Studies in Mechanobiology, Tissue Engineering and Biomaterials

Volume 10

Series Editor

Amit Gefen, Ramat Aviv, Israel

For further volumes:
http://www.springer.com/series/8415

Liesbet Geris
Editor

Computational Modeling in Tissue Engineering

Springer

Editor
Liesbet Geris
Biomechanics Research Unit
University of Liège
Liège
Belgium

ISSN 1868-2006 ISSN 1868-2014 (electronic)
ISBN 978-3-642-32562-5 ISBN 978-3-642-32563-2 (eBook)
DOI 10.1007/978-3-642-32563-2
Springer Heidelberg New York Dordrecht London

Library of Congress Control Number: 2012951135

Printed on acid-free paper

Springer is part of Springer Science+Business Media (www.springer.com)

To Kristof, Robbe and Soetkin, for making my life complete

Preface

This volume focuses on the recent advances in a domain that is situated at the interface of engineering and biomedical sciences and is an excellent example of the added value of integrative research: computational tissue engineering. After two decades of 'traditional' tissue engineering research with a strong focus on the biology of the tissue engineered construct and biomaterial development, the engineering side of the field is finally becoming exploited to its fullest potential. A common engineering approach when designing any kind of product or manufacturing process, from food to the chemical to the automotive industry, is to use in silico models of the product and/or the manufacturing process. These models, based on physical, mechanical, or (bio)chemical laws/equations and/or experimental data, are used in order to minimize the variability, optimize the overall process, and increase the quantity and quality of the final result. Accordingly, the chapters in the book fall under the following categories: computational tools for product design, computational tools for process design, and computational tools for the study of in vivo processes. The underlying tissue engineering applications will vary from blood vessels to trachea to cartilage and bone. For the chapters describing examples of the first two areas, the main focus is on (the optimization of) mechanical signals, mass transport, and fluid flow encountered by the cells in scaffolds and bioreactors as well as on the optimization of the cell population itself. In the chapters describing modeling contributions in the third area, the focus will shift toward the biology, the complex interactions between biology and the micro-environmental signals, and the ways in which modeling might be able to assist in investigating and mastering this complexity. The chapters cover issues related to (multiscale/multiphysics) model building, training and validation, but also discuss recent advances in scientific computing techniques that are needed to implement these models as well as new tools that can be used to experimentally validate the computational results.

The research groups which contributed to this book, from 10 different countries—mostly in USA/Canada and Europe—are all very well known internationally for their fundamental contributions to the field of computational tissue engineering, and the book as a whole therefore represents the state of the art of research in computational tissue engineering. The book also nicely demonstrates how computational tissue engineering research not only benefits from but also contributes to advances in other fields such as biomedical sciences, applied mathematics and computer sciences, material sciences, nanotechnology and bio-informatics. Faculty and students with interest in these fields should find this book an excellent source of information indicating where computational tissue engineering research brought us so far, and what are some of the open questions that wait to be answered.

Finally, I would like to take this opportunity to pay my tributes to Christian Oddou (1938–2011) who passed away unexpectedly last year. Only shortly before his death we had been discussing his contribution to this book which would have been, without a doubt, a valuable one. His warmth, kindness, passion and enthusiasm for anything related to biomechanics and computational tissue engineering will surely be missed.

Prof. Dr. Liesbet Geris

Contents

In Vivo, In Vitro, In Silico: Computational Tools for Product and Process Design in Tissue Engineering

Liesbet Geris

Abstract This chapter aims to provide an introduction to how engineering tools in general and computational models in particular can contribute to advancing the tissue engineering (TE) field. After a description of the current state of the art of TE, the developmental engineering paradigm is briefly discussed. Subsequently an overview is provided of different model categories that focus on different aspects of TE. These categories consists of the models that focus on either the TE product, the TE process or the in vivo results obtained after implantation. Generally, in all these models the aim is firstly to understand the biological process at hand and secondly to design strategies in silico to enhance the desired in vitro or in vivo behaviour. Finally, the need for quantification and parameter determination is discussed along with the computational tools and models that can be used to design the thereto required experiments in the most intelligent way.

1 Introduction

The presence of the word engineering in "Tissue Engineering", the interdisciplinary field combining biomedical and engineering sciences in the search for replacements of diseased/non-functional organ (parts) by manufactured living implants that support functional tissue regeneration [1], has not yet fulfilled its true potential. Ever since the

L. Geris (✉)
Biomechanics Research Unit, University of Liège,
Chemin des Chevreuils 1 B52/3, 4000 Liège, Belgium
e-mail: liesbet.geris@ulg.ac.be

L. Geris
Prometheus, Division of Skeletal Tissue Engineering Leuven,
KU Leuven, Herestraat 49, box 813, 3000 Leuven, Belgium

Stud Mechanobiol Tissue Eng Biomater (2013) 10: 1–15
DOI: 10.1007/8415_2012_144

Published Online: 22 August 2012

official introduction of the field in 1987, the research focus has been predominantly on the biology of the tissue construct and biomaterial development. Apart from the latter, the engineering input has remained limited to technical aspects such as bioreactor and biosensor development, and automation. Although these aspects are needed when bringing the biological processes from bench to bed side, they cover only a small part of what the engineering sciences have to offer the TE field.

One key issue the TE field in general is struggling with is the lack of quantity and quality of the generated products [2, 3]. Protocols and procedures followed in the lab are mainly established based on trial and error, requiring a huge amount of manual interventions and without clear early time-point quality criteria to guide the process. This also makes these processes very hard to scale up to industrial manufacturing levels as can be appreciated from the limited number of companies that have survived the first decade of TE [3]. Overall, there is a lack of intelligent process design in the current TE field. Over the last few years, a number of leading labs [4–8] in the TE field have realized that the trial and error approach is not a good way to obtain products that can meet the quality standards of international regulatory bodies such as EMA or FDA. It was proposed to return to an approach inspired by nature's own regenerative and developmental processes [4–8] called developmental engineering. Much the same as aspired in any manufacturing process, developmental processes are robust, multistage, observable, controllable, path-dependent and autonomous. A common engineering approach when designing any kind of manufacturing process, from the food over the chemical to the automotive industry, is to use *in silico* models of the product and/or the manufacturing process, based on physical, mechanical or (bio)chemical laws/equations and/or experimental data, in order to minimize the variability, increase the quality and optimize the overall process. *In silico* models can, amongst others, help to identify key regulating parameters of the manufacturing process and extrapolate early time point information to predict final product behaviour.

This chapter will briefly introduce a number of modelling techniques and applications related to the design of TE products and processes. It will start by introducing the novel paradigm of developmental engineering that has recently emerged in the TE field. Subsequently various examples of how computational tools can assist in the development of robust and reliable products and *processes* are discussed, many of which will come back in the other chapters of this book.

2 Developmental Engineering: A New Paradigm for Tissue Engineering

Over the last few years, several leading labs [4–8] in the TE field have proposed to return to an approach inspired by nature's own regenerative and developmental processes [4–8] called developmental engineering as many of the characteristics of developmental processes are desirable in process design. Firstly, the developing embryo has the ability to cope with a wide variety of external perturbations, i.e. it is a robust system. This robustness of developmental processes would allow the in vitro process to be impervious to a wide scale of external perturbations, which are often unavoidable in an artificial environment. Additionally developmental processes exhibit a multistage character, meaning cells will differentiate in characteristic stages, with a distinct morphology and marker genes. Accordingly, the in vitro process could be divided into a series of sequential subprocesses, each corresponding to a specific stage in developmental biology. This would make the process highly observable, by for example determining expression of certain marker genes, and highly controllable, since growth factors could be added when the cells are at a stage where they are competent to respond to them. Another concept advocating the use of in vitro developmental processes is path-dependence, or the dependence of one developmental stage on the previous ones. This means that the optimal conditions of the successive stage are provided by its predecessors. These conditions consequently do not have to be incorporated in the process design and will make the process more autonomous. An example of path-dependence can be found in endochondral ossification, where first a cartilage anlage is formed which creates optimal conditions for the invading ossification front [9]. Furthermore, some intermediate tissue forms have a great robustness thanks to intrinsic factors, allowing them to be treated as individual modules. The regulative, self-controlled behaviour exhibited by these modules will likely lead to a high product consistency. Several modular forms appear during development, including cellular modules like cartilage condensations and multicellular modules with a spatially extended and heterogeneous cell population like the growth plate.

The growth plate is a developmental centre that integrates many signalling pathways in order to regulate the patterning and growth of the skeleton. As a cell progresses throughout the growth plate, going from the long bone's epiphysis towards the diaphysis, its shape and function changes drastically [10]. At the epiphysis, a pool of small round chondrocytes makes up the resting zone. These cells differentiate into more rapidly proliferating flat chondrocytes, forming proliferative columns. The resting and proliferating chondrocytes secrete structural proteins, such as collagen type II, that form a hyaline cartilage matrix. Towards the diaphysis, chondrocytes differentiate further into prehypertrophic, secreting Indian Hedgehog (Ihh), and thereafter hypertrophic chondrocytes [11]. Hypertrophic chondrocytes remodel the cartilage matrix into a calcifying matrix comprising primarily collagen type X (Col-X). At terminal differentiation, the cells will induce invasion and resorption of the hypertrophic cartilage as well as the start of vascularisation by excreting proteins like Matrix Metallopeptidase 13 (MMP13) and Vascular Endothelial Growth Factor (VEGF) [12]. The evolution the chondrocytes undergo is reminiscent of the developmental process of endochondral ossification, indicating these events can be recapitulated using adult stem cells [13–17]. Indeed, implantation of articular chondrocytes (mixed with osteoblasts) in mice has been shown to result in formation of a structure similar to that of the growth plate [18].

Lenas et al. [4–8] extensively discuss how the growth plate developmental process can be used as a template for robust and reliable bone TE processes. Proof-of-concept for the basic idea that aggregates or constructs stem cells of (embryonal and postnatal bone marrow derived) after in vitro differentiation along the chondrogenic lineage (all the way up to hypertrophy) will result in bone formation after implantation in vivo has been delivered by several groups [19–21].

3 Computational Tools for Product and Process Design in Tissue Engineering

Engineered products will only be a viable and competitive alternative to upcoming off-the-shelf innovations in regenerative medicine if they are manufactured with reproducible properties, a prerequisite for consistent clinical outcomes. This important target is mainly challenged by the intrinsic variability in the behaviour of human cells from different batches or donors as well as by the sensitivity of cells to perturbations in the culture environment [22].

Although research-oriented systems are generally too complex, user-unfriendly, unsafe and expensive for direct use in clinical applications, their underlying principles could nevertheless lay a solid foundation for more clinically compliant manufacturing systems. This will require not only the optimization of the TE product itself but also the identification of only the most essential processes, culture parameters and construct parameters that must be monitored and controlled to standardise production and provide meaningful quality and traceability data, at the same time minimise risks, costs and user complexity [22].

Below we describe a number of computational tools are being used to optimize the design of both TE products and processes as well as the in vivo result. These tools use a variety of modelling techniques building on physical, mechanical or (bio)chemical laws/equations and/or experimental data. The models can range from mechanistic (hypothesis-driven, white box) to phenomenological (data-driven, black box) and, depending on the specific application, range from the gene/protein level over the cellular level up to the tissue/organ level. Figure 1 shows an overview figure of different models that have been developed in the field of bone tissue engineering by the research group of the author. The overview in this chapter is by no means exhaustive but serves to illustrate different aspects of product and process design. The reader is referred to the other chapters of this book that provide a more in depth review on a number of the aspects mentioned here.

3.1 Computational Tools for Product Design

There are a number of aspects on the design of TE products where computational models can make and already have made contributions, not only in the way the carrier structure (when dealing with combination products or biomaterials only) is designed but also the way cells are processed prior to implantation.

Obvious aspects of scaffold design include their structural, mass transport and mechanical properties. Several chapters in this book describe the design and

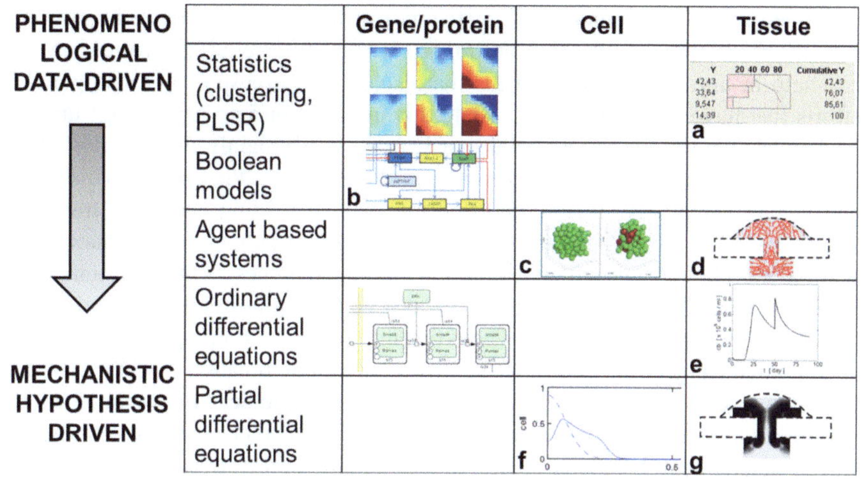

Fig. 1 Overview of the different models developed in the research group of the author of this chapter. Models can be classified according to the origin of their development (phenomenological vs. mechanistic) or according to the length and time scales of the processes they describe (gene/protein, cell, tissue/organ). **a** Roberts et al. [23]; **b** Kerkhofs et al. [24]; **c** Geris et al. [25]; **d** Peiffer et al. [26]; **e** Carlier et al. [27]; **f** Geris et al. [28]; **g** Geris et al. [29, 30]

characterisation of hydrogels, frequently used as carrier structure in tissue engineering, but each focusing on a different aspect. Israelowitz et al. [31] argue that in order to define the correct position of e.g. collagen in the fibre network arrangement of extracellular matrix, which is important to determine its tensile strength, an optimized tertiary structure of the protein needs to be characterized. They provide an introduction into the different methods that are currently used to determine protein conformation in silico. On a higher length scale, Nekouzadeh et al. [32] describe the development of a mechanical model to design and evaluate engineered tissues and/or carrier structures (such as hydrogels) that serve a mechanical role. An important component of such models is often viscoelasticity, or the dependence of mechanical response on loading rate and loading history. In a great number of biological and bio-artificial tissues the passive tissue force (or stress) relates to changes in tissue length (or strain) in a nonlinear viscoelastic manner. Choosing and fitting nonlinear viscoelastic models to data for a specific tissue can be a computational challenge. Nekouzadeh et al. [32] describe the range of such models (focusing in particular on adaptive quasi-linear viscoelastic models), criteria for selecting amongst them, and computational and experimental techniques needed to fit these to uniaxial data. Additional to structural mechanics, mass transport is an important property of hydrogels which can influence the behaviour of cells encompassed in these hydrogels in various ways. Lambrechts et al. [33] provide a thorough overview of how consumption and production of soluble medium components gives rise to gradients inside hydrogels and how mass transport related phenomena can shape these gradients. The authors focus on the combined use of experiments and mathematical modelling and describe not only how the simulation results play an important role in generating information that can help in unravelling mechanisms that drive solute transport but also genuine efforts that have been taken to translate this information into real TE set-ups.

Another type of carrier material that is often used, mainly in musculoskeletal tissue engineering, is a (macro-) porous scaffold. Similar to the hydrogels discussed above, these scaffolds allow supporting mechanical loading and mass transport. Olivares and Lacroix [34] review the computational methods applied to characterize scaffold morphology and simulate different biological processes in and around these scaffolds. These processes include cell seeding, cell migration, cell proliferation, cell differentiation, vascularisation, oxygen consumption, mass transport and/or scaffold degradation. Song et al. [35] describe how using a combination of computational fluid dynamics and finite element analysis allows to predict flow regimes within scaffolds and to optimize flow rates to deliver mechanical cues during cell seeding and subsequent cell behaviour. They furthermore demonstrate how computational modelling can be used to optimize spatiotemporal mechanical cue delivery and mechanically modulated biochemical gradients through optimization of scaffold geometry, material behaviour and mechanical properties.

Besides the scaffold's physical properties, also its chemical properties (e.g. its release properties) can have a substantial influence on the overall behaviour of the TE construct. Chemicals released form the scaffolds can either be dissolved components of the scaffold material itself (e.g. the release of soluble calcium from calcium-phosphate-collagen scaffolds) or substances that were added to the scaffold structure for delivery in vivo and have a specific biological function (e.g. controlled release of growth factors). Mathematical models have been developed describing these release processes and have been applied to determine in silico optimal scaffolds for a variety of biomedical problems, e.g. Carlier et al. [27] and many others.

For TE products that include (or are solely consisting of) a cellular component, models have been developed to investigate aspects ranging from storage over proliferation and selection to implantation strategies. Cincotti and Fadda [36] describe a model of the cryopreservation process of cell suspensions, a critical step in tissue engineering. The model is based on bio-physical properties and takes into account size distribution of the cell population. After validation, the authors have used their model to investigate the effect cell size distribution on system behaviour under various operating conditions showing that under commonly used operating conditions, intra-cellular ice formation may be lethal for the largest cells in the population. In addition to cell size, cell populations are also heterogeneous in various other functional and molecular aspects. Galle et al. [37] review the most recent results on heterogeneity in mesenchymal stem cells (MSC) and introduce a mathematical framework that approaches MSC heterogeneity on the single cell level. This framework is capable of describing the impact of MSC heterogeneity on in vitro expansion and differentiation and can be used to investigate MSC adaptation to changing environments and the cell's intrinsic control of state fluctuations. Prior to implantation, the quality of the cells needs to be assessed in order to guarantee a safe and effective therapy. As this quality check should preferably be non-invasive, complete (meaning all cells and not only a sample should be checked), real-time and predictive of further clinical therapeutic effect, conventional cell biology techniques are ruled out. Sasaki et al. [38] discuss the potential of image-based quality assessment by implementing machine learning models to connect biological phenomena with the measurements. After storage and quality assurance of the cell population, the timely administration of the appropriate concentrations of cells in the correct location is another crucial point where computational modelling can be an interesting tool. Geris et al. [39] have investigated the administration of MSCs in and at the fracture site of atrophic non-unions by means of a computational model and have corroborated their simulation results through comparison with the results from a pilot experiment (which was based on the in silico predictions).

3.2 Computational Tools for Process Design

Nature uses a very complex system of regulatory mechanisms compounded by a huge amount of redundancy. Systems biology and bio-informatics are just beginning to unlock the huge amount of information that is hidden within the human genome. From this huge amount of info a limited number of functional regulators (targets or markers) needs to be distilled that are indicative of the progress of the biological process in vitro and can hence be used to control the TE manufacturing process. These regulators are not necessarily restricted to biological parameters but can also be properties of the carrier structure and culture environment. These regulators are part of an intricate network that is too complex to be interpreted without the help of *in silico* modelling.

An additional challenge when dealing with biological processes is how to extract knowledge on these regulators from the relatively few process states that can be measured on-line [40]. In this context, the monitoring and control of bioreactor systems will be crucial at the research stage of product development, in order to identify these key regulators and to establish standardised production methods [41]. A mathematical model of the process is a cornerstone for modern control approaches such as model-based predictive controllers (MPC). Therefore, a complete design of an automatic bioreactor system should include the development of a good model, which should be complete enough to fully capture the process dynamics at interest and should also be capable of allowing the predictions to be calculated but at the same time, it should be intuitive and permit theoretic analysis [42].

Various types of models can be used as long as they allow accurate predictions of the most important process output(s) and are compact enough to be implemented in the bioreactor system. In many control applications black box models are used (e.g. impulse response models, step response models, transfer function models, state space models, neural networks, etc.) that describe the process under consideration based on data of dynamic experiments (dynamic data-based models). They have the advantage that they are compact, allow accurate predictions of the process behaviour and are easy to implement in a model-based control framework. However, an important drawback of these models is that they are not based on knowledge of the system and as such are difficult to interpret in a biological way. At the other end of the spectrum, there are mechanistic models (white box) that are knowledge based and as a result often (much) more complex than data-based models. A hybrid (grey box) approach has been developed that combines the advantages of both the dynamic data-based modelling approach and the mechanistic modelling approach into a so called data-based mechanistic (DBM) modelling approach [43]. DBM models can be developed in different ways, but one commonly used approach is to start from available mechanistic models that then will be reduced in complexity by applying sensitivity analysis and principal component approaches (e.g. [44, 45]). Parameters of the reduced order model structure can be estimated in a time varying way by using e.g. a recursive instrumental variable estimation method using data from dynamic experiments with the bioreactor system to be controlled.

Mechanistic (white box) models of in vitro bioreactor processes have been repeatedly proposed in the literature. O'Dea et al. [46] provide an overview of the models that use continuum modelling techniques to investigate how the different underlying processes interact to produce functional tissues for implantation in cell-seeded porous scaffolds. They aim to demonstrate how a combination of mathematical modelling, analysis and in silico computation, undertaken in collaboration with experimental studies, may lead to significant advances in the understanding of the fundamental processes regulating biological tissue growth and the optimal design of in vitro methods for generating replacement tissues that are fully functional. Raimondi et al. [47] discuss, also for cell-seeded porous scaffolds, the need for and the advances in the use of multiphysics and multiscale mathematical models. They describe various possible approaches to couple biomass growth, medium flow and mass transport in a single model. Furthermore they discuss recent advances in scientific computing techniques that are needed to implement these multiscale/multiphysics models as well as new tools that can be used to experimentally validate the computational results.

Besides the control of the bioreactor process, the design of the bioreactor set-up itself can play a major role obtaining the desired results. Bjork et al. [48] use computational models focusing on the dissolved oxygen transport to design bioreactor set-ups for engineered vascular tissues that improve transport, particularly by perfusion of medium through the interstitial space by transmural flow. Their computational models, supported by empirical data, specifically investigated designs that would eliminate nutrient gradients evident during static culture methods, in order to develop more uniform engineered vascular tissues which would lead to improved mechanical properties of the resulting construct.

3.3 Computational Tools for the Study of the In Vivo Process

Although according to the developmental engineering concept [4–8], the establishment of robust modular tissue intermediates in vitro should lead to the desired high-quality outcome in vivo, the effect of the in vivo environment is an important unknown that requires thorough experimental investigation. From the obtained experimental results mechanisms of actions can be proposed and subsequently mathematical models can be used to translate these mechanisms of action into a coherent set of mathematical equations. These equations form a quantitative spatio-temporal framework of interrelated biological variables and sub-processes, providing a dynamic and comprehensive overview of the entire repair process. As such, the mathematical models can help in interpreting the in vivo data (by establishing causal relations) on the one hand and generating new hypotheses on in vivo outcome (by running in silico experiments) on the other hand, in this way adding to the design and optimization of TE products and processes.

A myriad of models has been proposed in the literature describing various pathologies and in vivo regenerative processes. Watton et al. [49] have developed a fluid–solid-growth model to simulate the evolution of abdominal aortic aneurysms. The model uses a realistic constitutive model of the arterial wall accounting for a wide number of lower scale structures and processes. With the help of this model they were able to predict e.g. the development of tortuosity that accompanies abdominal aortic aneurysm enlargement. Besides providing a basis for further investigation and elucidation of the aetiology of aneurysm formation, the computational framework can also be applied to aid the design and optimisation of tissue engineered vascular constructs. In the field of bone regeneration Geris et al. [50] have reviewed the existing models of fracture healing, dividing these models into bioregulatory (fracture healing guided by biological stimuli), mechanoregulatory (fracture healing guided by mechanical stimuli) and mechanobioregulatory models (fracture healing guided by mechanical and biological stimuli). Nagel and Kelly [51] adapted a well-known mechanoregulatory model to explicitly account for the influence of oxygen tension on tissue differentiation. They furthermore discuss the effects of incorporating the tissue architecture during skeletal regeneration as well as the variability of the process. Reina-Romo et al. [52] discuss the importance of angiogenesis on both bone regeneration and TE. They describe the role of the vascular network in these processes as well as the most recent in silico models simulating the vascular network within bone constructs. They analyse discrete as well as continuum approaches from a computational perspective.

As mentioned above, simulation of the behaviour of a TE construct after implantation is another crucial aspect in the optimisation of TE products and processes. Lemon et al. [53] have developed a mathematical model of the regeneration of a tissue-engineered trachea seeded with cells in situ, in order to study the biological processes (e.g. stenosis) taking place after implantation for various designs of the TE construct (different cell seeding strategies). They provide an in depth discussion on the obstacles that are encountered when trying to formulate a faithful model of (any kind of) biological product or process. Furthermore they investigate how a simplified mathematical model that omits much detail of the biology can be of use for studying regeneration of a TE construct, using their model of a tissue-engineered trachea as an example.

4 Discussion

As shown above, computational tools have already investigated a wide variety of products and processes in the tissue engineering field. Whereas in the early models the distance between the computer and the bench was quite substantial, integration of (biological) experiments and simulation efforts are increasing. It has become evident that imaginative and refined experimental strategies based on genetics, imaging, quantitative and biophysical approaches, combined with the exploration of the fullest potential of mathematical modelling are necessary to understand cellular and developmental biology. The increased attention for this integrative approach can be appreciated from the initiatives that have been and are being taken by large funding agencies to promote this research, e.g. the Quantissue network [54] (funded by ESF-RNP) and the Physiome [55]/Virtual Physiological Human initiatives [56] (funded by agencies worldwide). The potential of this integrative research has already been demonstrated in a number of biomedical fields [57–62]. For example, Faratian et al. [57] successfully used a systems biology approach to stratify patients for personalized therapy in cancer and provided further compelling evidence that a particular biomarker, appropriately measured in the clinical setting, could refine clinical decision making in patients treated with a specific therapy. In developmental biology, Von Dassow and co-workers [59, 60] showed by means of a computational model that the drosophila segment polarity genes constitute a robust developmental module. The simulation results provided important insights into the overall dynamics of the gene network and highlighted mechanistic details that require further experimental research.

With the increasing demand for more quantitative models, there is also an increasing attention for the determination of relevant parameter sets [63–65]. Precise measurements of the different parameter values is in almost all cases impossible, either due to the fact that not all parameters represent physical processes (even when dealing with mechanistic or white box models) or because the physical property cannot be measured without altering the process. An example of the latter is the use of in vitro experiments to determine properties of in vivo processes. Classical system identification techniques, typically used in grey and black box approaches, will determine the parameter values as to fit the model to the system it is intended to describe. Depending on the system at hand and on the available experimental information, estimation theory or neural networks are commonly used concepts. Additionally, engineering concepts such as the design of experiments and optimal experimental design are finding their way into the biomedical sciences to increase the amount of information that can be retrieved from experiments while reducing the number of experimental runs required to obtain this information. Alternatively, or better yet concomitantly, to finding appropriate parameter values based on experimental results, many modellers apply techniques to investigate the impact of the chosen parameter values on their simulation results by means of sensitivity analyses. Sensitivity analyses appear under many different forms. The most frequently used technique is the one-at-the-time (OAT) analysis where only one parameter is altered (e.g. [66] to give but one example). This provides information on the main effects of this parameter but it does not provide any information on the combined effects or the interactions between different parameters. Design of experiment techniques have been successfully applied to mathematical models to overcome the limitations of the OAT technique [27, 67].

In the above discussion on the optimal way to determine parameter values for quantitative models, a completely different point of view is taken by a number of researchers. Gutenkunst and co-workers argue against the focus on optimizing experimental design to best constrain model parameters with collective fits as discussed above, particularly in cases when the understanding of a system is tentative and incomplete. An important consideration underlying their point of view is the question of how we should deal with uncertainties in the data [68], in the fitting of parameters, and in resulting predictions. Brown et al. rigorously explored one source of uncertainty in their model of growth-factor signalling in PC12 cells; their analysis considered not just the set of parameters that best fit the data but a statistical sampling of all parameter sets that fit the data [69, 70]. Like in many other systems [71], the space of parameter sets that could fit the data was vast. Perhaps surprisingly, some predictions were still very well constrained even in the face of this enormous parameter uncertainty. Brown et al. found a striking 'sloppy' pattern in the sensitivity of their model to parameter changes; when plotted on a logarithmic scale, the sensitivity eigenvalues were roughly evenly spaced over many decades. This sloppy nature was then further investigated by Gutenkunst and others [72–74]. Even though sloppiness is not unique to biological systems, it is particularly relevant to biology [75] because the collective behaviour of most biological systems is much easier to measure in vivo than the values of individual parameters. Using sloppy parameter analysis, concrete predictions can be extracted from models long before their parameters are even roughly known [70], and when a system is not already well-understood, it can be more profitable to design experiments to directly improve predictions of interesting system behaviour [76] rather than to improve estimates of parameters.

5 Conclusion

In conclusion, this chapter has provided an overview of how computational modelling could contribute to advancing the tissue engineering field. Regardless of whether the models focus on the product, the process or the in vivo results, the aim is always to try to understand the biological process and to design strategies in silico to enhance the desired in vitro or in vivo behaviour. Finally, if models are to be applied in a quantitative way, experiments need be designed as to feed the models in the most intelligent way. Also here computational tools and models can play an important role.

Acknowledgments The author gratefully acknowledges funding from the Special Research Fund of the University of Liège (FRS.D-10/20), the Belgian National Fund for Scientific Research (FNRS) grant FRFC 2.4564.12 and the European Research Council under the European

Union's Seventh Framework Programme (FP7/2007–2013)/ERC grant agreement n°279100. The authors also gratefully acknowledges collaborators from the Biomechanics Research Group of the University of Liège and from Prometheus, the KU Leuven Research and Development Division of Skeletal Tissue Engineering (www.kuleuven.be/Prometheus) for their invaluable input.

References

1. Langer, R., Vacanti, J.P.: Tissue engineering. Science **260**, 920–926 (1993)
2. Meijer, G., de Bruijn, J.D., Koole, R., van Blitterswijk, C.A.: Cell-based bone tissue engineering. PLoS Med. **4**, e9 (2007)
3. Archer, R., Williams, D.: Why tissue engineering needs process engineering. Nat. Biotechnol. **23**, 1353–1355 (2005)
4. Ingber, D.E., Mow, V.C., Butler, D., Niklason, L., Huard, J., Mao, J., Yannas, I., Kaplan, D., Vunjak-Novakovic, G.: Tissue engineering and developmental biology: going biomimetic. Tissue Eng. **12**, 3265–3283 (2006)
5. Lenas, P., Moos, M., Luyten, F.P.: Developmental engineering: a new paradigm for the design and manufacturing of cell-based products. Part II: from genes to networks: tissue engineering from the viewpoint of systems biology and network science. Tissue Eng. Part B Rev. **15**(4), 395–422 (2009)
6. Lenas, P., Moos, M., Luyten, F.P.: Developmental engineering: a new paradigm for the design and manufacturing of cell-based products. Part I: from three-dimensional cell growth to biomimetics of in vivo development. Tissue Eng. Part B Rev. **15**(4), 381–394 (2009)
7. Lenas, P., Luyten, F.: An emerging paradigm in tissue engineering: from chemical engineering to developmental engineering for bioartificial tissue formation through a series of unit operations that simulate the in vivo successive developmental stages. Ind. Eng. Chem. Res. **50**(2), 482–522 (2011)
8. Lenas, P., Luyten, F.P., Doblare, M., Nicodemou-Lena, E., Lanzara, A.E.: Modularity in developmental biology and artificial organs: a missing concept in tissue engineering. Artif. Organs **35**(6), 656–662 (2001)
9. Kronenberg, H.M.: PTHrP and skeletal development. Ann. N. Y. Acad. Sci. **1068**, 1–13 (2006)
10. Kronenberg, H.M.: Developmental regulation of the growth plate. Nature **423**, 332–336 (2003)
11. Solomon, L.A., Bérubé, N.G., Beier, F.: Transcriptional regulators of chondrocyte hypertrophy. Birth Defects Res. C Embryo Today **84**, 123–130 (2008)
12. Burdan, F., Szumi, J., Korobowicz, A., Farooquee, R., Patel, S., Patel, A., Dave, A., Szumi, M., Solecki, M., Klepacz, R., Dudka, J.: Morphology and physiology of the epiphyseal growth plate. Folia Histochem. Cytobiol. **47**, 5–16 (2009)
13. Luyten, F.P., Dell'Accio, F., De Bari, C.: Skeletal tissue engineering: opportunities and challenges. Best Pract. Res. Clin. Rheumatol. **15**(5), 759–770 (2001)
14. Shapiro, F.: Bone development and its relation to fracture repair. The role of mesenchymal osteoblasts and surface osteoblasts. Eur. Cell Mater. **15**, 53–76 (2008)
15. Reddi, A.H.: Morphogenesis and tissue engineering of bone and cartilage: inductive signals, stem cells, and biomimetic materials. Tissue Eng. **6**(4), 351–359 (2000)
16. Hunziker, E.B.: Articular cartilage repair: basic science and clinical progress. A review of the current status and prospects. Osteoarthr. Cartil. **10**(6), 432–463 (2002)
17. Ferguson, C., Alpern, E., Miclau, T., Helms, J.A.: Does adult fracture repair recapitulate embryonic skeletal formation? Mech. Dev. **87**(1–2), 57–66 (1999)
18. Alsberg, E., Anderson, K.W., Albeiruti, A., Rowley, J.A., Mooney, D.J.: Engineering growing tissues. Proc. Natl. Acad. Sci. U S A **99**, 12025–12030 (2002)

19. Jukes, J.M., Both, S.K., Leusink, A., Sterk, L.M., van Blitterswijk, C.A., de Boer, J.: Endochondral bone tissue engineering using embryonic stem cells. Proc. Natl. Acad. Sci. U S A **105**(19), 6840–6845 (2008)
20. Scotti, C., Tonnarelli, B., Papadimitropoulos, A., Scherberich, A., Schaeren, S., Schauerte, A., Lopez-Rios, J., Zeller, R., Barbero, A., Martin, I.: Recapitulation of endochondral bone formation using human adult mesenchymal stem cells as a paradigm for developmental engineering. Proc. Natl. Acad. Sci. U S A **107**(16), 7251–7256 (2010)
21. Weiss, H.E., Roberts, S.J., Schrooten, J., Luyten, F.P.: A semi-autonomous model of endochondral ossification for developmental tissue engineering. Tissue Eng. Part A (2012). doi:10.1089/ten.tea.2011.0602
22. Martin, I., Smith, T., Wendt, D.: Bioreactor-based roadmap for the translation of tissue engineering strategies into clinical products. Trends Biotechnol. **27**, 495–502 (2009)
23. Roberts, S.J., Geris, L., Kerckhofs, G., Desmet, E., Schrooten, J., Luyten, F.P.: The combined bone forming capacity of human periosteal derived cells and calcium phosphates. Biomaterials **32**(19), 4393–4405 (2011)
24. Kerkhofs, J., Roberts, S.J., Luyten, F.P., Van Oosterwyck, H., Geris, L.: Relating the chondrocyte gene network to growth plate morphology: from genes to phenotype. PLoS ONE **7**(4), e34729 (2012)
25. Geris, L., Van Liedekerke, P., Smeets, B., Tijskens, E., Ramon, H.: A cell based modelling framework for skeletal tissue engineering applications. J. Biomech. **43**(5), 887–892 (2010)
26. Peiffer, V., Gerisch, A., Vandepitte, D., Van Oosterwyck, H., Geris, L.: A hybrid bioregulatory model of angiogenesis during bone fracture healing. Biomech. Model. Mechanobiol. **10**(3), 383–395 (2011)
27. Carlier, A., Chai, Y.C., Moesen, M., Theys, T., Schrooten, J., Van Oosterwyck, H., Geris, L.: Designing optimal calcium phosphate scaffold-cell combinations using an integrative model based approach. Acta Biomater. **7**(10), 3573–3585 (2011)
28. Geris, L., Ashbourn, J.M.A., Clarke, T.: Continuum-level modelling of cellular adhesion and matrix production in aggregates. Comput. Method Biomech. Biomed. Eng. **14**(5), 403–410 (2011)
29. Geris, L., Gerisch, A., Vander Sloten, J., Weiner, R., Van Oosterwyck, H.: Angiogenesis in bone fracture healing: a bioregulatory model. J. Theor. Biol. **251**(1), 137–158 (2008)
30. Geris, L., Vander Sloten, J., Van Oosterwyck, H.: Connecting biology and mechanics in fracture healing: an integrated mathematical modeling framework for the study of nonunions. Biomech. Model. Mechanobiol. **9**(6), 713–724 (2010)
31. Israelowitz, M., Rizvi, S.W.H., Weyand, B., Gille, C., von Schroeder, H.P.: Protein modeling and surface folding by limiting the degrees of freedom (2012). doi:10.1007/8415_2012_141
32. Nekouzadeh, A., Genin, G.M.: Adaptive quasi-linear viscoelastic modeling (2012). doi:10.1007/8415_2012_142
33. Lambrechts, D., Schrooten, J., Van de Putte, T., Van Oosterwyck, H.: Computational modeling of mass transport and its relation to cell behavior in tissue engineering constructs (2012). doi:10.1007/8415_2012_139
34. Olivares, A.L., Lacroix, D.: Computational methods in the modeling of scaffolds for tissue engineering (2012). doi:10.1007/8415_2012_136
35. Song, M.J., Dean, D., Knothe Tate, M.L.: Computational modeling of tissue engineering scaffolds as delivery devices for mechanical and mechanically modulated signals (2012). doi:10.1007/8415_2012_138
36. Cincotti, A., Fadda, S.: Modeling the cryopreservation process of a suspension of cells: the effect of a size distributed cell population (2012). doi:10.1007/8415_2012_134
37. Galle, J., Hoffmann, M., Krinner, A.: Mesenchymal stem cell heterogeneity and ageing in vitro—a model approach (2012). doi:10.1007/8415_2012_116
38. Sasaki, H., Matsuoka, F., Yamamoto, W., Kojima, K., Honda, H., Kato, R.: Image-based cell quality assessment: modeling of cell morphology and quality for clinical cell therapy (2012). doi:10.1007/8415_2012_132

39. Geris, L., Reed, A.A., Vander Sloten, J., Simpson, A.H., Van Oosterwyck, H.: Occurrence and treatment of bone atrophic non-unions investigated by an integrative approach. PLoS Comput. Biol. **6**(9), e1000915 (2010)
40. Karim, M.N., Hodge, D., Simon, L.: Data-based modeling and analysis of bioprocesses: some real experiences. Biotechnol. Prog. **19**(5), 1591–1605 (2003)
41. Wendt, D., Riboldi, S.A., Cioffi, M., Martin, I.: Potential and bottlenecks of bioreactors in 3D cell culture and tissue manufacturing. Adv. Mater. **21**(32–33), 3352–3367 (2009)
42. Camacho, E.F., Bordons, C.: Model Predictive Control. Springer-Verlag, London (1999)
43. Young, P.: Data-based mechanistic modeling, generalised sensitivity and dominant mode analysis. Comput. Phys. Commun. **117**, 113–129 (1999)
44. Moore, B.C.: Principal component analysis in linear systems: controllability, observability and model reduction. IEEE Trans. Autom. Control **26**(1), 17–32 (1981)
45. Huang, Z., Chu, Y., Hahn, J.: Model simplification procedure for signal transduction pathway models: an application to IL-6 signaling. Chem. Eng. Sci. **65**(6), 1964–1975 (2010)
46. O'Dea, R.D., Byrne, H.M., Waters, S.L.: Continuum mathematical modelling of in vitro tissue engineering: a review (2012). doi:10.1007/8415_2012_140
47. Raimondi, M.T., Causin, P., Lagana, M., Zunino, P., Sacco, R.: Multiphysics computational modeling in cartilage tissue engineering (2012). doi:10.1007/8415_2012_112
48. Bjork, J.W., Safonov, A.M., Tranquillo, R.T.: Oxygen transport in bioreactors for engineered vascular tissues (2012). doi:10.1007/8415_2012_133
49. Watton, P.N., Huang, H., Ventikos, Y.: Multi-scale modelling of vascular disease: abdominal aortic aneurysm evolution (2012). doi:10.1007/8415_2012_143
50. Geris, L., Vander Sloten, J., Van Oosterwyck, H.: In silico biology of bone modeling and remodeling—regeneration. Philos. Trans. R. Soc. A **367**(1895), 2031–2053 (2009)
51. Nagel, T., Kelly, D.J.: Computational mechanobiology in cartilage and bone tissue engineering: from cell phenotype to tissue structure (2012). doi:10.1007/8415_2012_131
52. Reina-Roma, E., Valero, C., Borau, C., Rey, R., Javierre, E., Gomez-Benito, M.J., Dominguez, J., Garcia-Aznar, J.M.: Mechanobiological modelling of angiogenesis impact on tissue engineering and bone regeneration (2012). doi:10.1007/8415_2012_111
53. Lemon, G., King, J.R., Macchiarini, P.: Mathematical modeling of regeneration of a tissue-engineered trachea (2012). doi:10.1007/8415_2012_145
54. http://quantissue.eu/. Accessed 19 June 2012
55. http://en.wikipedia.org/wiki/Physiome. Accessed 19 June 2012
56. http://ec.europa.eu/information_society/activities/health/research/fp7vph/index_en.htm. Accessed 19 June 2012
57. Faratian, D., Goltsov, A., Lebedeva, G., Sorokin, A., Moodie, S., Mullen, P., Kay, C., Um, I.H., Langdon, S., Goryanin, I., Harrison, D.J.: Systems biology reveals new strategies for personalizing cancer medicine and confirms the role of PTEN in resistance to trastuzumab. Cancer Res. **69**(16), 6713–6720 (2009)
58. Edelman, L.B., Eddy, J.A., Price, N.D.: In silico models of cancer. Wiley Interdiscip. Rev. Syst. Biol. Med. **2**(4), 438–459 (2010)
59. von Dassow, G., Meir, E., Munro, E.M., Odell, G.M.: The segment polarity network is a robust developmental module. Nature **406**(6792), 188–192 (2000)
60. Von Dassow, G., Odell, G.M.: Design and constraints of the Drosophila segment polarity module: robust spatial patterning emerges from intertwined cell state switches. J. Exp. Zool. **294**(3), 179–215 (2002)
61. Eldar, A., Dorfman, R., Weiss, D., Ashe, H., Shilo, B.Z., Barkai, N.: Robustness of the BMP morphogen gradient in Drosophila embryonic patterning. Nature **419**(6904), 304–308 (2002)
62. Tsai, T.Y., Choi, Y.S., Ma, W., Pomerening, J.R., Tang, C., Ferrell Jr, J.E.: Robust, tunable biological oscillations from interlinked positive and negative feedback loops. Science **321**(5885), 126–129 (2008)
63. Faller, D., Klingmuller, U., Timmer, J.: Simulation methods for optimal experimental design in systems biology. Simulation **79**, 717–725 (2003)

64. Zak, D.E., Gonye, G.E., Schwaber, J.S., Doyle III, F.J.: Importance of input perturbations and stochastic gene expression in the reverse engineering of genetic regulatory networks: insights from an identifiability analysis of an in silico network. Genome Res. **13**, 2396–2405 (2003)
65. Gadkar, K.G., Varner, J., Doyle III, F.J.: Model identification of signal transduction networks from data using a state regulator problem. IEEE Syst. Biol. **2**, 17–30 (2005)
66. Geris, L., Gerisch, A., Maes, C., Van Oosterwyck, H., Carmeliet, G., Weiner, R., Vander Sloten, J.: Mathematical modeling of fracture healing in mice: comparison between experimental data and numerical simulation results. Med. Biol. Eng. Comput. **44**(4), 280–289 (2006)
67. Isaksson, H., van Donkelaar, C.C., Huiskes, R., Yao, J., Ito, K.: Determining the most important cellular characteristics for fracture healing using design of experiments methods. J. Theor. Biol. **255**(1), 26–39 (2009)
68. Cumming, G., Fidler, F., Vaux, D.L.: Error bars in experimental biology. J. Cell Biol. **177**, 7–11 (2007)
69. Brown, K.S., Sethna, J.P.: Statistical mechanical approaches to models with many poorly known parameters. Phys. Rev. E **68**, 021904 (2003)
70. Brown, K.S., Hill, C.C., Calero, G.A., Myers, C.R., Lee, K.H., et al.: The statistical mechanics of complex signaling networks: nerve growth factor signaling. Phys. Biol. **1**, 184–195 (2004)
71. Brodersen, R., Nielsen, F., Christiansen, J.C., Andersen, K.: Characterization of binding equilibrium data by a variety of fitted isotherms. Eur. J. Biochem. **169**, 487–495 (1987)
72. Gutenkunst, R.N., Waterfall, J.J., Casey, F.P., Brown, K.S., Myers, C.R., Sethna, J.P.: Universally sloppy parameter sensitivities in systems biology models. PLoS Comput. Biol. **3**(10), 1871–1878 (2007)
73. Gutenkunst, R.N., Casey, F.P., Waterfall, J.J., Myers, C.R., Sethna, J.P.: Extracting falsifiable predictions from sloppy models. Ann. N. Y. Acad. Sci. **1115**, 203–211 (2007)
74. Ashbourn, J.M., Miller, J.J., Reumers, V., Baekelandt, V., Geris, L.: A mathematical model of adult subventricular neurogenesis. J. R. Soc. Interface (2012). doi:10.1098/rsif.2012.0193
75. Daniels, B.C., Chen, Y.J., Sethna, J.P., Gutenkunst, R.N., Myers, C.R.: Sloppiness, robustness, and evolvability in systems biology. Curr. Opin. Biotechnol. **19**(4), 389–395 (2008)
76. Casey, F.P., Baird, D., Feng, Q., Gutenkunst, R.N., Waterfall, J.J., et al.: Optimal experimental design in an epidermal growth factor receptor signalling and down-regulation model. IET Syst. Biol. **1**, 190–202 (2007)

Part I
Computational Tools For Product Design

Protein Modelling and Surface Folding by Limiting the Degrees of Freedom

Meir Israelowitz, Birgit Weyand, Syed W. H. Rizvi, Christoph Gille and Herbert P. von Schroeder

Abstract One aspect of tissue engineering represents modelling of the extracellular matrix of connective tissue as the fiber network arrangement of the matrix determines its tensile strength. In order to define the correct position of the e.g. collagen in a structure, an optimized tertiary structure must be characterized. Existing approaches of protein models consider random packing of rigid spheres. We propose an alternative strategy to model protein structure by focusing on the folding. Our model considers (a) segments of amino-acid peptides or beads, (b) hydrogen bond distances, and (c) the distance geometry as functional components rather than minimizing distances between the centers of atoms. We reduced the molecular volume by using concepts from low dimensional topology, such as *braids* and surfaces, via differential geometry. A *braid* group maintains the continuity of a sequence while the spatial minimization is performed, and guarantees the continuity during the process. We have applied this approach to different examples of known protein sequences using ab initio protocols of ProteoRubix Systems™. Sequence files of three different proteins types were tested and modeled by ProteoRubix™ and compared to models derived by other methods. ProteoRubix™

M. Israelowitz · S. W. H. Rizvi · C. Gille · H. P. von Schroeder (✉)
Biomimetics Technologies Inc, Toronto, Canada
e-mail: herb.vonschroeder@uhn.ca

B. Weyand
Department of Plastic, Hand and Reconstructive Surgery, Hannover Medical School, Hannover, Germany

C. Gille
Institute of Biochemistry, Charite Universitsaetsmedizin, Berlin, Germany

H. P. von Schroeder
University Health Network, Toronto, Canada

H. P. von Schroeder
University of Toronto, Toronto, Canada

Stud Mechanobiol Tissue Eng Biomater (2013) 10: 19–45
DOI: 10.1007/8415_2012_141

Published Online: 1 July 2012

created near-identical models with minimal computational load. This model can be expanded to large, multi-molecular network structures.

1 Introduction

Tissue Engineering is an interdisciplinary field in the cross-section of biology, medicine, chemistry, physics, material sciences, engineering sciences and informatics. One aspect of tissue engineering represents modelling of extracellular matrix of connective tissue. The mechanical properties of the extracellular matrix are defined by its structure and composition. Collagen, for example, is the most abundant extracellular matrix protein in connective tissue, and is composed of three polypeptide strands each with 1,000 amino acids that are assembled to form a fibril [31]. The fiber network arrangement of the collagen matrix determines its tensile strength. In order to define the correct position of the collagen in a structure, an optimized tertiary structure must be characterized.

With the purpose of modelling a protein such as collagen, energy calculations must consider six degrees of freedom for each single atom. The calculations include hydrogen bonding and van der Walls forces, and moreover, because the interactions of the amino acids are both hydrophobic and hydrophilic forces in an aqueous environment, these calculations must also be considered in time [27]. The time for protein folding takes place in milliseconds to minutes [68]. In the case of collagen or other large proteins, there are assistance proteins, the so-called chaperones such as P52, which support protein folding and tertiary structure by holding initial amino acids for assembly of the total protein [61].

Protein modelling is considered to be a nondeterministic polynomial (NP) time problem that is regarded as inherently difficult if its solution requires significant resources, regardless of the algorithm that is used to solve the problem [45]. In order to solve a NP problem, we need to consider a program that makes 2^n operations before halting, since exponential-time algorithms might be unusable from the practical point of view. Collagen is an exceptionally large protein, but even for a small number of molecules (e.g. $n = 100$) and 10^{12} computer operations per second, a program would run for about 4×10^{10} years to solve a problem ([70] and Appendix 1) and therein lies the challenge for protein modelling.

Protein modelling is generally based on the fact that a structure is striving for the lowest state of energy based on the Lennard-Jones potential that defines energy as a function of the distances of the minimizations. Protein modelling uses free Gibbs energy to obtain the lowest state of energy. Currently, there are eight major methods used to denote protein conformation spaces or lowest state of energy:

1.1 *Molecular Dynamics* in which every coordinate position of the atoms contained within the amino acids in the sequence is taken into consideration. This method calculates the gradient or steep-out slope to obtain an energy minimum [69, 75].

1.2 The *Monte Carlo* method uses computational algorithms that rely on repeated random sampling techniques to find a global minimum. The *Monte Carlo* method takes a sample of the averages on the confirmation path; however, it is possible for the path to enter into local minima that are difficult to differentiate [13, 89].

1.3 *Smith's Microfibril model* calculates the conformation energy by finding the difference between the random state and the final conformation. The difference between the two is the objective function [17, 43, 67].

1.4 *Probabilistic techniques* based on Bayesian inference have been developed by Garnier et al. [30] as information theory. The method takes into account the probability of each amino acid having a particular secondary structure, and also considers the assumptions and probabilities of each structure contributing to that of its neighbours. The method is roughly 65 % accurate and is dramatically more successful in predicting alpha helices than beta sheets, which it frequently erroneously predicted as loops or disorganized regions [52].

1.5 *Neural Networks* use training sets of solved structures to identify common sequence motifs associated with particular arrangements of secondary structures. These methods are over 70 % accurate in their predictions, although beta strands are still often under-predicted due to the lack of three-dimensional structural information that would allow assessment of hydrogen bonding patterns that can promote formation of the extended conformation required for the presence of a complete beta sheet [52].

1.6 *Support Vector Machine* (SVM) is a method which analyzes data and recognizes patterns that are used for classification. SVM takes a set of input data and predicts, for each given input, which of two possible classes forms the input. SVM has proven particularly useful for predicting the locations of turns, which are difficult to identify with statistical methods data [60]. The SVM requirement for relatively small training sets has also been cited as an advantage to avoid over-fitting to existing structures. The major limitation for machine learning techniques is the attempt to predict more fine-grained local properties of proteins, such as multiple back bone dihedral angle in unassigned regions [87].

1.7 *Genetic Algorithm* (GA) is population of strings, which encode candidate solutions to optimize a problem to evolve a better solution. The fundamental *Genetic Algorithm* problems approach, due protein modelling interactions, are elaborate, hence *Genetic Algorithm* evaluations for complex problems are often the most prohibitive. *Genetic Algorithms* do not scale well with complexity, meaning that with increasing numbers of elements there is an exponential increase in the search space size. In general, the *Genetic Algorithm* has a tendency to optimize to local minima, and therefore becomes globally limited [79].

1.8 Ab initio models have been used exclusively to model binding sites of metals to proteins [80] or use statistical methods to find a secondary structures

[10, 35, 71]. *Ab initio* (from the beginning) refers to the beginning of an amino acid sequence.

There are several major limitations with all of these models. In general, all of the model simulations are restricted by computational time and cost. To minimize a structure, there is a need to consider every change of position in the configuration of a peptide or protein. In the case of establishing global minima, all possible energies need to be tested which can become an exponentially large problem [13, 36, 74, 81]. Further, existing models do not consider the dynamic behavior of proteins in an aqueous environment since they cannot consider water molecules [28, 38, 57, 76] primarily due to computer limitations [7, 12]. Secondary structure formation also depends on several other factors [90] including for example accessibility of residues to a solvent [47], the protein structural class [22], and even the organism from which the proteins are obtained [49]. These factors can further confound modelling paradigms.

While X-ray crystallography is used to validate a protein structure, it only shows an aspect of a structure in that it does not map the real aqueous environment of the protein that affects protein interactions and structure. In addition, X-ray crystallography does not describe the actual interactions that occur during the folding process which is a further limitation [82]. In contrast, nuclear magnetic resonance (NMR) can be used to show structure in real time in an aqueous environment, but so far this method is limited to 70 amino acids [63].

Based on the limitations of existing protein modelling techniques, other approaches need to be considered; for example, if the *volume and area* of the protein can be optimized, this optimization can be used to calculate primary but also tertiary structures. The method presented uses the optimal volumes to bring a structure closest to the lowest energy state.

2 Model

The polypeptide chain of a protein can be represented by a low-dimensional topology structure called a *braid* group [1, 2, 8]. It is defined as the union of arc lengths forming a string, or *braid* that can readily model the fundamental properties of the peptide backbone [48, 53]. Rather than considering a peptide as a linear sequence of amino acid residues, the peptide bonds of the protein chain form the arc lengths of our *braid*. A peptide chain thus consists of a series of rigid arc lengths carrying various substitute groups. Each arc length runs from the oxygen to carbon of the amide bond to, but not including the next peptide carbonyl carbon [9]. Folding the polypeptide chain into different conformations simply results in changing the relative orientation of these arc lengths. Although this grouping does not follow the biosynthetic pattern, it limits orientation changes to movements about the rotating C–CO bond given by Φ and Θ as shown in Fig. 1. Next, the volume of the chain can be minimized [57]. The volume, given by a protein, can be

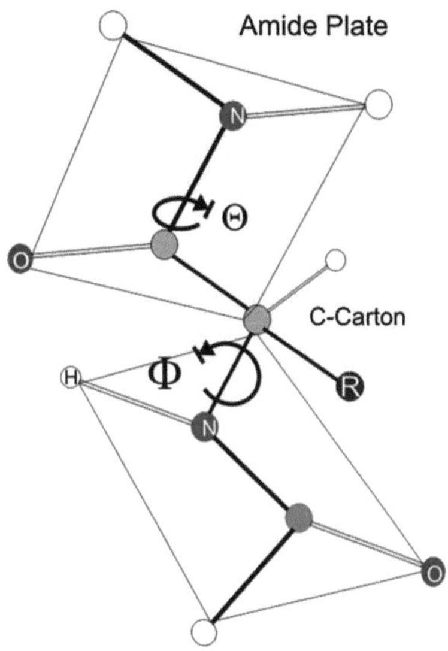

Fig. 1 Rotations around peptide groups. Two planar peptide groups are shown in this illustration. The only reasonably free movements are rotations around the C_α–ON bond (measured as Φ) and the C_α–OC bond (measured as Θ). By convention, Φ is 180° and Θ is 90° in the conformation shown and increase, as indicated, in the clockwise direction when viewed from C_α

thought of as a backbone with additional groups attached to it. The C–N backbone is not straight because the bonds are not collinear. For example, carbon forms single bonds that are spaced equally apart from each other form a tetrahedron angle (109.5°) [83] rather than straight chains in the case of the residues.

With the distances between the C–N backbone being non-collinear, the angles between directional vectors are not parallel. From this, the groups are replaced with an outline of the atoms centered on the backbone so that we have strings of *beads* (though the bead shape is not round [21]). The lacing of the *beads* is the backbone of the protein is shown in Fig. 2a. Each of the amino acids has bonds that can rotate. In most cases there are two bonds that rotate. The R groups (amino acid side chains) can take one of several states. In the case of proline there is only one free rotating bond (to the H) [28], this is handled by an error function that adds a large penalty to the optimization function such that the bond will stay at the optimal angle (Fig. 2b). Generally, all bonds are considered to be of fixed length and only rotation is allowed. The angles are the parameter. The volume and surfaces are the results. The simplex method requires only functions given by the objective function [41, 55]. The bond lengths never change, the only change occurs in the two angles per residue in our configuration search [85].

The rotation of non-collinear bonds allows the molecule to twist, similar to a Rubik's cube puzzle toy where a set of angles are joined by rotating joints. This rotation allows the protein to take a shape [56]. The molecule can be twisted to nearly any shape, but the proper shape is achieved by optimization of an objective function. The objective function mirrors an energy function.

The method used for the geometrical model is *trilateration*, which is a type of measurement that determines a point by using the geometry of spheres, circles, or triangles. Unlike triangulation, which uses the measurement of angles to determine

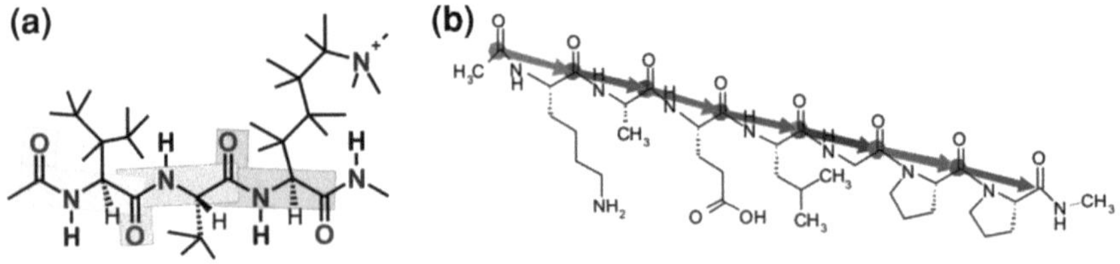

Fig. 2 **a** Beads are created by lacing the amino acid along the N–O backbone to construct an initial stage of the protein model. **b** Illustration of the optimized orientation given by the arrows of the minimization along the N–O backbone

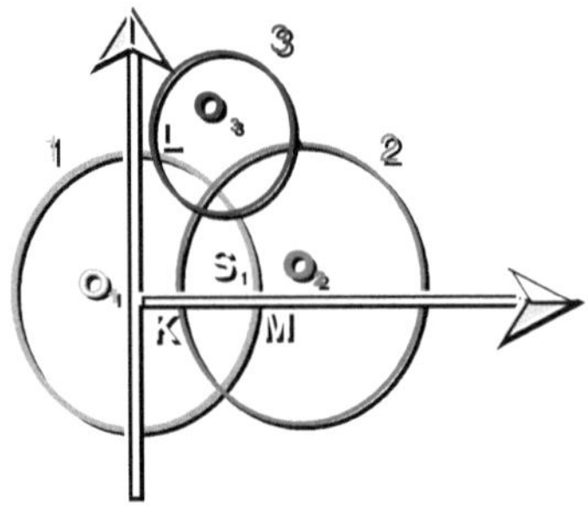

Fig. 3 Creating the surface for the bead building. The three overlapping/intersecting spheres with centers O_1, O_2 and O_3 form a new object and the net area is given by $A = \left[\frac{1}{2}(A_{O_1} + A_{O_2} + A_{O_3})\right]$ – Internal Area

location, trilateration uses measures of distance. In two-dimensional geometry, the radii of circles can sometimes be used to find the location of a point. Hence, the surface is calculated from the intersection of the surfaces (atomic radii) of the atoms in the residue [64]. For one single sphere the volume is $V = {}^{4}\!/_{3}\pi r^3$ and surface is $S = 4\pi r^2$. For two spheres the volume is $V = V_1 - V_2 - V_L$ and the surface is $S = S_1 + S_2 - S_L$ (V_1, V_2 are volume one and volume two; S_1, S_2 are surface one and surface two; V_L, S_L are overlapping).

The model requires enough volume for the distances and surfaces. The surfaces are determined by allowing the 3 overlapping and intersecting spheres of centers O_1, O_2 and O_3 to form a new object with defined volume. Its net area will be given by:

$$A = \left[\frac{1}{2}(\mathrm{A}_{O_1} + \mathrm{A}_{O_2} + \mathrm{A}_{O_3})\right] - \text{Internal Area}$$

where the internal area is given by (Fig. 3):

$$\text{Area}_{123} = \int_{L}^{K} \text{Area}(S_1)dS_1 + \int_{K}^{M} \text{Area}(S_1)dS_1 + \int_{M}^{L} \text{Area}(S_1)dS_1 \text{ and } 0<K<M$$

We obtain the minimal free energy by minimizing the volume which we have accounted for by the surface areas as calculated above. Note that more than three overlapping spheres are improbable for atoms in molecules [20]. The objective function for the minimization is given by:

$$\begin{aligned}\text{Objective} &= \text{volume(volumeweighing)} \\ &\quad - \sum \text{surfacearea}_{12}(\Phi, \Theta)\text{hydrophobicity}_1\text{hydrophobicity}_2\end{aligned}$$

The volume weights are proportional to the amount of energy needed to move the R from cyclohexane to water (0 is neutral, -1 is hydrophilic and 1 is hydrophobic), using the surface of the whole amino acid, rather than just the R group [65]. Each residue has a surface area with a known hydrophobicity [12]. The summation is over each set of residues that are touching (i.e. adjacent to each other). The surface area is the common surface area between the residues. This term will tend towards having hydrophobic residues together and hydrophilic ones together, but will avoid having hydrophilic next to hydrophobic residues.

There is also a volume term that minimizes the size of the molecule [11, 58]. This volume is given as the volume enclosed by the surface wetted by a solvent molecule with a $\sim$1.4 Å radius. The model is based on changing the distance and the angles and are adjusted to the distances. If we do not know the structure, we can calculate in parallel the α-coil and β-sheet simultaneously forming the braid used to obtain a globular or complex structure.

2.1 Constraints in the Standard Braid Theory

Prohibiting the braids from incidental intersection with themselves or with other braids is properly observed (rule of “no intersection”) in this application to keep the modeled peptide chains from overlapping each other (Appendix 2). The simple arc length model has been expanded to address the finite volume occupied by each amino acid residue [48]. While keeping the length and direction of the arc lengths constant, each segment is expanded into a *bead* enveloping the remainder of its amino acid residues. Each *bead* interacts with at most two other *beads*, and the intersection of any two sequential beads is a single point [39]. A *braid* now represents the peptide chain, which is a collection of *beads* (Appendix 3). Various peptide conformations can now be treated as changes in the relative orientation between pairs of *beads*. For large, single-chain proteins this is a significantly simplified approach to molecular modelling (Appendix 4; Appendix 5). The power of this approach is seen when considering protein structures composed of multiple peptides, such as the collagen triple helix. The model presented in this section includes the collagen fibril as a three stranded *braid*.

2.2 Orientation of Protein Structures

Collagen is a 3_{10} helix which contains 3.6 residues per turn and has 10 atoms in the ring which is formed by making the hydrogen bond three residues up the chain [32]. The distance is determined by H bonds that lie parallel to the helix, and the carbonyl groups pointing along the axis in the opposite direction [66]. The opposite direction of nitrogen and the carbonyl will set the preference distance and define the α-helix. Since the direction is measured from the carbonyl, the distance between each turn is 3.6 residues [3, 26, 77, 86].

A protein β-sheet orientation is symmetric. The β-sheet is measured from the nitrogen terminal to the carbonyl terminal. The residue, carbonyl, and nitrogen are on the same side [5]. The inter-strand symmetric amide proton is the donor of the hydrogen bond to the carbonyl. For an anti-parallel orientation the exchange is perpendicular. This is not the case for a parallel orientation. The distance between residues is $\sim$0.347 nm for anti-parallel and $\sim$0.325 nm for parallel pleated sheet. Parallel β-sheets tend to be more regular than anti-parallel β-sheets [73]. The range of angles Φ and Θ angles for the peptide bonds in parallel sheets is much smaller than that for anti-parallel sheets [72]. Parallel sheets are typically large structures and sheets that are composed of less than five strands are rare [24]. Anti-parallel sheets however typically consist of only a few strands [50].

Parallel sheets characteristically distribute hydrophobic side chains on both sides of the sheet, while anti-parallel sheets are usually arranged with all their hydrophobic residues on one side of the sheet [57]. This required an alteration of hydrophilic and hydrophobic residues in the primary structure of peptides involved in anti-parallel β-sheets because alternate side chains project to the same side of the sheet [37].

In general the N–H and the C=O (each with an individual dipole moment) need to be in the same plane to create a large net dipole for the structure whether it is an α, β or 3_{10} structure [46].

3 Computer Program

A computer program was developed to implement area minimization [14, 41]. The program accepts Protein Data Base (PDB) files or just amino acid sequences as input.

Input Parameters The data from the PDB or the amino acid sequence are randomized to simulate pre-coiling of the structure. The data are *relaxed* [23, 42, 54] and flattened by randomly given the Θ and Φ bonds a twist [4]. All the atoms stay in relative position inside of a bead [44]. The energy minimization is determined by the *downhill simplex method*, which is a heuristic search method and which converges to a non-stationary point. The *objective function technique* is implemented in multiple-space. The method uses a simplex which is a generalization of a triangle or tetrahedron and can be used for line segments or triangles [55].

Output The output consists of a molecular coordinate system data file that is exported to RasMol (Molecular Visualization, Freeware) molecular graphics visualization tool for export to a GIF three-dimensional image. Five amino acid sequences for five proteins, including collagen, were tested using the minimizer and compared to their published three-dimensional structures for confirmation. The deviation between the ProteoRubix^{TM} model and the known model was determined.

4 Results

We considered the following examples of available Protein Data Base (PDB) files in three categories of proteins for the minimization process: 3_{10} coil (i.e. collagen consists of 3 coils and the coil turns every 10 amino acids); α-coil; and, globular structure protein. The minimization program was used to create three-dimensional structures of five proteins and was compared with data from the original protein crystallized structures. A correlation was achieved for each of the proteins as described below. For illustration purposes, *dumps* (i.e. the folding process export as data files for visualization) were added between the stages such that twisted regions *unfold*. The data from the original crystallized protein structures were compared to the data obtained after minimization. Different structural protein types were minimized by using the objective function and compared to known structures using a sequencer alignment.

4.1 3_{10} Coil Protein

For 1BBF glycoprotein, the source of the original reference structure was a theoretical model [56]. Figure 4a shows the matching superposition of the original file given by the green strand and the minimized construct shown by the purple backbone line along the green strand 1BBF-1. Figure 5 shows the sequence alignment given by Clustal-W. The sequences of the three coils are superimposed and differentiated by Coil (chain) A, Coil B and Coil C; note that the superposition shows the alignment between the minimized file and the original file. Figure 6 shows the local and global root mean square difference (RMSD) of the α-carbon and the back bone with distances in Angstroms in the structural alignment of the

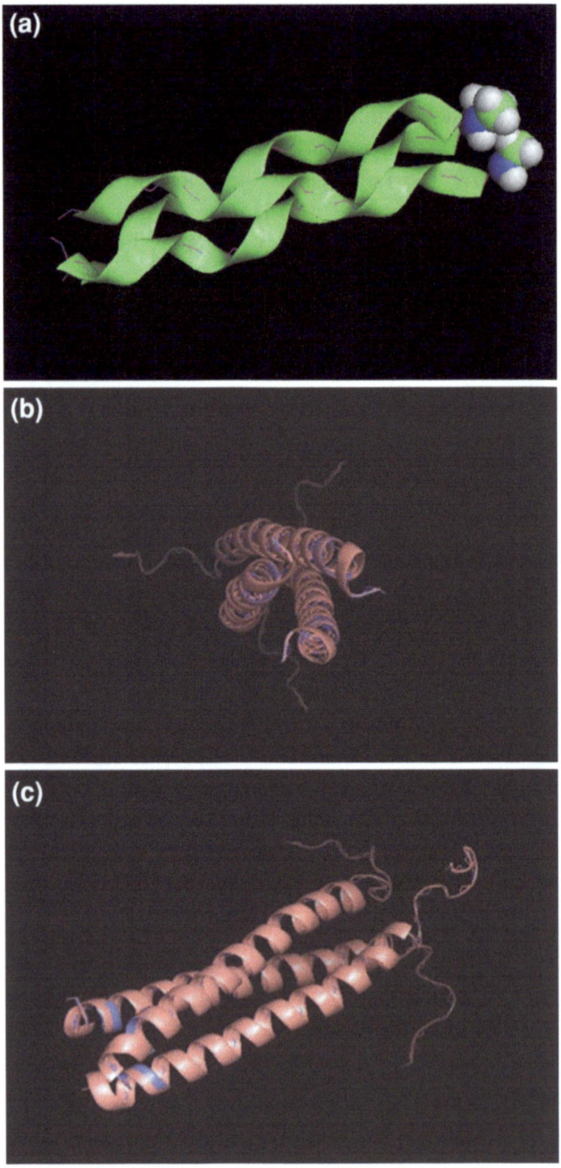

Fig. 4 Correlation between known protein structures and the minimized files. **a** Superposition of 1BBF collagen protein shown by the *green* is the known original and the *blue–green line* is minimized file showing excellent fit between the two models. **b** The *frontal view* of the superimposed files of original (*pink*) and the minimized (*purple*) models (with side chains) for 1AQ5 chicken cartilage matrix protein. Near-identical structural correlation was seen. **c** *Side view* of the superposition; of note are the tips of the coils that show the same alignment

superposition of the known and the modeled protein. The heavy atoms (one with a significantly higher atomic scattering factor than the others) are derivative structure factor amplitudes or changes along RMSD per molecule. The non-bonding energies used in order to minimize modelling of inter-molecular interactions can be integrated to the trilateration model [40].

CLUSTAL W Multiple Sequence Alignment

1BBF model chain A	**-PPGPPGPPGPPG**
PDBB model chain A	**-PPGPPGPPGPPG**
1BBF model chain C	**PGPPGPPGPPGP—**
PDBB model chain C	**PGPPGPPGPPGP—**
1BBF model chain B	**--GPPGPPGPPGPP-**
PDBB model chain B	**--GPPGPPGPPGPP-**

Fig. 5 The Clustal-W sequence alignment denoted in *colour* show identical matched regions of alignment between the original known 1BBF collagen file and the "PDBB" minimized file

Local RMSD-1BBF

	Alpha Carbons	Back Bone	Heavy	All
RMSD	0.148	0.127	0.158	0.193

Global RMSD-1BBF

	Alpha Carbons	Back Bone	Heavy	All
RMSD	0.148	0.127	0.158	0.193

Fig. 6 Root Mean Square Difference (RMSD) resulting from the superposition of the 1BBF collagen and the minimized model. The RMSD considers local and global solutions for the C_α, the back bone and the heavy portion of the atoms (i.e. every atom in the molecule, with exception of the hydrogen). The distances are in Angstroms

4.2 α-Coil Protein

1AQ5 is a coil–coil cartilage matrix protein of *Gallus gallus* (chicken). Based on NMR [84], Fig. 4b and c show frontal and side superposition of the known original file and the minimized file with excellent alignment. Figure 7 shows Clustal-W superposition of sequence of the two files with complete alignment. The adjoining coils are the sequence matched in the superposition and the others unmatched by Clustal-W. Figure 8 shows the local and global RMSD of the α-carbon and the back bone. The local minimization gives the RMSD per amino acid per carbon bone and global minimization gives the RMSD of the sequence. The RMSD compares atoms and the total number back bones of the PDBs. Figure 9 shows a graph of RMSD per amino acid.

CLUSTAL W multiple sequence alignment for 1AQ5

1AQ5_model_1_chain_A
GSHMEEDPCECKSIVKFQTKVEELINTLQQKLEAVAKRIEALENKII
1AQ5_model_1_chain_C
GSHMEEDPCECKSIVKFQTKVEELINTLQQKLEAVAKRIEALENKII
PDBB-AQ5_model_default_chain_A
GSHMEEDPCECKSIVKFQTKVEELINTLQQKLEAVAKRIEALENKII
PDBB-AQ5_model_default_chain_B
GSHMEEDPCECKSIVKFQTKVEELINTLQQKLEAVAKRIEALENKII
PDBB-AQ5_model_default_chain_C
GSHMEEDPCECKSIVKFQTKVEELINTLQQKLEAVAKRIEALENKII
1AQ5_model_1_chain_B
GSHMEEDPCECKSIVKFQTKVEELINTLQQKLEAVAKRIEALENKII

Fig. 7 The Clustal-W for the alignment between the original sequence of 1AQ5 and the minimized structure labeled as PDBB-1AQ5; *colors* identify each chain and matching of the files

Local RMSD-1AQ5

	Alpha Carbons	Back Bone	Heavy	All
RMSD	0.02	0.03	0.03	0.12
Atoms	47	188	303	414

Global RMSD-1AQ5

	Alpha Carbons	Back Bone	Heavy	All
RMSD	0.02	0.03	0.03	0.12
Atoms	47	188	303	414

Fig. 8 The RMSD calculations resulting from the superposition of the 1AQ5 and the minimized model. The RMSD considers local and global solutions for the C_{α}, the back bone and the heavy portion of the atoms (i.e. every atom in the molecule, with exception of the hydrogen). The distances are in Angstroms

4.3 Globular Proteins

1CQD, a globular protein, is a hydrolase from *Zingiber officinale* (ginger). Choi et al. [19] studied the structure via X-ray diffraction. Figure 10a shows the superposition of the original known file (green) with the minimized file (purple). Figure 11 shows the Clustal-W superposition of the sequence of the two files with 1,294 total number of amino acids. The colors identify the individual chains and the similar strands that are shown together. The superimposed structures of the PDB and the minimize file match. Figure 12 shows the local and global RMSD of the α-carbon and back bone. The local gives the RMSD per amino acid per carbon bone and global gives the RMSD of the sequence. The RMSD compare atom and the total number back bones of the PDBs. Figure 13 shows the graph of RMSD per residue and number of residues.

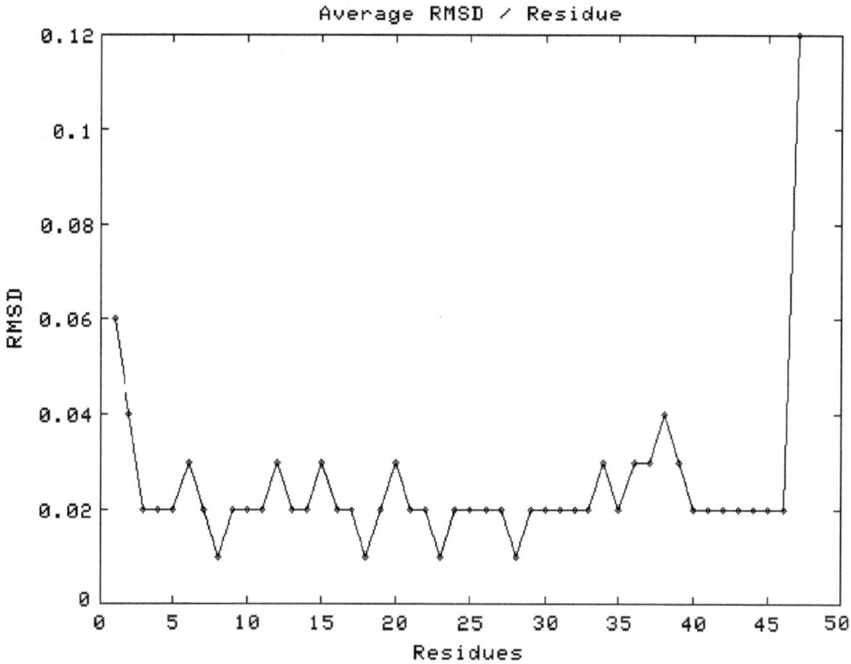

Fig. 9 The RMSD per residue, with the spikes showing the difference between the original file and the minimized file for 1AQ5

1AQP, a blood clotting protein from *Bos Taurus* (cattle), has been modeled in 1997 via X-ray diffraction [6]. Figure 10b shows the superposition of the known model and the minimized model. The grey color shows the superimposed model matching the original file. Figure 10c shows ribbons to depict the protein and the minor differences between the original file in yellow and the minimized file in gray; the distances of the differences are at ~9 Å. Figure 14 shows the local and global RMSD of the α-carbon and Back Bone. The local gives the RMSD per amino acid per carbon bone and the global gives the RMSD of the sequence of 124 residues.

To further validate our methods, we compared the predicted structures with the respective PDB entries using the alignment program STRAP [29, 33, 34] freely obtainable from http://3d-alignment.eu/. To compare two given 3D-structures of proteins, STRAP uses the method of 3D-superposition. STRAP implements several different back-ends for computation. Here TM-align [88] was used to perform a rigid superposition. TM-align moves one of the two models in space until both structures coincide as best as possible. There is a trade off between a low RMSD and a high percentage of assigned C-alpha positions. The result is a translation vector and rotation matrix. STRAP processes the result. It determines the RMSD which is a measure for the dissimilarity of two structures, derives a (multiple) sequence alignment and visualizes the result in Pymol [25].

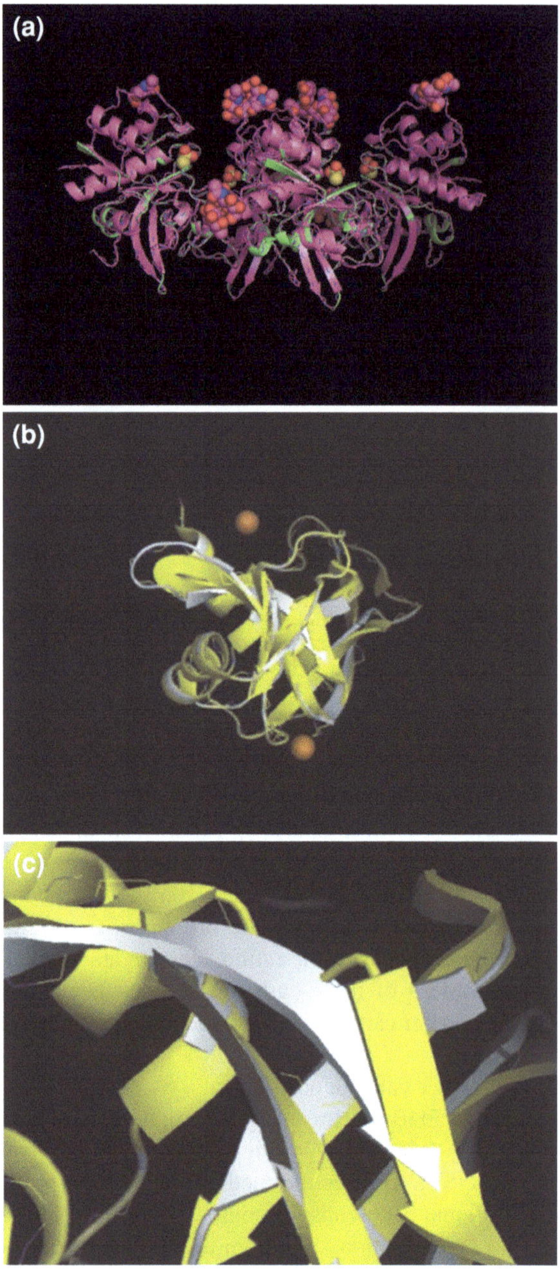

Fig. 10 Correlation between known protein structures and the minimized files. **a** The superposition of known (*green*) and minimized (*purple*) models for 1CQD globular protein from ginger showing high match pattern between the two structures. **b** The superposition of the known original model (*yellow*) and the minimized model (*grey*) for 1AQP blood clotting protein from cattle, illustrating the matched regions and locations. **c** Zoomed image of the superposition at 9 Angstroms

The analysis was done using SuperPose software [51]. Minimizer output was in the format of PDB. STRAP software was used for the PDB alignment and Clustal-W software (Clustal-W: Multiple Alignment for DNA or Proteins, Freeware) [78] was

CLUSTAL W (1.83) multiple sequence alignment for 1CQD

```
PDBB_model_default_chain_A   LPDSIDWRENGAVVPVKNQGGCGSCWAFSTVAAVEGINQIVTGDLISLSE
PDBB_model_default_chain_B   LPDSIDWRENGAVVPVKNQGGCGSCWAFSTVAAVEGINQIVTGDLISLSE
PDBB_model_default_chain_C   LPDSIDWRENGAVVPVKNQGGCGSCWAFSTVAAVEGINQIVTGDLISLSE
1CQD_model_default_chain_D   LPDSIDWRENGAVVPVKNQGGCGSCWAFSTVAAVEGINQIVTGDLISLSE
PDBB_model_default_chain_D   LPDSIDWRENGAVVPVKNQGGCGSCWAFSTVAAVEGINQIVTGDLISLSE
1CQD_model_default_chain_C   LPDSIDWRENGAVVPVKNQGGCGSCWAFSTVAAVEGINQIVTGDLISLSE
1CQD_model_default_chain_B   LPDSIDWRENGAVVPVKNQGGCGSCWAFSTVAAVEGINQIVTGDLISLSE
1CQD_model_default_chain_A   LPDSIDWRENGAVVPVKNQGGCGSCWAFSTVAAVEGINQIVTGDLISLSE
**************************************************

PDBB_model_default_chain_A   QQLVDCTTANHGCRGGWMNPAFQFIVNNGGINSEETYPYRGQDGICNSTV
PDBB_model_default_chain_B   QQLVDCTTANHGCRGGWMNPAFQFIVNNGGINSEETYPYRGQDGICNSTV
PDBB_model_default_chain_C   QQLVDCTTANHGCRGGWMNPAFQFIVNNGGINSEETYPYRGQDGICNSTV
1CQD_model_default_chain_D   QQLVDCTTANHGCRGGWMNPAFQFIVNNGGINSEETYPYRGQDGICNSTV
PDBB_model_default_chain_D   QQLVDCTTANHGCRGGWMNPAFQFIVNNGGINSEETYPYRGQDGICNSTV
1CQD_model_default_chain_C   QQLVDCTTANHGCRGGWMNPAFQFIVNNGGINSEETYPYRGQDGICNSTV
1CQD_model_default_chain_B   QQLVDCTTANHGCRGGWMNPAFQFIVNNGGINSEETYPYRGQDGICNSTV
1CQD_model_default_chain_A   QQLVDCTTANHGCRGGWMNPAFQFIVNNGGINSEETYPYRGQDGICNSTV
**************************************************

PDBB_model_default_chain_A   NAPVVSIDSYENVPSHNEQSLQKAVANQPVSVTMDAAGRDFQLYRSGIFT
PDBB_model_default_chain_B   NAPVVSIDSYENVPSHNEQSLQKAVANQPVSVTMDAAGRDFQLYRSGIFT
PDBB_model_default_chain_C   NAPVVSIDSYENVPSHNEQSLQKAVANQPVSVTMDAAGRDFQLYRSGIFT
1CQD_model_default_chain_D   NAPVVSIDSYENVPSHNEQSLQKAVANQPVSVTMDAAGRDFQLYRSGIFT
PDBB_model_default_chain_D   NAPVVSIDSYENVPSHNEQSLQKAVANQPVSVTMDAAGRDFQLYRSGIFT
1CQD_model_default_chain_C   NAPVVSIDSYENVPSHNEQSLQKAVANQPVSVTMDAAGRDFQLYRSGIFT

1CQD_model_default_chain_B   NAPVVSIDSYENVPSHNEQSLQKAVANQPVSVTMDAAGRDFQLYRSGIFT
1CQD_model_default_chain_A   NAPVVSIDSYENVPSHNEQSLQKAVANQPVSVTMDAAGRDFQLYRSGIFT
**************************************************

PDBB_model_default_chain_A   GSCNISANHALTVVGYGTENDKFWIVKNSWGKNWGESGYIRAERNIENP
PDBB_model_default_chain_B   GSCNISANHALTVVGYGTENDKFWIVKNSWGKNWGESGYIRAERNIENP
PDBB_model_default_chain_C   GSCNISANHALTVVGYGTENDKFWIVKNSWGKNWGESGYIRAERNIENP
1CQD_model_default_chain_D   GSCNISANHALTVVGYGTENDKFWIVKNSWGKNWGESGYIRAERNIENP
PDBB_model_default_chain_D   GSCNISANHALTVVGYGTENDKFWIVKNSWGKNWGESGYIRAERNIENP
1CQD_model_default_chain_C   GSCNISANHALTVVGYGTENDKFWIVKNSWGKNWGESGYIRAERNIENP
1CQD_model_default_chain_B   GSCNISANHALTVVGYGTENDKFWIVKNSWGKNWGESGYIRAERNIENP
1CQD_model_default_chain_A   GSCNISANHALTVVGYGTENDKFWIVKNSWGKNWGESGYIRAERNIENP
*************************************************

PDBB_model_default_chain_A   DGKCGITRFASYPVKK
PDBB_model_default_chain_B   DGKCGITRFASYPVKK
PDBB_model_default_chain_C   DGKCGITRFASYPVKK
1CQD_model_default_chain_D   DGKCGITRFASYPVKK
PDBB_model_default_chain_D   DGKCGITRFASYPVKK
1CQD_model_default_chain_C   DGKCGITRFASYPVKK
1CQD_model_default_chain_B   DGKCGITRFASYPVKK
1CQD_model_default_chain_A   DGKCGITRFASYPVKK
```

Fig. 11 The Clustal-W file chains are identified by *colors* and show alignment matches between 1CQD (control file) and PDBB (minimized output)

used for the sequence between the known structure and the minimized structure. The STRAP software followed the structural trace through to the C-termini and takes the C-termini of both files to compare. No regions were ignored or skipped (i.e., even loops were carefully considered and aligned). The alignment was constructed with the primary aim of maximizing the aligned positions between structures, provided that there was a rational basis for the alignment. Excellent alignment was achieved.

5 Discussion

The implementation of the *ab intio* model as a geometry solution was considered since other approaches for protein modelling were fraught with difficulty due mainly to insufficient computation resources. The complexity of the protein

Local RMSD -1CQD

	Alpha Carbons	Back Bone	Heavy	All
RMSD	0.01	0.01	0.21	0.21
Atoms	216	864	1294	1294

Global RMSD -1CQD

	Alpha Carbons	Back Bone	Heavy	All
RMSD	0.01	0.01	0.21	0.21
Atoms	216	864	1294	1294

Fig. 12 The RMSD calculations for superposition of the 1CQD with the minimized model. The RMSD considers local and global solutions for the C_{α}, the back bone and the heavy portion of the atoms. The number of residues that were considered was 1,294

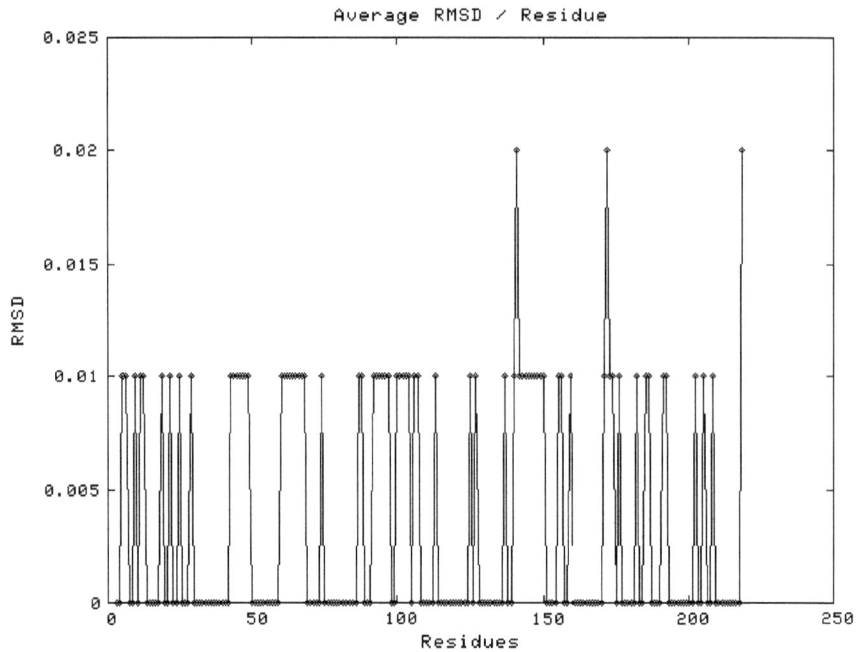

Fig. 13 RMSD per residue. The *spikes* show the difference between the original file and the minimize file for 1CQD

folding, and also the numbers of interactions needed to give a solution, leads to the NP problem. Furthermore, the actual environmental conditions such as interaction of protein in an aqueous environment increase the complexity of this challenge.

Local RMSD-1AQP

	Alpha Carbons	Back Bone	Heavy	All
RMSD	0.01	0.01	0.21	0.21
Atoms	124	496	785	929

Structure	Residues
PDBA	1-124
PDBB	1-124

Global RMSD-1AQP

	Alpha Carbons	Back Bone	Heavy	All
RMSD	0.01	0.01	0.02	0.01
Atoms	124	496	785	929

Structure	Residues
PDBA	1-124
PDBB	1-124

Fig. 14 The RMSD calculations from the superposition of the 1AQP with the minimized model for 929 residues

Based on our alternative approach, we were able to achieve accurate models for known proteins based on PDB files or amino acid sequences. The alignments were shown between the modeled and known structures via Clustal-W. The alignment differences given by the standard deviation (RMSD) showed minimal differences.

To summarize, the peptide *braid* provided the mathematical foundation that described the model and the three dimensional structure. The union arc-lengths for the model were given by the dipole moment (plane) of the N–H and the C=O. The differential geometry was used to create the algorithm for the projections in the rotations of the angles and the surface of the structure.

The *minimizer* calculated complex structures based the interactions between each of the 20 types of residues. This only gives 210 possible interactions and includes the interactions between the solute with the sodium or other charged ions. The residue interaction was obtained by assigning properties to the surface charges and hydrophobicity.

For the examples considered, we used Clustal-W and STRAPS for the alignment and RMSD for the verification. The simulations assumed residue interactions but ignored charges and sulfur bridges that are not systematic (all high by an eV or two). To make the model more physiological, we added the solute component, and assumed that the structure was formed in plain water with no ionic charge, and a size of ~ 1.8 Å. As an example, if we add a charged item to the solute, this did influence charge attraction by shielding the charged residues. If the charged species interacted at the site of hydrogen or other ions, the ionic size would alter the protein size and additional terms would be required to account for this based on species size. These changes can readily be accomplished within the software and hence numerous further applications are possible.

One limitation that we encountered with our approach was an objective function that required readjustment when distances approached gapping in the protein structure that were over 10 %. A solution for this would be to add the functionality of AMBER software (Assistance Model Building with Energy Refinement) where the energy for the distance is predicted and can therefore control or refine the distance [16, 18, 59, 62]. The added approach does not consider the atoms, but simply the distance, and can deal with large or small sequences.

6 Conclusion

We present a new model to solve three-dimensional modelling of proteins that can create supra structures within tissues. By considering amino acid peptides as beads, hydrogen bond distances, the surface and distance geometry as functional components, and by focusing on folding, the Ab initio protocols can successfully model complex proteins in three dimensions with minimal computational load.

Acknowledgments The authors thank: Isabella Verdinelli and Lauren Ernst for the discussions related to the model and Troy Wymore from the Pittsburgh Supercomputing Center for his suggestions in verifying the models and a special thanks to Dr. Alex Cohen from ProteoRubix Systems who developed the minimizer and for discussion and development of ProteoRubix™ software for the Geometry Modelling. Chris Holm for editing the manuscript.

Appendix 1

NP

Suppose the bead involves d dihedral angles. Let $\phi^* = (\phi_1^*, \ldots, \phi_n^*) \in [0, 360]^d$ be an optimal solution to the constrained optimization problem

(1.1) $\min_{\phi \in [0,360]^d} \{v(\phi : \phi \text{ is arotation about the bond i})\}$

Then there is a maximal number $n > 0$.

(1.2) The p^n hard problem of an exhaustive search over the angles $0 \leq \phi_i^1 < \cdots < \phi_i^P \leq 360$ to find an approximate optimizer $\overline{\phi}$ to ϕ^* may be possible for a modern computer.

(1.3) There is exactly one solution to (4) in $|\overline{\phi_i} - p, \overline{\phi_i} + p|$ which would be ϕ^* and can be approximated using a given constrained optimization algorithm (**B 1**).

The convergence ball for the constrained optimization algorithm provides a candidate for p in the proposition. Using this proposition, we can obtain an acceptable initial condition for a constrained optimization algorithm.

Fig. 15 The sequence [(a, b) is length of a single coil] gives an approximation to the polygon arc P; the length between two points (a, b), where D is segments of arc-length

Appendix 2

Boundary Determination to Prevent Overlap

A path is a one-dimensional sub-manifold M of R^3, so that, for any point $x \in M$ there is a local parameterization near x. $C^k (k \geq 2)$ denotes the curvature of the path and D denotes the coordinates identifying the path. The output of each iteration is a set of coordinates in three dimensions, $(D = x_1, x_2, \ldots, x_n)$ identifying a path. We denote by *length bond* is the polygonal arc around the path (Fig. 15). The curvature C^k and the arc-length are non-regular. Let $x = x(t)$, with $a \leq t \leq b$ and consider a partition [15]:

$a = t_0 < t_1 < \cdots < t_n = b$, of an interval (a, b).

The sequence (a, b) are the boundaries of a single coil) gives an approximation to the polygon arc C. As illustrated the length between two points (a, b), where D are segments of arc-length given by:

$$\lambda(D) = \sum_{j=1}^{n} D_j = \sum_{i=1}^{n} \|x_i - x_{i-1}\| = \sum_{i=1}^{n} \|x(t_i) - x(t_{i-1})\| \tag{2.1}$$

The arc-length can be bounded from above and from below. The upper bound is given by:

$$\rho_{+}(K, D) = \frac{1}{\lambda(D)} \sum_{(K \circ D_j) \cap D \neq \emptyset} \lambda(K \circ D_j) \tag{2.2}$$

And the lower bound is:

$$\rho_{-}(K, D) = \frac{1}{\lambda(D)} \sum_{(K \circ D_j) \subset D} \lambda(K \circ D_j) \tag{2.3}$$

where $\rho_+(K,D)$ is the ratio of the total measure of the set in the system K (is the volume minimization) so that the transformation $^\circ$ (*projection*) of the segments and the curve C give the lower and the upper bound(a,b).

$$b = \rho_{+} = \lim_{\lambda(D) \to \infty} \sup \rho_{+}(K, D) = \lim_{\lambda \to \infty} \operatorname{Sup} \rho_{\substack{+ \\ \lambda(D) \geq \lambda}} (K, D) \tag{2.4}$$

$$a = \rho_{-} = \lim_{\lambda(D) \to \infty} \inf \rho_{-}(K, D) = \lim_{\lambda \to \infty} \inf_{\lambda(D) \geq \lambda} \rho_{-}(K, D) \tag{2.5}$$

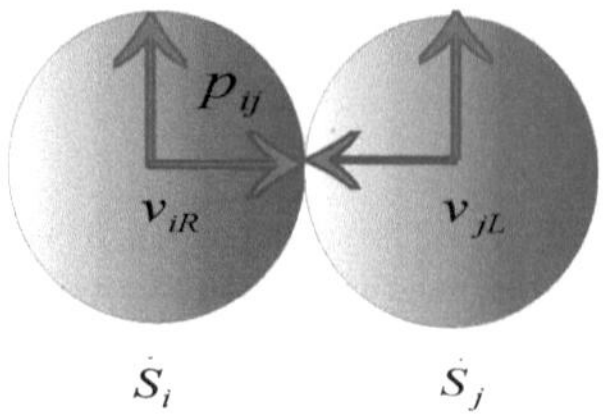

Fig. 16 The radii are chosen so that the intersection of the closure of any two beads $\bar{S}_i$ and $\bar{S}_j$ is a single point p_{ij},. The point p_{ij}, is the origin of a right and a left vector v_{iR}, v_{jL}

Hence the boundaries of C are given in (2.4) and in (2.5).

Appendix 3

The geometry structure of the protein is defined by a braid (see **B 2** for a description of a chain as a collection of *beads* forming a *braid*). The jth molecule of the chain is fitted to a conveniently shaped open *bead* Sj (see **B 3**) with is 0 center located at the center of the *bead* and the radius r_i has size such that the ith *bead* does not overlap with the jth *bead* when $i \neq j$.

The radii in Fig. 16 r_i are chosen so that the intersection of the closure of any two beads $\bar{S}_i$ and $\bar{S}_j$ is a single point p_{ij}, (see **B 3**). The point p_{ij}, is the origin of a right and a left vector v_{iR}, v_{jL}. In this process it is important to translate (projection) and rotate these vectors. The mathematics of this construction justifies geometry of the bead construction.

Appendix 4

With our model of collagen in mind, we next introduced the concept of the *braid* group. The *braid* was defined as the union of the backbones creating a string representing the amino acids. The collagen has three strands (as a group) or *coils* and each strand has a back bone, represented as the union of all points x (t_{i-1}, t_i) that are generated:

$$\text{Bonds} = \bigcup_{n=1}^{N} \{x(t_{i-1}, t_i)\} \tag{4.1}$$

A braid is a collection of *beads* for which two operators $(\circ, =)$ can be defined. The *bead* in the collection can be *projected* using least of the squares. Let B denote this collection of *beads*, so B = (braids), and (B,∘) is a group. We are checking the segments of the radius of bead of a single *braid*. The *enclosed volume* shrinks driven by minimization and through the homoeopathy is guaranty [2] (see **B 5**). We are modelling three coils, and their geometrical configuration has an equivalence class denoted by σ_i and σ_i^{-1}. A *braid* is equivalent and it is called *isotope* if the three coils cannot pass each other or themselves without intersecting [8] Fig. 17.

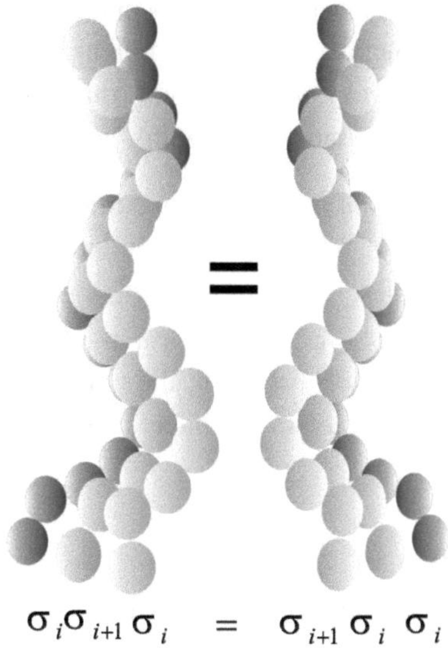

Fig. 17 A *braid* is equivalent and it is called *isotope* if the three coils cannot pass each other or themselves without intersecting

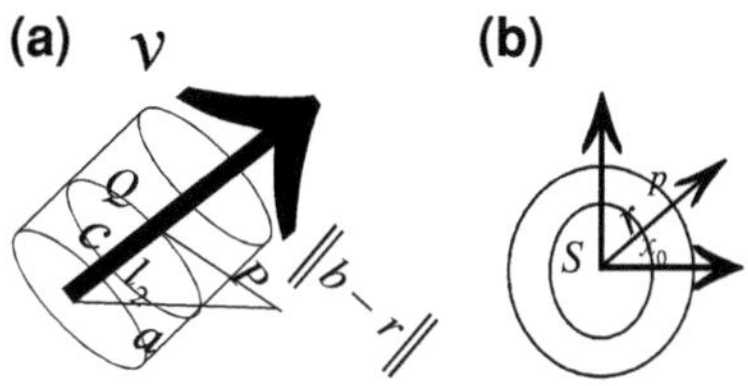

Fig. 18 The projection of the rotation of the rotation angles. The *enclosed volume* shrinks by the minimization and via homoeopathy which is guaranteed

$\sigma_i\sigma_{i+1}\sigma_i = \sigma_{i+1}\sigma_i\sigma_{i+1}$ if $1 \leq i \leq n-2$ [1]

Appendix 5

The distances of the projection to P is given by $\|b-r\|$, where $v(x-p)=0$ and by Pythagorean gives us that $b^2 = c^2 - a^2$, where, $a^2 = \left\| \frac{b}{\|b\|_2} \times (Q-P) \right\|_2^2$, $c^2 = \|Q-P\|_2^2$ or $\frac{b(Q-P)^2}{\|v\|_2}$ shown in Fig. 18a.

B 1

Let $\bar{\phi}$ the solution to $\phi^* = \left(\phi_1^*, \ldots, \phi_n^*\right) \in [0, 360]^d$ and d dihedral angle, $\phi_n^* = \sum^* \rightarrow N$, $1 \leq n \leq k$; Let q, r be polynomial such $\phi_n^*(I) \leq q(|I|)$, where I is the instance of the angle in our problem. Then test instance construction system for all the angles of our problem $(TICA)$, then $P = NP$.

Conversion We know that $\phi^* = (\phi_1^*, \ldots, \phi_n^*) \in [0, 360]^d$ is the optimal solution, where d dihedral angles then $\delta = \frac{1}{2} \min_{\phi \in [0,360]^d} \{v(\phi : \phi \text{ is a rotation about the bond } i)\}$ n is the maximum number of angles, $n > 0$ and $\delta > 0$. Let $\in\ > 0$ be given. Where ϕ^* is continuous, there is a point $p \in \phi^*$, $\phi^* \leq \frac{1}{2}\phi(p)$ where implies $|\overline{\phi_i} - p, \overline{\phi_i} + p| < \in$ and $v(p) \leq \frac{1}{2}\phi(p)$, we have $|\overline{\phi_i} - p, \overline{\phi_i} + p| \leq \phi^* + |v(p)| < \delta + \frac{1}{2}\phi(p) < \in$

Uniqueness using the existence and uniqueness theorem, we know that ϕ^* is continuous, in the interval $|\overline{\phi_i} - p, \overline{\phi_i} + p|$ then converges.

B 2

D is said to be covering itself if $\bigcup_j D_j \supset D$ and each elements of at least one of D belongs to d_j. The system D_j is packing if $D_i \cap D_j = \emptyset (i \neq j)$, $\bigcup_j D_j \supset D$

If two sets $D_1, D_2, \ldots$ have the same elements in common then each element $D_1, D_2, \ldots$ belong to D.

B 3

Each segment can be treated as open *beads,* as such the coordinates belong to a set X and for any point $p \subset D_j$ and $\delta = D_j$ where the measure is positive.

So, the definition of the *bead* is:

$$D = \{x{:}d(p,x) < \delta\}$$

B 4

Let *A* and B be a disjoint convex set in a convex space, then

$A = \{x{:}(x - D_i)^2 < r_i\}$ and $B = \{x{:}(x - D_j)^2 < r_j\}$, the distance is given by: $dis(D_i, D_j) = r_i + r_j$. The closure of B is given by $\overline{B} = \{x{:}(x - D_j)^2 \leq r_j\}$ then $A \cap \overline{B} = \emptyset$.

A is an open set by construction. A & B are the convex hull, also by construction, and then:

$\exists l(x) = a$ if

$$x \in A\ l(x) \le a$$
$$x \in B\ l(x) \ge a$$

where a is

$$v_j = (a - D_j)^2$$
$$v_i = (a - D_i)^2$$

where

$$D_i = dist(a - v_{Ri}) \text{ and } D_j = dist(a - v_{Lj})$$

B 5

Let $x \in S(r,x_0)$, $S \in \Re^n$ and $x_0 \neq 0$ i.e. $p(x) = x_0 + r\frac{X}{\|X\|}$ (Fig. 18b) then; $r = \|p - x_0\| = \left\| x_0 - r\frac{X}{\|X\|} - x_0 \right\| = \frac{r}{\|X\|}\|x\| = r$ hence $p \in S(r,x_0)$ [2].

References

1. Adams, C., Hildebrand, M., Weeks, J.: Hyperbolic invariants of knots and links. Trans. Amer. Math. Soc. **1**, 1–56 (1991)
2. Adams, J.F.: Vector fields on spheres. Ann. Math. **75**, 603–632 (1962)
3. Andrew, C.D., Penel, S., Jones, G.R., Doig, A.J.: Stabilizing nonpolar/polar side-chain interactions in the α-helix. Proteins Struct. Funct. Genetics **45**, 449–455 (2001)
4. Arge, E., Bruaset, A.M., Langtangen, H.P.: Modern Software tools for Scientific Computing. Birkhauser Press, Boston (1997)
5. Baker, E.N., Hubbard, R.E.: Hydrogen bonding in globular proteins. Prog. Biophys. Mol. Biol. **179**, 97–177 (1984)
6. Balakrishnan, R., Ramasubu, N., Varughese, K., Parthasathy, R.: Crystal structure of the cooper and nickel complexes of RNAase A: metal-induced interprotein interactions and identification of a novel cooper binding motif. Proc. Natl. Acad. Sci. U. S. A. **94**, 9620–9625 (1997)
7. Biegler, T.F., Mumenthaler, C., Wener, B.: Folding proteins by energy minimization and Montecarlo simulations with hydrophobic surface area potentials. J. Mol. Model. **1**, 1–10 (1995)
8. Birman, J.S.: Recent developments in Braid and Link theory. Math. Intell. **13**, 52–60 (1991)
9. Blaney, J.M., Dixon, J.S.: Distance Geometry in Molecular Modelling. CRC Press, Boca Raton (1993)
10. Bonneau, R., Tsai, J., Ruczinski, I., Chivian, D., Rohl, C., Strauss, C.E., Baker, D.: Rosetta in CASPA 4: progress in ab initio protein structure prediction. Proteins **5**, 119–126 (2001)
11. Bondi, A.: van der Waals volumes and radii. J. Phys. Chem. **68**, 441–451 (1964)
12. Bryngelson, J.D., Billing, E.M.: Interatomic interactions to protein structure. Rev. Comput. Chem. **5**, 84 (1995)
13. Bussemaker, H.J., Thirumalia, D., Bhattacharjee, J.K.: Thermodynamic stability of folding protein against mutation. Phys. Rev. Let. **79**, 3530–3533 (1997)
14. Campbell, P.G., Cohen, A.P., Ernst, L.A., Ernsthausen, J., Farkas, D.L., Galbraith, W., Israelowitz, M.: US Patent Application. USPTO Patent Application Number 20030216867 (2003)

15. Carmo, M.P.: Differential Geometry of Curves and surfaces. Prentice-Hall, Englewood Cliffs (1976)
16. Case, D.A., Cheatham III, T.E., Darden, T., Gohlke, H., Luo, R., Merz Jr, K.M., Onufriev, A., Simmerling, C., Wang, B., Woods, R.: The Amber biomolecular simulation programs. J. Comput. Chem. **26**, 1668–1688 (2005)
17. Chen, J.M., Kung, C.E., Feairheller, S.H., Brown, E.M.: An energetic evaluation of a "Smith" collagen microfil model. J. Protein Chem. **10**(5), 535–551 (1991)
18. Cheatham III, T.E., Young, M.A.: Molecular dynamics simulation of nucleic acids: successes, limitations and promise. Biopolymers **56**, 232–256 (2001)
19. Choi, H.K., Laursen, R.A., Allen, K.N.: The 2.1 Å structure of a cysteine protease with proline specificity from ginger rhizome, Zingiber officinale. Biochemistry **38**, 11624–11633 (1999)
20. Connolly, M.L.: Computation of molecular volume. J. Am. Chem. Soc. **107**, 1118–1124 (1985)
21. Connolly, M.L.: Adjoin volumes. J. Math. Chem. **15**, 339–352 (1994)
22. Costantini, S., Colonna, G., Facchiano, A.M.: Amino acid propensities for secondary structures are influenced by the protein structural class. Biochem. Biophys. Res. Commun. **342**, 441–451 (2006)
23. Cramer, C.J.: Essentials of computational chemistry: theories and models. Wiley, West Sussex (2004)
24. Das, A.K., Cohen, P.W., Barford, D.: The structure of the tetratricopeptide repeats of protein phosphatase 5: implications for TPR-mediated protein–protein interactions. EMBO J. **17**, 1192–1199 (1998)
25. DeLano, W.L.: The PyMOL molecular graphics system on World Wide Web (2002). http://www.pymol.org
26. Eaton, W.A., Munoz, V., Thompson, P.A., Henry, E.R., Hofrichter, J.: Kinetics and dynamics loops, α-helices, β-haipins, and fast-folding proteins. Acc. Chem. Res. **31**, 745–753 (1998)
27. Eğe, S.: Organic Chemistry, pp. 18–71. D.C. Heath and Company, Lexington (1984)
28. Fernandez, A., Sinanoglu, O.: Denaturation of proteins in methanol/water mixtures. Biophys. Chem. **21**, 163–164 (1985)
29. Frömmel, C., Gille, C., Goede, A., Gröpl, C., Hougardy, S., Nierhoff, T., Preissner, R., Thimm, M.: Accelerating screening of 3D protein data with a graph theoretical approach. Bioinformatics **19**, 2442–2447 (2003)
30. Garnier, J., Osguthorpe, D.J., Robson, B.: Analysis of the accuracy and implications of simple methods for predicting the secondary structure of globular proteins. J. Mol. Biol. **120**, 97–12 (1978)
31. Garrett, R., Grisham, C.M.: Biochemistry, p. 150. Brooks/Cole, Belmont (2005)
32. Gibson, K.D., Scheraga, H.A.: An algorithm for packing multistrand polypeptide structures by energy minimization. J. Comput. Chem. **15**, 1414–1428 (1994)
33. Gille, C., Lorenzen, S., Michalsky, E., Frömmel, C.: KISS for STRAP: user extensions for a protein alignment editor. Bioinformatics **19**, 2489–2491 (2003)
34. Gille, C.: Structural interpretation of mutations and SPNs using STRAP-NT. Protein Sci. **15**, 208–210 (2006)
35. Gong, H., Porter, L.L., Rose, G.: Counting peptide-water hydrogen bonds in unfolded proteins. Protein Sci. **574**, 417–427 (2011)
36. Havel, T.F.: An evaluation of computational strategies for use in the determination of protein structure from distance constraints obtained by nuclear magnetic resonance. Prog. Biophys. Mol. Biol. **56**, 43–78 (1991)
37. Hill, B.R., Raleigh, D.P., Lombardi, A., Degrado, W.F.: De novo design of helical bundles as models for understanding protein folding and function. Acc. Chem. Res. **33**, 745–754 (2000)
38. Hummer, G., Garde, S., García, A.E., Paulaitis, M.E., Pratt, L.R.: The pressure dependence of hydrophobic interactions is consistent with the observed pressure denaturation of proteins. Proc. Natl. Acad. Sci. U. S. A. **95**, 1522–1555 (1998)
39. Hunt, A.J., Gittes, F., Howard, J.: The force exerted by a kinesin molecule against a viscous load. Biophys. J. **67**, 766–781 (1994)

40. Israelowitz, M., Rizvi, S.W.H., Kramer, J., von Schroeder, H.P.: Computational modelling of type I Collagen fibers to determine the extracellular matrix structure of connective tissues. Protein Eng. Des. Sel. **18**, 329–335 (2005)
41. Jacoby, S.L.S., Kowalik, J.S., Pizzo, J.T.: Interactive methods for nonlinear optimization problems. Prentice-Hall, Englewood Cliffs (1972)
42. Tang, C.J.K., Alexandrov, V.: Relaxed Monte Carlo linear solver. In: Alexandrov, V.N., Dongarra, J.J., Juliano, B.A., Tan, C.J.K. (eds.) Lecture Notes in Computer Science, vol. 2073, p. 1289. Springer, Heidelberg (2001)
43. King, G., Brown, E.M., Chen, J.M.: Computer model of a bovine type I collagen microfribil. Protein Eng. **1**, 43–49 (1996)
44. Kuntz, I.D., Thomason, J.F., Oshiro, C.M.: Distance geometry. In: Openheimer N.J., James T.L. (eds.) Methods in Enzymology, vol. 177, pp. 159–204. Academic press, New York (1989)
45. Lyngsø, R.B., Pedersen, C.N.: RNA pseudoknot prediction in energy-based models. J. Comput. Biol. **7**, 409–427 (2000)
46. Liu, W., Chou, K.: Prediction of protein secondary structure content. Protein Eng. **12**, 1041–1050 (1999)
47. Macdonald, J.R., Johnson Jr, W.C.: Enviromental features are important in determining protein secondary structure. Protein Sci. **10**, 1172–1177 (2001)
48. Maritan, A., Micheletti, C., Triovato, A., Banava, J.B.: Optimal shapes of compact strings. Nature **406**, 287–290 (2000)
49. Marashi, S.A., Behrouzi, R., Pezeshk, H.: Adaptation of proteins to different environments: a comparison of proteome structural properties in *Bacillus subtilis* and *Escherichia coli*. J. Theor. Biol. **244**(1), 127–132 (2007)
50. MacCallum, P.H., Poet, R., Milner-White, J.E.: Coulombic interaction between partially charged main-chain atoms hydrogen-bonded to each other influence the confirmations of α–helices and antiparallel β–sheet. A new method for analysing the forces between hydrogen bonding groups in proteins includes all the coulombic interactions. J. Mol. Biol. **248**, 361–373 (1995)
51. Maiti, R., von Domseleear, G.H., Zang H., Wisshart DS.: Super pose: a simple server sophisticated structural superposition. Nucleic Acid Res. **32**, W590–W594 (2004)
52. Mount, D.M.: Bioinformatics: Sequence and Genome Analysis, vol. 2. Cold Spring Harbor Laboratory Press, Cold Spring Harbor (2004)
53. More, J.J., Wu, Z.: Issues in large scale global minimization. In: Biegler, L.T., Coleman, T. F., Conn, A.R., Santosa, F.N. (eds.) Large-scale optimization with applications, Part III, p. 99. Springer-Verlag, New York (1997)
54. Morgan, D., Ceder, G., Curtarolo, S.: High-throughput and data mining with ab initio methods. Meas. Sci. Technol. **16**, 296–301 (2005)
55. Nelder, J.A., Mead, R.: A simplex method for function minimization. Comput. J. **7**, 308–313 (1965)
56. Nemethy, G., Gibson, K.D., Palmer, K.A., Yoon, C.N., Paterlini, G., Zagari, A., Rumsey, S., Scheraga, H.A.: Energy parameters in polypeptides. 10. Improved geometric parameters and nonbonded interactions for use in the ECEPP/3 algorithm, with application to proline-containing peptides. J. Phys. Chem. **96**, 6472–6484 (1992)
57. Nicholls, A., Sharp, K.A., Honig, B.: Protein folding and association: insights from the interfacial and thermodynamic properties of hydrocarbons. Proteins **11**, 281–296 (1991)
58. Pavanï, R., Ranghino, G.: A method to compute the volume of a molecule. Comput. Chem. **6**, 133–135 (1982)
59. Pearlman, D.A., Case, D.A., Caldwell, J.W., Ross, W.R., Cheatham III, T.E., DeBolt, S., Ferguson, D., Seibe,l G., Kollman, P.: AMBER, a computer program for applying molecular mechanics, normal mode analysis, molecular dynamics and free energy calculations to elucidate the structures and energies of molecules. Comput. Phys. Commun. **91**, 1–41 (1995)
60. Pham, T.H., Satou, K., Ho, T.B.: Support vector machines for prediction and analysis of beta and gamma-turns in proteins. J. Bioinform. Comput. Biol. **3**, 343–358 (2005)

61. Polanowska-Grabowska, R., Simon Jr, C.G., Shabanowitz, J., Hunt, D.F., Gear, A.R.L.: Platelet adhesion to collagen under flow causes dissociation of a phosphoprotein complex of heat-shock proteins and protein phosphatase 1. Blood **90**, 1516–1526 (1997)
62. Ponder, J.W., Case, D.A.: Force fields for protein simulations. Adv. Protein Chem. **66**, 27–85 (2003)
63. Ochsebein, F., Gilquin, B.: NMR for protein analysis. CLEFS CEA **56**, 52–55 (2008)
64. Ott, R., Bijma, J., Hemleben, C.: A computer method for estimating volume and surface areas of complex structure consisting of overlapping spheres **16**, 83–98 (1992)
65. Radzicka, A., Wolfeden, R.: Comparing the polarities of the amino acids: side-chain distribution coefficients between the vapor phase, cyclohexane, 1-octanol, and neutral aqueous solution. Biochemistry **27**, 1664–1670 (1988)
66. Ramachandra, G.N., Ramakrishnan, C., Sasisekharan, V.: Stereochemistry of polypeptide chain configurations. J. Mol. Biol. **7**, 95–99 (1963)
67. Raspanti, M.: Different architectures of collagen fibrils enforce different fibrillogenesis mechanisms. J. Biomed. Sci. Eng. **3**, 1169–1174 (2010)
68. Rojnuckarin, A., Santae, K., Shankar, S.: Brownian dynamics simulations of protein folding: Access to milliseconds time scale and beyond. PNAS **68**, 4288–4292 (1998)
69. Roux, B.: Perspective in molecular dynamics and computational method. J. Cell Biol. **135**, 547–548 (2010)
70. Sanjeev, A., Barak, B.: Computational Complexity: A Modern Approach, pp. 50–59. Cambridge University Press, Cambridge (2009)
71. Sanchez, R., Pieper, U., Mirkovic, N., de Bakker, P.I.W., Wittenstein, E., Sali, A.: MODBase, a database of annotated comparative protein structure models. Nucleic Acids Res. **28**, 250–253 (2000)
72. Sharman, G.J., Searle, M.S.: Cooperative interaction between the three strands of a designed antiparallel β-sheet. J. Am. Chem. Soc. **120**, 5291–5300 (1998)
73. Schenck, H.L., Gelmman, S.H.: Use of a designed tripled-stranded antiparallel β-sheet to probe β–sheet cooperativity in aqueous solution. J. Am. Chem. Soc. **120**, 4869–4870 (1998)
74. Shakhnovich, E.I., Farztdinov, G., Gutin, A.M., Karplus, M.: Protein folding bottlenecks: a lattice Monte Carlo simulation. Phys. Rev. Lett. **67**, 1665–1668 (1991)
75. Shreraga, H., Gibson, K.D.: An algorithm for packing regular multistrand polypeptide structures by energy minimization. J. Comput. Chem. **15**, 1414–1428 (1994)
76. Sinaglou, O.: Microscopic surface tension down to molecular dimensions and microthermodynamic surface areas of molecules or clusters. J. Chem. Phys. **1**, 463–468 (1981)
77. Srinavasan, R.: Helix length distribution from protein crystallographic data. Indian J. Biochem. Biophys. **13**, 192–193 (1976)
78. Thompson, J.D., Higgins, D.G., Gibson, T.J.: CLUSTAL W: improving the sensitivity of progressive multiple sequence alignment through sequence weighting, position-specific gap penalties and weight matrix choice. Nucleic Acids Res. **22**, 4673–4680 (1994)
79. Ting, C.-K.: On the mean convergence time of multi-parent genetic algorithms without selection. In: Freitas, A.A., Bentley, P.J., Johnson, C.G., Timmis, J. (eds.) Advances in artificial life, p. 403 Springer-Verlag, Berlin (2005)
80. Tiraboschi, G., Gresh, N., Giessner-Prettre, C., Pedersen, L.G., Deerfield II, D.W.: A joint ab initio and molecular mechanics investigation of polycoordinated Zn(II) complexes with model hard and soft ligands. Variations of the binding energy and of its components with the number and the charges of the ligands. J. Comput. Chem. **21**, 1011–1039 (2000)
81. Torda, A.E., Van Gunsteren, W.F.: Molecular modelling using nuclear magnetic resonance data. In: Lipkowitz, K.B., Boy, D.B. (eds.) Reviews in Computational Chemistry, vol 3, pp. 143–172. VCH Publishers, New York (1992)
82. Voelz, V.A., Bowman, G.R., Beauchamp, K., Pande, V.S.: Molecular simulation of ab initio protein folding for a millisecond folder NTL9 (1–39). J. Am. Chem. **132**, 1526–1528 (2010)
83. Wade, L.G.: Structure and Stereochemistry of Alkanes: Organic Chemistry, 6th edn, pp. 103–122. Pearson Prentice Hall, Upper Saddle River (2006)

84. Wiltscheck, R., Kammerer, R.A., Dames, S.A., Schulthess, T., Blommers, M.J., Engel, J., Alexandrescu, A.T.: NMR assignments and secondary structure of the coiled coil trimerization domain from cartilage matrix protein in oxidized and reduced forms. Protein Sci. **6**, 1734–1745 (1997)
85. William, P., Teukolsky, S.A., Vetterling, W.T., Flannery, B.P.: Numerical Recipes in Fortran, 2nd edn., pp. 312–326. Cambridge University Press, Cambridge (1992)
86. Word, J.M., Lovell, S.C., LaBean, T.H., Taylor, H.C., Zalis, M.E., Presley, B.K., Richardson, J.S., Richardson, D.C.: Visualizing and quantifying molecular goodness-of-fit: small-probe contact dots with explicit hydrogens. J. Mol. Biol. **285**, 1711–1733 (1999)
87. Zimmermann, O., Hansmann, U.H.: Support vector machines for prediction of dihedral angle regions. Bioinformatics **22**, 3009–3015 (2006)
88. Zhang, Y., Skolnick, J.: TM-align: a protein structure alignment algorithm based on the TM-score. Nucleic Acids Res. **33**(7), 2302–2309 (2005)
89. Zhang, X.Q., Jansem, A.P.: Kinetic Monte Carlo method for simulating reactions in Solution. Phy. Rev. E Stat. Nonlin. Softw. Matter Phys. **82**, 046704 (2010)
90. Zhong, L., Johnson Jr, W.C.: Environment affects amino acid preference for secondary structure. PNAS **89**, 4462–4465 (1992)

Adaptive Quasi-Linear Viscoelastic Modeling

Ali Nekouzadeh and Guy M. Genin

Abstract Engineered tissues are often designed to serve a mechanical role. The design and evaluation of such tissues requires a mechanical model. An important component of such models is often viscoelasticity, or the dependence of mechanical response on loading rate and loading history. In a great number of biological and bio-artificial tissues the passive tissue force (or stress) relates to changes in tissue length (or strain) in a nonlinear viscoelastic manner. Choosing and fitting nonlinear viscoelastic models to data for a specific tissue can be a computational challenge. This chapter describes the range of such models, criteria for selecting amongst them, and computational and experimental techniques needed to fit these to uniaxial data. The chapter begins with Fung's quasi-linear viscoelastic (QLV) model, which is nearly a standard first model to try for non-linear viscoelastic tissues. The chapter then describes the two major limitations of the Fung QLV model, and presents approaches for overcoming these. The first limitation is accuracy: the Fung QLV model imposes a severe set of restrictions on constitutive behavior, and a generalized form of the Fung QLV model is needed in many cases. The second limitation is that the Fung QLV model is cumbersome computationally, especially for calibration experiments. The Adaptive QLV model is far simpler to calibrate and provides greater flexibility than the Fung QLV model. The Adaptive QLV model extends linear viscoelastic models to incorporate nonlinearity using a principle different from that of the Fung QLV model: it adapts nonlinearity according to the instantaneous level of strain. The Adaptive QLV model can be used in simple or generalized form. The chapter concludes with a

---

A. Nekouzadeh
Cardiac Bioelectricity and Arrhythmia Center, Department of Biomedical Engineering, Washington University in Saint Louis, Saint Louis, MO, USA

G. M. Genin (✉)
Department of Mechanical Engineering and Materials Science, Washington University in Saint Louis, Saint Louis, MO, USA
e-mail: gening@seas.wustl.edu

Stud Mechanobiol Tissue Eng Biomater (2013) 10: 47–83
DOI: 10.1007/8415_2012_142

Published Online: 19 August 2012

series of test protocols for calibrating QLV models along with the associated calibration procedures, using the nonlinear viscoelastic behavior of reconstituted collagen tissue as an example. The Adaptive QLV model is not only simpler to calibrate but also more accurate in predicting the mechanical response of the reconstituted collagen tissue.

1 Nonlinear Viscoelasticity in Biological Systems

Mechanics plays a role in the function of nearly every biologic tissue. Tissues such as bone and tendons serve roles that are predominantly mechanical. Mechanical forces play a vital role throughout growth and development [1–6]. Even in tissues and protein structures whose primary roles are not mechanical, mechanical responses have important implications for cellular physiology, cellular pathologies, and tissue injury [7–15].

Similarly, mechanics is important in the functionality of engineered, bio-artificial tissues. In the engineering of replacements and grafts for soft tissues, [16], bone [17], and their attachments [18], mechanical models are needed to identify the degree to which the replacement mimics the natural tissue. In the engineering of tissue microenvironments that serve to guide the differentiation of stem cells, the dynamic responses of extracellular matrix proteins such as reconstituted collagen determine the mechanical environments of cells (e.g., [19]). Here, mechanical models are needed to design microenvironments of appropriate dynamic stiffness. In engineered tissues that serve as three-dimensional scaffolds for probing cellular biophysics, mechanical models are required for delineating cellular responses from those of the engineered tissue as a whole and for estimating cellular forces from motion of markers within an engineered tissue [20–24]. Strain can be measured easily in such systems, but estimation of stresses requires knowledge of the mechanical responses, or constitutive law, of the extracellular matrix [25–27].

The constitutive laws of biological and bio-artificial tissues usually depend upon how quickly the tissues are loaded, and upon how they were loaded previously. Rapid length increases, over a physiologic range, are usually met with greater mechanical resistance than are more gradual length increases. If stretched and then held in the stretched condition, the isometric force needed to retain a tissue in its stretched configuration usually decreases or "relaxes" over time from the peak value reached immediately following the stretch. From a mechanical perspective, the materials that exhibit such a behavior are considered viscoelastic.

While many biological and bio-artificial tissues exhibit linear viscoelastic behavior for sufficiently small loading increments, their viscoelastic force response to an elongation of $\Delta l_1 + \Delta l_2$ is often different than the sum of viscoelastic responses to elongations Δl_1 and Δl_2, meaning that their viscoelasticiy is nonlinear. The constitutive relations for such a material, which are mathematical relations between force and displacement or between stress and strain, are nonlinear viscoelastic. The focus of this chapter is phenomenological constitutive models that can be used for computational prediction of such nonlinear viscoelastic behavior.

Before delving into these, a word is needed about an alternative approach that is *not* the focus of this chapter: the micromechanical approach, in which continuum constitutive laws are constructed from knowledge of the microstructures and deformation mechanisms of a tissue. For a number of biological tissues, some aspects of nonlinearity are understood in terms of the underlying deformation mechanisms of the protein structure. An example is the mechanics of tendons and ligaments, in which the uncrimping of type I collagen likely underlies non-linear stiffening of the tissue at low strains [28–30]. Lanir developed the first approach to developing tissue-level constitutive relations for such tissues by averaging fiber-level models over an orientation distribution of fibers [31–35], and important extensions to this approach have been made through studies on planar collagenous tissues [36], and through studies of related systems such as the anterior cruciate ligament [37], the tendon enthesis [38, 39], and reconstituted collagen matrices [22–24, 40–43]. Such models offer clear advantages over phenomenological models in a great number of situations, but require detailed advance knowledge of tissue microstructure and deformation mechanisms [44].

1.1 Overview of Phenomenological Nonlinear Viscoelastic Frameworks and Their Limitations

Phenomenological models involve frameworks that can be fit to the response of a material with limited information about the material's make-up. The simplest class of phenomenological models for viscoelastic behavior is linear viscoelasticity. Linear viscoelasticity can be described completely in terms of a relaxation function and a constant modulus of elasticity.

The problem is more complicated for nonlinear viscoelastic behavior, with numerous approaches available. Any nonlinear viscoelastic model can be considered a subset of the most general nonlinear viscoelastic form derived by Coleman and Noll [45, 46]. This formulation involves a summation of three different terms: elastic stress, linear pure viscoelastic stress, and nonlinear pure viscoelastic stress. The constitutive law of a nonlinear material can also be modeled mathematically using nonequilibrium thermodynamics through two approaches: functional thermodynamics and state-variable thermodynamics [47]. In these approaches, scalar response quantities like free energy are modeled as functionals of the strain (or stress) tensor, and stress (or strain) is calculated using functional differentiation of those scalar quantities.

The most commonly used phenomenological model to describe the nonlinear viscoelastic behavior of biological tissues is a quasi-linear viscoelastic (QLV) model proposed by Fung [48]. Although highly specific in its applicability, the attractive feature of the Fung QLV model is its simplicity: it identifies a class of quasi-linearity that is appropriate for many tissues. For many other biological and bio-artificial tissues, the Fung QLV model cannot achieve the desired accuracy. The literature contains many examples of these problems, and many elegant solutions [49–56]. A simple example of such problems in the authors' work comes from uniaxial testing of reconstituted type I collagen specimens: the Fung QLV model, when fitted to the mechanical response of such a specimen to a rapid stretch, was unable to predict with reasonable accuracy the subsequent mechanical responses of the same or identical specimens to stretches applied at other rates, or in other sequences [55, 56].

A major limitation of the predictive capability of the Fung QLV model is the model's assumption that the "reduced relaxation function," which determines the influence of historical loads on the current level of stress, is the same at all levels of tissue stretch. This is not valid for many biological and bio-artificial tissues. One way to overcome this limitation is the approach of a "Generalized Fung QLV" model, in which this constraint is lifted, and the reduced relaxation function is permitted to vary with the degree of stretch of the tissue [55, 56]. In the specific example of reconstituted collagen, the Generalized Fung QLV model succeeds in predicting the nonlinear viscoelastic behavior of reconstituted collagen with far more accuracy, and passes the elementary test of predicting the responses to various ramp stretches when fitted to stretch-and-hold tests.

A second limitation, of both the Fung QLV model and the Generalized Fung QLV model, is that they are tedious computationally. The stress response to a specific stretch input must be calculated numerically from a convolution integral, unless the stretch input is an instantaneous step. These numerical convolutions complicate the optimization algorithms needed to fit these models to a specific material.

The focus of this chapter is a relatively new class of nonlinear viscoelastic models, with a specific implementation called the Adaptive QLV model. This new nonlinear viscoelastic model is considered quasi-linear because the dependence of the stress on the strain history is calculated through a linear convolution integral as in the Fung QLV model. However, the Adaptive QLV model incorporates the nonlinearity into the linear viscoelastic model based on a principle that differs from that of the Fung QLV model. This change not only adds the flexibility needed to model a broader range of biological and bio-artificial tissues, but also simplifies calibration tremendously, often without the need to evaluate the convolution integral numerically. To determine the stress at a specific time, the Adaptive QLV model adapts its nonlinear function according to the instantaneous strain. At any instant, only the strain history is required, and the history of the nonlinearity is not considered. In contrast, in the Fung QLV model stress at any instant depends on the history of both the nonlinear function and the strain. This chapter reviews quasi-linear viscoelastic approaches to modeling of tissues, with special emphasis on the Adaptive QLV model and its ability to predict the nonlinear viscoelastic behavior of reconstituted collagen gels.

2 Nonlinear and Quasi-Linear Viscoelasticity

The most general framework for nonlinear viscoelasticity considers the stress (or force) as a function of the entire history of the straining (or elongation) of a material [57]. Such a model is too general for any practical application. Imposing some symmetry constraints and the principle of fading memory, Coleman and Noll derived a general nonlinear viscoelastic model that can be approximated by a Taylor-like series about a zero strain history [45, 46]. In this approximation, stress is calculated in terms of first and higher order integrals of the strain history. Up to the first order integral, the constitutive law is linear; it becomes nonlinear when second and higher order integrals are included in the approximation. Alternative integral series were derived by Rivlin and Spencer [58, 59] and by Pipkin and Rogers [60, 61]. In the models of Pipkin, the stress is calculated in terms of the time derivative of the strain history rather than the strain history itself.

The single-integral approximation of the Pipkin-Rogers model can be rewritten as the Fung QLV for one-dimensional viscoelastic modeling of biological tissues. The central feature of the Fung model is that the dependence of the force response on the stretch history can be obtained from a linear convolution integral, which allows the nonlinear model to retain all of the benefits of linearity. Fung incorporated nonlinearity into a linear viscoelastic constitutive law by replacing strain with a nonlinear function of strain. The resultant viscoelastic model is linear with respect to a pure function of strain instead of strain itself, and hence "quasi-linear."

The term "quasi-linear viscoelastic" has since become more general, and encompasses any nonlinear viscoelastic constitutive law in which loading history dependence can be modeled by a linear convolution integral or a summation of linear convolution integrals [55]. A feature of such QLV models is that the stress and strain (or force and displacement) are related by an intermediate variable that separates the nonlinearity from the viscoelasticity. An example of this is the "elastic stress" in the Fung QLV model, which serves as an intermediate variable that allows use of a linear convolution integral to derive the stress. This is described in detail in the next section.

2.1 *The Fung QLV Model*

In a linear viscoelastic model, stress is calculated in terms of strain history using a linear convolution integral as:

$$\sigma(t) = \int_{-\infty}^{t} G(t-\tau)\ \frac{d\varepsilon(\tau)}{d\tau}\, d\tau \tag{1}$$

where σ is the uniaxial stress and ε is the corresponding strain. $G(t)$ is the unit step relaxation function, which is the recorded stress in response to a unit step elongation of the tissue. Qualitatively, $G(t-\tau)$ represents the diminishing effect of the strain state at a time τ before the current time, t, on the current stress, $\sigma(t)$.

The Fung QLV model incorporates nonlinearity into Eq. (1) by replacing strain with a nonlinear function of strain. The stress state is then calculated by the following convolution integral:

$$\begin{aligned}\sigma(t) &= \int_{-\infty}^{t} g(t-\tau)\,\frac{d\sigma^{(e)}(\varepsilon(\tau))}{d\tau}\,d\tau \\ &= \int_{-\infty}^{t} g(t-\tau)\,\frac{d\sigma^{(e)}(\varepsilon)}{d\varepsilon}\,\frac{d\varepsilon(\tau)}{d\tau}\,d\tau\end{aligned} \tag{2}$$

where $g(t)$ is called the "reduced" relaxation function, which is the unit step relaxation function normalized by its initial value (so that $g(0) = 1$), and $\sigma^{(e)}(\varepsilon)$ is a function of strain called the "elastic stress." In this model, the nonlinearity is included within the elastic tangent stiffness term, $d\sigma^{(e)}/d\varepsilon$.

The "reduced" relaxation function $g(t)$ represents the shape of the normalized unit step relaxation curve, which is the decay in isometric stress level following an instantaneous stretch. For different tissues and depending on the shape of the experimentally recorded reduced relaxation functions, $g(t)$ may be an exponential, polynomial, logarithmic or any other mathematical function. A common choice of $g(t)$ is the sum of several exponential terms:

$$g(t) = a_o + \sum_{i=1}^{M} a_i e^{-t/\tau_i}, \tag{3}$$

where each of the M exponential terms (each time constant τ_i and corresponding amplitude a_i) can be associated with a series combination of a spring and a dashpot (a Maxwell element), and a_o can be associated with a spring.

Fung [39] also observed that many biological tissues are well modeled by a reduced relaxation function with a continuous spectrum of time constants:

$$g(t) = \frac{1 + \int_0^\infty S(\tau)e^{-t/\tau}d\tau}{1 + \int_0^\infty S(\tau)d\tau}, \tag{4}$$

where $S(\tau)$ dictates the distribution of time constants. As did Neubert in his earlier model of rubber viscoelasticity [62], Fung adopted the form $S(\tau) = c/\tau$, where c is a constant; this provides $g(t)$ with a uniform spectrum of time constants, as is appropriate for many tissues that exhibit a logarithmic decay in isometric force following a stretch [22].

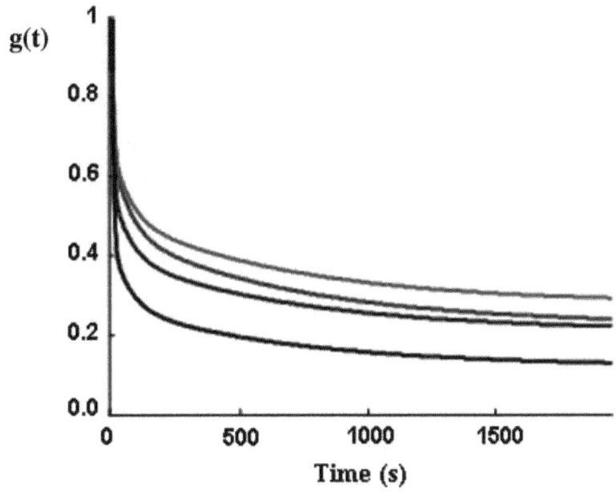

Fig. 1 Four normalized relaxation functions for a reconstituted collagen tissue specimen, at four different initial strains. The relaxation function of these tissues is clearly a function of initial strain, meaning that the Fung QLV model is inadequate for modeling this tissue. Each curve represents a 3-term exponential fit to experimentally recorded data, as reported in [55]

2.2 *The Generalized Fung QLV Model*

The Fung QLV model requires that isometric stress relaxation following an instantaneous stretch be proportional to a single reduced relaxation function $g(t)$, regardless of the level to which a tissue has been strained prior to the application of the instantaneous stretch, and regardless of the amplitude of the instantaneous stretch. This behavior is inconsistent with that observed in many biological and bio-artificial tissues. An example, shown in Fig. 1, is of reconstituted collagen stretched to four different (nearly) instantaneous levels of strain. The Fung QLV model requires that $g(t)$ be independent of strain level, and that the four curves shown in Fig. 1 be a single curve; however, the four curves are significantly different. This limits accuracy of the Fung QLV model in representing the mechanical characteristics of these and other biological and bio-artificial tissues [50, 54, 55].

To overcome this limitation the Fung QLV model has been generalized by allowing the reduced relaxation function to vary with the tissue strain level:

$$\sigma(t) = \int_{-\infty}^{t} g^G(t-\tau;\varepsilon)\,\frac{d\varepsilon(\tau)}{d\tau}\,d\tau. \tag{5}$$

In this Generalized Fung QLV model, the reduced relaxation function $g^G(t;\varepsilon)$ is considered as a weighted summation of multiple shape functions; and for each shape function there is a distinct nonlinear function of strain that determines its weight in the relaxation curve. For an exponential representation of the reduced relaxation function analogous to Eq. (3), $g^G(t;\varepsilon)$ can be written to depend explicitly on both time and strain:

$$g^G(t;\varepsilon) = A_o(\varepsilon) + \sum_{i=1}^{M} A_i(\varepsilon)e^{-t/\tau_i(\varepsilon)} \tag{6}$$

in which the strain-dependent amplitudes $A_i(\varepsilon)$ of the relaxation modes correspond to strain-dependent time constants $\tau_i(\varepsilon)$. Note that the Fung QLV model is recovered by restricting $A_i(\varepsilon)$ and $\tau_i(\varepsilon)$ to be independent of strain. Following the Fung QLV model (Eq. (2)), the stress–strain relation for the Generalized Fung QLV can be written as the sum of the convolution integrals associated with the different shape functions:

$$\sigma(t) = \sigma_o(\varepsilon(t)) + \sum_{i=1}^{M} \int_{-\infty}^{t} g_i(t-\tau)\, A_i(\varepsilon(t)) \frac{d\varepsilon(\tau)}{d\tau}\, d\tau \tag{7}$$

where $\sigma_o(t)$ is the steady state tissue stress for a given tissue strain, $\varepsilon(t)$, and is taken out of the convolution integral to simplify model calibration. $g_i(t)$ is the ith shape function and is considered usually as an exponential term (i.e. $\exp(-t/\tau_i)$). $A_i(\varepsilon)$is the nonlinear function of strain and is equivalent to the derivative of the elastic stress in the Fung QLV model (i.e. $d\sigma^{(e)}(\varepsilon)/d\varepsilon$). Since the steady state tissue stress is taken out of the convolution integral, all shape functions should approach zero as time approaches infinity.

2.3 The Adaptive QLV Model

The Adaptive QLV model differs from the Fung and Generalized Fung QLV models in that it uses an alternative approach to incorporate nonlinearity into a linear viscoelastic model. This alternative approach simplifies the calibration procedure significantly, and is conceptually simple as well because it enables representation of the resultant QLV model in terms of an infinite array of Maxwell elements (springs and dashpots).

In the Adaptive QLV model, stress and strain are related through an intermediate variable termed the viscoelastic strain, $V^{(\varepsilon)}(t)$, and through a linear convolution integral as:

$$\begin{aligned} \sigma(t) &= k(\varepsilon(t)) V^{(\varepsilon)}(t) \\ V^{(\varepsilon)}(t) &= \int_{-\infty}^{t} g(t-\tau) \frac{d\varepsilon(\tau)}{d\tau} d\tau \end{aligned} \tag{8}$$

where $k(\varepsilon)$ is a pure nonlinear function of strain and $g(t)$ is a reduced relaxation function that can be expressed as a sum of exponentials with different time constants. $V^{(\varepsilon)}(t)$ represents the dependence of the stress on the history of straining. Nonlinearity enters the model through $k(\varepsilon)$, which converts the strain history (viscoelastic strain) to stress through a simple multiplication.

The Adaptive QLV model can be written in generalized form, as with the Generalized Fung QLV model, by assuming different nonlinear behavior for different shape functions:

$$\begin{aligned} \sigma(t) &= \sigma_o(\varepsilon(t)) + \sum_{i=1}^{M} k_i(\varepsilon(t)) V_i^{(\varepsilon)}(t) \\ V_i^{(\varepsilon)}(t) &= \int_{-\infty}^{t} g_i(t-\tau)\frac{d\varepsilon(\tau)}{d\tau}d\tau, \quad i = 1, 2, \ldots, M \end{aligned} \tag{9}$$

where $\sigma_o(\varepsilon)$ is a pure function of strain representing the long-term elastic part of the response. Each $g_i(t)$ could be any relaxation function such that $g_i(0) = 1$ and $g_i(\infty) = 0$. $g_i(t)$ is termed a shape function rather than a reduced relaxation function because the relaxation curves are a weighted summation of $g_i(t)$ functions. Usually the shape functions are chosen to be exponential terms (i.e., $g_i(t) = e^{-t/\tau_i}$).

The main advantage of the Adaptive QLV model is simplicity of calibration. The Adaptive QLV model involves a convolution of the relaxation function with the strain function rather than with the elastic stress as in the Fung QLV model. The strain function is prescribed, and can be chosen to be any function. In contrast, the elastic stress depends on the mechanical characteristics of the tissue and needs to be calibrated along with the relaxation function. Calibrating only one unknown function (the relaxation function) through the convolution integral is far simpler than calibrating two unknown functions through the convolution integral. For a great many displacement inputs, the convolution integral (Eq. (9)) of the Adaptive QLV model can be found in closed analytical form. Some particularly useful examples of these are described in the next section.

2.3.1 Representation of the Adaptive QLV Model with Maxwell Elements

When the shape functions $g_i(t)$ are represented by exponential terms, the Adaptive QLV model can be represented in terms of parallel Maxwell elements, although such elements need not exist physically [55]. Viewing the Adaptive QLV model in this way allows for physical insight into the underlying principles, and is thus the focus of this section. These elements involve a set of nonlinear springs and dashpots where the spring stiffnesses and dashpot coefficients are functions of overall tissue strain and not their individual strain (Fig. 2). For each element i:

$$\begin{cases} \dot{V}_i^{(\varepsilon)} + \frac{V_i^{(\varepsilon)}}{\tau_i(\varepsilon)} = \dot{\varepsilon} \\ \sigma_i = k_i(\varepsilon(t)) V_i^{(\varepsilon)}(t) \end{cases} \tag{10}$$

where:

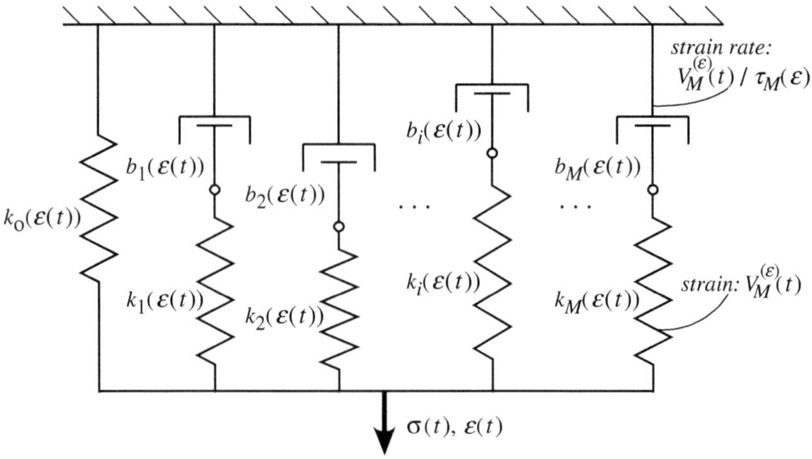

Fig. 2 Spring-dashpot representation of the Adaptive QLV model, as reported in [55]

$$\tau_i(\varepsilon) = \frac{b_i(\varepsilon)}{k_i(\varepsilon)}. \tag{11}$$

For the elastic response (the single spring element with no dashpot):

$$\sigma_o = k_o(\varepsilon) = \sigma_o(\varepsilon). \tag{12}$$

A great simplification results by requiring both the spring stiffness and dashpot coefficient of a single Maxwell element to be proportional to the same nonlinear function of strain:

$$\begin{cases} k_i(\varepsilon) = k_i\psi_i(\varepsilon) \\ b_i(\varepsilon) = b_i\psi_i(\varepsilon) \end{cases}. \tag{13}$$

Each time constant τ_i is then independent of strain:

$$\tau_i(\varepsilon) = \frac{b_i(\varepsilon)}{k_i(\varepsilon)} = \frac{b_i}{k_i} = \tau_i. \tag{14}$$

The consequence is that the first order differential equations of Eq. (10) become linear and their solution, which can be calculated from a linear convolution integral, exists in closed form for many strain functions, including a constant rate strain:

$$\dot{V}_i^{(\varepsilon)} + \frac{V_i^{(\varepsilon)}}{\tau_i} = \dot{\varepsilon} \Rightarrow V_i^{(\varepsilon)}(t) = \int_{-\infty}^{t} e^{-(t-\xi)/\tau_i} \frac{d\varepsilon(\xi)}{d\xi} d\xi \quad i = 1, 2, \ldots, M. \tag{15}$$

The total stress takes the form of Eq. (9):

$$\sigma(t) = \sigma_o(t) + \sum_{i=1}^{M} k_i(\varepsilon(t)) V_i^{(\varepsilon)}(t). \tag{16}$$

2.3.2 Physical Interpretation of the Adaptive QLV Model

We conclude this section with a qualitative example that highlights the physical basis of the Adaptive QLV model, and how this differs from that of the Fung QLV model [55]. The example follows the approach of Fung [63] and treats a strain history involving instantaneous jumps to a possibly new strain level e_n at each time increment $n\Delta t$, where Δt is a constant time interval:

$$\varepsilon_s(t) = \sum_n e_n u(t - n\Delta t) \tag{17}$$

where e_n is the value of strain function at time $n\Delta t$ and $u(t)$ is a unit step function. This strain history represents a series of incremental stepwise strains. Therefore, for a linear viscoelastic material, the stress can be written as a summation of incremental relaxation curves corresponding to the incremental strains:

$$\sigma_{Linear}(t) = \sum_n e_n y(t - n\Delta t), \tag{18}$$

where $y(t)$ is the stress response to a unit step strain.

In the Fung QLV model, the incremental relaxation function that starts at time $n\Delta t$ is scaled by a function of strain at time $n\Delta t$ to include the nonlinearity:

$$\sigma_{Fung}(t) = \sum_n e_n f_1(\varepsilon(n\Delta t)) y(t - n\Delta t) \tag{19}$$

In the Adaptive QLV model, the incremental relaxation function that starts at time $n\Delta t$ is scaled by a function of the current total strain at time t:

$$\sigma_{Adaptive}(t) = \sum_n e_n f_2(\varepsilon(t)) y(t - n\Delta t) \tag{20}$$

The difference between the Fung and Adaptive QLV models in this example is that the way a strain increment at time $n\Delta t$ influences the current stress. In the Fung QLV model, this influence is locked in at the time that the strain increment occurs, whereas in the Adaptive QLV model the influence adjusts based upon the current value of strain; the "Adaptive" in fact refers to this continuous adaptation of the relaxation function. As will be shown in the following sections, this adaptation provides for improved predictions in a number of biological and bio-artificial tissues. Additionally, the model affords a highly simplified calibration.

3 Calibration Protocols for QLV Models

In this section we review four common test protocols employed to calibrate quasi-linear viscoelastic constitutive models and describe procedures for calibrating the parameters of the Adaptive QLV, Generalized Fung QLV, and, when significantly different, the Fung QLV models to data obtained following these protocols. All tests are one dimensional and involve data acquired by monitoring the tissue stretch and resistive force over time (Fig. 3), and all assume loading rates sufficiently small that inertial effects could be neglected (e.g., [64]). The test specimen is assumed to be in a fully relaxed state prior to application of load, so that the history of prior loading of the specimen does not affect its subsequent mechanical response measurably. Axial strain, ε, is calculated in each case from axial stretch, Δl, through:

$$\varepsilon(t) = (l(t) - l_o)/l_o, \tag{21}$$

where l_o is the initial (usually zero-force) length of the tissue. Uniaxial stress, σ, is calculated from axial force, f, through:

$$\sigma(t) = f(t)/S_o, \tag{22}$$

where S_o is the initial cross-sectional area of the tissue. Alternatively, one may consider and calibrate the QLV models as representative of the constitutive law between the force and the relative displacement of the ends of a specimen rather than between the stress and the strain.

3.1 Relaxation Tests with Multiple Amplitudes

An ideal relaxation test involves an instantaneous (stepwise) stretch of a tissue, followed by an isometric holding of the tissue at this level of stretch. The force response of a viscoelastic material to such a stretch is a rapid increase in resistive force followed by a gradual force relaxation over time (Fig. 3a). Force relaxation data should be measured until the force approaches a final steady state value. Each relaxation test is defined by the initial tissue strain, ε, (or displacement) and the amplitude $\Delta\varepsilon$ of the stepwise strain (or displacement) increment. Before a relaxation test, the tissue force should have reached the steady value associated with the initial strain. This is required for the subsequent relaxation data to represent only the current relaxation test.

The goal here is to characterize the strain dependence of nonlinear viscoelastic behavior. One suitable protocol is to perform a series of relaxation tests of different amplitudes, all from the same initial length, with ample time for relaxation of the specimen back to its initial configuration between tests. In this protocol the initial length the tissue is ideally its stress free reference configuration, but this is not always possible.

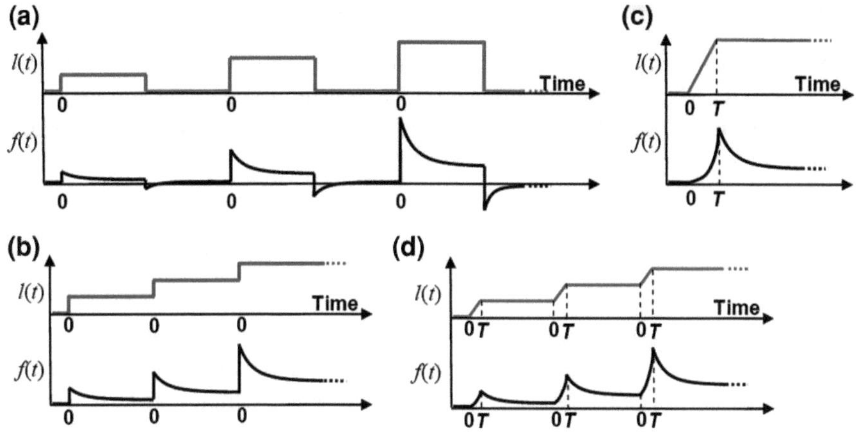

Fig. 3 Schematics describing four common test protocols for calibration of QLV models. Tissue strain is plotted in gray and tissue stress is plotted in black. **a** Relaxation tests with multiple amplitudes. **b** Incremental relaxation tests. **c** Single large amplitude ramp-and-hold test. **d** Incremental ramp-and-hold tests

The following notation is used to define the fitting procedure for such a series of relaxation tests. $R_n(t)$ is the time history of stress recorded in the nth relaxation test, whose strain amplitude is $\Delta\varepsilon_n$ ($n = 1, 2,\ldots, N$). The strain history for the nth relaxation test, $\varepsilon_n(t)$, involves a step function $u(t)$, in which each test is treated as being performed at time $t = 0$ on a specimen that has never before been stretched:

$$\varepsilon_n(t) = \Delta\varepsilon_n u(t). \tag{23}$$

3.1.1 Adaptive QLV Model

The first step in fitting the Adaptive QLV model is to calculate the time course of each viscoelastic strain in the nth relaxation test using Eq. (9):

$$V_{ni}^{(\varepsilon)}(t) = \int_{-\infty}^{t} g_i(t-\tau)\Delta\varepsilon_n \frac{du(\tau)}{d\tau} d\tau = \int_{-\infty}^{t} g_i(t-\tau)\Delta\varepsilon_n \delta(\tau)\, d\tau = \Delta\varepsilon_n g_i(t), \tag{24}$$

where $\delta(t)$ is Dirac's delta function. Substituting into the first line of Eq. (9):

$$\sigma_n(t) = \sigma_o(\Delta\varepsilon_n) + \Delta\varepsilon_n \sum_{i=1}^{M} k_i(\Delta\varepsilon_n) g_i(t), \tag{25}$$

The first term in Eq. (25), $\sigma_o(\varepsilon)$, can be fit from the steady state value of $R_n(t)$ by noting that all of the shape functions approach zero over time:

$$\sigma_o(\Delta\varepsilon_n) = R_n(t \to \infty). \tag{26}$$

Next, the values of k_i can be calibrated for each strain level tested. If the intent is to model the viscoelastic properties of a tissue using some predetermined shape functions g_i, then for each relaxation test the values of k_i are chosen to minimize I_n defined as:

$$I_n = \int_{-\infty}^{+\infty} (R_n(t) - \sigma_n(t))^2 dt \quad n = 1, 2, \ldots, N. \tag{27}$$

where $\sigma_n(t)$ is the model prediction of the relaxation stress using Eq. (25). To increase accuracy, the shape functions may instead be expressed in terms of unknown parameters that can also be calibrated. For example, we may assume $M = 3$ shape functions of the form $g_i(t) = \exp(-t/\tau_i)$ in which the three time constants τ_1, τ_2 and τ_3 are unknown parameters that should be calibrated to experimental data. Here, because the model uses the same shape functions to model all relaxation tests, all of the relaxation curves should be calibrated at once. The calibration procedure is then to determine all k_i values and the parameters of the shape functions (e.g., τ_1, τ_2 and τ_3) by minimizing I defined as:

$$I = \sum_{n=1}^{N} \int_{-\infty}^{+\infty} \left(\frac{R_n(t) - \sigma_n(t)}{R_n(0)} \right)^2 dt \tag{28}$$

With the values of the functions $k_i(\varepsilon)$ calibrated at several strain levels $\Delta\varepsilon_n$, an appropriate interpolation method may be used to determine these functions for all values of strain.

3.1.2 Generalized Fung QLV Model

Substituting for strain from Eq. (23) into the Fung QLV model (Eq. (2)) or the Generalized Fung QLV model (Eq. (7)), the model prediction for stress in the nth relaxation test is of the following form:

$$\begin{aligned}\sigma_n(t) &= \sigma_o(\varepsilon(t)) + \sum_i \int_{-\infty}^{t} g_i(t-\tau)\, A_i(\varepsilon(\tau))\Delta\varepsilon_n \delta(\tau)\, d\tau \\ &= \sigma_o(\Delta\varepsilon_n) + \sum_i A_i(\Delta\varepsilon_n)\Delta\varepsilon_n g_i(t)\end{aligned} \tag{29}$$

In the Fung and Generalized Fung QLV models, the nonlinear functions are determined for the tissue strain at the beginning of the test (time $t = 0$). However, at the beginning of the test the strain is discontinuous. In deriving Eqs. (29), the strain at time zero is assumed to be the strain at time 0^+ (the final strain) rather than the strain at time 0^- (the initial strain). Assuming the initial strain to be the strain at time zero forces these models to behave linearly for all relaxation tests starting from the same initial strain. Note that this problem does not exist for the Adaptive QLV model because the Adaptive QLV model incorporates nonlinear functions in terms of the instantaneous strain, which is the final strain level of a relaxation test.

Using the predictions of Eq. (29), σ_o can be calibrated using Eq. (26). Subsequently, the A_i functions can be calibrated at each $\Delta\varepsilon_n$ so that the integral I_n given by Eqs. (27) or the integral I given by Eq. (28) is minimized for known or parametric shape functions, respectively. As with the Adaptive QLV model, interpolation may be used for intermediate straining for the case of the Generalized Fung QLV model. Note that this procedure would apply as well for the fitting of the Fung QLV model to any of the N individual tests, but only a single parameter set would result. If these parameters are dependent upon strain, then the Fung QLV model is not valid and the Generalized Fung QLV model must be used.

3.2 Incremental Relaxation Tests

Incremental relaxation testing is a faster protocol to assess the nonlinear behavior of tissues and calibrate QLV models. The time savings arises from the reduced number of force relaxation intervals required: in the previous protocol the specimen must be shortened back to its initial length following each test, and a waiting period is required before performing any subsequent relaxation tests. This wait should be long enough to ensure that any viscous effect associated with the shortening back to the initial specimen length have settled, and should be of a duration equal to that of the relaxation test. To avoid these wait periods the relaxation tests may be performed in an incremental manner with a series of small (and usually equal amplitude) stretches that are performed with no unloading between the stretch increments (Fig. 3b). In this protocol the initial length of each relaxation test is the final length from the previous test. For equal strain increments of amplitude $\Delta\varepsilon$ on a specimen that is strain free before the application of the first increment, the initial strain in the nth relaxation test is $(n-1)\Delta\varepsilon$. The strain function can therefore be written:

$$\varepsilon_n(t) = \begin{cases} (n-1)\Delta\varepsilon, & t<0 \\ n\Delta\varepsilon, & t>0 \end{cases}. \tag{30}$$

3.2.1 Adaptive QLV Model

Substituting the strain function from Eq. (30) into Eq. (9) the viscoelastic strains in the nth relaxation test are:

$$V_{ni}^{(\varepsilon)}(t) = \int_{-\infty}^{t} g_i(t-\tau)\Delta\varepsilon\delta(\tau)d\tau = \Delta\varepsilon g_i(t), \tag{31}$$

and the stress is:

$$\sigma_n(t) = \sigma_o(n\Delta\varepsilon) + \Delta\varepsilon \sum_{i=1}^{M} k_i(n\Delta\varepsilon)g_i(t). \tag{32}$$

With $R_n(t)$ again denoting the experimentally recorded time course for the relaxation of stress in the nth relaxation test, the function $\sigma_o(\varepsilon)$ can be calibrated as:

$$\sigma_o(n\Delta\varepsilon) = R_n(t \to \infty). \tag{33}$$

Then, the values of $k_i(n\Delta\varepsilon)$ may be calibrated so that the integrals I_n in Eqs. (27) are minimized for known shape functions, or the integral I in Eq. (28) is minimized for parametric shape functions. As before, the functions k_i at intermediate strain levels may be estimated by interpolation.

3.2.2 Generalized Fung QLV Model

In the Generalized Fung QLV model, the predicted stress for the input strain function of Eq. (30) is:

$$\sigma_n(t) = \sigma_o(n\Delta\varepsilon) + \sum_{i} A_i(n\Delta\varepsilon)g_i(t)\,. \tag{34}$$

$\sigma_o(\varepsilon)$ can be calibrated using Eq. (33), and the functions A_i can be calibrated at $n\Delta\varepsilon$ so that the integrals I_n in Eqs. (27) are minimized for known shape functions, or the integral I in Eq. (28) is minimized for parametric shape functions. As before, interpolation may be used for intermediate strains.

3.3 Single, Large Amplitude Ramp-and-Hold Test

The ideal stepwise stretch is the preferred protocol for testing a viscoelastic material. However, in many practical applications performing an ideal stepwise stretch is not possible because of challenges associated with stretching the tissue sufficiently fast, preventing tissue damage, avoiding significant wave motion, and logging data with sufficient speed. The ramp-and-hold test is an alternative test protocol that usually generates more accurate data compared with the stepwise relaxation tests.

In a ramp-and-hold protocol the tissue length is increased at a constant rate to its final value during the ramp-loading phase, and kept constant to allow the tissue force to relax to its steady state value during the hold relaxation phase (Fig. 3 c). The strain rate in the ramp-loading phase can be adjusted to prevent tissue damage and mechanical waves during the loading [64]. Unlike the stepwise stretch tests (relaxation tests) in which the tissue experiences only two levels of strain (the initial strain and the final strain), the ramp loading tests expose the tissue to a continuous range of strain over the loading ramp. Therefore, the ramp loading stress reflects the tissue characteristics for all strains between the initial and final strain. As will be described, this can be a benefit or a drawback. On the one hand, this can complicate model fitting, but on the other, a single large amplitude ramp-and-hold may be sufficient to calibrate a QLV model to the viscoelastic properties of a tissue.

We again assume that initial configuration of the tissue is its unstretched, zero strain, zero strain history configuration. In the ramp-loading phase of a single large strain ramp-and-hold test, the tissue strain increases by $\Delta\varepsilon$ at a constant rate over T seconds. In the hold-relaxation phase, then the tissue is maintained at its final strain ($\Delta\varepsilon$) until the stress relaxes to its steady state value. The strain function in this protocol can be written:

$$\varepsilon_n(t) = \begin{cases} 0, & t<0 \\ \frac{\Delta\varepsilon}{T}t, & 0<t<T \\ \Delta\varepsilon, & t>T \end{cases}. \tag{35}$$

We denote the experimentally recorded stress during the ramp-loading phase ($0 \leq t \leq T$) as $P(t)$, and that recorded during the hold-relaxation phase ($t \geq$ T) as $H(t)$.

3.3.1 Adaptive QLV Model

A single ramp-and-hold test does not provide sufficient information to calibrate the generalized form of the Adaptive QLV model presented in Eq. (9). However, this protocol does provide sufficient information for calibration of the simple form of the Adaptive QLV model presented in Eq. (8). Substituting the strain from Eq. (35) into Eq. (8), the viscoelastic strain may be calculated as:

$$V^{(\varepsilon)}(t) = \begin{cases} \int_0^t g(t-\tau)\frac{\Delta\varepsilon}{T}d\tau = \frac{\Delta\varepsilon}{T}\gamma(t), & 0<t<T \\ \int_0^T g(t-\tau)\frac{\Delta\varepsilon}{T}d\tau = \frac{\Delta\varepsilon}{T}(\gamma(t)-\gamma(t-T)), & t>T \end{cases} \tag{36}$$

where $\gamma(t)$ is the integral of $g(t)$ defined as:

$$\gamma(t) = \int_0^t g(\tau)d\tau. \tag{37}$$

The stress may be written:

$$\sigma(t) = \begin{cases} \frac{\Delta\varepsilon}{T} k\left(\frac{\Delta\varepsilon}{T} t\right)\gamma(t) = \sigma_P(t), & 0 < t < T \\ \frac{\Delta\varepsilon}{T} k(\Delta\varepsilon)(\gamma(t) - \gamma(t-T)) = \sigma_H(t), & t > T \end{cases} \tag{38}$$

where $\sigma_P(t)$ and $\sigma_H(t)$ are the predictions of the time variation of stress over the ramp-loading and hold-relaxation phases, respectively.

Here we use the experimental stress during the hold phase to calibrate the parameters of the reduced relaxation function, $g(t)$. Parameters of the reduced relaxation functions are those that minimize the following integral:

$$I = \sum_n \int_T^{+\infty} \left(\frac{\sigma_H(t)}{\sigma_H(T)} - \frac{H(t)}{H(T)}\right)^2 dt. \tag{39}$$

Usually, if $g(t)$ is known parametrically $\gamma(t)$ can be determined parametrically as well. Once the $g(t)$ function and the associated $\gamma(t)$ function are determined, the nonlinear function $k(\varepsilon)$ may be calculated algebraically using the experimentally recorded stress during the ramp-loading phase:

$$k\left(\frac{\Delta\varepsilon}{T} t\right) = \frac{T}{\Delta\varepsilon} \frac{\sigma_P(t)}{\gamma(t)}. \tag{40}$$

3.3.2 Fung QLV Model

Substituting the strain from Eq. (35) into the Fung QLV model given by Eq. (2) yields the following prediction for stress:

$$\sigma(t) = \begin{cases} \frac{\Delta\varepsilon}{T} \int_0^t g(t-\tau) \frac{d\sigma^{(e)}\left(\frac{\Delta\varepsilon}{T}\tau\right)}{d\varepsilon} d\tau = \sigma_P(t), 0 < t < T \\ \frac{\Delta\varepsilon}{T} \int_0^T g(t-\tau) \frac{d\sigma^{(e)}\left(\frac{\Delta\varepsilon}{T}\tau\right)}{d\varepsilon} d\tau = \sigma_H(t), \ t > T \end{cases} \tag{41}$$

Unlike the Adaptive QLV model, no simple procedure exists for calibration the Fung QLV model to a ramp-and-hold test, and calibration requires far more computation. The challenge is that the $\sigma^{(e)}(\varepsilon)$ function, an unknown function, appears inside the convolution integral in both the ramp stress and the hold stress. $g(t)$ and $\sigma^{(e)}(\varepsilon)$ may be calibrated by trial and error and by testing different choices of these functions. Sophisticated optimization algorithms may be used to reduce the number of trials by identifying and subsequently choosing functions based on previous trails, and indeed several functional choices of $g(t)$ and $\sigma^{(e)}(\varepsilon)$ are common in the literature [52, 53, 65, 66]. The optimization procedure requires computing the convolution integrals of Eq. (41) for numerous choices of $g(t)$ and $\sigma^{(e)}(\varepsilon)$ functions, optimizing the prediction of Eq. (41) to find the parameters of these functions. Additional optimization is required for choosing amongst the fits to find optimum functions and parameters.

3.4 Incremental Ramp-and-Hold Protocol

Although a single, large amplitude ramp-and-hold test is a simple and fast protocol for calibrating a QLV model to a biological or bio-artificial tissue, the fit that results may be oversimplified. The drawback is that a single, large amplitude ramp-and-hold test might not contain sufficient information to characterize the way that the relaxation function evolves with tissue length: such a protocol is appropriate only if a tissue is well-modeled by a single reduced relaxation function that is valid for all tissue lengths. The incremental ramp-and-hold protocol (Fig. 3d) is an alternative that generates more information than either the single large amplitude ramp-and-hold protocol or the incremental relaxation protocol.

The incremental ramp-and-hold protocol is similar to the incremental relaxation protocol, except that the stepwise stretches are replaced by ramp stretches. The strain function for the nth ramp-and-hold test can be written:

$$\varepsilon_n(t) = \begin{cases} (n-1)\Delta\varepsilon, & t<0 \\ (n-1)\Delta\varepsilon + \frac{\Delta\varepsilon}{T}t, & 0<t<T \\ n\Delta\varepsilon, & t>T \end{cases} \tag{42}$$

where T is the duration of ramp loading in each increment. We denote the experimentally recorded stress during the nth ramp-loading phase ($0 \leq t \leq T$) as $P_n(t)$, and that recorded during nth hold-relaxation phase ($t \geq$ T) as $H_n(t)$.

3.4.1 Adaptive QLV Model

Substituting the strain from Eq. (42) into Eq. (9), the viscoelastic strains in the nth relaxation test are:

$$V_{ni}^{(\varepsilon)}(t) = \begin{cases} \int\limits_0^t g_i(t-\tau)\frac{\Delta\varepsilon}{T}d\tau = \frac{\Delta\varepsilon}{T}\gamma_i(t), & 0<t<T \\ \int\limits_0^T g_i(t-\tau)\frac{\Delta\varepsilon}{T}d\tau = \frac{\Delta\varepsilon}{T}(\gamma_i(t)-\gamma_i(t-T)), & t>T \end{cases} \tag{43}$$

where, analogous to the previous section, $\gamma_i(t)$ is the integral of $g_i(t)$:

$$\gamma_i(t) = \int_0^t g_i(\tau)d\tau. \tag{44}$$

The stress can be written:

$$\sigma_n(t) = \begin{cases} \sigma_o\left((n-1)\Delta\varepsilon + \frac{\Delta\varepsilon}{T}t\right) \\ \qquad + \frac{\Delta\varepsilon}{T}\sum_i k_i\left((n-1)\Delta\varepsilon + \frac{\Delta\varepsilon}{T}t\right)\gamma_i(t) = \sigma_{Pn}(t), & 0<t<T \\ \sigma_o(n\Delta\varepsilon) \\ \qquad + \frac{\Delta\varepsilon}{T}\sum_i k_i(n\Delta\varepsilon)(\gamma_i(t) - \gamma_i(t-T)) = \sigma_{Hn}(t), & t>T \end{cases} \tag{45}$$

where $\sigma_{Pn}(t)$ and $\sigma_{Hn}(t)$ are the predictions of the time variation of stress over the nth ramp-loading and nth hold-relaxation phases, respectively.

As before, $\sigma_o(\varepsilon)$ can be calibrated at each strain level $n\Delta\varepsilon$ through:

$$\sigma_o(n\Delta\varepsilon) = H_n(t \to \infty). \tag{46}$$

For known shape functions, the values of $k_i(n\Delta\varepsilon)$ may be calibrated using the hold-relaxation data of the nth ramp-and-hold test so that the integrals I_n are each minimized:

$$I_n = \int_T^{+\infty} (H_n(t) - \sigma_{Hn}(t))^2 dt. \tag{47}$$

For parametric shape functions, the parameters of the shape functions and the values of $k_i(n\Delta\varepsilon)$ for all n may be calibrated so that the following integral is minimized:

$$I = \sum_n \int_T^{+\infty} \left(\frac{H_n(t) - \sigma_{Hn}(t)}{H_n(T)}\right)^2 dt. \tag{48}$$

As before, the values of the functions k_i at intermediate stains may be estimated by interpolation.

An advantage of this protocol amongst the four considered in this chapter is that it affords an additional set of data with which to check or refine the predictive capability of a QLV model: once the k_i functions have been determined for all strains using a fitting of the hold-relaxation data, the fitted model may be used to predict the ramp-loading stress histories σ_{Pn}. These predictions can be compared with the measured data to assess the predictive capability of the fitted constitutive model.

This fitting procedure is far simpler than that needed for fitting the Fung QLV and Generalized Fung QLV models. However, it is more complicated than that for the Adaptive QLV model in a single, large amplitude ramp-and-hold test. In the incremental ramp-and-hold test, the variation of the relaxation function with tissue length can be modeled. This precludes the use the exceptionally simple fitting procedure from the single, large amplitude ramp-and-hold protocol (Eq. (40)) to determine the k_i functions at intermediate stains $((n-1)\Delta\varepsilon < \varepsilon < n\Delta\varepsilon)$.

3.4.2 Generalized Fung Model

Substituting the strain function from Eq. (41) into Eq. (9), stress history for the nth relaxation test is:

$$\sigma_n(t) = \begin{cases} \sigma_o\left((n-1)\Delta\varepsilon + \frac{\Delta\varepsilon}{T}t\right) \\ \qquad + \frac{\Delta\varepsilon}{T}\sum\limits_i \int\limits_0^t g_i(t-\tau)A_i\left((n-1)\Delta\varepsilon + \frac{\Delta\varepsilon}{T}\tau\right)d\tau = \sigma_{Pn}(t), \quad 0<t<T \\ \sigma_o(n\Delta\varepsilon) \\ \qquad + \frac{\Delta\varepsilon}{T}\sum\limits_i \int\limits_0^T g_i(t-\tau)A_i\left((n-1)\Delta\varepsilon + \frac{\Delta\varepsilon}{T}\tau\right)d\tau = \sigma_{Hn}(t), \quad t>T \end{cases} \tag{49}$$

The nonlinear unknown functions $A_i(\varepsilon)$ appear inside the convolution integral in both the ramp-loading stress history $\sigma_{Pn}(t)$ and the hold-relaxation stress history $\sigma_{Hn}(t)$. Therefore, both $\sigma_{Pn}(t)$ and $\sigma_{Pn}(t)$ depend on the functions $A_i(\varepsilon)$ for all the strains between $(n-1)$ $\Delta\varepsilon$ and $n\Delta\varepsilon$. This complicates the optimization procedure as in the case of the Fung model for the single large amplitude ramp-and-hold protocol. Similarly, the optimization protocol is based on several trial-and-error steps. However, because in the incremental ramp-and-hold protocol the incremental strain $\Delta\varepsilon$ is small, we may approximate the $A_i(\varepsilon)$ functions as piecewise linear. For the case of a stress-free, fully relaxed specimen at zero strain, we may assume all $A_i(0)$ to be zero. In the nth ramp-and-hold test, A_i may be written as a linear function in terms of a single parameter, a slope m_{ni}, as:

$$A_i\left((n-1)\Delta\varepsilon + \frac{\Delta\varepsilon}{T}t\right) = A_i((n-1)\Delta\varepsilon) + m_{ni}\frac{\Delta\varepsilon}{T}t, \tag{50}$$

where $A_i((n-1)\,\Delta\varepsilon)$ was determined in the $(n-1)$th ramp-and-hold test. For known shape functions, the parameters m_{ni} may be determined from data in the nth ramp-and-hold test by minimizing the integral I_n given in Eq. (47). For parametric shape functions, the parameters of the shape functions and all the m_{ni} parameters may be calibrated so that the integral I given by Eq. (48) is minimized.

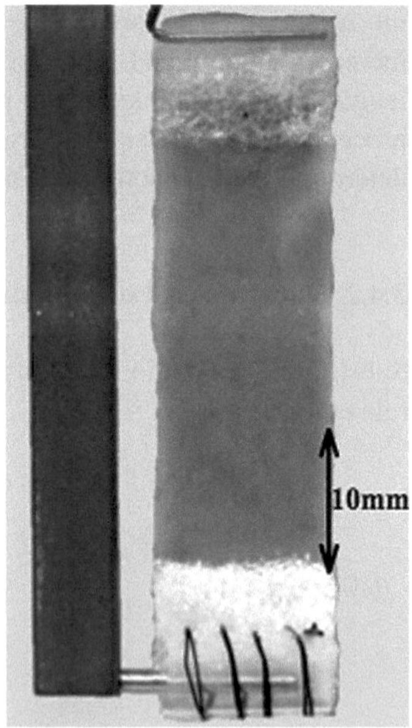

Fig. 4 Reconstituted collagen test specimen mounted on the two parallel bars of the testing apparatus [55]. The upper bar was fixed and connected to a force transducer. The lower bar was movable, with motion controlled by a computer to apply a specified displacement strain function

4 Example of the Calibration and Evaluation of Nonlinear Viscoelastic Models: Reconstituted Collagen

We present here complete examples of characterizing a bioartificial tissue using quasi-linear viscoelastic methods. The example is the stress–strain relation of reconstituted type I collagen gel specimens, used in our laboratories as a scaffold for bioartifical tissues containing prescribed populations of cells [67–69].

A number of protocols for these tissue constructs can be found in the literature [70, 71]. The specimens studied in this section were synthesized from 4 mg/ml rat-tail type I collagen stock solution (Millipore, Inc., Billerica, MA) in 0.02 M acetic acid, with pH brought to 7.4 using sodium hydroxide. The solution was poured into rectangular molds and incubated at 37 °C for 15 h as it gelled through non-covalent cross-linking of collagen molecules. Final specimen dimensions were 30 mm long × 10 mm wide × 3 mm thick. Molds contained fabric attachments that were folded and stitched over plastic tubes to facilitate attachment to testing bars (Fig. 4). The testing apparatus consisted of two horizontal bars, one suspended from a force transducer and the other attached to a sliding element controlled by a stepper motor through a micrometer. The micro-stepping driver was programmable and allowed prescription of specific displacement functions.

4.1 Test Protocol

Constitutive models were fit to data from an incremental ramp-and-hold protocol. Four tests were performed on each specimen, as described in [55]. Each ramp increment involved a 2.0 mm stretch over $T = 20$ s; each hold lasted 2000 s, sufficient for specimens to relax to an almost constant level of isometric force. Force was recorded at 10 Hz. Before the first test, the loading bars were adjusted so that the stress was zero at the initial (zero strain) length of the specimen. After the final ramp-and-hold increment, the loading bars were returned to their pre-test separation, and the specimen was allowed to relax for 2000 s.

The predictions of the models were thereafter checked against data from a single, large amplitude ramp-and-hold protocol. The same specimens were stretched 8.0 mm at a constant rate over $T = 10$ s [55].

4.2 Calibration

Here we present detailed examples of how to calibrate the Adaptive QLV model and Generalized QLV model to the incremental ramp-and-hold test data. Strain and stress were calculated from experimentally measured displacement and force using Eqs. (21) and (22). Three exponential shape functions were chosen:

$$g_1(t) = e^{-t/\tau_1},\ g_2(t) = e^{-t/\tau_2},\ g_3(t) = e^{-t/\tau_3}, \tag{51}$$

where τ_1, τ_2 and τ_3 are the three parameters of these shape functions that should be calibrated. The incremental strain in each test was:

$$\Delta\varepsilon = \frac{\Delta l}{l_o} = \frac{2}{30} = 0.0667, \tag{52}$$

and the strain function in the nth test ($n = 1, 2, 3$, or 4) was:

$$\varepsilon_n(t) = \begin{cases} 0.0667(n-1) + \frac{0.0667}{20}t, & 0 < t < 20 \\ 0.0667n, & t > 20 \end{cases}, \tag{53}$$

so that the four final strains following the four incremental tests were 0.0667, 0.1333, 0.2000 and 0.2667.

In the plot of the experimentally recorded stress data for the four incremental tests (Fig. 5) [55], the first 20 s (ramp-loading phase) is expanded 5 times in the time domain for clarity, and the initial stress (stress at the beginning of each ramp loading), the maximum stress (stress at the end of each ramp loading, and at the beginning of the subsequent hold relaxation), and the final stress (at the end of each hold relaxation) are labeled in each test. Following the fitting procedures described above, we calibrated the QLV models using the hold relaxation time courses of stress, H_n, recorded during the four incremental tests.

4.2.1 Adaptive QLV Model

According to Eq. (46) and using the experimental recordings, the elastic part of stress, σ_o, was 1, 6, 19 and 38 Pascal at 4 strains of 0.0667, 0.1333, 0.2000 and

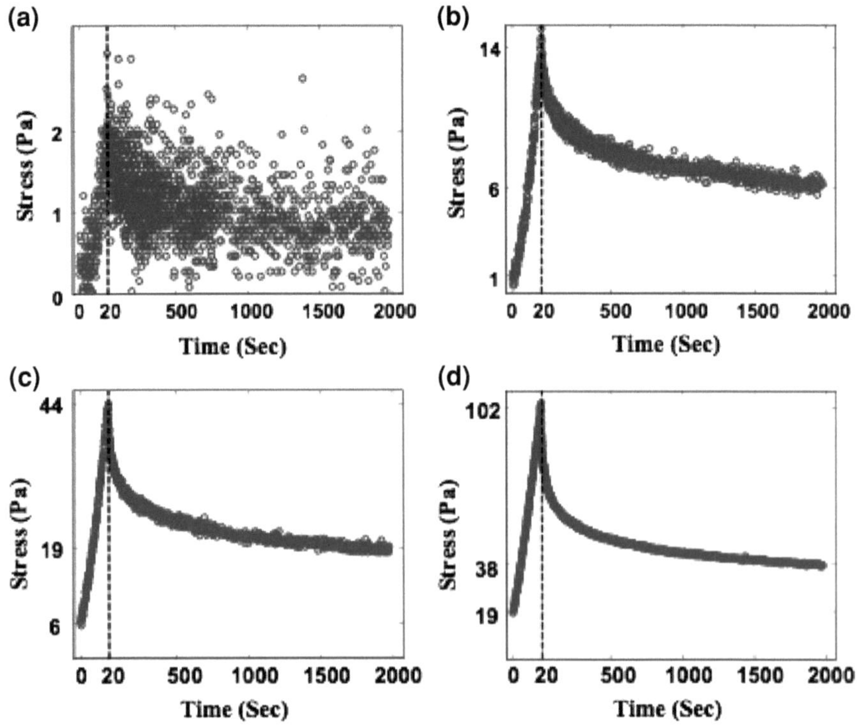

Fig. 5 Experimentally recorded stress histories for four incremental ramp-and-hold tests performed on a single reconstituted collagen specimen [55]. The nominal strain increment is 0.0667 in all tests. The ramp portion is expanded in time scale for clarity. **a** Initial strain of 0; **b** Initial strain of 0.0667; x **c** Initial strain of 0.1333; **d** Initial strain of 0.2000

0.2667, respectively. As the stress was zero initially, $\sigma_o(\varepsilon = 0) = 0$. Calibration k_1, k_2 and k_3 required calculation of the model prediction of the hold stresses. For the g_i functions given by Eq. (51), the G_i functions are:

$$G_1(t) = \tau_1\left(1 - e^{-t/\tau_1}\right),\ G_2(t) = \tau_2\left(1 - e^{-t/\tau_2}\right),\ G_3(t) = \tau_3\left(1 - e^{-t/\tau_3}\right). \quad (54)$$

Substituting these into Eq. (45), the hold stress is:

$$\sigma_{Hn}(t) = \sigma_o(0.0667n) + \frac{0.0667}{20}\sum_{i=1}^{3} k_i(0.0667n)\,\tau_i\left(e^{20/\tau_i} - 1\right)e^{-t/\tau_i}. \quad (55)$$

The hold-relaxation stress prediction is, in this case, an exponential function with the same exponents as the reduced relaxation functions:

$$\sigma_{Hn}(t) = \sigma_o(0.0667n) + \sum_{i=1}^{3} c_i(0.0667n)\,e^{-t/\tau_i}, \quad (56)$$

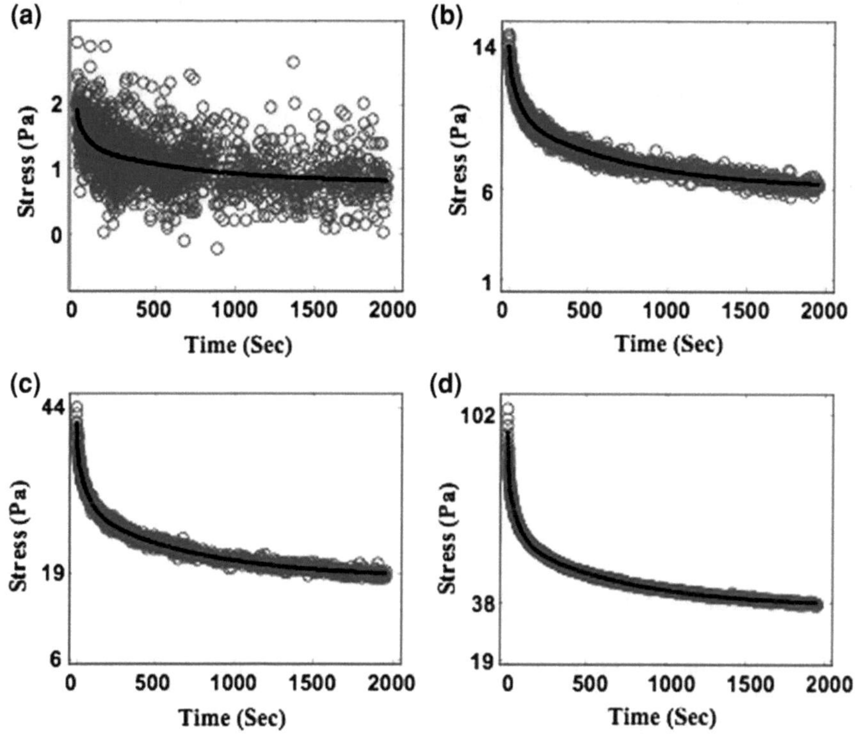

Fig. 6 Three time constant exponential fits (*black curves*) to hold-relaxation stress data (*gray circles*) for the Adaptive QLV model. Optimum time constants: 5.5, 66 and 699 s. **a** Initial strain of 0; **b** Initial strain of 0.0667; **c** Initial strain of 0.1333; **d** Initial strain of 0.2000

where

$$c_i(0.0667n) = \frac{0.0667}{20} k_i(0.0667n)\left(e^{20/\tau_i} - 1\right), \quad i = 1, 2, 3. \tag{57}$$

Therefore, the time constants τ_1, τ_2 and τ_3 could be calibrated simply by calibrating a three time constant exponential function (with variable amplitudes) to all of the hold-relaxation stress histories simultaneously. This is an interesting feature of the Adaptive QLV model that simplifies the calibration significantly for exponential shape functions.

All four hold relaxation stresses were fitted with three-time constant exponential curves by minimizing the integral I given in Eq. (48). The resultant fitted curves (Fig. 6), with σ_o, c_i and τ_i as shown in Table 1, was an excellent fit of the hold stress time history data in all four tests. The functions k_i calculated for these strains using Eq. (57) are listed in Table 2. A cubic interpolation was used to determine the σ_o and k_i functions for strains from 0 to 0.2667 (Fig. 7) [55]. Knowing the three parameters of the shape functions: τ_1, τ_2 and τ_3, and σ_o and k_i functions for all strains the Adaptive QLV model was fully calibrated.

Table 1 Optimum time constants and coefficients of exponential fits to hold-relaxation stress history data for the Adaptive QLV model

ε	$\tau_0 = \infty$ (Pa)	$\tau_1 = 5.5$ s (Pa)	$\tau_2 = 66$ s (Pa)	$\tau_3 = 699$ s (Pa)
0.0667	$\sigma_o = 0.8$	$c_1 = 0.2$	$c_2 = 0.4$	$c_3 = 0.6$
0.1333	$\sigma_o = 6.0$	$c_1 = 1.6$	$c_2 = 2.8$	$c_3 = 4.1$
0.2000	$\sigma_o = 18.3$	$c_1 = 6.6$	$c_2 = 8.4$	$c_3 = 10.7$
0.2667	$\sigma_o = 36.9$	$c_1 = 23.2$	$c_2 = 20.7$	$c_3 = 23.2$

Table 2 Fitted values of k_i for the Adaptive QLV model at the final strains of incremental ramp-and-hold tests

ε	$\tau_1 = 5.5$ s (Pa)	$\tau_2 = 66$ s (Pa)	$\tau_3 = 699$ s (Pa)
0.0667	$k_1 = 12$	$k_2 = 7$	$k_3 = 9$
0.1333	$k_1 = 91$	$k_2 = 49$	$k_3 = 62$
0.2000	$k_1 = 371$	$k_2 = 146$	$k_3 = 163$
0.2667	$k_1 = 1300$	$k_2 = 361$	$k_3 = 354$

4.2.2 Generalized Fung QLV Model

As with the Adaptive QLV model, fitting of the Generalized Fung QLV model began with function σ_o. This, again, has values of 1, 6, 19 and 38 Pa at strains of 0.0667, 0.1333, 0.2000 and 0.2667, respectively. Calibration of the values of the functions $A_1(\varepsilon)$, $A_2(\varepsilon)$, and $A_3(\varepsilon)$ at each strain level required calculation of the model predictions of the hold stresses. Substituting the strain and shape functions from Eqs. (53) and (51) into Eq. (49), the hold relaxation stress in Generalized Fung model is:

$$\sigma_{Hn}(t) = \sigma_o(0.0667n) + \frac{0.0667}{20}\sum_{i=1}^{3}\int_0^T A_i\left(0.0667(n-1) + \frac{0.0667}{20}\tau\right)e^{-(t-\tau)/\tau_i}d\tau \tag{58}$$

To simplify the calibration we assumed a piecewise linear approximation of the unknown $A_i(\varepsilon)$ functions as given in Eq. (50). For piecewise linear $A_i(\varepsilon)$ functions the hold stress can be rewritten as:

$$\sigma_{Hn}(t) = \sigma_o(0.0667n) + \frac{0.0667}{20}\sum_{i=1}^{3}\int_0^T \left(A_i(0.0667(n-1)) + m_{ni}\frac{0.0667}{20}\tau\right)e^{-(t-\tau)/\tau_i}d\tau \tag{59}$$

or, equivalently, as:

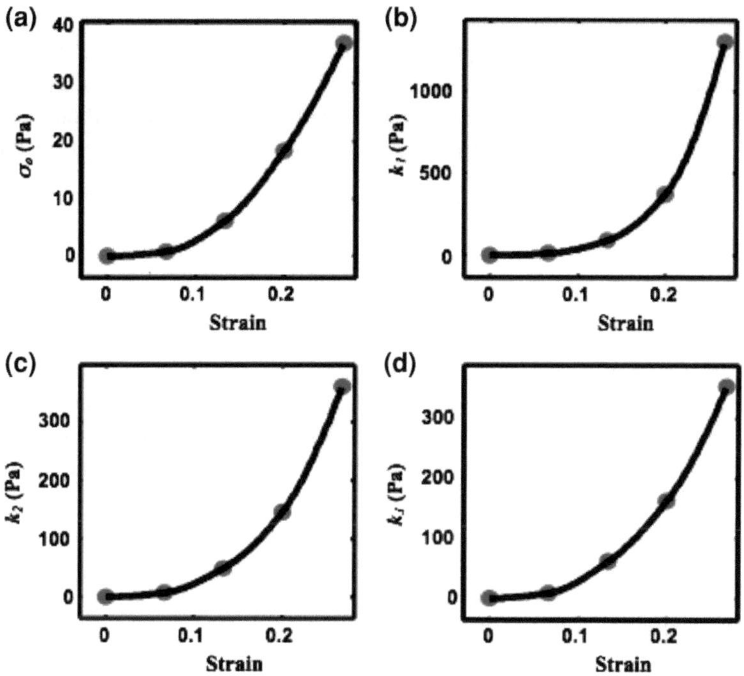

Fig. 7 Elastic part of the stress (**a**) and three nonlinear functions k_1 (**b**), k_2 (**c**) and k_3 (**d**) for the fitted Adaptive QLV model [55]. *Gray circles* are calibrated values based on experimental data and *black curves* are cubic spline interpolations

$$\sigma_{Hn}(t) = \sigma_o(0.0667n) + \frac{0.0667}{20}\sum_{i=1}^{3} A_i(0.0667(n-1))\tau_i\left(e^{T/\tau_i} - 1\right) e^{-t/\tau_i} + \left(\frac{0.0667}{20}\right)^2 \sum_{i=1}^{3} m_{ni}\tau_i^2\left(1 - e^{T/\tau_i} + \frac{T}{\tau_i}e^{T/\tau_i}\right)e^{-t/\tau_i}. \quad (60)$$

Because stress is zero at zero strain, we may assume $A_i(0) = 0$. Thus, using Eq. (50), we may write:

$$A_i(n\Delta\varepsilon) = \Delta\varepsilon\sum_{v=1}^{n} m_{ni}. \quad (61)$$

This allowed Eq. (60) to be rewritten in terms of τ_i and m_{ni}.
These parameters were calibrated so that the integral I given by Eq. (48) was minimized. The fitted curves for the Generalized QLV model with pricewise linear approximations of the A_i functions (Fig. 8) again produced an excellent fit to the hold stress in all four tests (Fig. 9); optimum time constants and slopes are listed in Table 3. The optimum time constants were very close to those found for the Adaptive QLV model. Indeed, one may use a different fitting procedure (that

applicable to exponential shape functions) to obtain the same time constants as those found for the Adaptive QLV model, which results in slightly different m_{ni} coefficients. Note that the functions A_i of the Generalized Fung QLV model differ fundamentally from the k_i functions of the Adaptive QLV model: although they are on the same order of magnitude, one is placed inside the convolution integral, while the other is placed outside.

4.3 Assessment of Predictions

The predictive ability of a model may be examined against experimental data that was not used for calibration. As described above, the incremental ramp-and-hold protocol always provides such data for model evaluation. The subsequent single large amplitude ramp-and-hold test, which occurred at a different strain rate, provides additional data for model evaluation. In this section we provide examples of model calibration using both sets of data.

4.3.1 Incremental Ramp Stresses

The ramp loading stresses predicted by the Adaptive QLV model are:

$$\sigma_{Pn}(t) = \sigma_o\left(0.0667(n-1) + \frac{0.0667}{20}t\right) + \frac{0.0667}{20}\sum_i k_i\left(0.0667(n-1) + \frac{0.0667}{20}t\right)\gamma_i(t), \qquad (62)$$

where G_i functions are given in Eq. (54), and those predicted by the Generalized Fung QLV model are:

$$\sigma_{Pn}(t) = \sigma_o\left(0.0667(n-1) + \frac{0.0667}{20}t\right) + \frac{0.0667}{20}\sum_{i=1}^{3}\int_0^t A_i\left(0.0667(n-1) + \frac{0.0667}{20}t\right)e^{-(t-\tau)/\tau_i}d\tau. \qquad (63)$$

Using the model parameters calibrated to the hold-relaxation data, the model prediction for the ramp stress histories were calculated for each incremental test and compared to experimental data (Fig. 10) [55]. Both the Adaptive QLV and Generalized Fung QLV models could reasonably predict the ramp stresses during incremental testing. The predictions of these two calibrated QLV models were slightly different for the ramp loading stress, although both models reproduced the hold-relaxation data almost identically. The Adaptive QLV model exhibited a slightly larger upward curvature for the ramp stresses compared with the Generalized Fung QLV model.

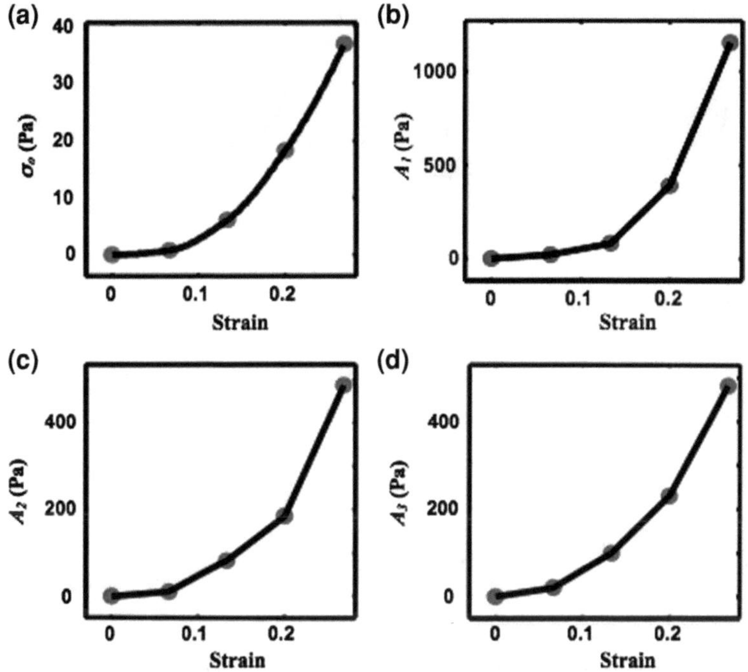

Fig. 8 Elastic part of the stress (**a**) and three nonlinear functions: A_1 (**b**), A_2 (**c**) and A_3 (**d**) for Generalized Fung QLV model. Gray circles are calibrated values based on experimental data and black curves are incrementally linear interpolations

4.3.2 Single, Large Amplitude Ramp-and-Hold Test

The final test of the model fits came from the single, large amplitude ramp-and-hold test, performed at a loading rate different from that used for model calibration. The ramp loading and hold relaxation stresses of the tissue specimen in the large amplitude ramp-and-hold test (Fig. 11a and b, respectively) show that, for the first 3 s of the ramp loading, the tissue stress remains almost zero, while the nominal strain increased (linearly) to about 0.08. In these reconstituted collagen specimens, this is an indicator of plastic deformation that increased the initial zero stress length of the specimens over the course of the series of incremental ramp-and-hold tests that preceded the single, large amplitude ramp-and-hold experiment presented in Fig. 11. The models were adapted for this inelastic deformation by increasing the zero stress length of the tissue for both elastic and viscous components of the stress by plastic strain increments $\Delta\varepsilon_i$. These additional parameters were calibrated using the hold relaxation stress in the large amplitude ramp-and-hold test without altering the already calibrated shape functions or nonlinear functions. We were able to calibrate these plastic strains for an excellent fit of the hold relaxation data in both the Adaptive QLV and Generalized Fung QLV models (Fig. 11d and f, respectively).

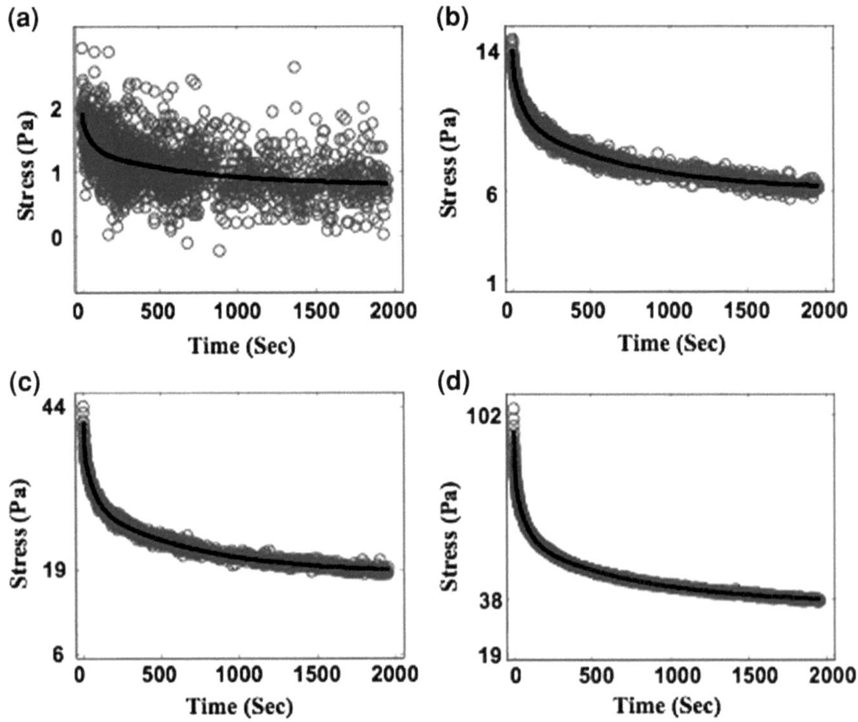

Fig. 9 Three time constant exponential fit of the Generalized Fung QLV model (*black curves*) to the hold relaxation stress history data (*gray circles*). Optimum time constants were: 6.6, 61 and 611 s. **a** Initial strain of 0; **b** Initial strain of 0.0667; **c** Initial strain of 0.1333; **d** Initial strain of 0.2000

Table 3 Optimum time constants and associated slopes of piecewise linear A_i functions for the Generalized Fung QLV model. Data are presented at the final strain levels of the strain increments

ε	$\tau_1 = 6.6$ s (Pa)	$\tau_2 = 61$ s (Pa)	$\tau_3 = 611$ s (Pa)
0.0667	$m_{11} = 309$	$m_{12} = 148$	$m_{13} = 310$
0.1333	$m_{21} = 878$	$m_{22} = 1072$	$m_{23} = 1191$
0.2000	$m_{31} = 4662$	$m_{32} = 1551$	$m_{33} = 1948$
0.2667	$m_{41} = 11491$	$m_{42} = 4544$	$m_{43} = 3767$

Taking into account the plastic deformation, the ramp-loading stress histories were computed in both models and plotted against the experimental recordings (Fig. 11c and e). The Adaptive QLV model has a reasonable fit to the ramp stress (maximum deviation of 2.5 Pa.) while the Generalized Fung model slightly overestimates the ramp stress (maximum deviation of 7 Pa.). The Adaptive QLV model predicts a ramp-loading stress history with a greater upward curvature than does the Generalized Fung QLV. This indicates that the nonlinearity has more significant effect during the ramp loading.

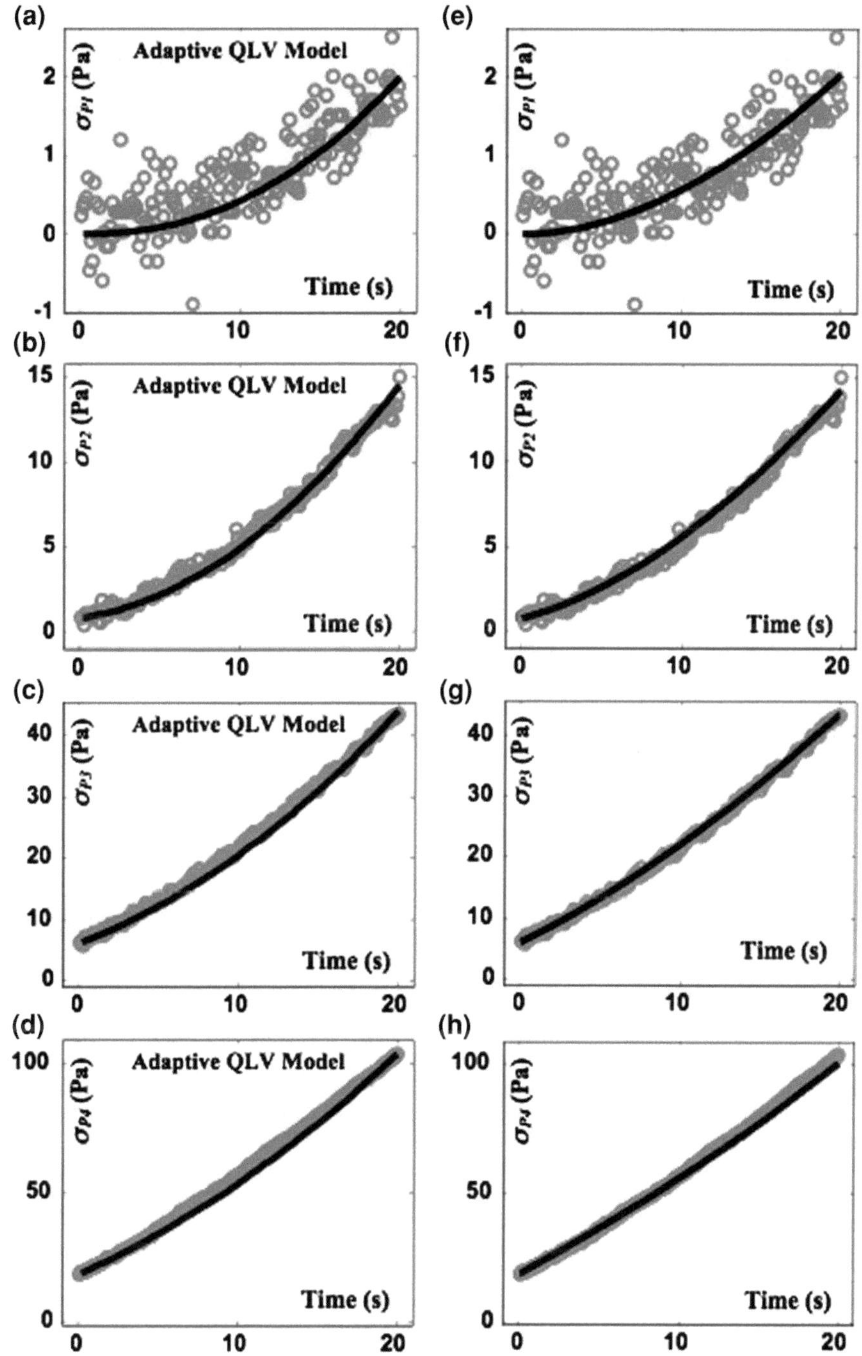

Fig. 10 Model predictions (*black curves*) for the incremental ramp-loading stress histories, plotted against the experimentally recorded stress histories (*gray circles*) [55]. **a–d** are predictions of the Adaptive QLV model prediction for tests with final strains of 0.0667, 0.1333, 0.2000 and 0.2667, respectively. **e–h** are predictions of the Generalized Fung QLV model for tests with final strains of 0.0667, 0.1333, 0.2000 and 0.2667, respectively

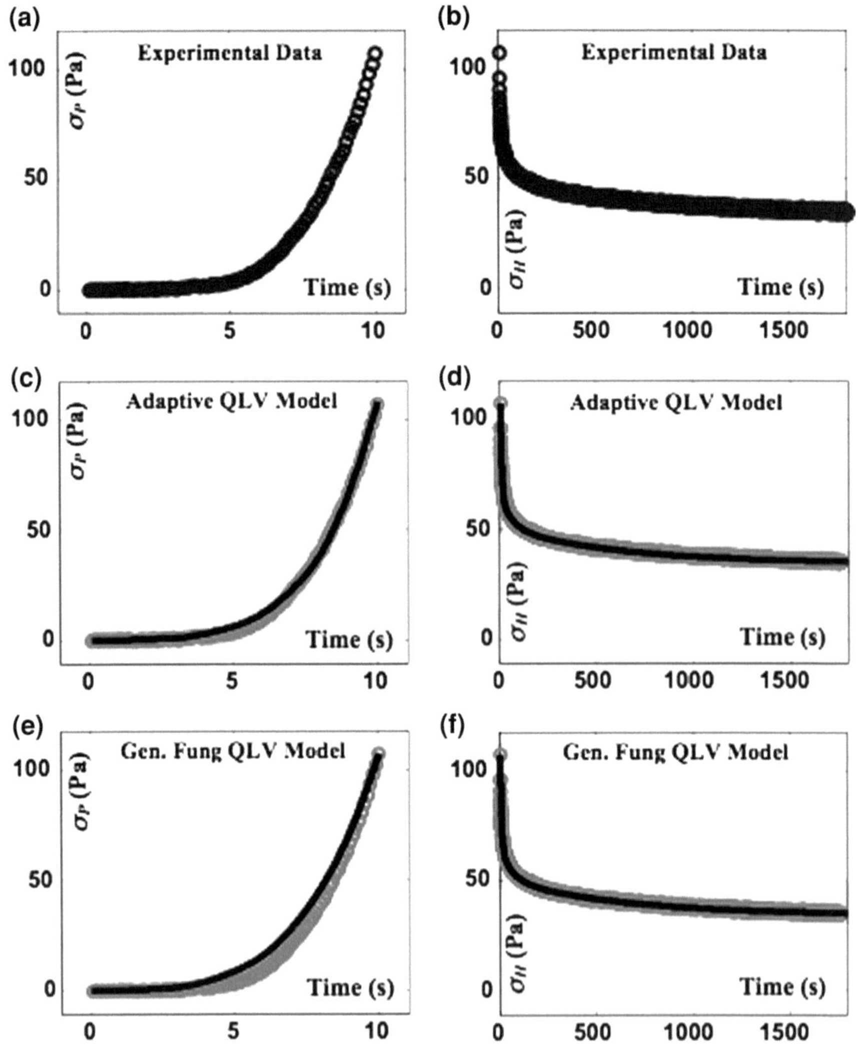

Fig. 11 Measured (*gray circles*) and predicted (*black lines*) ramp-loading and hold-relaxation stress histories for a reconstituted collagen specimen during a large amplitude ramp-and-hold test that followed a series of incremental ramp-and-hold tests [55]. **a** Measured stress during the ramp-loading phase. **b** Measured stress during hold-relaxation phase. **c** Prediction of the Adaptive QLV model for ramp stress (*black curve*) compared to measured stress (*gray circles*). **d** Calibration of the plastic strain increments for the Adaptive QLV model (*black curve*) compared to measured hold-relaxation stress (*gray circles*). **e** Prediction of the Generalized Fung QLV model for ramp stress (*black curve*) compared to measured stress (*gray circles*). **f** Calibration of the plastic strain increments for the Generalized Fung QLV model (*black curve*) compared to measured hold-relaxation stress (*gray circles*)

4.4 Concluding Remarks

We conclude with thoughts on selecting a viscoelastic model that is appropriate for a specific material and application.

The simplest approach by far is to use a linear viscoelastic model, with the assumption of small strain. Nearly every viscoelastic material can be modeled using linear viscoelasticity over a range of conditions, and the first assessment to make is whether this range of conditions encompasses those in which the material will be studied. Geometric linearity becomes an increasingly poor assumption with increasing strain. As a rule of thumb, it is a poor approximation for strain levels above 10–15 % in cases in which a constitutive model is to be applied to conditions that differ substantially from those used to fit the constitutive law. In the example used in this chapter, strain levels exceeded this range substantially, but accounting for geometric nonlinearity was not needed for comparing amongst identical uniaxial stretches applied to a single specimen. However, if these data were to be applied to predict the response of a tissue to a large strain, multiaxial stress state, an appropriate pullback to the reference configuration would need to be applied to the fitted models prior to generalizing to a 3D constitutive law. For this, we refer the reader to the texts of Gurtin et al. [72] and of Bower [73].

The validity of an assumption of material viscoelasticity can be assessed in a number of ways. Qualitatively, a good indication from a ramp-loading test that a nonlinear model is required is a force–displacement that is concave-up: all linear viscoelastic models yield predictions that are linear or concave-down. If such curvature is significant over strain rates and amplitudes of interest, a nonlinear viscoelastic model must be used.

If a nonlinear viscoelastic model is required, quasi-linear viscoelasticity is a great place to start. The Fung QLV model is often fully adequate if a single reduced relaxation function is appropriate for all levels of tissue stretch. An example of how to check this is the set of tests shown in Fig. 1. If, as in Fig. 1, the reduced relaxation function varies with the degree that a tissue is stretched, a different framework is required.

The Fung QLV model is easily extended through the Generalized Fung QLV model described in this chapter to allow for reduced relaxation functions that vary with strain. However, all of the tedious computation associated with the Fung QLV model is compounded when fitting the Generalized Fung QLV model. The Adaptive QLV model is far simpler than either of these models to fit, and has an especially simple form for many standard ramp-and-hold tests.

The principle by which nonlinearity is incorporated into the Adaptive QLV model differs fundamentally from that of the Fung and Generalized Fung QLV models. The consequence of this is that the choice of constitutive model cannot be based upon the simplicity of the Adaptive QLV model alone. Instead, tests must be performed to assess the suitability of model predictions. The incremental ramp-and-hold protocol always provides data for such assessment, and in the case of the reconstituted collagen specimens discussed in this chapter the Generalized Fung

QLV and Adaptive QLV models were both suitable when compared against these data. An additional single, large amplitude ramp-and-hold test performed at a different loading rate showed that, for this particular tissue, both models provided suitable predictions. However, minor differences in the two model predictions were also evident, and careful study would be needed to determine model suitability for loading rates outside of those tested.

In summary, the Adaptive QLV model provides a simple-to-fit nonlinear alternative to the Fung and Generalized Fung QLV models. Because each of these three models rests upon a set of assumptions, application of any of these models to a particular tissue requires careful testing over the range of strains and strain rates that are of interest for a particular application.

Acknowledgments This work was supported in part by the National Institutes of Health (HL079165) and by the Johanna D. Bemis trust.

References

1. Taber, L.A.: Biomechanics of growth, remodeling, and morphogenesis. Appl. Mech. Rev. **48**, 487 (1995)
2. Taber, L.A.: Biomechanics of cardiovascular development. Ann. Rev. Biomed. Eng. **3**(1), 1–25 (2001)
3. Varner, V.D., Taber, L.A.: Not just inductive: a crucial mechanical role for the endoderm during heart tube assembly. Development **139**(9), 1680–1690 (2012)
4. Varner, V.D., Voronov, D.A., Taber, L.A.: Mechanics of head fold formation: investigating tissue-level forces during early development. Development **137**(22), 3801–3811 (2010)
5. Zamir, E.A., Czirok, A., Cui, C., Little, C.D., Rongish, B.J.: Mesodermal cell displacements during avian gastrulation are due to both individual cell-autonomous and convective tissue movements. Proc. Natl. Acad. Sci. USA **103**(52), 19806–19811 (2006)
6. Thomopoulos, S., Genin, G.M., Galatz, L.M.: The development and morphogenesis of the tendon-to-bone insertion - what development can teach us about healing. J. Musculoskelet. Neuronal. Interact. **10**(1), 35–45 (2010)
7. Bayly, P.V., Cohen, T.S., Leister, E.P., Ajo, D., Leuthardt, E.C., Genin, G.M.: Deformation of the human brain induced by mild acceleration. J. Neurotrauma. **22**(8), 845–856 (2005)
8. Cohen, T.S., Smith, A.W., Massouros, P.G., Bayly, P.V., Shen, A.Q., Genin, G.M.: Inelastic behavior in repeated shearing of bovine white matter. J. Biomech. Eng. **130**(4), 044504 (2008)
9. Elson, E.L., Fried, E., Dolbow, J.E., Genin, G.M.: Phase separation in biological membranes: integration of theory and experiment. Annu. Rev. Biophys. **39**, 207–226 (2010)
10. Massouros, P.G., Genin, G.M.: The steady-state response of a Maxwell viscoelastic cylinder to sinusoidal oscillation of its boundary. Proc. Roy. Soc. Lond. A **464**(2089), 207–221 (2008)
11. Nekouzadeh, A., Pryse, K.M., Elson, E.L., Genin, G.M.: Stretch-activated force shedding, force recovery, and cytoskeletal remodeling in contractile fibroblasts. J. Biomech. **41**(14), 2964–2971 (2008)
12. Nekouzadeh, A., Rudy, Y.: Continuum molecular simulation of large conformational changes during ion-channel gating. PLoS One **6**(5), e20186 (2011)
13. Qiu, H., Zhu, Y., Sun, Z., Trzeciakowski, J.P., Gansner, M., Depre, C., Resuello, R.R., Natividad, F.F., Hunter, W.C., Genin, G.M., Elson, E.L., Vatner, D.E., Meininger, G.A., Vatner, S.F.: Vascular smooth muscle cell stiffness as a mechanism for increased aortic stiffness with aging. Circ. Res. **107**(5), 615–619 (2010)

14. Sabet, A.A., Christoforou, E., Zatlin, B., Genin, G.M., Bayly, P.V.: Deformation of the human brain induced by mild angular head acceleration. J. Biomech. **41**(2), 307–315 (2008)
15. Silva, J.R., Pan, H., Wu, D., Nekouzadeh, A., Decker, K.F., Cui, J., Baker, N.A., Sept, D., Rudy, Y.: A multiscale model linking ion-channel molecular dynamics and electrostatics to the cardiac action potential. Proc. Natl. Acad. Sci. USA **106**(27), 11102–11106 (2009)
16. Nerurkar, N.L., Elliott, D.M., Mauck, R.L.: Mechanical design criteria for intervertebral disc tissue engineering. J. Biomech. **43**(6), 1017–1030 (2010)
17. Grayson, W.L., Chao, P.H., Marolt, D., Kaplan, D.L., Vunjak-Novakovic, G.: Engineering custom-designed osteochondral tissue grafts. Trends Biotechnol. **26**(4), 181–189 (2008)
18. Yang, P.J., Temenoff, J.S.: Engineering orthopedic tissue interfaces. Tissue Eng. Part B Rev. **15**(2), 127–141 (2009)
19. Thomopoulos, S., Das, R., Birman, V., Smith, L., Ku, K., Elson, E.L., Pryse, K.M., Marquez, J.P., Genin, G.M.: Fibrocartilage tissue engineering: the role of the stress environment on cell morphology and matrix expression. Tissue Eng Part A **17**(7–8), 1039–1053 (2011)
20. Nekouzadeh, A., Silva, J.R., Rudy, Y.: Modeling subunit cooperativity in opening of tetrameric ion channels. Biophys. J. **95**(7), 3510–3520 (2008)
21. Marquez, J.P., Elson, E.L., Genin, G.M.: Whole cell mechanics of contractile fibroblasts: relations between effective cellular and extracellular matrix moduli. Philos. Transact. A Math. Phys. Eng. Sci. **368**(1912), 635–654 (2010)
22. Marquez, J.P., Genin, G.M., Pryse, K.M., Elson, E.L.: Cellular and matrix contributions to tissue construct stiffness increase with cellular concentration. Ann. Biomed. Eng. **34**(9), 1475–1482 (2006)
23. Marquez, J.P., Genin, G.M., Zahalak, G.I., Elson, E.L.: Thin bio-artificial tissues in plane stress: the relationship between cell and tissue strain, and an improved constitutive model. Biophys. J. **88**(2), 765–777 (2005)
24. Marquez, J.P., Genin, G.M., Zahalak, G.I., Elson, E.L.: The relationship between cell and tissue strain in three-dimensional bio-artificial tissues. Biophys. J. **88**(2), 778–789 (2005)
25. Butler, J.P., Tolić-Nørrelykke, I.M., Fabry, B., Fredberg, J.J.: Traction fields, moments, and strain energy that cells exert on their surroundings. Am. J. Physiol-Cell Physiol. **282**(3), C595–C605 (2002)
26. Legant, W.R., Miller, J.S., Blakely, B.L., Cohen, D.M., Genin, G.M., Chen, C.S.: Measurement of mechanical tractions exerted by cells in three-dimensional matrices. Nat. Methods **7**(12), 969–971 (2010)
27. Oliver, T., Jacobson, K., Dembo, M.: Traction forces in locomoting cells. Cell Motil. Cytoskelet. **31**(3), 225–240 (1995)
28. Kastelic, J.: A structural mechanical model for tendon crimping. J. Biomech. **13**, 887–893 (1980)
29. Stouffer, D.C., Butler, D.L., Hosny, D.: The relationship between crimp pattern and mechanical response of human patellar tendon-bone units. J. Biomech. Eng. **107**(2), 158–165 (1985)
30. Hansen, K.A., Weiss, J.A., Barton, J.K.: Recruitment of tendon crimp with applied tensile strain. J. Biomech. Eng. **124**(1), 72–77 (2002)
31. Lanir, Y.: Two dimensional mechanical properties of mammalian skin- II. Experimental results. J. Biomech. **7**, 171–182 (1974)
32. Lanir, Y.: Structure-function relations in mammalian tendon: the effect of geometrical nonuniformity. J. Bioeng. **2**(1–2), 119–128 (1978)
33. Lanir, Y.: A structural theory for the homogeneous biaxial stress-strain relationship in flat collagenous tissues. J. Biomech. **12**, 423–436 (1979)
34. Lanir, Y.: A microstructure model for the rheology of mammalian tendon. J. Biomech. Eng. **102**(4), 332–339 (1980)
35. Lanir, Y.: Constitutive equations for fibrous connective tissues. J. Biomech. **16**(1), 1–12 (1983)
36. Sacks, M.S.: Incorporation of experimentally-derived fiber orientation into a structural constitutive model for planar collagenous tissues. J. Biomech. Eng. **125**(2), 280–287 (2003)
37. De Vita, R., Slaughter, W.S.: A structural constitutive model for the strain rate-dependent behavior of anterior cruciate ligaments. Int. J. Solids Struct. **43**(6), 1561–1570 (2006)

38. Genin, G.M., Kent, A., Birman, V., Wopenka, B., Pasteris, J.D., Marquez, P.J., Thomopoulos, S.: Functional grading of mineral and collagen in the attachment of tendon to bone. Biophys. J. **97**(4), 976–985 (2009)
39. Thomopoulos, S., Marquez, J.P., Weinberger, B., Birman, V., Genin, G.M.: Collagen fiber orientation at the tendon to bone insertion and its influence on stress concentrations. J. Biomech. **39**(10), 1842–1851 (2006)
40. Raghupathy, R., Barocas, V.H.: A closed-form structural model of planar fibrous tissue mechanics. J. Biomech. **42**(10), 1424–1428 (2009)
41. Raghupathy, R., Barocas, V.H.: Generalized anisotropic inverse mechanics for soft tissues. J. Biomech. Eng. **132**(8), 081006 (2010)
42. Raghupathy, R., Witzenburg, C., Lake, S.P., Sander, E.A., Barocas, V.H.: Identification of regional mechanical anisotropy in soft tissue analogs. J. Biomech. Eng. **133**(9), 091011 (2011)
43. Zahalak, G.I., Wagenseil, J.E., Wakatsuki, T., Elson, E.L.: A cell-based constitutive relation for bio-artificial tissues. Biophys. J. **79**(5), 2369–2381 (2000)
44. Hollander, Y., Durban, D., Lu, X., Kassab, G.S., Lanir, Y.: Constitutive modeling of coronary arterial media: comparison of three model classes. J. Biomech. Eng. **133**, 061008 (2011)
45. Coleman, B.D., Noll, W.: Foundations of linear viscoelasticity. Rev. Mod. Phys. **33**(2), 239 (1961)
46. Coleman, B.D., Noll, W.: The thermodynamics of elastic materials with heat conduction and viscosity. Arch. Rat. Mech. Anal. **13**(1), 167–178 (1963)
47. Schapery, R.: Nonlinear viscoelastic solids. Int. J. Solids Struct. **37**(1), 359–366 (2000)
48. Fung, Y.C.: Ch 7: stress-strain-history relations of soft tissues in simple elongation. In: Fung, Y.C., Perrone, N., Anliker, M. (eds.) Biomechanics: Its Foundations and Objectives, pp. 181–208. Prentice-Hall, San Diego (1972)
49. Quaia, C., Ying, H.S., Nichols, A.M., Optican, L.M.: The viscoelastic properties of passive eye muscle in primates. I: Static forces and step responses. PLoS One **4**(4), e4850 (2009)
50. Quaia, C., Ying, H.S., Optican, L.M.: The viscoelastic properties of passive eye muscle in primates. II: testing the quasi-linear theory. PLoS One **4**(8), e6480 (2009). doi:10.1371/journal.pone.0006480
51. Quaia, C., Ying, H.S., Optican, L.M.: The viscoelastic properties of passive eye muscle in primates. III: force elicited by natural elongations. PLoS One **5**(3), e9595 (2010)
52. Hingorani, R.V., Provenzano, P.P., Lakes, R.S., Escarcega, A., Vanderby Jr, R.: Nonlinear viscoelasticity in rabbit medial collateral ligament. Ann Biomed Eng **32**(2), 306–312 (2004)
53. Provenzano, P., Lakes, R., Corr, D., Vanderby, R.: Application of nonlinear viscoelastic models to describe ligament behavior. Biomech Model. Mechanobiol. **1**(1), 45–57 (2002)
54. Provenzano, P., Lakes, R., Keenan, T., Vanderby Jr, R.: Nonlinear ligament viscoelasticity. Ann. Biomed. Eng. **29**(10), 908–914 (2001)
55. Nekouzadeh, A., Pryse, K.M., Elson, E.L., Genin, G.M.: A simplified approach to quasi-linear viscoelastic modeling. J. Biomech. **40**(14), 3070–3078 (2007)
56. Pryse, K.M., Nekouzadeh, A., Genin, G.M., Elson, E.L., Zahalak, G.I.: Incremental mechanics of collagen gels: new experiments and a new viscoelastic model. Ann. Biomed. Eng. **31**(10), 1287–1296 (2003)
57. Truesdell, C., Noll, W., Antman, S.S.: The non-linear field theories of mechanics, vol. 3. Springer, New York (2004)
58. Rivlin, R.: Further remarks on the stress-deformation relastions for isotropic materials. J. Rat. Mech. Anal. **4**, 681–702 (1955)
59. Spencer, A., Rivlin, R.: The theory of matrix polynomials and its application to the mechanics of isotropic continua. Arch. Rat. Mech. Anal. **2**(1), 309–336 (1958)
60. Pipkin, A.: Small finite deformations of viscoelastic solids. Rev. Mod. Phys. **36**(4), 1034 (1964)
61. Pipkin, A., Rogers, T.: A non-linear integral representation for viscoelastic behaviour. J. Mech. Phys. Solids **16**(1), 59–72 (1968)
62. Neubert, H.K.P.: A simple model representing internal damping in solid materials. Aeronaut. Q **14**, 171–182 (1963)
63. Fung, Y.C.: Biomechanics: mechanical properties of living tissues, 2nd edn. Springer, New York (1993)

64. Nekouzadeh, A., Genin, G.M., Bayly, P.V., Elson, E.L.: Wave motion in relaxation-testing of nonlinear elastic media. Proc. Roy. Soc. Lond. A **461**, 1599–1626 (2005)
65. Chun K.J, Hubbard R.P.: Development of a reduced relaxation function and comparison of stress relaxation for anatomically paired tendons. J. Mech. Sci. Technol. **23** (7):1893–1898 (2009). doi:10.1007/s12206-009-0504-3
66. Duenwald, S.E., Vanderby, R., Lakes, R.S.: Constitutive equations for ligament and other soft tissue: evaluation by experiment. Acta. Mech. **205**(1–4), 23–33 (2009)
67. Wagenseil, J.E., Elson, E.L., Okamoto, R.J.: Cell orientation influences the biaxial mechanical properties of fibroblast populated collagen vessels. Ann. Biomed. Eng. **32**(5), 720–731 (2004)
68. Wagenseil, J.E., Wakatsuki, T., Okamoto, R.J., Zahalak, G.I., Elson, E.L.: One-dimensional viscoelastic behavior of fibroblast populated collagen matrices. J. Biomech. Eng. **125**(5), 719–725 (2003)
69. Wakatsuki, T., Kolodney, M.S., Zahalak, G.I., Elson, E.L.: Cell mechanics studied by a reconstituted model tissue. Biophysical. J. **79**(5), 2353–2368 (2000)
70. Thomopoulos, S., Fomovsky, G.M., Holmes, J.W.: The development of structural and mechanical anisotropy in fibroblast populated collagen gels. J. Biomech. Eng. **127**(5), 742–750 (2005)
71. Barocas, V.H., Girton, T.S., Tranquillo, R.T.: Engineered alignment in media equivalents: magnetic prealignment and mandrel compaction. J. Biomech. Eng. **120**(5), 660–666 (1998)
72. Gurtin, M.E., Fried, E., Anand, L.: The Mechanics and Thermodynamics of Continua. Cambridge University Press, Cambridge (2009)
73. Bower, A.F.: Applied Mechanics of Solids. CRC, New York (2009)

Computational Modeling of Mass Transport and Its Relation to Cell Behavior in Tissue Engineering Constructs

Dennis Lambrechts, Jan Schrooten, Tom Van de Putte and Hans Van Oosterwyck

Abstract Effective recapitulation of extracellular matrix properties into a Tissue Engineering strategy is strongly involved with the need for a proper transport environment. Consumption and production of soluble medium components gives rise to gradients which influence cell behavior in various ways. Understanding how transport related phenomena can shape these gradients is targeted in this chapter by the combined use of experiments and mathematical modeling. An overview of different models is given that describe solute transport and its relation to specific cell behavior. From the simulation results important information can be extracted which help to unravel mechanisms that drive solute transport. Finally we describe the genuine efforts that have been taken to translate this information into real tissue engineering setups (e.g., optimization of culture conditions and controlled-release of growth factors).

D. Lambrechts (✉) · J. Schrooten
Department of Metallurgy and Materials Engineering, KU Leuven, Kasteelpark Arenberg 44-box 2450, 3001 Leuven, Belgium
e-mail: Dennis.Lambrechts@mech.kuleuven.be

D. Lambrechts · J. Schrooten · H. Van Oosterwyck
Prometheus, Division of Skeletal Tissue Engineering Leuven, KU Leuven, Herestraat 49-box 813, 3000 Leuven, Belgium

T. Van de Putte
TiGenix NV, Haasrode Researchpark 1724, Romeinse straat 12 box 2, 3001 Leuven, Belgium

H. Van Oosterwyck
Biomechanics Section, KU Leuven, Celestijnenlaan 300C-box 2419, 3001 Leuven, Belgium

Stud Mechanobiol Tissue Eng Biomater (2013) 10: 85–105
DOI: 10.1007/8415_2012_139

Published Online: 7 June 2012

1 Introduction

The functionality of extracellular matrix (ECM) in organ tissues is much broader than just structural support for cells that reside within. Cells are able to interact with ECM, both biochemically and biophysically, from structural remodeling induced by cell proteolytic activity [72] to stem cell lineage specification via straining of linked proteins [30]. The ECM also provides an important storage space for signaling molecules, a feature which appeared to be crucial in tissue morphogenesis [61, 121]. Morphogenesis is mainly driven by the gradients in signaling molecules (i.e. morphogens) which can arise from differences in morphogen diffusivity [79, 127] or from interstitial fluid flow-induced asymmetry in morphogen distribution upon enzymatic release from the matrix [50]. Understanding the mass transport principles which underlie the formation and maintenance of morphogen gradients is therefore fundamental to understand how these gradients will direct tissue patterning.

It should be clear that recapitulating fundamental ECM properties in a tissue engineering (TE) context relates closely to the aim of establishing a proper mass transport environment. From basic nutrients to signaling molecules, concentration gradients might exist for any soluble medium component that is consumed or produced by the cells [42]. The effects they can elicit on cell behavior are numerous and have proven to be a function of absolute concentrations, the range of operation and slope [46]. To measure such gradients, the use of biosensors [1] and tracer molecules [127] has been previously reported. Experimental quantification is however not always straightforward and can even become too challenging for more complex (in vivo-like) setups. A powerful tool that makes quantification easier and can predict gradient magnitudes for the even most complex situations [86], lies in the combined use of experiments and mathematical modeling [24].

Mathematical models help in establishing relations and insights between evolving solute profiles and specific cell behavior [80]. Such models have proven their applicability in unraveling important mechanisms and dynamics of experimental observations [25, 37]. Their usability ranges from the establishment of numerical interactions between influencing parameters [130] to optimization of culture conditions for nutrient transport [111] and modeling-based TE carrier design [16].

In light of the design of biomaterial carriers, special attention should be attributed to the environmental remodeling abilities of a cell. Triggered by their proteolytic activity cells can break down ECM components for migration or modify tissue architecture in response to biophysical or biochemical forces [97]. These changes have however important consequences on the transport and activity of autocrine and paracrine signaling molecules, both directly and indirectly [115]. Implementation of structural biomaterial remodeling in mathematical models has contributed to a better comprehension of its active role in cell signaling [120]. Translating these remodeling principles in a TE strategy has led to the development of biomaterials crosslinked by enzyme-degradable peptide sequences which allows for a spatiotemporal control of their degradation properties [71, 100]. Models of in vivo regeneration processes (e.g. fracture healing [40]) will thereby give crucial information on the timing and location of remodeling and hence also proteolytic events. This information can be effectively implemented in treatment strategies aiming to provide continuous mechanical support to the fracture, by matching scaffold degradation with new tissue generation [105], and allows the cells to re-establish a properly working signaling environment.

In this book chapter we aim to elucidate how the key actors that govern mass transport in a TE carrier (i.e., biomaterial, cells, culture environment and solutes) can influence the overall functioning of this carrier. This will be described in terms of specific cell behavior that is provoked under defined concentrations and gradients of soluble factors. The influence of the carrier components will be captured in a series of continuum parameters that are used to formulate the mass transport problem in a mathematical landscape. Finally for each component illustrations are given on how to exploit this information in the optimization of culture conditions and the rational design of setups used for TE applications.

2 General Mass Transport in Carriers

Solute transport in biomaterial carriers (e.g., hydrogels and macro- or microporous scaffolds) used for TE applications is generally governed by passive diffusion. Diffusive transport as a primary transport mechanism in carriers can however put major constraints on the remodeling capabilities of cells that reside within this material [42] and hence also on new tissue formation. Since in this setting of dynamic tissue architecture and composition the transport of solutes with large molecular weight is most strongly affected [9], important modulations in cell signaling can be expected [120]. However also the transport of small molecules, such as oxygen, can be impeded as cells grow and new tissue is produced which gives rise to an imbalance between solute uptake and supply [25].

For this reason bioreactor systems have been developed which try to overcome fundamental limitations that are associated with diffusive mass transport. A wide variability in bioreactor configurations exists that enhance mass transport in and to the carrier, either by direct perfusion/compression or indirect perfusion/mixing [82]. Since movement and exercise are important driving forces for the body's interstitial fluid flow [120], special attention will be given here to the influence of mechanical carrier loading on solute transport.

Based on the previous discussion we introduce the general equation for mass conservation,

$$\frac{\partial C_i(x,y,z,t)}{\partial t} = -\nabla \cdot \vec{\Gamma}_i(x,y,z,t) + R_i(x,y,z,t)$$

where C_i is the concentration of the solute of interest; $\vec{\Gamma}_i$ is the mass flux; R_i is a reaction term which accounts for consumption, production, degradation, or binding of solute i to the matrix; and t is time. The mass flux due to molecular diffusion is proportional to the gradient in solute concentration ($\vec{\Gamma}_i = -D_{ij}\nabla C_i$), while convective transport is driven by the velocity field $\vec{v}$ ($\vec{\Gamma}_i = C_i\vec{v}$) [6]. Substitution of both terms into the previous equation gives (in the case of an incompressible medium),

$$\frac{\partial C_i(x,y,z,t)}{\partial t} = D_{ij}(x,y,z,t)\nabla^2 C_i(x,y,z,t) - \vec{v}(x,y,z,t)\cdot\nabla C_i(x,y,z,t) + R_i(x,y,z,t)$$

where $\vec{v}$ is the solute velocity vector; D_{ij} is the diffusion coefficient of the solute in solvent j; and ∇^2 is the Laplacian operator. This general equation applies for most biomaterial setups used in tissue engineering and its constitutive transport parameters can be determined either from experiments or from theoretical formulations.

2.1 *Diffusion*

Experimental quantification of solute diffusion rates through a carrier have been performed in well-controlled release kinetics experiments and by fitting analytical solutions to Fick's diffusion law [16, 25]. Also well established are fluorescence techniques to measure dispersal of fluorescently labeled target molecules, such as Fluorescence Recovery After Photobleaching (FRAP) [11], photoactivation [99], photoconversion [47] or photoswitching [3] of these fluorescent molecules. Major advantage of the latter methods is that they are less time-consuming as compared to release kinetics [11] and also have the ability to record local differences in solute diffusivity, which have been shown to result from structural matrix heterogeneities [118].

Alternatively, solute diffusion rates for a specific carrier matrix can be estimated from existing literature values. Reported values are obtained either for diffusion in free solution or for a given solute carrier combination. Based on microscale structural differences, several authors have formulated relationships which try to explain discrepancies in diffusion coefficients between often seemingly equal matrices [55, 57, 62, 95]. These models take into account the steric interactions of solutes diffusing through the matrix. The interplay between structural matrix features and effective solute diffusivity can also be described in terms of the hydrodynamic obstructions of diffusing solute. Correlations have resulted from this approach using techniques of volume averaging, which requires the use of periodic structural models [128, 129], the effective-medium approximation, which does not impose any restrictions on the structural model but is at the cost of a reduced validity and reliability of the predictions [18, 103] or by using a random walk approach [118].

2.2 Convection

Convective mass transport through a porous medium can be described using the averaged equations as formulated by Darcy or Brinkman [27]. These equations express a relation between medium velocity and the applied pressure gradient which is governed by the permeability, a factor that characterizes the influencing matrix properties. Experimental setups are described in literature that can measure different carrier permeabilities, either by applying a constant pressure or a constant flow [69, 91].

Several models are available that provide a link between structural carrier properties and permeability [49, 53]. Their application range is however constrained to approximations at the macroscopic scale since they depend on physical and geometric idealizations of the microporous carrier [120]. From the theory of mixtures and based on experimental results, it was found that the effects of induced fluid flows (which are rather low) on solute transport in the carrier are most significant for solutes with large molecular weight [35, 111].

2.3 Compression-Induced Mass Transport

Dynamic compression of a carrier combines matrix compaction with interstitial fluid transport [33]. Augmented solute transport associated with this convective fluid transport will therefore at the same time be restricted resulting from a decrease in matrix diffusivity. This compression-induced loss in diffusivity can either be measured experimentally with FRAP [38, 66] or estimated from structural diffusivity relations such as in Mackie and Meares [73, 74, 92], which assumes a high dependency of matrix diffusivity on fluid volume fraction. An interesting alternative would be the coupling of diffusive transport with structural deformation at the microscale level [117].

The effect of unconfined compression on enhanced solute transport has previously been formulated in mathematical terms using the theory of incompressible mixtures [5, 89]. In agreement with experimental observations, it was established that compression frequency and solute molecular weight are both decisive factors for the extent of compression-enhanced solute transport [32, 83]. As a major conclusion from these studies it was shown that mechanical carrier stimulation can significantly improve the transport of larger molecules (from glucose to large signaling molecules). The mechanism which underlies this phenomenon is found in the dual action of small convective flows and the increased peripheral solute gradient during dynamic loading [83].

In the next sections we will give an overview on how solute molecules with different molecular weight can influence specific cell behavior and how the transport of these molecules is influenced by the cellular carrier components. Solutes of interest range from small molecules (e.g., oxygen and glucose) to large molecules (e.g., growth factors).

3 Oxygen

The main mechanism by which cells acquire their energy is through oxidative phosphorylation [2]. In this process oxygen serves as an oxidizing agent that facilitates the flux of electrons through progressively lower energy states, which allows for a large extraction of free energy used to synthesize adenosine triphosphate (ATP) molecules. Apart from energy production, oxygen has proven to be a potent modulator of cell behavior that can change cellular phenotype [56], stimulate matrix production by the cells [94] or induce angiogenesis by the release of angiogenic factors [26]. Molecular oxygen however has a low solubility in culture medium and is rapidly consumed by the cells in order to meet their continuous energy demands. These factors make soluble oxygen very prone to become depleted during culture [42].

To what extent in time and space oxygen might become depleted within a carrier is not only regulated by the intrinsic mass transport properties of the carrier alone. A major influence comes from the cells themselves. This includes the cellular demand for dissolved oxygen, expressed by the cellular oxygen uptake rate (OUR), which is known to be controlled by many factors.

Firstly, cells harvested from distinct tissue types in the body can have significant differences in OUR [109]. Secondly, the availability of oxygen to the cells is a strong determinant of mitochondrial respiration. When cells are exposed to oxygen tensions below a critical value, the redox state of cytochrome oxidase or the respiration rate itself is partially limited [12]. This effect can be captured by a Michaelis–Menten kinetic [34],

$$Q(x,y,z,t) = Q_{\max}\frac{c_{O_2}(x,y,z,t)}{K_q + c_{O_2}(x,y,z,t)}$$

where Q is the oxygen uptake rate (mol cell^{-1} h^{-1}); Q_{max} is the maximal OUR (mol cell^{-1} h^{-1}); and K_q the oxygen tension at half of the maximal consumption rate (mol m^{-3}). Both kinetic parameters were furthermore shown to be dependent upon specific cell-material interactions [45, 80]. This relation could have important consequences related to biomaterial choice and cell remodeling behavior.

Underlying the total drop in oxygen tension inside the carrier is the effective number of metabolically active cells. Cell growth inside a biomaterial carrier can be modeled in many ways. This ranges from simple linear or piece-wise linear relationships with available nutrient concentrations [84] to more detailed models such as the modified Contois equation,

$$P_C(x,y,z,t) = c_{cell}\left[\frac{A_{cell}c^n_{O_2}(x,y,z,t)}{K_C c_{cell}\lambda\varepsilon^n_{cell} + c^n_{O_2}(x,y,z,t)} - d(x,y,z,t)\right]$$

where P_i denotes the cell growth kinetic function (cell m^{-3} day^{-1}); c_{cell} is the cell density (cell m^{-3}); A_{cell} is the homogeneous growth rate (day^{-1}); K_C is the modified Contois saturation constant; ε_{cell} is the cell volume fraction (V_{cell}/V); V_{cell}

is the averaging volume of the cell phase (m^3); V is the averaging volume (m^3); λ is a cell conversion factor (mol cell^{-1}); d is the death rate (day^{-1}); and n is a semi-empirical parameter. This equation inhibits cell growth in case of an overpopulation of available carrier space [19]. The Moser equation restricts cell growth in a direct relation to the available oxygen concentration [88, 104],

$$P_{Mr}(x,y,z,t) = c_{cell}\left[\frac{A_{cell}c^n_{O_2}(x,y,z,t)}{K_q\varepsilon^n_{cell} + c^n_{O_2}(x,y,z,t)} - d(x,y,z,t)\right]$$

The Moser equation reduces to the Monod equation for n equal to one. This equation couples the OUR directly to the cell growth rate as,

$$P_{Md}(x,y,z,t) = c_{cell}\left[\left(\frac{A_{cell,\max}}{Y_{CO_2}} + m_{cell}\right)\frac{c_{O_2}(x,y,z,t)}{K_q\varepsilon_{cell} + c_{O_2}(x,y,z,t)} - d(x,y,z,t)\right]$$

where $A_{cell,max}$ is the maximal specific cell growth rate (day^{-1}); Y_{CO2} is the yield of cells per unit oxygen (cells mol^{-1}); m_{cell} is the maintenance coefficient (mol cell^{-1} day^{-1}), the minimum oxygen consumption required to keep the cells alive.

The main difference between linear and non-linear systems is that the former will produce a significant region of uniform proliferation, a phenomenon that is rarely observed in practice [70]. This heterogeneity can also implicitly be implemented using a custom-defined function derived from experimental data [25, 36].

Finally new matrix production and carrier remodeling can also directly alter oxygen delivery to the cells, though this effect is more pronounced for larger solutes [9, 13, 94].

In the following paragraphs we will give a brief overview on how these oxygen models can be efficiently applied to tackle some specific problems a tissue engineer could encounter. Static in vitro culture of tissue substitutes generally gives rise to heterogeneous cell growth, especially when substitute dimensions exceed a critical size [25, 80]. Enhanced proliferation of cells in the peripheral regions and coupled increases in oxygen uptake, have thereby been speculated as factors determining the incidence and severity of tissue hypoxia and associated cell death [10]. To test this hypothesis and gain improved understanding of the mechanisms which underlie these observations, mathematical models have been developed describing the interactions between oxygen tension and cell density. Effectively applying this strategy Demol et al. presented a model to describe in vitro behavior of human periosteum derived cells cultured inside a fibrin hydrogel construct (Fig. 2) [25].

Necessary model input parameters were derived from dedicated in vitro experiments that allowed to assess cell proliferation, the influence of oxygen tension on cell death and proliferation, and the diffusivity of oxygen in fibrin. As the constructs were cultivated over a period of 14 days, a significant region of dead cells in the construct center could be detected which progressively expanded outwards with longer cultivation times (Fig. 3a). The observation was accompanied by the formation of a multilayered cell sheet which had an average thickness

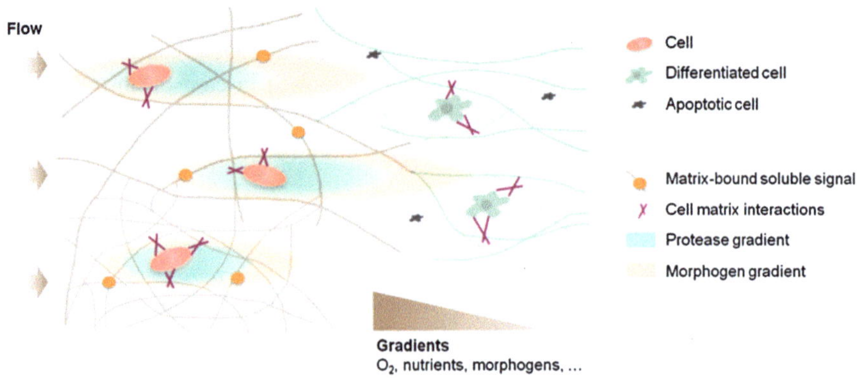

Fig. 1 Fibrous matrix remodeling is triggered by cell proteolytic activity. Transport of various solutes drives specific cell behavior during the different stages of new tissue formation (e.g., differentiation and apoptosis) and is strongly influenced by changes in carrier components (e.g., new matrix production by differentiated cells or release of matrix-bound soluble factors). Adapted from [50]

of about 40 μm and was present at the construct periphery (Fig. 3b). To account for this in the model a heterogeneous proliferation rate was implemented, with fast cell proliferation at the construct surface-its value being determined by means of a dedicated experiment-and slow proliferation inside the construct.

The model predicted the highest cell density in the outer layer (i.e., multilayered cell sheet), which was about 30 times higher than the maximum density in the rest of the gel (Fig. 4a). This led to the highest volumetric oxygen consumption rate and the highest gradient in oxygen tension in this outer layer. (Fig. 4b). However, this was not the primary cause of hypoxic regions detected within the hydrogel center, as the surface region only accounted for 2.3 % of the total decrease in oxygen tension compared to a decrease of 16.5 % in the rest of the hydrogel. This finding was further confirmed in a model parameter sensitivity analysis in which (reasonable) changes in cell sheet thickness did not have a significant influence on the occurrence of tissue hypoxia. On the contrary, variations in cellular oxygen consumption and oxygen diffusivity in the hydrogel region had a major effect on total construct hypoxia, making both governing factors for design and upscaling of static in vitro cultured hydrogel-based constructs.

As a second application we consider the development of TE pancreatic substitutes. In this field the use of mouse insulinoma βTC3 cells has since long been investigated for the long-term treatment of insulin dependent diabetes mellitus (IDDM) [28]. Reports on the insulin secretory capacities of these cells have indicated a strong reduction in secretion as the available oxygen tension drops below 10 μM [98]. This observation has therefore strong implications for the functioning of the substitute in vitro and in vivo, a problem that has been assessed using diffusion–reaction models for oxygen transport in the carrier [44]. By combining experimental measurements with the steady-state solutions of a

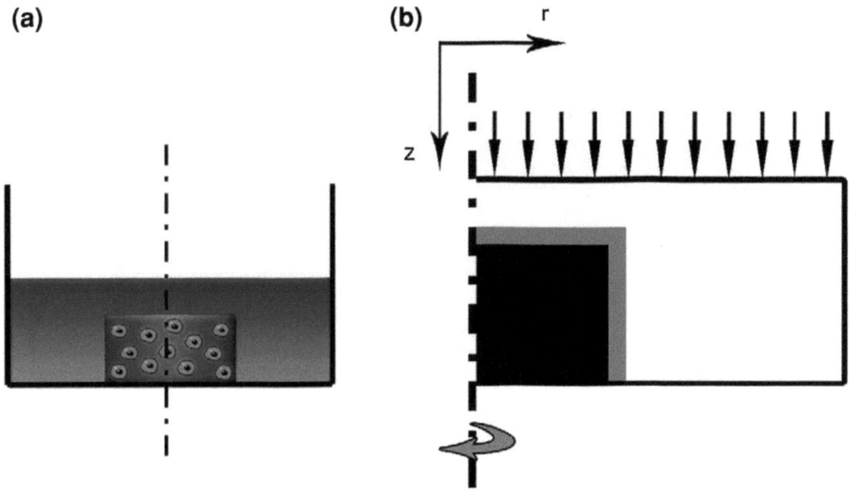

Fig. 2 **a** Schematic representation of cells cultured in a fibrin hydrogel. **b** Geometry of axisymmetric model (half of vertical cross-section) which consists of three distinct regions: cells encapsulated in the hydrogel (*black*), cells on the hydrogel surface (*grey*, not shown proportionally) and culture medium (*white*). Figure reprinted from [25]

mathematical model, these authors aimed at providing a methodology for rapid evaluation of the substitute performance. Experimental input data was obtained from ^{19}F nuclear magnetic resonance imaging on a dissolved perfluorocarbon emulsion inside the carrier. This gave a single averaged read-out of dissolved oxygen tension which showed a unique relation with the number of viable cells within the carrier, for a given environmental oxygen tension. Spatiotemporal evolution of both oxygen and viable cell density were estimated from the mathematical model using the average oxygen tension as a fitting parameter. Application of this approach therefore allowed for quick measurements on substitute functioning. Design and optimization of experimental configurations for tissue substitute cultivation and remodeling is a third important application where mathematical modeling can be of great value. Starting from this principle comparative numerical studies of bioreactor setups for in vitro tissue construct cultivation have led to an improvement of cultivation regimes and construct designs [111, 130]. These studies have indicated that reaching a critical cell mass inside the construct in vitro would require the use of a perfusion culture setup. Enhancing the transport of small molecules (such as oxygen) in perfusion systems would however require relatively high perfusion speeds. This regime demands a change in carrier geometry or biomaterial properties (e.g., stiffer matrix) in order to avoid permanent damage to the carrier. In this optimization process special attention should however be attributed to the presence of multiple soluble components, since changing the transport properties of a given solute might have important consequences for other solute transport [14, 85]. Such altered solute transport

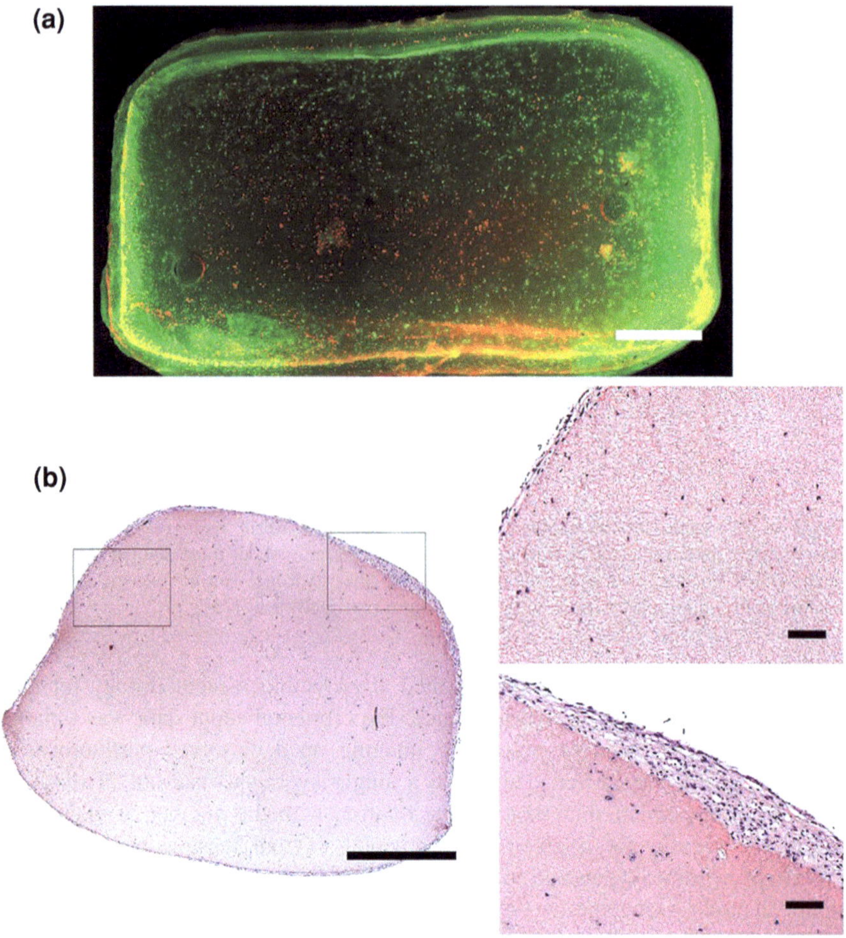

Fig. 3 **a** LIVE/DEAD viability images of human periosteum derived cells cultured for 14 days in fibrin hydrogels (bar = 1 mm). **b** H & E staining on histological sections of a fibrin construct cultured for 21 days (bar is 1 mm for the left image and 100 μm in the two images at higher magnification). Figure reprinted from [25]

could originate from differences in molecular size as described in previous sections, but could also be attributed to differences in cellular response, which will be described in the next sections.

Another strategy aiming to overcome these limitations of inadequate solute mass transport is to induce de novo synthesis of a vascular network or sprouting of existing vessels using the body as an in vivo bioreactor system [116]. The use of mathematical models has been proposed in this context as a tool for the intelligent design of scaffold structures and implantation techniques [22]. The experimental in vivo model investigated by these authors consisted of hepatocyte-seeded PLGA

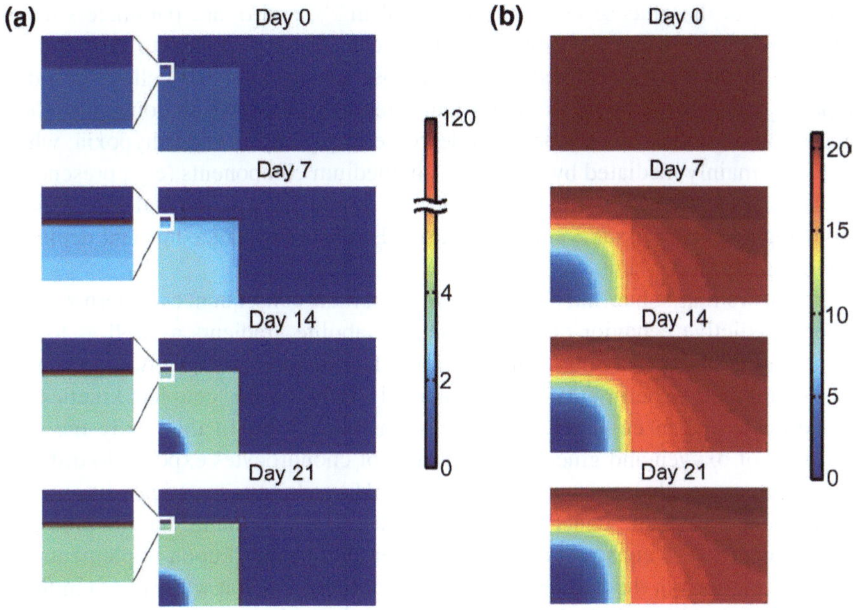

Fig. 4 Cell density (**a**) and oxygen tension (**b**) inside cell-seeded fibrin hydrogels as predicted by the mathematical model. Cell density is expressed in 10^6 cells/ml and oxygen tension in percentage. Figure reprinted from [25]

foam scaffolds with an integrated arteriovenous loop [15]. Based on the results of a computational model describing tissue growth and oxygen diffusion in the scaffold, they argued the need for a heterogeneous seeding pattern in which a small tissue biopsy of the desired tissue would be placed close to the blood vessel loop. This system would better tune angiogenesis and cell outgrowth, reducing the incidence of deficient oxygen transport to the cells and hence improve tissue functionality.

4 Nutrients and Metabolites

Cellular energy metabolism is not only dependent on the presence of a single solute, but on the entire environment of nutrients and metabolites [7, 110, 113]. Reported mechanisms of this interdependent behavior include the stimulation of OUR of cells exposed to low glucose conditions, known as the Crabtree effect [20, 51, 52, 96]. This relationship can be expressed in mathematical terms using an exponential decay function [130],

$$OUR = a_1 + a_2 e^{-c_{glc}/a_3}$$

where c_{glc} is the glucose concentration (mol m^{-3}); and a_i are parameters determined by curve fitting. A similar equation can also be applied to describe the increase in glycolysis (defined by the glucose uptake rate) for cells exposed to hypoxic conditions, known as the Pasteur effect [81, 93, 108]. In contrast to these studies other authors have reported a decrease in glycolysis under hypoxia, which would be mainly mediated by differences in medium components (e.g., presence of oxidants) [43, 67, 68, 130]. These effects should receive special attention in modeling avascular tissues, and hence also TE substitutes, since nutrient depletion could readily occur here.

It follows that solute interactions have important consequences in terms of the model predictive behavior of nutrient and metabolite gradients as well as for the optimization of experimental configurations. To illustrate this we give an example of a combined experimental and numerical study of the coupled kinetics for chondrocytes in an engineered cartilage construct [130]. In this study the relationships of oxygen and glucose uptake rates of chondrocytes exposed to different oxygen tensions, glucose concentrations and pH levels was investigated as well as the influence of lactate concentration on pH level. It was found that the predicted cell viability in the construct center was generally enhanced upon implementation of these relations in the numerical model, an observation that was most significant at high cell densities [130]. This interesting cell behavior predicted *in silico* could hence be indicative of a cellular rescue mechanism.

5 Signaling Molecules

Growth factors serve important roles as signaling molecules during development [46, 90, 119] and regeneration [17, 123]. Most of these signaling molecules need to travel through the surrounding extracellular environment either locally (autocrine signaling) or to more distant locations (paracrine and endocrine signaling) in order to exert their influence on cell behavior [112]. In the following paragraphs we will show how a cell is able to interfere with this transport process and direct this signaling reaction to provoke its intended effect. More specifically the example of cellular signaling in the growth plate will be discussed.

Bone elongation occurs through the action of endochondral ossification, which is driven primarily by the differentiation rate of proliferating chondrocytes into hypertrophic chondrocytes within the growth plate [64]. At a molecular level this process is strongly controlled by two paracrine signaling factors, Indian hedgehog (Ihh) and parathyroid hormone-related protein (PTHrP) [65, 126]. While Ihh coordinates chondrocyte proliferation, differentiation and osteoblast differentiation, PTHrP mainly keeps the proliferating chondrocytes in a proliferative phase [65, 124, 126]. Heparan sulphate (HS) complexes associated with proteoglycans in the ECM of the growth plate serve important roles in morphogenesis by providing binding sites for specific signaling molecules [8, 48]. This specificity depends on the sulfation pattern of the heparan sulphate chains and allows for interactions with several members of the Hedgehog, transforming growth factor-β (TGFβ), bone morphogenetic protein (BMP), Wingless (Wnt/Wg), and fibroblast growth factor (FGF) families [31, 48].

Simple mathematical models have been developed expressing the total width between the early hypertrophic zone and the perichondrium using coupling equations of PTHrP and Ihh concentrations [125]. As concentrations of both signaling molecules can be modulated by various transport-related factors (e.g. mechanical compression, solute binding, changes in diffusivity due to matrix degradation) the relative importance of these factors on rates of proliferation and hypertrophy can be rapidly assessed using this model.

In a second model bone growth and morphogenesis is described by differences in spatial distribution and proliferation rates of proliferative and hypertrophied chondrocytes [37]. Underlying this growth process were the regulatory capacities of the Ihh and PTHrP spatial concentration distributions, that were modeled by a set of reaction–diffusion equations [21, 75, 78]. In order to obtain a physiological growth pattern the ratio of diffusion coefficients for both signaling molecules was bound to certain criteria. The influences of these morphogen diffusion rates on characteristics of the growth plate were acknowledged already earlier in studies of the skeletal disorder Exostosin (EXT1) [54, 63]. There it was shown that in EXT1 mutations, expressing reduced amounts of HS, the range of Ihh signaling within the growth plate was increased giving rise to an extended proliferative zone.

Calcification of the matrix surrounding the hypertrophic chondrocytes in the growth plate triggers the invasion of blood vessels from the metaphyseal bone [58]. This capillary invasion is mediated by the expression of vascular endothelial growth factor (VEGF) in hypertrophic chondrocytes [39]. Binding of VEGF to ECM components has thereby been implicated as a possible requisite for cellular autocrine signaling, giving rise to amplified VEGF gradients that are able to direct capillary morphogenesis [50, 76]. The mechanism underpinning this gradient amplification results from the combined action of a small interstitial fluid flow, biasing the secreted protease distribution, and the distribution of liberated VEGF molecules, influenced by both protease distribution and convective flows [35].

The use of growth factors (such as VEGF) in controlled-release systems has been widely proposed for TE strategies aiming at the regeneration of damaged or diseased tissues [29, 101, 102]. Given the short half-life and residence of free growth factors in solution, controlled-release strategies hold great promise providing a means to protect these factors from degradation and internalization [77, 105]. Though such systems can deliver signaling molecules in a time- and space-controlled manner, the lack of detailed knowledge on in vivo growth factor concentrations and possible interfering behavior of administered compounds complicates rational decisions on the required growth factor concentrations [60]. For such applications we can however greatly take advantage of the use of numerical models. In this way a modeling-based design approach was proposed for the controlled delivery of VEGF in a mouse model of hindlimb ischemia [16]. Using a reaction diffusion model to predict VEGF distribution in vivo, a layered scaffold design was proposed to deliver VEGF in a spatial concentration gradient where it is able to both initiate and spatially control angiogenesis. Regulating spatial VEGF presentation increased hindlimb blood flow which was reflected in a reduced incidence of limb necrosis.

Other more detailed numerical models of in vivo regeneration processes could equally well be applied for the rational design of controlled delivery systems [40]. Such models have the added advantage of testing the efficacy of a certain treatment strategy (such as controlled-delivery systems) *in silico*, hence helping researchers to identify the most promising strategies and having the potential to significantly reduce experimental costs [41].

6 Discussion and Conclusions

Proper functionalizaton of carriers used for TE applications, requires a profound understanding of the mechanisms that drive solute transport to and from the active cell units. These solutes can be as small as oxygen, essential for cellular nutrition, or as large as protein complexes, which allow cells to communicate with each other or probe their environment [122]. The cellular actions they can elicit range from basic cell survival and division to the organized patterning of cells into tissues (i.e., morphogenesis).

Mimicking the normal in vivo solute transport environment of a cell and optimization of culture conditions is complicated by the various mechanisms which underlie these transport processes. We have shown that this variability can be induced by differences in molecular size between solutes, differences in synthesis and uptake of solutes by the cell, interactions at a cell level between various nutrients and metabolites (but also signaling molecules [4, 106]), solute interactions with the surrounding matrix and many others. Systems that involve such high degrees of complexity can however greatly benefit from mathematical modeling, as we have shown in this chapter.

Choosing an optimal model and setting an appropriate level of detail is strongly determined by the extent of construct remodeling that is taking place, the availability and type of experimental data, and spatial resolution [107]. Focus in this chapter has been mainly set on single scale (continuum) models which is ascribed to their abundant availability and their low computational costs. If we however want to recreate interactions at multiple levels of organization with respect to space and time, more attention should be attributed to the integrative properties of the model [114], a strategy that has been defined in literature as multiscale modeling. Crucial to the success of a multiscale modeling framework, is to provide a consistent cross-scale linkage interface between the models at each biological level [23]. We have shown the importance of such an interface in the context of diffusive transport within TE carriers, by providing a means to correlate microstructural matrix properties to solute diffusion rates using for example a volume averaging technique [129].

Finally it should be mentioned that diffusion–reaction mechanisms are not only prominent for solute transport inside the carrier, also inside a cell these mechanisms are important for proper cell functioning (for example in signal propagation [59]). Information generation by the use of solute gradients is accordingly a widely conserved mechanism in biology, operating at multiple scales of organization [87]. This scale-invariant principle holds great promise by extending the applicability of both models and imaging techniques that were described within this chapter.

References

1. Acosta, M.A., Ymele-Leki, P., Kostov, Y.V., Leach, J.B.: Fluorescent microparticles for sensing cell microenvironment oxygen levels within 3D scaffolds. Biomaterials **30**(17), 3068–3074 (2009). doi:10.1016/j.biomaterials.2009.02.021
2. Alberts, B., Johnson, A., Lewis, J., Raff, M., Roberts, K., Walter, P.: Molecular Biology of the Cell, 4th edn. Garland Science, New York (2007)
3. Ando, R., Mizuno, H., Miyawaki, A.: Regulated fast nucleocytoplasmic shuttling observed by reversible protein highlighting. Science **306**(5700), 1370–1373 (2004). doi:10.1126/science.1102506. 306/5700/1370[pii]
4. Arany, Z., Foo, S.Y., Ma, Y.H., Ruas, J.L., Bommi-Reddy, A., Girnun, G., Cooper, M., Laznik, D., Chinsomboon, J., Rangwala, S.M., Baek, K.H., Rosenzweig, A., Spiegelman, B.M.: HIF-independent regulation of VEGF and angiogenesis by the transcriptional coactivator PGC-1 alpha. Nature **451**(7181), 1008–U1008 (2008). doi:10.1038/Nature06613
5. Armstrong, C.G., Lai, W.M., Mow, V.C.: An analysis of the unconfined compression of articular-cartilage. J. Biomech. Eng. Trans. ASME **106**(2), 165–173 (1984)
6. Beard, D.A., Qian, H.: Chemical Biophysics: Quantitative Analysis of Cellular Systems. Cambridge Texts in Biomedical Engineering. Cambridge University Press, New York (2008)
7. Bibby, S.R.S., Jones, D.A., Ripley, R.M., Urban, J.P.G.: Metabolism of the intervertebral disc: effects of low levels of oxygen, glucose, and pH on rates of energy metabolism of bovine nucleus pulposus cells. Spine **30**(5), 487–496 (2005)
8. Bishop, J.R., Schuksz, M., Esko, J.D.: Heparan sulphate proteoglycans fine-tune mammalian physiology. Nature **446**(7139), 1030–1037 (2007). doi:10.1038/Nature05817
9. Blum, J.J., Lawler, G., Reed, M., Shin, I.: Effect of cytoskeletal geometry on intracellular diffusion. Biophys. J. **56**(5), 995–1005 (1989). doi:10.1016/S0006-3495(89)82744-4. S0006-3495(89)82744-4[pii]
10. Brown, D.A., MacLellan, W.R., Laks, H., Dunn, J.C., Wu, B.M., Beygui, R.E.: Analysis of oxygen transport in a diffusion-limited model of engineered heart tissue. Biotechnol. Bioeng. **97**(4), 962–975 (2007). doi:10.1002/bit.21295
11. Brown, E.B., Wu, E.S., Zipfel, W., Webb, W.W.: Measurement of molecular diffusion in solution by multiphoton fluorescence photobleaching recovery. Biophys. J. **77**(5), 2837–2849 (1999)
12. Brown, G.C.: Control of respiration and Atp synthesis in mammalian mitochondria and cells. Biochem. J. **284**, 1–13 (1992)
13. Bursac, P.M., Freed, L.E., Biron, R.J., Vunjak-Novakovic, G.: Mass transfer studies of tissue engineered cartilage. Tissue Eng. **2**(2), 141–150 (1996). doi:10.1089/ten.1996.2.141
14. Cartmell, S.H., Porter, B.D., Garcia, A.J., Guldberg, R.E.: Effects of medium perfusion rate on cell-seeded three-dimensional bone constructs in vitro. Tissue Eng. **9**(6), 1197–1203 (2003)
15. Cassell, O.C., Morrison, W.A., Messina, A., Penington, A.J., Thompson, E.W., Stevens, G.W., Perera, J.M., Kleinman, H.K., Hurley, J.V., Romeo, R., Knight, K.R.: The influence of extracellular matrix on the generation of vascularized, engineered, transplantable tissue. Ann. N. Y. Acad. Sci. **944**, 429–442 (2001)

16. Chen, R.R., Silva, E.A., Yuen, W.W., Brock, A.A., Fischbach, C., Lin, A.S., Guldberg, R.E., Mooney, D.J.: Integrated approach to designing growth factor delivery systems. FASEB J. **21**(14), 3896–3903 (2007). doi:10.1096/fj.06-7873com
17. Chen, Y., Whetstone, H.C., Lin, A.C., Nadesan, P., Wei, Q.X., Poon, R., Alman, B.A.: Beta-catenin signaling plays a disparate role in different phases of fracture repair: implications for therapy to improve bone healing. PLoS Med **4**(7), 1216–1229 (2007). doi:10.1371/journal.pmed.0040249. ARTN e249
18. Clague, D.S., Phillips, R.J.: Hindered diffusion of spherical macromolecules through dilute fibrous media. Phys. Fluids **8**(7), 1720–1731 (1996)
19. Contois, D.E.: Kinetics of bacterial growth: relationship between population density and specific growth rate of continuous cultures. J. Gen. Microbiol. **21**, 40–50 (1959)
20. Crabtree, H.G.: Observations on the carbohydrate metabolism of tumours. Biochem. J. **23**(3), 536–545 (1929)
21. Crampin, E.J., Hackborn, W.W., Maini, P.K.: Pattern formation in reaction-diffusion models with nonuniform domain growth. Bull. Math. Biol. **64**(4), 747–769 (2002). doi:10.1006/bulm.2002.0295
22. Croll, T.I., Gentz, S., Mueller, K., Davidson, M., O'Connor, A.J., Stevens, G.W., Cooper-White, J.J.: Modelling oxygen diffusion and cell growth in a porous, vascularising scaffold for soft tissue engineering applications. Chem. Eng. Sci. **60**(17), 4924–4934 (2005). doi:10.1016/j.cea.2005.03.051
23. Dallon, J.C.: Multiscale modeling of cellular systems in biology. Curr. Opin. Colloid Interface Sci. **15**(1–2), 24–31 (2010). doi:10.1016/j.cocis.2009.05.007
24. Dehmelt, L., Bastiaens, P.I.H.: Spatial organization of intracellular communication: insights from imaging. Nat. Rev. Mol. Cell Biol. **11**(6), 440–452 (2010). doi:10.1038/Nrm2903
25. Demol, J., Lambrechts, D., Geris, L., Schrooten, J., Van Oosterwyck, H.: Towards a quantitative understanding of oxygen tension and cell density evolution in fibrin hydrogels. Biomaterials **32**(1), 107–118 (2011). doi:10.1016/j.biomaterials.2010.08.093
26. Dor, Y., Porat, R., Keshet, E.: Vascular endothelial growth factor and vascular adjustments to perturbations in oxygen homeostasis. Am. J. Physiol. Cell Physiol. **280**(6), C1367–C1374 (2001)
27. Durlofsky, L., Brady, J.F.: Analysis of the Brinkman equation as a model for flow in porous-media. Phys. Fluids **30**(11), 3329–3341 (1987)
28. Efrat, S., Linde, S., Kofod, H., Spector, D., Delannoy, M., Grant, S., Hanahan, D., Baekkeskov, S.: Beta-cell lines derived from transgenic mice expressing a hybrid insulin gene-oncogene. Proc. Natl. Acad. Sci. U. S. A. **85**(23), 9037–9041 (1988)
29. Elisseeff, J., McIntosh, W., Fu, K., Blunk, T., Langer, R.: Controlled-release of IGF-I and TGF-beta 1 in a photopolymerizing hydrogel for cartilage tissue engineering. J. Orthop. Res. **19**(6), 1098–1104 (2001)
30. Engler, A.J., Sen, S., Sweeney, H.L., Discher, D.E.: Matrix elasticity directs stem cell lineage specification. Cell **126**(4), 677–689 (2006). doi:10.1016/j.cell.2006.06.044. S0092-8674(06)00961-5[pii]
31. Esko, J.D., Selleck, S.B.: Order out of chaos: assembly of ligand binding sites in heparan sulfate. Annu. Rev. Biochem. **71**, 435–471 (2002). doi:10.1146/annurev.biochem.71.110601.13545
32. Evans, R.C., Quinn, T.M.: Dynamic compression augments interstitial transport of a glucose-like solute in articular cartilage. Biophys. J. **91**(4), 1541–1547 (2006). doi:10.1529/biophysj.105.080366
33. Evans, R.C., Quinn, T.M.: Solute convection in dynamically compressed cartilage. J. Biomech. **39**(6), 1048–1055 (2006). doi:10.1016/j.jbiomech.2005.02.017. S0021-9290(05)00127-2[pii]
34. Fassnacht, D., Portner, R.: Experimental and theoretical considerations on oxygen supply for animal cell growth in fixed-bed reactors. J. Biotechnol. **72**(3), 169–184 (1999). S0168165699001297[pii]
35. Fleury, M.E., Boardman, K.C., Swartz, M.A.: Autologous morphogen gradients by subtle interstitial flow and matrix interactions. Biophys. J. **91**(1), 113–121 (2006). doi:10.1529/biophysj.105.080192

36. Galban, C.J., Locke, B.R.: Analysis of cell growth kinetics and substrate diffusion in a polymer scaffold. Biotechnol. Bioeng. **65**(2), 121–132 (1999)
37. Garzon-Alvarado, D.A., Garcia-Aznar, J.M., Doblare, M.: A reaction-diffusion model for long bones growth. Biomech. Model. Mechan. **8**(5), 381–395 (2009). doi:10.1007/s10237-008-0144-z
38. Gefen, A., Cornelissen, L.H., Gawlitta, D., Bader, D.L., Oomens, C.W.: The free diffusion of macromolecules in tissue-engineered skeletal muscle subjected to large compression strains. J. Biomech. **41**(4), 845–853 (2008). doi:10.1016/j.jbiomech.2007.10.023. S0021-9290(07)00471-X[pii]
39. Gerber, H.P., Vu, T.H., Ryan, A.M., Kowalski, J., Werb, Z., Ferrara, N.: VEGF couples hypertrophic cartilage remodeling, ossification and angiogenesis during endochondral bone formation. Nat. Med. **5**(6), 623–628 (1999)
40. Geris, L., Gerisch, A., Maes, C., Carmeliet, G., Weiner, R., Vander Sloten, J., Van Oosterwyck, H.: Mathematical modeling of fracture healing in mice: comparison between experimental data and numerical simulation results. Med. Biol. Eng. Comput. **44**(4), 280–289 (2006). doi:10.1007/s11517-006-0040-6
41. Geris, L., Schugart, R., Van Oosterwyck, H.: In silico design of treatment strategies in wound healing and bone fracture healing. Philos. Trans. A Math. Phys. Eng. Sci. **368**(1920), 2683–2706 (2010). doi:10.1098/rsta.2010.0056. 368/1920/2683[pii]
42. Griffith, L.G., Swartz, M.A.: Capturing complex 3D tissue physiology in vitro. Nat. Rev. Mol. Cell Biol. **7**(3), 211–224 (2006). doi:10.1038/Nrm1858
43. Grimshaw, M.J., Mason, R.M.: Bovine articular chondrocyte function in vitro depends upon oxygen tension. Osteoarthr. Cartil. **8**(5), 386–392 (2000). doi:10.1053/joca.1999.0314. S1063-4584(99)90314-X[pii]
44. Gross, J.D., Constantinidis, I., Sambanis, A.: Modeling of encapsulated cell systems. J. Theor. Biol. **244**(3), 500–510 (2007). doi:10.1016/j.jtbi.2006.08.012. S0022-5193(06)00365-1[pii]
45. Guaccio, A., Borselli, C., Oliviero, O., Netti, P.A.: Oxygen consumption of chondrocytes in agarose and collagen gels: a comparative analysis. Biomaterials **29**(10), 1484–1493 (2008). doi:10.1016/j.biomaterials.2007.12.020
46. Gurdon, J.B., Bourillot, P.Y.: Morphogen gradient interpretation. Nature **413**(6858), 797–803 (2001)
47. Gurskaya, N.G., Verkhusha, V.V., Shcheglov, A.S., Staroverov, D.B., Chepurnykh, T.V., Fradkov, A.F., Lukyanov, S., Lukyanov, K.A.: Engineering of a monomeric green-to-red photoactivatable fluorescent protein induced by blue light. Nat. Biotechnol. **24**(4), 461–465 (2006). doi:10.1038/Nbt1191
48. Hacker, U., Nybakken, K., Perrimon, N.: Heparan sulphate proteoglycans: the sweet side of development. Nat. Rev. Mol. Cell Biol. **6**(7), 530–541 (2005). 10.1038/nrm1681]nrm1681[pii]
49. Happel, J.: Viscous flow relative to arrays of cylinders. AIChE J. **5**(2), 174–177 (1959)
50. Helm, C.L.E., Fleury, M.E., Zisch, A.H., Boschetti, F., Swartz, M.A.: Synergy between interstitial flow and VEGF directs capillary morphogenesis in vitro through a gradient amplification mechanism. Proc. Natl. Acad. Sci. U. S. A. **102**(44), 15779–15784 (2005). doi:10.1073/pnas.0503681102
51. Heywood, H.K., Bader, D.L., Lee, D.A.: Rate of oxygen consumption by isolated articular chondrocytes is sensitive to medium glucose concentration. J. Cell. Physiol. **206**(2), 402–410 (2006). doi:10.1002/Jcp.20491
52. Heywood, H.K., Knight, M.M., Lee, D.A.: Both superficial and deep zone articular chondrocyte subpopulations exhibit the crabtree effect but have different basal oxygen consumption rates. J. Cell. Physiol. **223**(3), 630–639 (2010). doi:10.1002/Jcp.22061
53. Higdon, J.J.L., Ford, G.D.: Permeability of three-dimensional models of fibrous porous media. J. Fluid Mech. **308**, 341–361 (1996)
54. Hilton, M.J., Gutierrez, L., Martinez, D.A., Wells, D.E.: EXT1 regulates chondrocyte proliferation and differentiation during endochondral bone development. Bone **36**(3), 379–386 (2005). doi:10.1016/j.bone.2004.09.025

55. Hrabe, J., Hrabetova, S., Segeth, K.: A model of effective diffusion and tortuosity in the extracellular space of the brain. Biophys. J. **87**(3), 1606–1617 (2004). doi:10.1529/biophysj.103.039495
56. Hunziker, E.B., Driesang, I.M.: Functional barrier principle for growth-factor-based articular cartilage repair. Osteoarthr. Cartil. **11**(5), 320–327 (2003). S1063458403000311[pii]
57. Johansson, L., Lofroth, J.E.: Diffusion and interaction in gels and solutions. 4. Hard-sphere Brownian dynamics simulations. J. Chem. Phys. **98**(9), 7471–7479 (1993)
58. Karsenty, G., Wagner, E.F.: Reaching a genetic and molecular understanding of skeletal development. Dev. Cell **2**(4), 389–406 (2002)
59. Kholodenko, B.N.: Cell-signalling dynamics in time and space. Nat. Rev. Mol. Cell Biol. **7**(3), 165–176 (2006). doi:10.1038/Nrm1838
60. Kirkpatrick, C.J., Fuchs, S., Unger, R.E.: Co-culture systems for vascularization—learning from nature. Adv. Drug Deliv. Rev. **63**(4–5), 291–299 (2011). doi:10.1016/j.addr.2011.01.009
61. Kleinman, H.K., Philp, D., Hoffman, M.P.: Role of the extracellular matrix in morphogenesis. Curr. Opin. Biotechnol. **14**(5), 526–532 (2003). S0958166903001186[pii]
62. Kosto, K.B., Deen, W.M.: Diffusivities of macromolecules in composite hydrogels. AIChE J. **50**(11), 2648–2658 (2004). doi:10.1002/Aic.10216
63. Koziel, L., Kunath, M., Kelly, O.G., Vortkamp, A.: Ext1-dependent heparan sulfate regulates the range of Ihh signaling during endochondral ossification. Dev. Cell **6**(6), 801–813 (2004)
64. Kronenberg, H.M.: Developmental regulation of the growth plate. Nature **423**(6937), 332–336 (2003). doi:10.1038/Nature01657
65. Lanske, B., Karaplis, A.C., Lee, K., Luz, A., Vortkamp, A., Pirro, A., Karperien, M., Defize, L.H.K., Ho, C., Mulligan, R.C., AbouSamra, A.B., Juppner, H., Segre, G.V., Kronenberg, H.M.: PTH/PTHrP receptor in early development and Indian hedgehog-regulated bone growth. Science **273**(5275), 663–666 (1996)
66. Leddy, H.A., Guilak, F.: Site-specific effects of compression on macromolecular diffusion in articular cartilage. Biophys. J. **95**(10), 4890–4895 (2008). doi:10.1529/biophysj.108.137752
67. Lee, R.B., Urban, J.P.: Evidence for a negative Pasteur effect in articular cartilage. Biochem. J. **321**(Pt 1), 95–102 (1997)
68. Lee, R.B., Urban, J.P.: Functional replacement of oxygen by other oxidants in articular cartilage. Arthritis Rheum. **46**(12), 3190–3200 (2002). doi:10.1002/art.10686
69. Levick, J.R.: Flow through interstitium and other fibrous matrices. Q. J. Exp. Physiol. CMS **72**(4), 409–438 (1987)
70. Lewis, M.C., MacArthur, B.D., Malda, J., Pettet, G., Please, C.P.: Heterogeneous proliferation within engineered cartilaginous tissue: the role of oxygen tension. Biotechnol. Bioeng. **91**(5), 607–615 (2005). doi:10.1002/Bit.20508
71. Lutolf, M.P., Hubbell, J.A.: Synthetic biomaterials as instructive extracellular microenvironments for morphogenesis in tissue engineering. Nat. Biotechnol. **23**(1), 47–55 (2005). doi:10.1038/Nbt1055
72. Lutolf, M.R., Weber, F.E., Schmoekel, H.G., Schense, J.C., Kohler, T., Muller, R., Hubbell, J.A.: Repair of bone defects using synthetic mimetics of collagenous extracellular matrices. Nat. Biotechnol. **21**(5), 513–518 (2003). doi:10.1038/Nbt818
73. Mackie, J.S., Meares, P.: The diffusion of electrolytes in a cation-exchange resin membrane. 1. Theoretical. Proc. R. Soc. Lond. Ser. A **232**(1191), 498–509 (1955a)
74. Mackie, J.S., Meares, P.: The diffusion of electrolytes in a cation-exchange resin membrane. 2. Experimental. Proc. R. Soc. Lond. Ser. A **232**(1191), 510–518 (1955b)
75. Madzvamuse, A.: Time-stepping schemes for moving grid finite elements applied to reaction-diffusion systems on fixed and growing domains. J. Comput. Phys. **214**(1), 239–263 (2006). doi:10.1016/j.jcp.2005.09.012
76. Maes, C., Carmeliet, G.: Vascular and nonvascular roles of VEGF in bone development. In: Ruhrberg, C. (ed.) VEGF in Development, pp. 79–90. Springer, New York (2008)
77. Mahoney, M.J., Krewson, C., Miller, J., Saltzman, W.M.: Impact of cell type and density on nerve growth factor distribution and bioactivity in 3-dimensional collagen gel cultures. Tissue Eng. **12**(7), 1915–1927 (2006)

78. Maini, P.K.: Using mathematical models to help understand biological pattern formation. C. R. Biol. **327**(3), 225–234 (2004). doi:10.1016/j.crvi.2003.05.006
79. Makarenkova, H.P., Hoffman, M.P., Beenken, A., Eliseenkova, A.V., Meech, R., Tsau, C., Patel, V.N., Lang, R.A., Mohammadi, M.: Differential interactions of FGFs with heparan sulfate control gradient formation and branching morphogenesis. Sci. Signal **2**(88), ra55 (2009). doi:2/88/ra55[pii]10.1126/scisignal.2000304
80. Malda, J., Rouwkema, J., Martens, D.E., le Comte, E.P., Kooy, F.K., Tramper, J., van Blitterswijk, C.A., Riesle, J.: Oxygen gradients in tissue-engineered PEGT/PBT cartilaginous constructs: measurement and modeling. Biotechnol. Bioeng. **86**(1), 9–18 (2004). doi:10.1002/Bit.20038
81. Marcus, R.E.: The effect of low oxygen concentration on growth, glycolysis, and sulfate incorporation by articular chondrocytes in monolayer culture. Arthritis Rheum. **16**(5), 646–656 (1973)
82. Martin, I., Wendt, D., Heberer, M.: The role of bioreactors in tissue engineering. Trends Biotechnol. **22**(2), 80–86 (2004). doi:10.1016/j.tibtech.2003.12.001
83. Mauck, R.L., Hung, C.T., Ateshian, G.A.: Modeling of neutral solute transport in a dynamically loaded porous permeable gel: Implications for articular cartilage biosynthesis and tissue engineering (vol. 125, 602, 2003). J. Biomech. Eng. Trans. ASME **126**(3), 392–392 (2004)
84. Mcelwain, D.L.S., Ponzo, P.J.: Model for growth of a solid tumor with nonuniform oxygen-consumption. Math. Biosci. **35**(3–4), 267–279 (1977)
85. Mizuno, S., Allemann, F., Glowacki, J.: Effects of medium perfusion on matrix production by bovine chondrocytes in three-dimensional collagen sponges. J. Biomed. Mater. Res. **56**(3), 368–375 (2001). doi:10.1002/1097-4636(20010905). 56:3<368:AID-JBM1105>3.0.CO;2-V[pii]
86. Mizutani, C.M., Nie, Q., Wan, F.Y.M., Zhang, Y.T., Vilmos, P., Sousa-Neves, R., Bier, E., Marsh, J.L., Lander, A.D.: Formation of the BMP activity gradient in the Drosophila embryo. Dev. Cell **8**(6), 915–924 (2005). doi:10.1016/j.devcel.2005.04.009
87. Moseley, J.B., Mayeux, A., Paoletti, A., Nurse, P.: A spatial gradient coordinates cell size and mitotic entry in fission yeast. Nature **459**(7248), 857–U858 (2009). doi:10.1038/Nature08074
88. Moser, H.: Structure and dynamics of bacterial populations maintained in the chemostat. Cold Spring Harb. Symp. Quant. Biol. **22**, 121–137 (1957)
89. Mow, V.C., Kuei, S.C., Lai, W.M., Armstrong, C.G.: Biphasic creep and stress relaxation of articular cartilage in compression? Theory and experiments. J. Biomech. Eng. **102**(1), 73–84 (1980)
90. Murry, C.E., Keller, G.: Differentiation of embryonic stem cells to clinically relevant populations: lessons from embryonic development. Cell **132**(4), 661–680 (2008). doi:10.1016/j.cell.2008.02.008
91. Ng, C.P., Swartz, M.A.: Fibroblast alignment under interstitial fluid flow using a novel 3-D tissue culture model. Am. J. Physiol. Heart Circ. Physiol. **284**(5), H1771–H1777 (2003). doi:10.1152/ajpheart.01008.2002
92. Nimer, E., Schneiderman, R., Maroudas, A.: Diffusion and partition of solutes in cartilage under static load. Biophys. Chem. **106**(2), 125–146 (2003). doi:10.1016/S0301-4622(03)00157-1
93. Obradovic, B., Carrier, R.L., Vunjak-Novakovic, G., Freed, L.E.: Gas exchange is essential for bioreactor cultivation of tissue engineered cartilage. Biotechnol. Bioeng. **63**(2), 197–205 (1999). doi:10.1002/(SICI)1097-0290(19990420). 63:2<197:AID-BIT8>3.0.CO;2-2[pii]
94. Obradovic, B., Meldon, J.H., Freed, L.E., Vunjak-Novakovic, G.: Glycosaminoglycan deposition in engineered cartilage: experiments and mathematical model. AIChE J. **46**(9), 1860–1871 (2000)
95. Ogston, A.G., Preston, B.N., Wells, J.D., Ogston, A.G., Preston, B.N., Snowden, J.M., Wells, J.D.: Transport of compact particles through solutions of chain-polymers. Proc. R. Soc. Lond. A Mat. **333**(1594), 297–316 (1973)
96. Otte, P.: Basic cell metabolism of articular cartilage. Manometric studies. Z. Rheumatol. **50**(5), 304–312 (1991)

97. Page-McCaw, A., Ewald, A.J., Werb, Z.: Matrix metalloproteinases and the regulation of tissue remodelling. Nat. Rev. Mol. Cell Biol. **8**(3), 221–233 (2007). doi:10.1038/nrm2125. nrm2125 [pii]
98. Papas, K.K., Long Jr, R.C., Constantinidis, I., Sambanis, A.: Effects of oxygen on metabolic and secretory activities of beta TC3 cells. Biochim. Biophys. Acta **1291**(2), 163–166 (1996). 0304-4165(96)00062-1[pii]
99. Patterson, G.H., Lippincott-Schwartz, J.: A photoactivatable GFP for selective photolabeling of proteins and cells. Science **297**(5588), 1873–1877 (2002)
100. Patterson, J., Martino, M.M., Hubbell, J.A.: Biomimetic materials in tissue engineering. Mater. Today **13**(1–2), 14–22 (2010)
101. Phelps, E.A., Landazuri, N., Thule, P.M., Taylor, W.R., Garcia, A.J.: Bioartificial matrices for therapeutic vascularization. Proc. Natl. Acad. Sci. U. S. A. **107**(8), 3323–3328 (2010). doi:10.1073/pnas.0905447107
102. Phillips, J.E., Burns, K.L., Le Doux, J.M., Guldberg, R.E., Garcia, A.J.: Engineering graded tissue interfaces. Proc. Natl. Acad. Sci. U. S. A. **105**(34), 12170–12175 (2008). doi:10.1073/pnas.0801988105. 0801988105 [pii]
103. Phillips, R.J., Deen, W.M., Brady, J.F.: Hindered transport of spherical macromolecules in fibrous membranes and gels. AIChE J. **35**(11), 1761–1769 (1989)
104. Picioreanu, C., van Loosdrecht, M.C., Heijnen, J.J.: A new combined differential-discrete cellular automaton approach for biofilm modeling: application for growth in gel beads. Biotechnol. Bioeng. **57**(6), 718–731 (1998). doi:10.1002/(SICI)1097-0290(19980320)57:6<718:AID-BIT9> . 3.0.CO;2-O [pii]
105. Place, E.S., Evans, N.D., Stevens, M.M.: Complexity in biomaterials for tissue engineering. Nat. Mater. **8**(6), 457–470 (2009). doi:10.1038/Nmat2441
106. Pugh, C.W., Ratcliffe, P.J.: Regulation of angiogenesis by hypoxia: role of the HIF system. Na.t Med. **9**(6), 677–684 (2003). doi:10.1038/nm0603-677nm0603-677 [pii]
107. Qutub, A.A., Mac Gabhann, F., Karagiannis, E.D., Vempati, P., Popel, A.S.: Multiscale models of angiogenesis. IEEE Eng. Med. Biol. Mag. **28**(2), 14–31 (2009). doi:10.1109/MEMB.2009.931791
108. Rajpurohit, R., Koch, C.J., Tao, Z., Teixeira, C.M., Shapiro, I.M.: Adaptation of chondrocytes to low oxygen tension: relationship between hypoxia and cellular metabolism. J. Cell. Physiol. **168**(2), 424–432 (1996). doi:10.1002/(SICI)1097-4652(199608). 168:2<424:AID-JCP21>3.0.CO;2-1[pii]10.1002/(SICI)1097-4652(199608)168:2<424:AID-JCP21>3.0.CO;2-1
109. Rolfe, D.F., Brown, G.C.: Cellular energy utilization and molecular origin of standard metabolic rate in mammals. Physiol. Rev. **77**(3), 731–758 (1997)
110. Sengers, B.G., Heywood, H.K., Lee, D.A., Oomens, C.W.J., Bader, D.L.: Nutrient utilization by bovine articular chondrocytes: A combined experimental and theoretical approach. J. Biomech. Eng. Trans. ASME **127**(5), 758–766 (2005). doi:10.1115/1.1993664
111. Sengers, B.G., van Donkelaar, C.C., Oomens, C.W.J., Baaijens, F.P.T.: Computational study of culture conditions and nutrient supply in cartilage tissue engineering. Biotechnol. Prog. **21**(4), 1252–1261 (2005). doi:10.1021/Bp0500157
112. Singh, A.B., Harris, R.C.: Autocrine, paracrine and juxtacrine signaling by EGFR ligands. Cell. Signal. **17**(10), 1183–1193 (2005). doi:10.1016/j.cellsig.2005.03.026
113. Soukane, D.M., Shirazi-Adl, A., Urban, J.P.G.: Computation of coupled diffusion of oxygen, glucose and lactic acid in an intervertebral disc. J. Biomech. **40**(12), 2645–2654 (2007). doi:10.1016/j.jbiomech.2007.01.003
114. Southern, J., Pitt-Francis, J., Whiteley, J., Stokeley, D., Kobashi, H., Nobes, R., Kadooka, Y., Gavaghan, D.: Multi-scale computational modelling in biology and physiology. Prog. Biophys. Mol. Biol. **96**(1–3), 60–89 (2008). doi:10.1016/j.pbiomolbio.2007.07.019. S0079-6107(07)00067-3 [pii]
115. Sternlicht, M.D., Werb, Z.: How matrix metalloproteinases regulate cell behavior. Annu. Rev. Cell Dev. Biol. **17**, 463–516 (2001)

116. Stevens, M.M., Marini, R.P., Schaefer, D., Aronson, J., Langer, R., Shastri, V.P.: In vivo engineering of organs: the bone bioreactor. Proc. Natl. Acad. Sci. U. S. A. **102**(32), 11450–11455 (2005). doi:10.1073/pnas.0504705102. 0504705102 [pii]
117. Stylianopoulos, T., Barocas, V.H.: Volume-averaging theory for the study of the mechanics of collagen networks. Comput. Method Appl. Mech. Eng. **196**(31–32), 2981–2990 (2007). doi:10.1016/j.cma.2006.06.019
118. Stylianopoulos, T., Diop-Frimpong, B., Munn, L.L., Jain, R.K.: Diffusion anisotropy in collagen gels and tumors: the Effect of fiber network orientation. Biophys. J. **99**(10), 3119–3128 (2010). doi:10.1016/j.bpj.2010.08.065
119. Sumi, T., Tsuneyoshi, N., Nakatsuji, N., Suemori, H.: Defining early lineage specification of human embryonic stem cells by the orchestrated balance of canonical Wnt/beta-catenin. Activin/Nodal BMP signal. Dev. **135**(17), 2969–2979 (2008). doi:10.1242/Dev.021121
120. Swartz, M.A., Fleury, M.E.: Interstitial flow and its effects in soft tissues. Annu. Rev. Biomed. Eng. **9**, 229–256 (2007). doi:10.1146/annurev.bioeng.9.060906.151850
121. Teleman, A.A., Strigini, M., Cohen, S.M.: Shaping morphogen gradients. Cell **105**(5), 559–562 (2001). S0092-8674(01)00377-4 [pii]
122. Tschumperlin, D.J., Dai, G.H., Maly, I.V., Kikuchi, T., Laiho, L.H., McVittie, A.K., Haley, K.J., Lilly, C.M., So, P.T.C., Lauffenburger, D.A., Kamm, R.D., Drazen, J.M.: Mechanotransduction through growth-factor shedding into the extracellular space. Nature **429**(6987), 83–86 (2004)
123. Tsuji, K., Bandyopadhyay, A., Harfe, B.D., Cox, K., Kakar, S., Gerstenfeld, L., Einhorn, T., Tabin, C.J., Rosen, V.: BMP2 activity, although dispensable for bone formation, is required for the initiation of fracture healing. Nat. Genet. **38**(12), 1424–1429 (2006). doi:10.1038/Ng1916
124. van der Eerden, B.C.J., Karperien, M., Wit, J.M.: Systemic and local regulation of the growth plate. Endocr. Rev. **24**(6), 782–801 (2003). doi:10.1210/Er.2002-0033
125. van Donkelaar, C.C., Huiskes, R.: The PTHrP-Ihh feedback loop in the embryonic growth plate allows PTHrP to control hypertrophy and Ihh to regulate proliferation. Biomech. Model. Mechanbiol. **6**(1–2), 55–62 (2007). doi:10.1007/s10237-006-0035-0
126. Vortkamp, A., Lee, K., Lanske, B., Segre, G.V., Kronenberg, H.M., Tabin, C.J.: Regulation of rate of cartilage differentiation by Indian hedgehog and PTH-related protein. Science **273**(5275), 613–622 (1996)
127. Williams, R.M., Zipfel, W.R., Tinsley, M.L., Farnum, C.E.: Solute transport in growth plate cartilage: in vitro and in vivo. Biophys. J. **93**(3), 1039–1050 (2007). doi:10.1529/biophysj.106.097675
128. Wood, B.D., Quintard, M., Whitaker, S.: Calculation of effective diffusivities for biofilms and tissues. Biotechnol. Bioeng. **77**(5), 495–516 (2002). 10.1002/bit.10075 [pii]
129. Wood, B.D., Whitaker, S.: Diffusion and reaction in biofilms. Chem. Eng. Sci. **53**(3), 397–425 (1998)
130. Zhou, S., Cui, Z., Urban, J.P.G.: Nutrient gradients in engineered cartilage: metabolic kinetics measurement and mass transfer modeling. Biotechnol. Bioeng. **101**(2), 408–421 (2008). doi:10.1002/Bit.21887

Computational Methods in the Modeling of Scaffolds for Tissue Engineering

Andy L. Olivares and Damien Lacroix

Abstract Tissue engineering uses porous biomaterial scaffolds to support the complex tissue healing process to fulfill two main functions: (1) to support mechanical loading and (2) to allow mass transport. Computational methods have been extensively applied to characterize scaffold morphology and to simulate different biological processes of tissue engineering. In addition, phenomena such a cell seeding, cell migration, cell proliferation, cell differentiation, vascularisation, oxygen consumption, mass transport or scaffold degradation can be simulated using computational methods. A review of the different methods used to model scaffolds in tissue engineering is described in this chapter.

1 Introduction

Tissue engineering is considered as "an interdisciplinary field that combines the knowledge and technology of cells, engineering materials, and suitable biochemical factors to create artificial organs and tissues or to regenerate damaged tissues" [1]. More recently, Williams defined tissue engineering as "the creation (or formation) of new tissue for the therapeutic reconstruction of the human body,

A. L. Olivares
Biomechanics and Mechanobiology,
Institute for Bioengineering of Catalonia,
Baldiri Reixac, 4, 08028 Barcelona, Spain
e-mail: aolivares@ibecbarcelona.eu

D. Lacroix (✉)
Department of Mechanical Engineering,
University of Sheffield, Mappin Street,
Sheffield, S1 3JD, UK
e-mail: D.Lacroix@sheffield.ac.uk

Stud Mechanobiol Tissue Eng Biomater (2013) 10: 107–126
DOI: 10.1007/8415_2012_136

Published Online: 11 March 2012

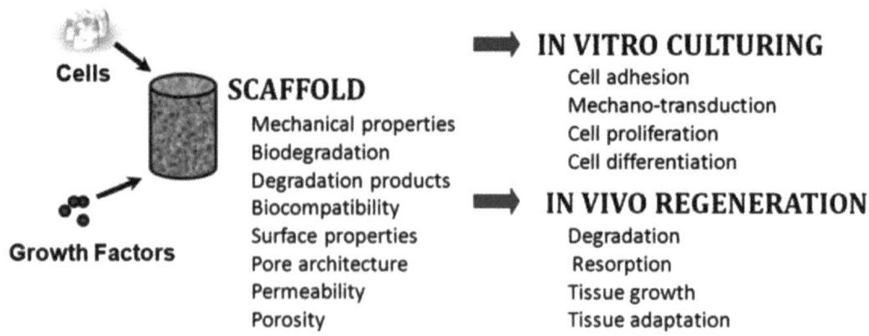

Fig. 1 Tissue engineering process using biomaterial scaffold (modified from Liu et al. [6])

by the deliberate and controlled stimulation of selected target cells through a systematic combination of molecular and mechanical signals" [2]. The process of creation of biological tissue needs an artificial substrate to guide the tissue and control the cell response to the supply of specific molecular and mechanical signals. "Scaffold" is the denomination for the temporal artificial support. Generally studies of biomaterial substrates are done in two-dimensions in order to evaluate the cell-biomaterial interactions. However, it is desired that the scaffold substitutes the defect or mimic the organs or tissues structures in a three-dimensional manner so that it ensures the functions of the damaged tissues.

In the last decade the advances made in tissue engineering and scaffold research have increased substantially [3]. These substantial changes in this scientific field should generate a higher quality of life of people. However, the main barrier found so far is related to the difficulties of translating scientific results into clinical applications. Another problem is related with the high complexity of the biological processes that are taking place and that are not well controlled or understood. The complexities and the use of the human cells as a cell source make tissue engineering an expensive process with usually a low reproducibility of the results at an industrial scale.

The requirements of a scaffold are multiple and different for each application. Nonetheless some common characteristics can be found such as: a good network of interconnecting pores, open channels capable to provide the oxygen and nutrients to the cells inside the scaffold, an easy removal of the waste product and a biocompatible material able to provide the appropriate mechanical strength and biodegradable properties. Some of the characteristics and phenomenon involved in tissue engineering are summarized in Fig. 1. The intrinsic biomaterial properties in relation to the scaffold architecture influence the affinity and response of the cells within the scaffold. During the cell culture it is desirable (a) to know the forces supported by the cells attached on the scaffold, (b) to study the distribution of cells after seeding, (c) to study how occur cell migration, proliferation, and differentiation, and (d) to study the influence of the mechanical stimuli on the cell response. This complexity is further increased under in vivo conditions where the biomaterial degradation and tissue formation, and the adaptation of a new tissue with the new blood vessel formation (angiogenesis) are combined.

Computational methods have been introduced in tissue engineering as tools to comprehend and predict the phenomenon occurring inside scaffolds. Initially, the cells are seeded on a porous scaffold that acts as a template to facilitate the formation of functional new tissue and organ. Some variables such as pore size, material type and fabrication process are known to influence the cell response [4]. A computational analysis of scaffold properties should consider the overall problem as a continuous two phase's problem: a solid bulk scaffold and a fluid medium inside the pores. The diversity of the methods used so far shows that different assumptions have been made to simplify the complex experimental or physiological conditions. Nevertheless the common aim of these studies is to predict, understand or determine the optimal culture conditions or scaffold morphology [5]. Mechanobiological concepts are also being used more commonly to explain how the cells sense the signals from the environment and how their response to them can affect the cell phenotype related processes. In this chapter, we will discuss some of the results obtained with the computational methods used to characterize the scaffold as an artificial structure to be used in tissue engineering applications.

2 Computational Structural Characterization of Scaffolds

Generally, the conventional scaffold fabrication techniques (salt leaching, phase separation or gas foaming) present difficulties to obtain a precise and repeatable scaffold microstructure in which pore interconnectivity or pore size are guaranteed [7–10]. Each technique is usually most suited to a specific biomaterial or scaffold and for a specific tissue engineering application. Porous scaffolds need to be characterized due to the irregular microstructures that give a different microenvironment to the cells attached onto the scaffold surface.

Novel techniques allow an initial computational control of fabrication using computer-aided design (CAD). The common name used for these techniques is rapid prototyping (RP). These techniques lead to better pore reproducibility of regular shape and better interconnection than the conventional methods. Regular scaffolds offer an easier understanding and optimization of diverse biological phenomena. However, once fabrication using RP techniques is made, it is necessary to characterize the regular pores because the accuracy of the fabrication method depends on the biomaterial properties, the RP techniques used and the conditions applied during the process [11, 12].

In both cases (conventional and RP techniques) it is useful to characterize the sample in a non destructive manner. Computationally it is possible to apply imaging techniques (micro-Computed Tomography and Nuclear Magnetic Resonance) and using reconstruction algorithms. Through scaffold reconstruction it is possible to know the pore interconnectivity, the overall porosity, the pore size distribution and the specific surface area. Some examples of scaffold structures reconstructed and characterized using computational techniques are shown in Fig. 2. The irregular distribution of pores of calcium phosphate (CaP) cement microstructure is shown in a cross section in Fig. 2a [9]. The reconstruction was

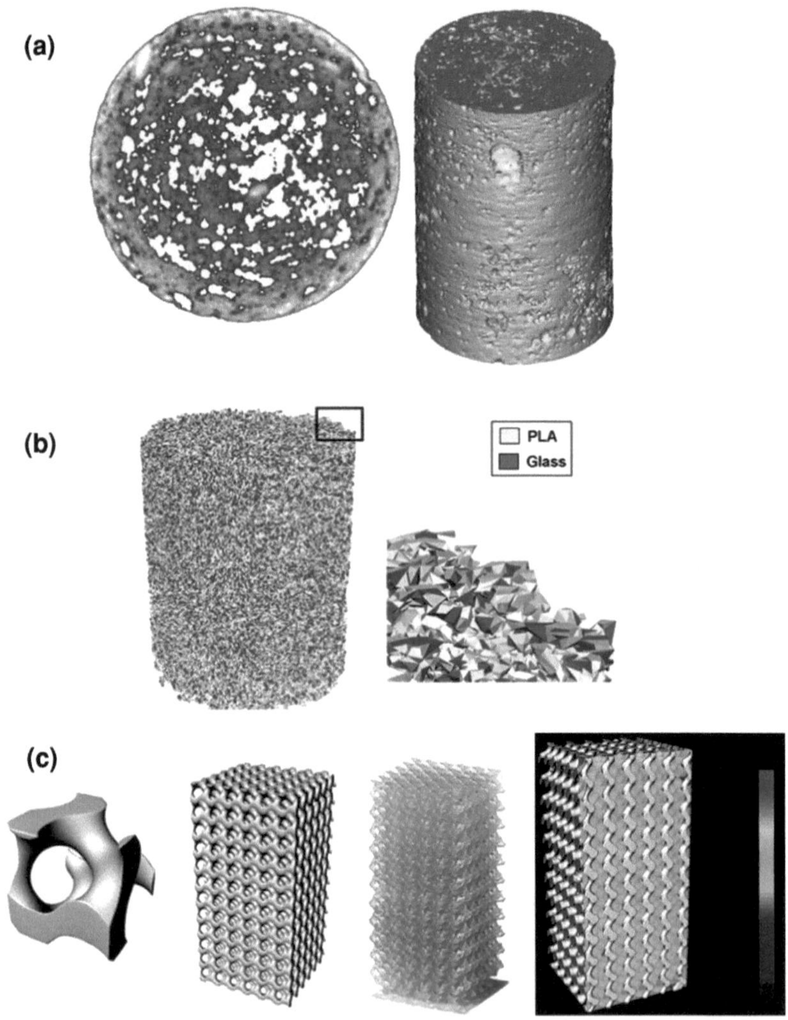

Fig. 2 Computational approaches for characterize of the scaffold. **a** (*left*) 2D cross section of micro-CT images of a non injected CaP cement and (*right*) 3D reconstruction in MIMIC (Materialize)of the CaP cement (from Lacroix et al. [9]). **b** Reconstructions of PLA-G5 (polylactide acid and glass) scaffold and representation of component in the finite element model (from Milan et al. [13]). **c** CAD-designs of the repeating gyroid unit cells assemble of $6 \times 6 \times 12$ unit cells, photo of build structures and μCT-scanning of the built structures (modified from Melchels et al. [14])

obtained by superimposing consecutive cross sections to form a three-dimensional reconstruction of the final sample (Fig 2a, right).

Usually, after this characterization based on the distributions, shape and size of the pores, the scaffold solid and internal volume of pores are computationally reconstructed to simulate different physical and cellular processes. For these reasons the reconstruction step is vital to obtain the correct characterization and numerical analysis. The second example (Fig. 2b) selected shows a volume of one cylindrical sample of a PLA (poly-lactide acid) polymer with glass composite biomaterial [13]. PLA and glass particles were identified using the gray values for each material allowing separation of the material properties for the computational analysis.

Until now it has been possible to obtain a complete structural characterization of irregular structure scaffolds. However, the heterogeneity of the geometry in non-regular scaffold minimizes the use of scaffold models since each sample is unique. Therefore it is difficult to know whether the modeled scaffold is representative of the real scaffolds used. For RP scaffolds the steps to follow are easier (Fig 2c). Initially the desired design (CAD) of pores is drawn, through a representative volume element (for example of a gyroid shape in Fig. 2c) and the complete scaffold (cylindrical volume) are designed [14]. After the scaffold fabrication using RP techniques (stereolithography in this example) is characterized by means of micro-CT method allowing comparison of the initial design with the final product.

3 Computational Evaluation of Mechanical Properties of Scaffold

3.1 Mechanical Properties Depending on the Scaffold Microstructure

The desired mechanical properties of porous scaffolds vary depending on the clinical applications. It is therefore desirable to be able to control and tune such properties on a specific case basis. The digital scaffold reconstruction methodology showed above offers the possibility to develop the scaffold structure maintaining the specific micro pores. The Finite Element Method (FEM) can be used to predict the mechanical properties of a scaffold. FEM is a numerical technique that gives approximate solutions through partial differential equations. Defining the problem from the geometry, the domain is divided in finite sub-domains called elements. Through an adequate mesh of elements, the bulk material properties and the loading conditions that correspond to the scaffold are applied. In tissue engineering, the FEM is used principally to determine the effective mechanical properties of the porous scaffold. For example, the method allows to calculate the effective Young's modulus E_f under compression (Eq. 1) dependent on the reaction force R computed at the nodes, the total cross section area of the scaffold A, and the axial strain applied $\varepsilon = \frac{\Delta l}{l}$ [6, 9].

$$E_f = \frac{R}{A\varepsilon} \tag{1}$$

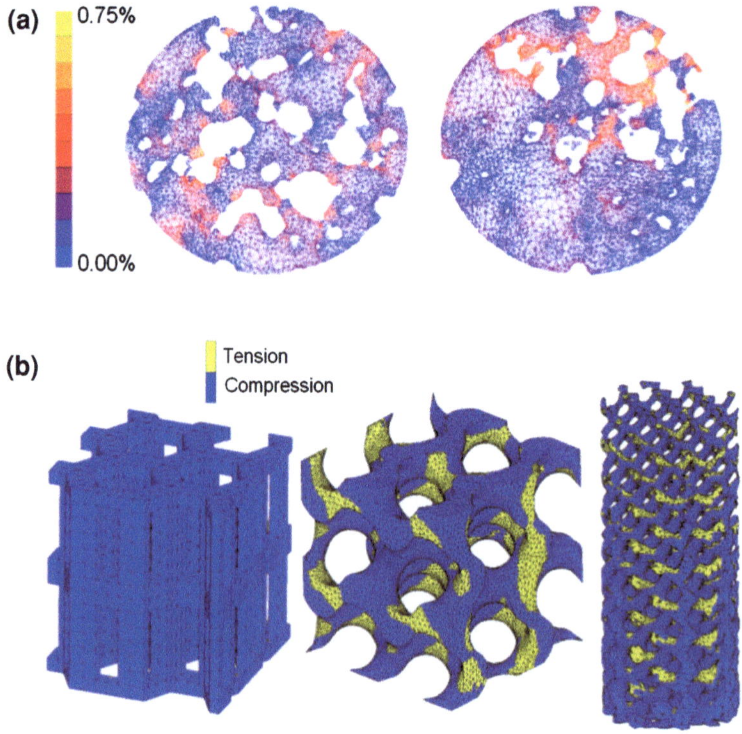

Fig. 3 **a** Examples of distribution of octahedral shear strain in two samples for CaP cement (*left*) and glass (*right*) scaffold morphologies (modified from Lacroix et al. [9]). **b** Major strain distribution in RP pores; the zones under tension or compression strain are delimited; prism hexagonal shape with 70% of porosity (*left*), gyroid pore structure with 70% of porosity (*middle*), and gyroid pore shape distributed gradually through the height of scaffold (global porosity equals 70%) (*right*) (data from Olivares et al. [5])

The mechanical parameters computed in the scaffold surface are considered as the stimuli or signals felt by the cells attached. Dependent on the stimuli magnitude (strain for example), the stimuli will alter cell proliferation or phenotype differentiation. Generally, for bone tissue engineering applications a compressive load is applied to determine the strength and the strain distribution in the scaffold wall surface (see Fig. 3). The effect of mechanical stimuli was studied through micro-FE models for macroporous CaP cement and for a porous glass ceramic scaffold. The octahedral shear strain distribution on a two dimensional section of both samples is shown in Fig. 3a; higher strains are found in areas close to the pores with magnitudes up to 0.75% for the application of a compressive strain of 0.5% [9]. The strain distribution throughout the section is quite inhomogeneous due to the inhomogeneous pore distribution.

Under a compressive load equivalent to a uniaxial strain of 0.5%, the structures shown in the Fig. 3b presented a higher proportion of material experiencing compressive strain than tensile strains. Hexagonal structures lead to lesser variability of major strain distributions (compressive strain from 0.25 to 0.75%) on the walls than the gyroid structures (from tensile value of 0.25% until compressive values of 1%) [5]. Thus, the stimuli not only is dependent on the uniform distribution of pores, but also depend on pore shape [6, 14].

3.2 Scaffold Degradation and Tissue Regeneration

Optimum scaffold degradation is related directly with the temporal mechanical properties needed to substitute the tissue damage. Few computational studies try to explore the scaffold implantation and in vivo response. One example is to correlate the relationship between the scaffold geometry with the resorption of the scaffold, but independently of its composition [15]. Another study based on multiscale simulation [16] shows that the degradation kinetic of polymer (PLGA, poly (lactic-co-glycolic acid) is fast and has a negative effect in the balance of tissue regeneration within scaffolds.

Another example selected here to explain the scaffold degradation was developed by Adachi et al. [17]. They proposed a framework to simulate bone regeneration, which includes the degradation rate. They evaluated the mechanical function in the bone regeneration process by changing the strain energy at the bone–scaffold system. Scaffolds with lattice-like and spherical pore structures were assessed (Fig. 4). The scaffold degradation was assumed to be due to hydrolysis, decreasingthe polymer molecular weight W, and therefore inducing a decrease in the scaffold Young's modulus E_S (Eq. 2). E_{S0} is the initial scaffold Young's modulus and W_0 is the initial molecular weight. The rate of degradation is affected by the morphology of the scaffold microstructure, and large surface areas accelerate the diffusion of water molecules into the bulk of the polymers.

$$E_s = E_{s0}\frac{w(t)}{w_0} \tag{2}$$

Bone formation and bone resorption are accomplished by osteoblastic and osteoclastic cellular activities respectively. New bone formation was modeled using the rate equation for trabecular surface remodeling (Eq. 3) based on the uniform stress hypothesis [18, 19]. Here, σ_c is the representative stress at point x_c on the bone or scaffold surface on which the osteoblasts form new bone matrix, and σ_d is the representative stress in the neighboring area around point $x_{c.}$ S denotes the surface; σ_r is the stress at neighboring point x_r within the sensing distance $l_L(l = |x_r - x_c|)$ and $w(l)[w(l) \geq 0\,(0 \leq l \leq l_L)]$ is the decaying weighting function.

$$\sigma_d = \frac{\int_S w(t)\sigma_r \mathrm{dS}}{\int_S w(t)\mathrm{dS}} \tag{3}$$

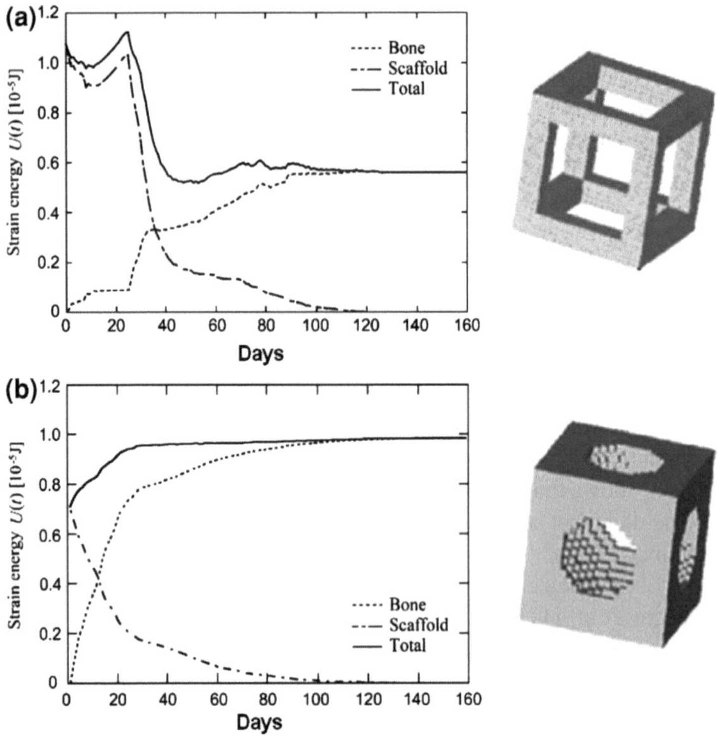

Fig. 4 Computational simulation of bone tissue regeneration that consists of scaffold degradation and new bone formation using the voxel finite element method. **a** Lattice-like structure with $1 = 1.6$ mm and **b** spherical pore structure with $d = 2.6$ mm (modified from Adachi et al. [17])

The changes in the total strain energy of the bone–scaffold system, *U(t)*, in the regeneration process are plotted in Fig. 4 for the lattice-like and spherical pore structures. A transition of the mechanical function between two structural components can be seen. For the lattice-like structure the crossing point occurs at 40 days, while for the spherical pore it occurs at 20 days. The scaffold pore structure design presents the highest influence in the bone regeneration process. Through the case presented, it was demonstrated that the optimal design variables of the scaffold can be determined by computational simulation of bone regeneration. However, the rate equations for new bone formation and scaffold degradation were derived on the basis of various simplifications and assumptions.

3.3 Homogenization Method for Scaffold Design

Studies of different mathematical cell units designed for RP methods were made to create a library of structures based on CAD design [20–22]. Porous scaffold design is a compromise between high mechanical function and high mass transport needs.

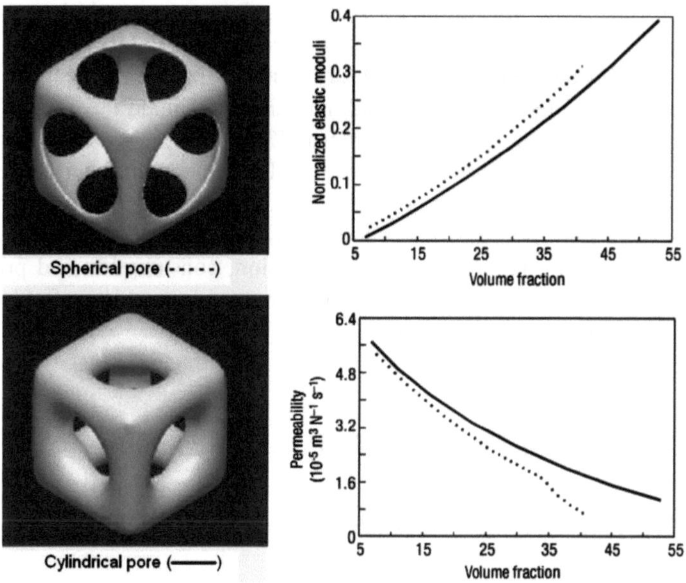

Fig. 5 Elastic modulus and permeability versus porosity on two pore cell units; spherical pore (*top, in plot with dashed line*) and cylindrical pore (*bottom, in plot with solid line*), modified from Hollister [23]

The homogenization theory is used to generate the multiscale equilibrium equations to compute effective properties of these unit cell designs [21, 22]. The study presented by Hollister [23] has been selected to illustrate to the reader the homogeneity method. The mechanical properties such as the macroscopic effective stiffness E^{macro} (Eq. 4) can be computed from the stiffness at a microscopic E^{micro} level depending of the strain tensor M and the volume of unit cell V. The mass transport (fluid inside the pores) is also homogenized focusing in the macroscopic permeability K (Eq. 5) computed through the average Stokes flow velocity v and pressure gradients [23].

$$E^{\text{macro}} = \frac{1}{|V_{\text{unit cell}}|} \int_V E^{\text{micro}} M \mathrm{d}V_{\text{unit cell}} \tag{4}$$

$$K^{\text{macro}} = \frac{1}{|V_{\text{unit cell}}|} \int_V v \mathrm{d}V_{\text{unit cell}} \tag{5}$$

Hollister [23] have found that an increment of the amount and disposition of material in a pore induces an increment of elastic properties, but with a reduction of the permeability (Fig. 5). Results demonstrate that the modulus increases as expected with volume fraction, and that the spherical pore leads to a stiffer scaffold. The permeability decreases as expected with volume fraction and cylindrical pore design allows fabricating a more permeable scaffold. Others studies illustrate how for different scaffold microstructure designs, different effective stiffnesses and permeabilities are computed [6, 14]. However, the importance of this method lies in the ability to maximize the permeability for cell migration and mass transport, always taking care to maintain the effective elastic properties similar to that of the natural tissue [24].

Homogenization is a powerful mathematical technique and can be applied in tissue engineering to optimize the process inside the scaffold. The study proposed by Shipley et al. [25] used this theory to optimize the oxygen, glucose and lactate transport through the maximum value of diffusion, consumption and production with optimal choice of cell density. So select the most favorable scaffold structure and the preference of perfuse direction.

4 Simulating Fluid Flow Within Porous Scaffolds

4.1 Computing Fluid Stimuli on the Scaffold Pores

In tissue engineering bioreactors are usually used as systems to provide the dynamic environment to the cells. The stimuli can be obtained from the scaffold surface deformations or from the fluid flow inside the pores. Mass transport within the pores is related with the cell seeding and with the possible response of the cells to the stimuli. Perfusion bioreactors are systems extensively used in tissue engineering, due to the capacity to lead to a uniform distribution of fluid over the scaffold [26–28]. Micro pore shapes in the scaffold have a higher influence in flow profile. Generally, the flow is simulated using computational fluid dynamic (CFD) method. In the specific case of direct perfuse flow, a laminar and Newtonian flow are simulated, computed by the Navier–Stokes equation (Eq. 6) and the continuity equation (Eq. 7), where u and ∇p are the fluid velocity and pressure gradient respectively.

$$\rho\vec{u}\cdot\nabla\vec{u} = -\nabla p + \eta\nabla^2\vec{u} \tag{6}$$

$$\nabla\cdot\vec{u} = 0 \tag{7}$$

In Fig. 6 results of CFD for irregular pores morphologies [13] and regular morphologies [5] are presented. Irregular morphology models were developed from the micro-CT images and the fluid volume inside the pore was modeled (Fig. 6a). The scaffold material property in itself was not accounted for in these studies, only the pore morphological characteristics influenced the results. Milan et al. [13] obtained a non homogeneous distribution of fluid flow due to the irregular interconnection between the pores. For an inlet velocity of 100 μm/s, the highest velocities (150–780 μm/s) were found in the center of the scaffold pores whereas the lowest velocities (0–50 μm/s) were found close to the pore walls (Fig. 6a).

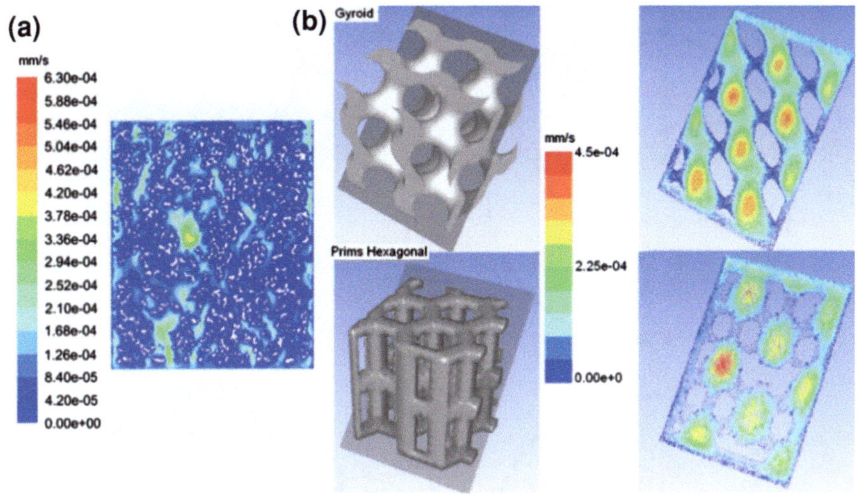

Fig. 6 Velocity profiles represented in a cross section under fluid flow perfusion (v = 100 μm/s). **a** Irregular scaffold morphology fabricated with PLA-G5 (poly-lactide acid and glass), model obtained from micro-CT (modified from Milan et al. [13]. **b** Representative volume element obtained from CAD of rapid prototyping scaffold. Gyroid shape with 70% of porosity (*top*), hexagonal prism with 70% of porosity (*down*) (modified from Olivares et al. [5])

RP can be an excellent fabrication technique to control the fluid stimuli within scaffolds. From the study developed by Olivares et al. [5], two different scaffold morphologies were selected (Fig. 6b). A gyroid structure was compared with a hexagonal straight prism. Pore interconnections of the hexagonal prism limited the accessibility of the fluid to some areas inside the scaffold. With an inlet velocity of 100 μm/s, the fluid flow distribution for the gyroid structure was in the range of 220–450 μm/s while for the hexagonal prism structure it was in the range of 220–300 μm/s.

4.2 Cell Seeding Simulation Using CFD Models

Cell seeding is a critical step in tissue engineering since it precedes the further steps for the in vitro and in vivo culture. In tissue engineering, the optimization of cell seeding and the comparison between different studies are problematic because cell seeding depends highly on the structure of the scaffolds, such as porosity, tortuosity, pore size and pore shape, the number of cells in suspension, etc. Recently, a system to control cell seeding through perfusion was proposed by combining a CFD study of a rapid prototyping porous scaffold with the perfusion fluid flow experimentation [12]. Gyroid pore design was used for the scaffold stereolithographic fabrication. The values of pore size distribution in an isotropic

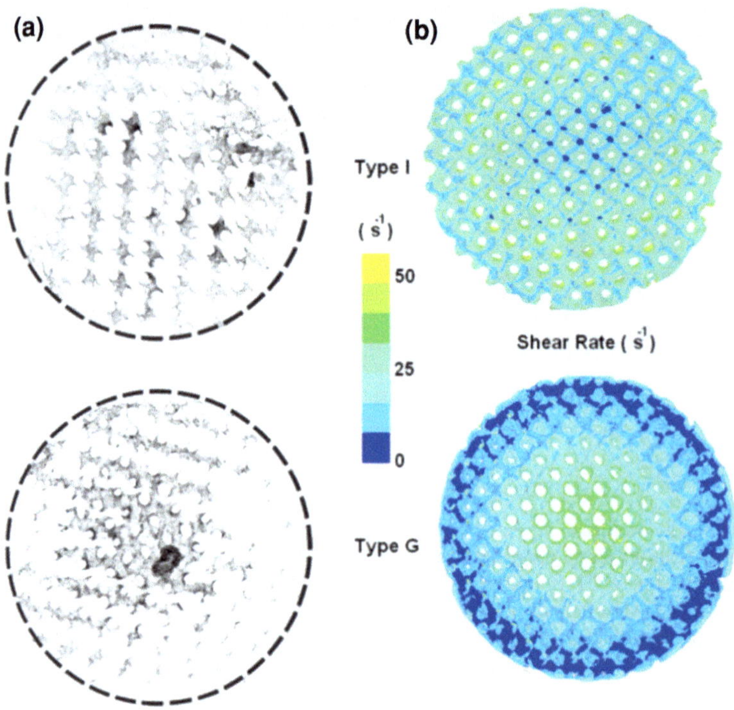

Fig. 7 Representative threshold z-stacks of confocal microscopy images (500 μm thickness) showing adhering cells after perfusion seeding. The fluid shear rates were measured in the scaffold surface through the CFD methods. **a** Type I. **b** Type G (from Melchels et al. [12])

scaffold (type I) were of 412 ± 13 μm with porosity of 62 ± 1% (Fig. 7a). In order to study the influence of pore size variation, a gradual variation of pore size (type G) was introduced where the scaffold pore size was ≈500 μm at the center and ≈250 μm at the periphery with a total porosity of 56 ± 3% [12]. Fluid volume was reconstructed from micro-CT images and represented the actual design of the perfusion chamber. These models assumed steady state conditions for maximum fluid flow while an alternant flow was applied in vitro when the cells are suspended in the medium. However, even with this restriction, the computational model demonstrated a close relation between distribution of maximum stimuli and final cell seeding distribution (Fig. 7).

When the distribution of pore size was uniform (Type I), the distribution of cells densities observed through confocal images after seeding was also homogeneous, in a similar behavior obtained numerically (Fig. 7a). In the gradient pore size (Type G), a gradient of shear stress was obtained in the cylindrical scaffold, which compared well with the gradual distribution of cells obtained after 16 h of seeding (Fig. 7b).

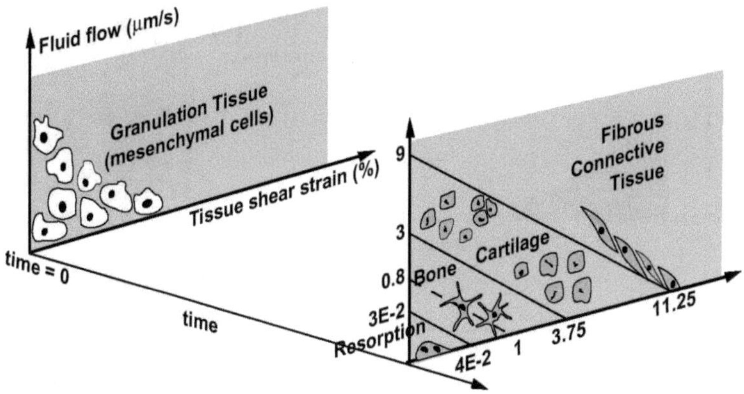

Fig. 8 Mechano-regulation model. The tissue phenotype is determined for each element dependent on its position in the mechano-regulation diagram (not drawn to scale) from Lacroix et al. [37])

5 Simulation of Mechano-Biological Concepts

5.1 Cell Differentiation Studied Through Separated Phases

A mechanobiological concept was proposed by Prendergast et al. [29] to explain the relation between mechanical stimuli with cell differentiation and tissue formation. This concept was applied on various applications such as implant–bone interface [30], fracture healing [31], osteochondral defect [32], bone chambers [30, 31, 33], and bone distraction [34, 35]. The first numerical application to tissue engineering of this concept was performed by Kelly and Prendergast [32] in which a mechanoregulation algorithm for tissue differentiation was used to determine the influence of scaffold material properties on chondrogenesis in a finite element model of an osteochondral defect.

The basis of the concept is that the stimulus S defined as a combination between the octahedral shear strain SS and the fluid velocity FF can produce some distortion of the cell and therefore affects the cell differentiative activity. In the stimuli equation (Eq. 8), a and b are empirical constants; $a = 0.0375$ and $b = 3\ \mu\text{ms}^{-1}$. If $S > 3$, then fibrous tissue differentiation occurs; if $3 > S > 1$, then cartilage differentiation occurs; if $1 > S > 0.267$, then immature bone differentiation occurs; and if $0.267 > S > 0.01$, then resorption occurs (Fig. 8).

$$S = \frac{\text{SS}}{a} + \frac{\text{FF}}{b} \tag{8}$$

One example to illustrate the use of the mechano-regulation model on scaffold analysis is the study performed by Sandino and Lacroix [36] on irregular CaP scaffold morphology where tissue differentiation was simulated for mechanical stimuli transmitted under compressive load and perfusion fluid. The properties of

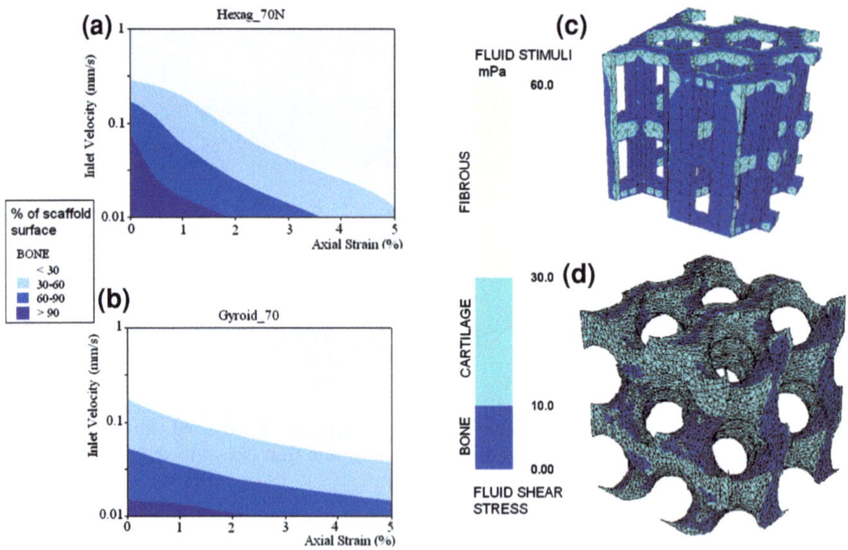

Fig. 9 Influence of scaffold morphology [**a, c** Hexagonal Prism, porosity 70%; **b, d** Gyroid, porosity 70%] in stimuli for tissue differentiation. **a, b** Optimization for conditions represented through the percentage of stimuli that correspond with bone tissue differentiation. **c, d** phenotypes representations in models distributed for fluid stimuli induced by 0.1 mm/s (modified from Olivares et al. [5])

the finite element (tissue) were changed depending on the stimuli. The fluid phase was the only phase to change (through a change of viscosity) to simulate the growth of tissues within the pores. The discretization of the fluid phase and the solid phase captured the discontinuity of mechanical stimuli affecting cells seeded within a scaffold over time. Using regular scaffolds with ideal morphology, Olivares et al. [5] concluded that the mechanoregulation diagram on these scaffolds was more sensitive to fluid flow changes than to solid strain changes. They studied the influence of the experimental conditions for eight different scaffold designs to create a bone formation optimization plot (Fig. 9a, b). In the same study, the authors related the scaffold design with the phenotype (Fig. 9c, d) using the mechano-regulation theory explained previously. For the same inlet fluid stimuli and porosity a scaffold with a hexagonal prism leads to increased bone formation areas compared with a gyroid structure.

5.2 *Lattice Point Approach*

The lattice formulation was introduced in a mechano-biological model by Perez and Prendergast [38] in order to include individual behavior of cell. A lattice is created within each finite element and is assumed with granulation tissue

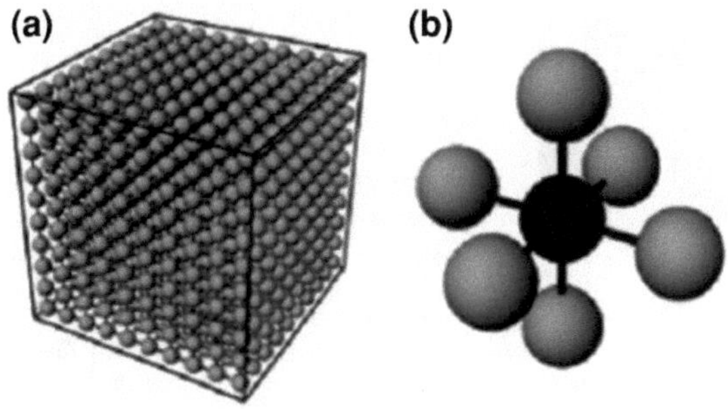

Fig. 10 **a** Lattice generated from granulation element geometry. **b** possible cell locations when migrating

properties (typical initial tissue that fill wounds). Each lattice is considered a region of space for both the cell and extracellular matrix (Fig. 10).

The lattice model has been used as an alternative to simulate cell proliferation and migration. The proliferation of a cell is initially assumed (in 3D) to be surrounded by six possible locations. First a new position is randomly selected from the surrounding locations (including its original position). In turn one of the remaining neighboring positions is then chosen for the daughter cell to occupy. In the event that the chosen location is already occupied, another position is chosen again at random. This process continues until either the simulation ends or all lattice positions are occupied. Recognizing that migration is a more rapid process, a new location for a migrating cell is chosen per iteration of the proliferation process. A lattice point can be occupied by a mesenchymal cell (MSC), a fibroblast, a chondrocyte, an osteoblast or an endothelial cell. The sequence of endothelial cells and the vessel growth direction, growth length and branching are defined through the lattice formulation as probabilistic functions.

The lattice concept was applied in tissue engineering by Byrne et al. [39] (Fig. 11). They studied the effect of important design properties such as scaffold porosity, degradation rate and scaffold mechanical properties, on the tissue formation process inside a regular structured bone scaffold. Initially the scaffold was assumed to be filled with granulation tissue and 1% of lattice points, chosen at random, were "seeded" with mesenchymal stem cells. Over time, the scaffold dissolved at a rate of 0.5% per iteration of the simulation, leaving space for the developing tissue. In this study, they were able to identify optimal scaffold properties that would lead to the highest amount of bone formation.

Studies of angiogenesis during tissue differentiation were performed by Checa and Prendergast [40] using the lattice model. The angiogenesis model was applied to a simplified scaffold for bone tissue engineering and the number of cells initially seeded into the scaffold was related to the rate of vascularization and the penetration of the vascular network. They showed that the initial cell seeding conditions had a significant effect on the vascularization of the scaffold.

In order to simulate also the angiogenesis phenomenon in CaP cement, Sandino et al. [41] applied the lattice approach in a scaffold with irregular morphology (Fig. 12). For the magnitudes of mechanical strain studied (0.5 and 1% of total

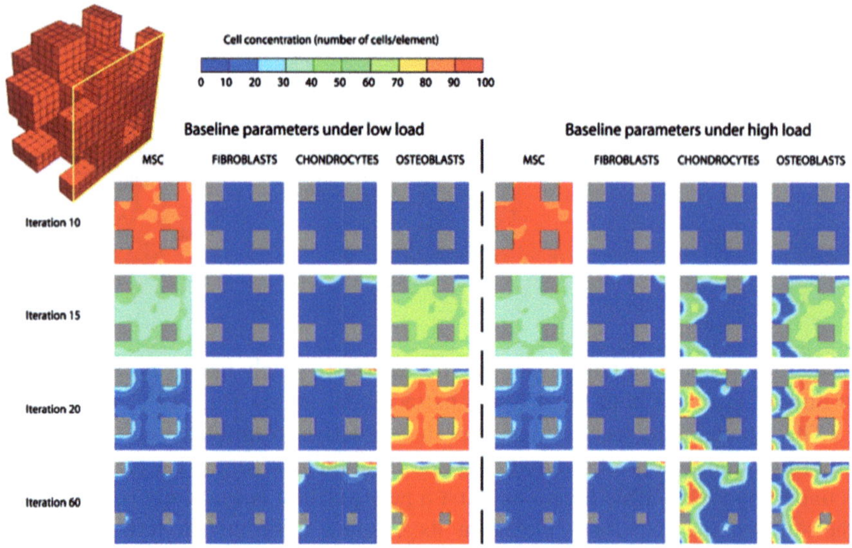

Fig. 11 Cells distribution in a simplified scaffold using lattice method. (from Byrne et al. [39])

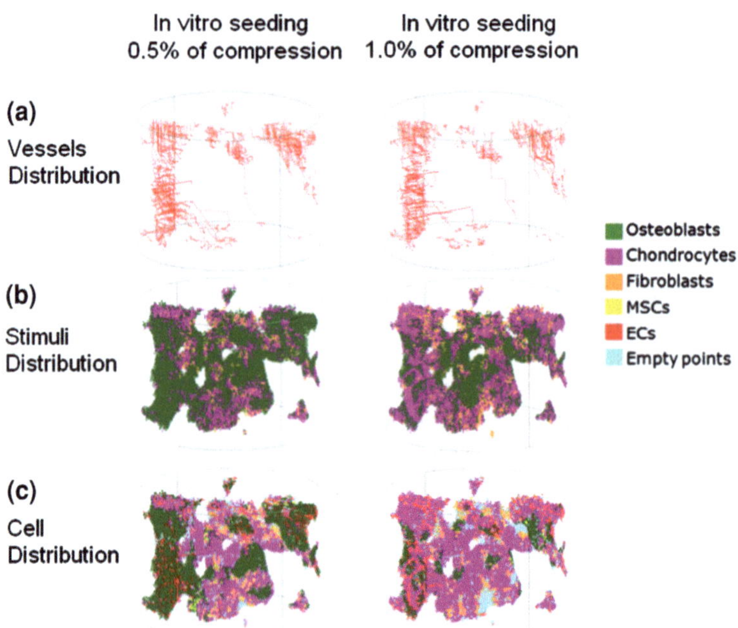

Fig. 12 Angiogenesis and mechano-regulatory stimuli to cell differentiation in simulation (0.5 and 1% of total strain and in vitro MSCs seeding) after 100 iterations are showed. **a** Vascular network. **b** Stimuli distribution. **c** Cell distribution (modified from Sandino et al. [41])

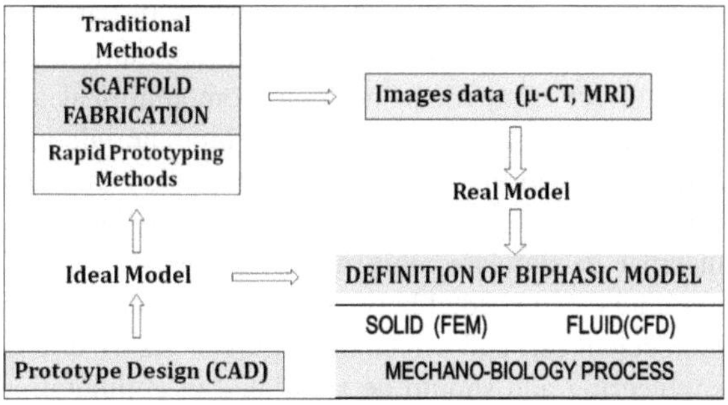

Fig. 13 Scheme representing the main methods applied for in tissue engineering scaffold

deformation), there was not a noticeable effect on angiogenesis. Similar vessel networks were observed in both cases of strain (Fig. 12). The angiogenic process was mostly driven by the actual porosity and permeability of the scaffold rather than by the load magnitude.

6 Conclusions and Future Trends

In this chapter, computational methods used in the modeling of scaffolds for tissue engineering were illustrated through some examples selected from the literature. Figure 13 shows a summary of the different schemes used to model scaffolds. Initially the characterization of the scaffold is based on image acquisition and treatment in order to obtain the morphological parameters. These parameters (porosity, pore shape, pore size) can be designed previously (CAD) when a rapid prototyping method is used in the fabrication process; but the real structure can be assessed using micro-CT scanning. In order to control the mechanical properties of the scaffold; displacement or load are applied to simulate the experiment condition and the effective stiffness can be computed. Generally, the FEM is applied in the characterization of the mechanical integrity of scaffold, including the possible simulation of scaffold degradation. The irregular scaffold morphology leads to heterogeneous distributions of stimuli induced from the solid phase. The best control of the mechanical stimuli can be obtained using RP scaffolds. Studies with RP scaffolds allow to develop methods such as homogenization were the optimization of the macro-scaffold properties are related with the micro pore properties.

For the evaluation of mass flow inside the scaffold, computational fluid dynamic (CFD) simulations can be used. Generally, the fluid shear stress is defined as the stimulus acting on the scaffold wall to characterize the fluid flow. In the literature different ranges of stimuli are found and a consensus must be found, but in principle using this methodology it is possible to control precisely the design of the size and shape of the scaffold pores. The mechano-biological phenomenon can also be simulated from the perspective that the porous scaffolds have two phases (fluid and solid), and that both phases will have an influence in the processes of seeding, migration, proliferation, differentiation, and cell death. In the future, a more direct comparison between computational and experimental results is needed, including differences between biomaterials in term of reaction with the cells and the possibility to discretise the cells (similar to the lattice approach) with cellular properties. Multiscale factors and chemical reaction can also be included in order to simulate oxygen and nutrient transport and consumption. More integration is also needed between the interrelations of biology, materials science and biomechanics. The effect of angiogenesis is crucial for the processes of tissue engineering and should be simulated in computational models.

Several studies [42, 43] have looked at the bone–implant interface when prosthesis is implanted. The formation of fibrous tissue, cartilage or bone depends mostly on the local loading conditions at that interface. Due to the intrinsic patient variability in terms of bone morphology, loading conditions and biological response, the patient specific modeling of such interface remains a challenge. In tissue engineering the same issues remain to be elucidated. However, the advances in in vitro techniques have enabled to control better the loading conditions and biological response, and to define precisely the morphological properties of the scaffold used. It is therefore believed that most of the work performed previously on the interaction of the implant with the host tissue can be transposed directly to the tissue engineering application. The use of multiscale modeling that span (1) from the molecular interaction of the scaffold coating with the molecules present to the surrounding medium, to (2) the cellular interaction with the scaffold surface, and to (3) the formation of extracellular matrix tissue will enable to unravel new knowledge of the interactions between material scaffolds and biological entities. As a conclusion the development of computational methods within tissue engineering opens many perspectives with the possibility to simulate the cellular processes that take place inside the scaffold and to optimize and maximize the formation of functional tissues.

References

1. Langer, R., Vacanti, J.P., Vacanti, C.A., Atala, A., Freed, L.E., Vunjak-Novakovic, G.: Tissue engineering: biomedical applications. Tissue Eng. **1**(2), 151–161 (1995)
2. Williams, D.F.: On the nature of biomaterials. Biomaterials **30**(30), 5897–5909 (2009)
3. Hollister, S.J.: Scaffold engineering: a bridge to where? Biofabrication **1**(1, 012001), 1–14 (2009)
4. Bohner, M., Loosli, Y., Baroud, G., Lacroix, D.: Commentary: deciphering the link between architecture and biological response of a bone graft substitute. Acta Biomater. **7**(2), 478–484 (2011)
5. Olivares, A.L., Marsal, E., Planell, J.A., Lacroix, D.: Finite element study of scaffold architecture design and culture conditions for tissue engineering. Biomaterials **30**(30), 6142–6149 (2009)
6. Liu, C., Xia, Z., Czernuszka, J.: Design and development of three-dimensional scaffolds for tissue engineering. Chem. Eng. Res. Des. **85**(7), 1051–1064 (2007)

7. Sandino, C., Planell, J.A., Lacroix, D.: A finite element study of mechanical stimuli in scaffolds for bone tissue engineering. J. Biomech. **41**(5), 1005–1014 (2008)
8. Cioffi, M., Boschetti, F., Raimondi, M.T., Dubini, G.: Modeling evaluation of the fluid-dynamic microenvironment in tissue-engineered constructs: a micro-CT based model. Biotechnol. Bioeng. **93**(3), 500–510 (2006)
9. Lacroix, D., Chateau, A., Ginebra, M.-P., Planell, J.A.: Micro-finite element models of bone tissue-engineering scaffolds. Biomaterials **27**(30), 5326–5334 (2006)
10. Milan, J.-L., Planell, J.A., Lacroix, D.: Simulation of bone tissue formation within a porous scaffold under dynamic compression. Biomech. Model. Mechanobiol. **9**(5), 583–596 (2010)
11. Saey, H., Hutmacher, D.W.: Application of micro CT and computation modeling in bone tissue engineering. Comput. Aided Des. **37**(11), 1151–1161 (2005)
12. Melchels, F.P.W., Tonnarelli, B., Olivares, A.L., et al.: The influence of the scaffold design on the distribution of adhering cells after perfusion cell seeding. Biomaterials **32**, 2878–2884 (2011)
13. Milan, J.-L., Planell, J.A., Lacroix, D.: Computational modelling of the mechanical environment of osteogenesis within a polylactic acid-calcium phosphate glass scaffold. Biomaterials **30**(25), 4219–4226 (2009)
14. Melchels, F.P.W., Bertoldi, K., Gabbrielli, R., Velders, A.H., Feijen, J., Grijpma, D.W.: Mathematically defined tissue engineering scaffold architectures prepared by stereolithography. Biomaterials **31**(27), 6909–6916 (2010)
15. Bohner, M., Baumgart, F.: Theoretical model to determine the effects of geometrical factors on the resorption of calcium phosphate bone substitutes. Biomaterials **25**(17), 3569–3582 (2004)
16. Sanz-Herrera, J.A., García-Aznar, J.M., Doblaré, M.: On scaffold designing for bone regeneration: a computational multiscale approach. Acta Biomater. **5**(1), 219–229 (2009)
17. Adachi, T., Osako, Y., Tanaka, M., Hojo, M., Hollister, S.J.: Framework for optimal design of porous scaffold microstructure by computational simulation of bone regeneration. Biomaterials **27**(21), 3964–3972 (2006)
18. Adachi, T., Kameo, Y., Hojo, M.: Trabecular bone remodelling simulation considering osteocytic response to fluid-induced shear stress. Philos. Trans. A Math. Phys. Eng. Sci. **368**(1920), 2669–2682 (2010)
19. Adachi, T., Tsubota, K.-ichi, Tomita, Y., Hollister, S.J.: Trabecular surface remodeling simulation for cancellous bone using microstructural voxel finite element models. J. Biomech. Eng. **23**(5), 403 (2001)
20. Cheah, C.M., Chua, C.K., Leong, K.F., Chua, S.W.: Development of a tissue engineering scaffold structure library for rapid prototyping. Part 1: Investigation and classification. Int. J. Adv. Manuf. Tech. **21**(4), 291–301 (2003)
21. Cheah, C.M., Chua, C.K., Leong, K.F., Chua, S.W.: Development of a tissue engineering scaffold structure library for rapid prototyping. Part 2: parametric library and assembly program. Int. J. Adv. Manuf. Tech. **21**(4), 302–312 (2003)
22. Hollister, S., Lin, C.: Computational design of tissue engineering scaffolds. Comput. Methods Appl. Mech. Eng. **196**(31–32), 2991–2998 (2007)
23. Hollister, S.J.: Porous scaffold design for tissue engineering. Nat. Mater. **4**(7), 518–524 (2005)
24. Hollister, S.J., Maddox, R.D., Taboas, J.M.: Optimal design and fabrication of scaffolds to mimic tissue properties and satisfy biological constraints. Biomaterials **23**(20), 4095–4103 (2002)
25. Shipley, R.J., et al.: Design criteria for a printed tissue engineering construct: a mathematical homogenization approach. J. Theor. Biol. **259**(3), 489–502 (2009)
26. Wendt, D., Stroebel, S., Jakob, M., John, G.T., Martin, I.: Uniform tissues engineered by seeding and culturing cells in 3D scaffolds under perfusion at defined oxygen tensions. Biorheology **43**(3–4), 481–488 (2006)
27. Wendt, D., Marsano, A., Jakob, M., Heberer, M., Martin, I.: Oscillating perfusion of cell suspensions through three-dimensional scaffolds enhances cell seeding efficiency and uniformity. Biotechnol. Bioeng. **84**(2), 205–214 (2003)

28. Santoro, R., Olivares, A.L., Brans, G., et al.: Bioreactor based engineering of large-scale human cartilage grafts for joint resurfacing. Biomaterials **31**(34), 8946–8952 (2010)
29. Prendergast, P.J., Huiskes, R., Søballe, K.: Biophysical stimuli on cells during tissue differentiation at implant interfaces. J. Biomech. **30**(6), 539–548 (1997)
30. Huiskes, R., Van Driel, W.D., Prendergast, P.J., Søballe, K.: A biomechanical regulatory model for periprosthetic fibrous-tissue differentiation. J. Mater. Sci. Mater. Med. **8**(12), 785–788 (1997)
31. Lacroix, D., Prendergast, P.J., Li, G., Marsh, D.: Biomechanical model to simulate tissue differentiation and bone regeneration: application to fracture healing. Med. Biol. Eng. Compu. **40**(1), 14–21 (2002)
32. Kelly, D.J., Prendergast, P.J.: Mechano-regulation of stem cell differentiation and tissue regeneration in osteochondral defects. J. Biomech. **38**(7), 1413–1422 (2005)
33. Geris, L., Vandamme, K., Naert, I., Vander Sloten, J., Duyck, J., Van Oosterwyck, H.: Application of mechanoregulatory models to simulate peri-implant tissue formation in an in vivo bone chamber. J. Biomech. **41**(1), 145–154 (2008)
34. Geris, L., Gerisch, A., Sloten, J.V., Weiner, R., Oosterwyck, H.V.: Angiogenesis in bone fracture healing: a bioregulatory model. J. Theor. Biol. **251**(1), 137–158 (2008)
35. Isaksson, H., et al.: Bone regeneration during distraction osteogenesis: mechano-regulation by shear strain and fluid velocity. J. Biomech. **40**(9), 2002–2011 (2007)
36. Sandino, C., Lacroix D.: A dynamical study of the mechanical stimuli and tissue differentiation within a CaP scaffold based on micro-CT finite element models. Biomech Model Mechanobiol. (2010)
37. Lacroix, D., Prendergast, P.J.: A mechano-regulation model for tissue differentiation during fracture healing: analysis of gap size and loading. J. Biomech. **35**(9), 1163–1171 (2002)
38. Pérez, M.A., Prendergast, P.J.: Random-walk models of cell dispersal included in mechanobiological simulations of tissue differentiation. J. Biomech. **40**(10), 2244–2253 (2007)
39. Byrne, D.P., Lacroix, D., Planell, J.A., Kelly, D.J., Prendergast, P.J.: Simulation of tissue differentiation in a scaffold as a function of porosity, Young's modulus and dissolution rate: application of mechanobiological models in tissue engineering. Biomaterials **28**(36), 5544–5554 (2007)
40. Checa, S., Prendergast, P.J.: A mechanobiological model for tissue differentiation that includes angiogenesis: a lattice-based modeling approach. Ann. Biomed. Eng. **37**(1), 129–145 (2009)
41. Sandino, C., Checa, S., Prendergast, P.J., Lacroix, D.: Simulation of angiogenesis and cell differentiation in a CaP scaffold subjected to compressive strains using a lattice modeling approach. Biomaterials **31**(8), 2446–2452 (2010)
42. Puleo, D.A., Nanci, A.: Understanding and controlling the bone–implant interface. Biomaterials **20**(23–24), 2311–2321 (1999)
43. Sanz-Herrera, J.A., García-Aznar, J.M., Doblaré, M.: Scaffold microarchitecture determines internal bone directional growth structure: a numerical study. J. Biomech. **43**(13), 2480–2486 (2010)

Computational Modeling of Tissue Engineering Scaffolds as Delivery Devices for Mechanical and Mechanically Modulated Signals

Min Jae Song, David Dean and Melissa L. Knothe Tate

Abstract In this chapter, we outline the use of computational modeling and novel experimental methods to develop tissue engineering scaffolds as delivery devices for exogenous and endogenous cues, including biochemical and mechanical signals, to drive the fate of mesenchymal stem cells (MSCs) seeded within. Tissue regeneration in mature organisms recapitulates de novo tissue generation during organismal development. This gave us the impetus to develop tissue engineering scaffolds that deliver mechanical and chemical cues intrinsic to the environment of cells during mesenchymal condensation, which marks the initiation of skeletogenesis during development. Cell seeding density and mode of achieving density (protocol) have been shown to effect dilatational (volume changing) stresses on stem cells and deviatoric (shape changing) stresses on their nuclei. Shear flow provides a practical means to deliver mechanical forces within scaffolds, resulting in both dilatational and deviatoric stresses on cell surfaces. Both spatiotemporal mechanical cue delivery and mechanically modulated biochemical gradients can be further honed through optimization of scaffold geometry and mechanical properties. We use computational fluid dynamics (CFD) coupled with finite element analysis (FEA) modeling to predict flow regimes within the scaffolds and optimize flow rates to simulate seeded cells. This chapter outlines to major advantages of using computational modeling to design and optimize tissue engineering scaffold geometry, material behavior, and tissue ingrowth over time.

M. J. Song · D. Dean · M. L. Knothe Tate (✉)
Case Western Reserve University, 10900 Euclid Avenue,
Cleveland, OH 44106, USA
e-mail: mkt4@case.edu

Stud Mechanobiol Tissue Eng Biomater (2013) 10: 127–143

DOI: 10.1007/8415_2012_138

Published Online: 7 June 2012

1 Introduction

During physiological activity, external loads in dynamic environments get transduced via the musculoskeletal system to the cells which build, maintain, and remodel musculoskeletal tissues. This loading of poroelastic, viscoelastic, and hyperelastic fluid-imbibed solid elements and complex fluids transduces dilatational mechanical stresses, which induce volume without shape changes, and deviatoric mechanical stresses, which induces shape without volume changes. In this way, ground forces transduced via the muscles, ligaments, tendons and bones are experienced as stresses at the tissue length scale (e.g., cortical or trabecular bone, tendon) and cellular length scale (e.g., osteocytes in the pericellular lacunocanalicular system, or tenocytes), respectively [6, 24, 33]. (Fig. 1) Osteocytes and tenocytes have a crucial role in sensing these mechanical signals through a putative feedback system that enables maintenance and remodeling of bone, respectively tendon, tissue structure and function in dynamic environments [9, 23].

Taking into account these typical examples of mechanoadaptation as a means to maintain structure–function relationships in tissues exposed to spatiotemporally dynamic mechanobiological environments, new strategies for engineering and manufacture of replacement tissues are incorporating biomimicry approaches to harness nature's smart biomaterial paradigms. The design and engineering of tissue engineering scaffolds has entered a new era, where such scaffolds are considered as much as delivery devices as structural and functional tissue replacements [2, 26]. To harness nature's paradigms, we aim to drive structure–function relationships at the tissue and organ length scales by delivering appropriate mechanical and chemical cues to cells. One approach to optimize scaffolds as delivery devices is to use predictive computational modeling as a powerful tool that "...help[s] us to prioritize which variables exert dominant effects on system behavior and thus which experiments are key to test predictions. [As such] predictive computational model[s] allow for the study of [tissue's smart, multiscale properties] without the imperative to carry out thousands of experiments," as summed up in a recent publication [25]. In this chapter, we review computational modeling of tissue engineering scaffolds as delivery devices for mechanical and mechanically modulated (biological and chemical) signals.

Computational models can predict and simulate the role of mechanical forces in cell differentiation, motility, adhesion, proliferation, and secretion of extracellular matrix proteins within tissue engineering scaffold environments. The computational method even helps to unravel the most enigmatic problems whose solutions are stymied by experimental or technological limitations. For example, experimental mechanical testing of the femur can elucidate boundary stresses and strains; in contrast, computational models can predict intrinsic mechanical loading distributions of the structure after experimental validation. With a given tissue engineering scaffold geometry, computational models can be used to control and optimize parameters to deliver mechanical stimuli to cells seeded within, in order to maximize the probability of achieving the targeted tissue manufacture and integration. (Fig. 2) [2].

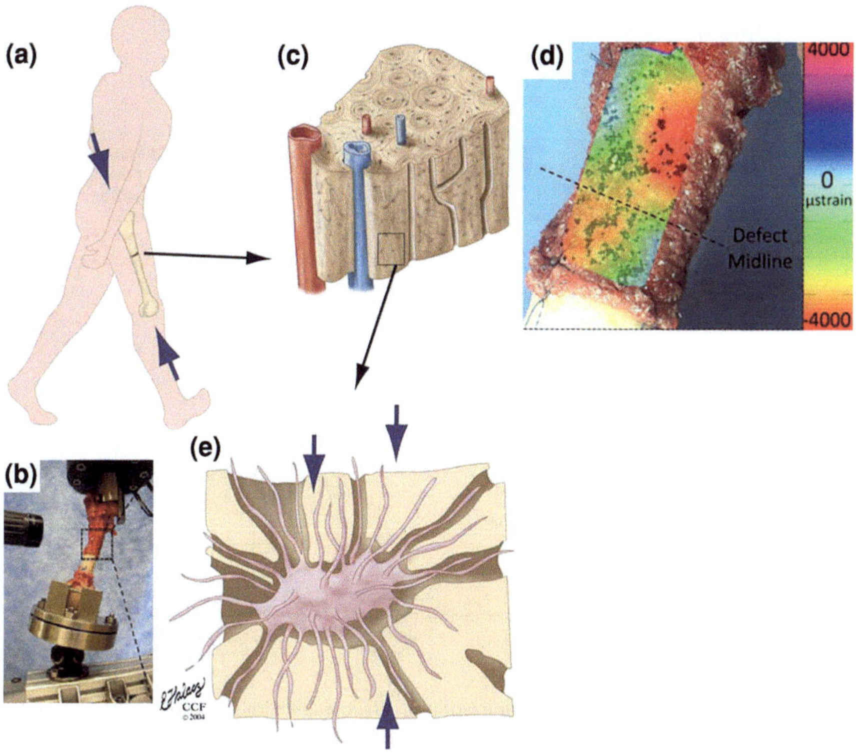

Fig. 1 Physiological loading of a patient or subject in a dynamic environment, from the organ length scale (**a**) in vivo and (**b**) ex vivo, tissue length scale (**c**) for cortical bone within the periosteum (**d**), and cellular length scale (pericellular, lacunocanalicular fluid space, (**e**) *Adapted from* [24]. **b** Ex vivo experimental set up for compressive loading of the femur to mimic stance shift in the first 2 weeks after one stage bone transport surgery. **d** Example of high definition optical strain mapping for periosteum surrounding a critical sized defect which is exposed ex vivo to mechanical loads mimicking stance shift after surgery (**b**). *Adapted from* [24, 33], used with permission

CFD models allow for study of flow induced forces at multiple length and time scales. Given the importance of the fluid environment for all cells of the body, this chapter emphasizes aspects of CFD for engineering and manufacture of tissues, using bone and treatment of bone defects as a case study.

2 Tissue Engineering Scaffold for Treatment of Critical Sized Bone Defects

Critical sized defects (CSDs) in bone are defined as the smallest defects that cannot heal spontaneously during the lifetime of a patient or study subject. Such critical sized defects are commonly caused after trauma, neurosurgical interventions, and reconstructive surgery of congenital abnormality, cancer, and infections. The

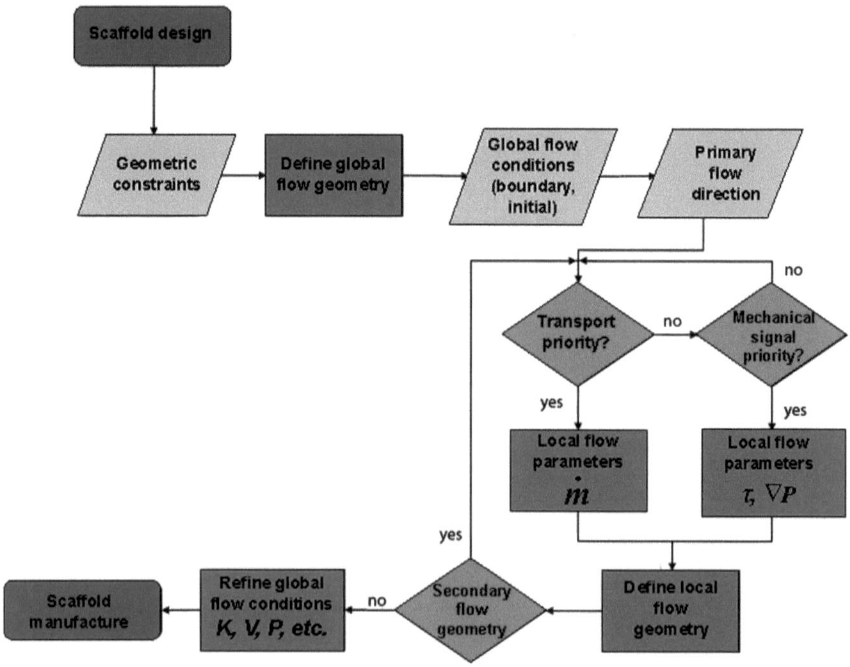

Fig. 2 Flow chart for design and optimization of tissue engineering scaffolds using CFD as a predictive tool. CFD multiphysics methods demonstrate chemical transport and delivery of mechanical signals within a given tissue engineering scaffold geometry, such as that shown in Fig. 3 where the primary flow direction is axial (along the length of the cylinder) and the secondary flow direction is orthogonal (transverse, defined by secondary flow geometry). *Adapted from* [2], used with permission. Please refer to online version of chapter for color version of the figure

current standard treatment option for such defects involves bone grafting, where graft is obtained from the patient himself (autograft), a bone bank (allograft), a graft substitute, or a structural implant or filler such as titanium or poly-methylmethacrylate (PMMA). Autografts are the gold standard treatment for CSDs due to their lack of immunogenicity; immune rejection is a critical complication associated with use of allograft. One complication associated with the use of autograft for CSDs, particularly of the cranium, is poor vascularization, which results in the need for reoperations and/or the removal of the implant [19]. In addition, packing of CSD with morcellized bone graft has been shown recently to retard the ingression of MSCs from the periosteum, when it is left in situ around the defect zone [29]. Furthermore, success of the surgery depends highly on both the size of the defect and the quality of its surrounding tissue [28]. Even with a sufficiently small defect that is surrounded by a healthy tissue bed, autografting *per se* is associated with risks including donor site morbidity and additional pains [19, 38].

Over the past several decades, tissue engineering has been developed as an alternative to tissue transplantation. For CSDs of long bone, a new surgical technique referred to as the one-stage bone transport procedure has been developed

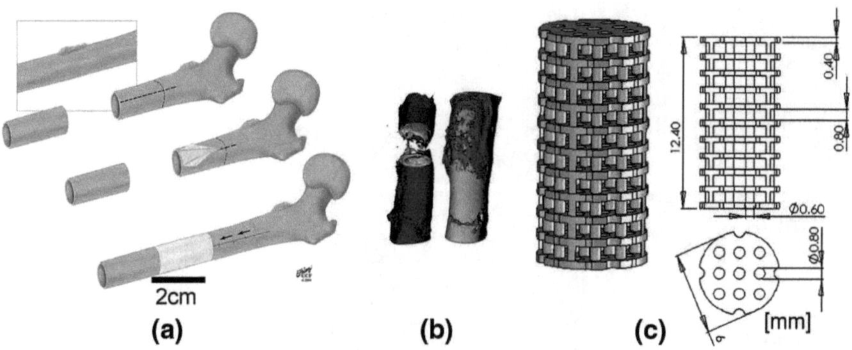

Fig. 3 Schematic of the one-stage bone-transport surgical procedure applied to the femur **a** and **b** three-dimensional (3D) micro-computed tomograph reconstruction of the critical sized defect control (unhealed after 16 weeks) and the defect surrounded by healthy periosteum in situ (healed completely after 16 weeks) [28], used with permission. **c** Computer Aided Design (CAD) drawings of 3D scaffold designs for treatment of CSD in the cranium [13, 18], adapted from [13] and used with permission . Please refer to online version of chapter for color version of the figure

recently (Fig. 3a, b) [28]. The technique results in woven bone regeneration in CSD zone within 2 weeks of surgery. Recent studies using a periosteum replacement implant *cum* delivery device have shown that incorporation of periosteal factors, including periosteum derived multipotent cells and periosteal strips without patent blood supply, around the defect zone can improve defect infilling compared to that observed with baseline controls [26]. For cranial CSDs, polymer scaffolds have been designed, some of which also incorporate MSCs (Fig. 3c) [13, 18]. This scaffold consists of polypropylene fumarate (PPF) mixed with a photoinitiator, which can be cross-linked by exposing it to a concentrated flood of UV light [13]. MSCs are seeded into the scaffold before implantation of this scaffold into cranial CSD zone. Broad interdisciplinary studies are in progress to elucidate mechanisms of tissue building, incorporating fundamentals of computational and experimental mechanics, polymer science, rapid prototyping, biochemistry, and stem cell mechanobiology [39, 41].

3 Mechanical Characteristics of Embryonic Stem Cells

Biophysical and biochemical cues define the local environment of the cell and play a key role in determining cell behavior including migration, proliferation, and differentiation; cumulatively, these cell behaviors result in de novo generation of tissue or bottom up tissue engineering. During embryonic development, cells respond to biophysical and biochemical signals to form the template of the complete organism. Defining the tissue template specifications to mimic the environment of the condensed mesenchyme (Fig. 4) during development allows

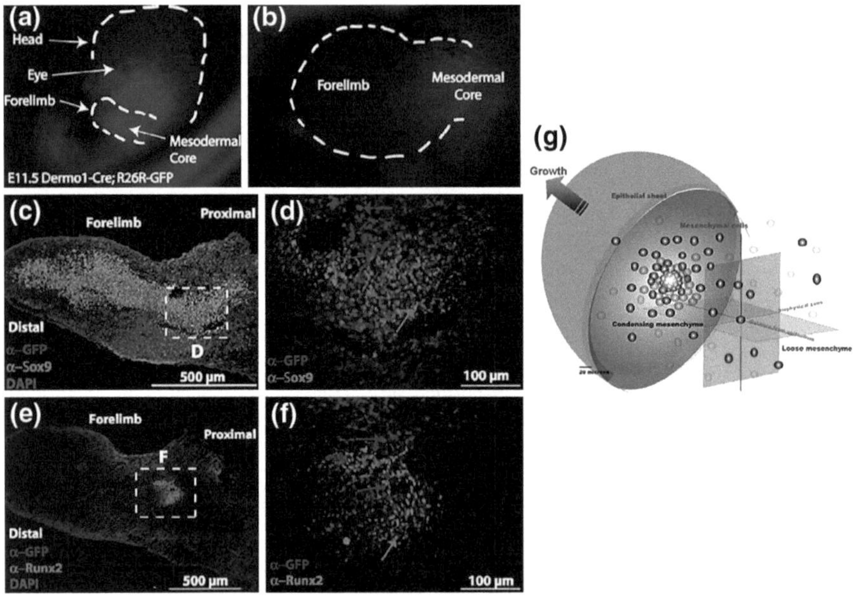

Fig. 4 Key extrinsic factors and their role in mesenchymal condensation, the first step of skeletogenesis. (**a**, **b**) The embryonic murine limbud is shown at E11.5, where the mesenchyme is bulbous in shape. Although possessing the same genetic code, lineage commitment by each cell is highly dependent on spatiotemporal signals including deviatoric and dilatational stress states as well as the biochemical *milieu*. For example, the expression of Sox9, a genetic marker of chondrogenesis (**c**, enlarged in **d**) is spatially distinct from that of Runx2, a genetic marker of osteogenesis (**e**, enlarged in **f**). *Adapted from* [27], used with permission

for exploitation of tissue scaffolds as delivery devices for exogenous and endogenous cues, including biochemical and mechanical signals, to drive the fate of MSC seeded within [27].

Recent studies demonstrate the promise of delivering spatiotemporally controlled mechanical cues to guide stem cell proliferation patterns [2, 14, 25] and lineage commitment [2–5, 12, 24–26, 34, 35, 46], essentially harnessing nature's approach to engineering tissues. Moreover, embryonic MSC exhibit 1000-fold greater mechanosensitivity than terminally differentiated cells (Fig. 5) [2, 12, 34, 35, 46]. Already half a century ago, Pauwels postulated that dilatational stresses such as hydrostatic or normal stresses cause differentiation of stem cells to chondrogenic lineage whereas deviatoric or shear stress guides stem cells toward ligamentous and tendonous phenotypes [37]. In vitro studies have shown that such mechanical loading significantly affected terminally differentiated cells as well as lineage commitment in undifferentiated multipotent stem cells [8, 11, 14, 15, 22, 42, 43]. Specifically, shear forces by fluid flow have been shown to affect baseline gene expression of Collagen type I, II, Runx2 and Sox 9, which are markers of mesenchymal condensation [12, 34, 35, 37, 46] (Fig. 6).

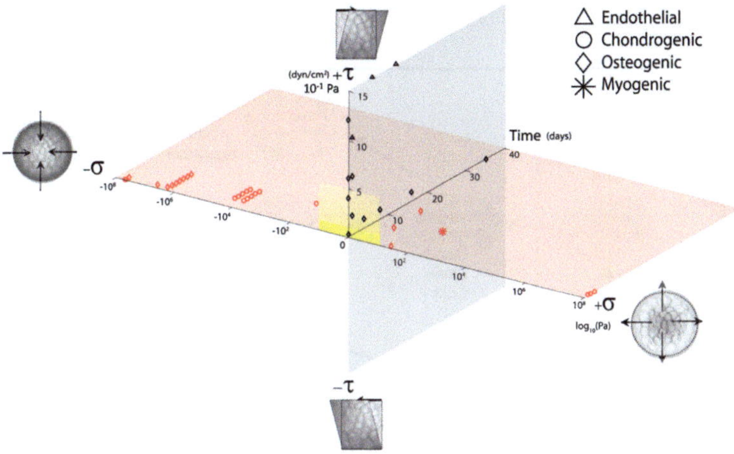

Fig. 5 Characteristic magnitudes and time domains of mechanical signals applied in studies of multipotent cell differentiation. *Red* data points dilatational stress and *black* data points deviatoric, i.e., shear stress. $-\sigma$ indicates hydrostatic compression and $+\sigma$ indicates tension depicted on a $\log_{10}$ scale in Pa. *Yellow* region shows dilatational and deviatoric stress ranges predicted to prevail during cell fate determination in utero. *Adapted from* [2, 41], used with permission

4 Stem Cell Morphology within Mechanical Environments

Stem cell shape has been shown to modulate fate or lineage commitment. Whereas flattened and spread shape directs stem cells toward osteogenesis, round and unspread shape directs stem cells toward adipogenesis. The RhoA-ROCK pathway, involving cytoskeletal remodeling of actin, has been shown to bridge stem cell shape changes to cell fate commitment [32].

The cytoskeleton is the main subcellular, mechanical support structure of the cell. Actin filaments and microtubules play a crucial role in the cellular response to external forces under fluid flow (Fig. 7). Recent studies show interesting relationships between stem cell shape and fate and the emergence of anisotropy in stem cells exposed to controlled dilatational and deviatoric stress environments. In these studies, larger cells at low cell density exhibit a more extensive cytoskeleton, as measured by amount of tubulin and actin expressed; in contrast, smaller cells at high cell density exhibit a less extensive cytoskeleton. Furthermore, expression of tubulin is more significantly affected by shear flow than expression of actin; this is particularly interesting, given tubulin's role in bearing compressive forces in the cell compared to actin, which withstands more tensile forces [12].

The role of nucleus shape change in fate determination is just beginning to be elucidated. A recent study demonstrated a significant relationship between nucleus shape and cell seeding protocol; namely, seeding at increasing target density changes the volume of the cell while changing the shape of the nucleus. Furthermore, changes in nucleus shape are significantly correlated to (fold) changes in

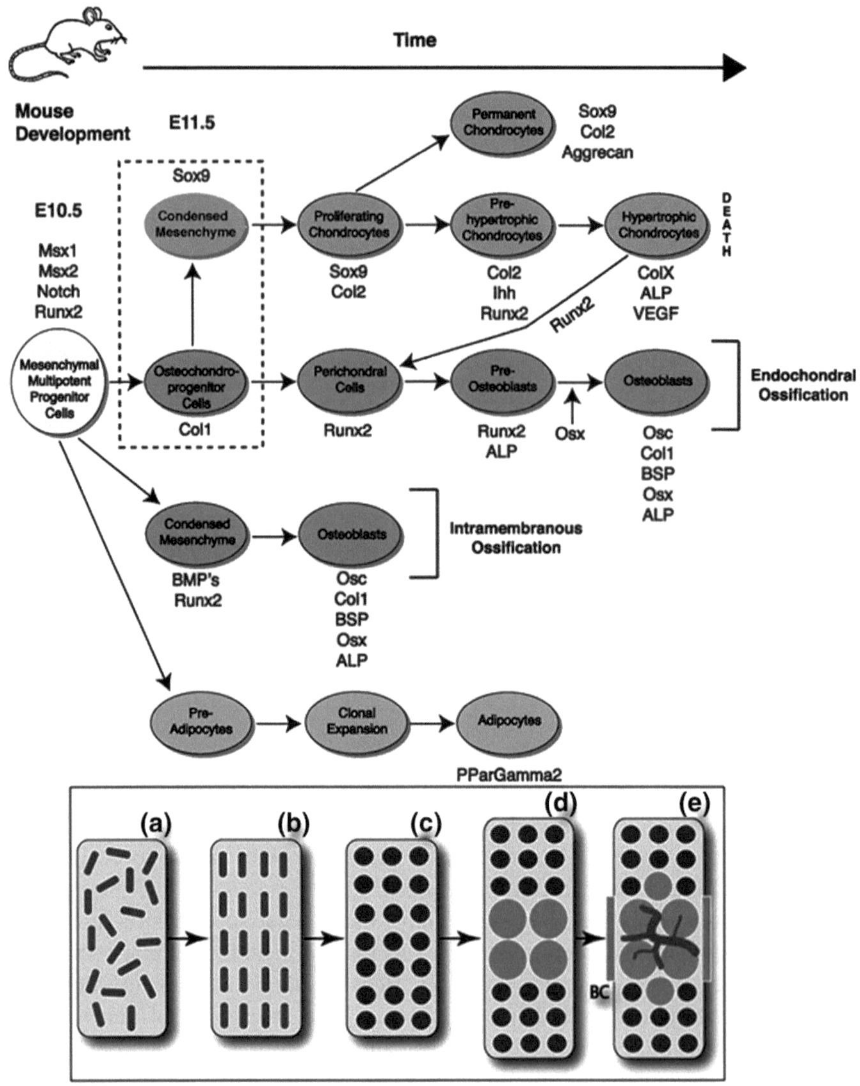

Fig. 6 Differentiation of mesenchymal stem cells into several lineages including chondrocytes (*orange* in online version of the figure), osteoblasts (*blue* in online version of the figure), and adipocytes (*green* in online version of the figure). (*Top*) Prior to mesenchymal condensation, upregulation of notch signals for chondrogenesis, which upregulates Sox9, one of the first transcription factors regulating endochondral ossification. (*Bottom*) In addition to direct formation of woven bone during intramembranous ossification (described in context of the one stage bone transport procedure), bone also forms via endochonral ossification, as shown schematically; cells (**a**) proliferate and organize into the condensed mesenchyme comprised of osteochondroprogenitor cells (**b**). The cells of the condensed mesenchyme differentiate into chondrocytes (**c**) and the cells at the center of the condensation stop proliferating and become hypertrophic (**d**). Perichondral cells adjacent to the hypertrophic chondrocytes differentiate into osteoblasts (**e**) and form the bone collar (**b**, **c**) as the invasion of blood vessels begins concomitant to continued osteoblast differentiation. *Adapted from* [27], used with permission. Please refer to online version of chapter for color version of the figure

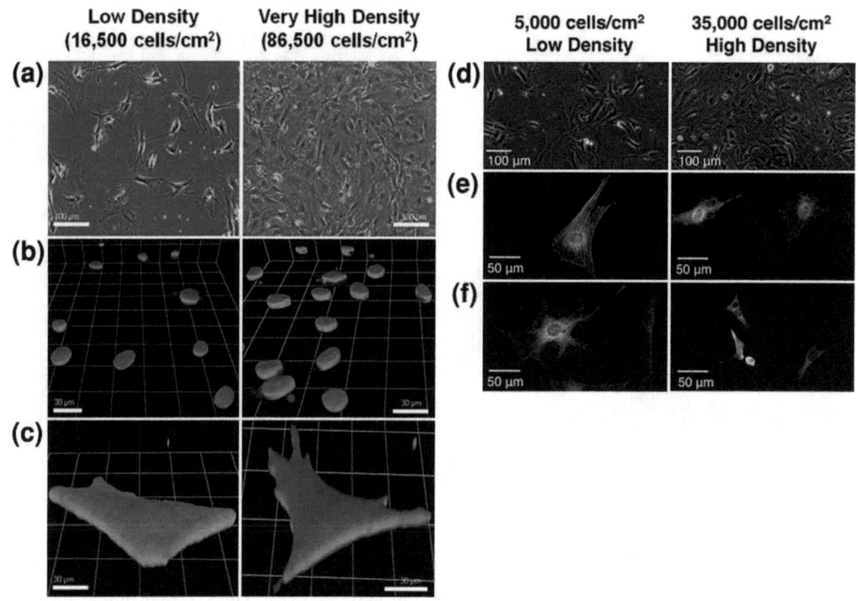

Fig. 7 Stem cell morphology changes at the subcellular length scale as a consequence of seeding at different densities and using different protocols to achieve density. **a**, **b**, **c** Nucleus shape changes **b** and cell shape changes **c** are attributable to target cell seeding densities and means of achieving density [12], used with permission. **d**, **e**, **f** The cytoskeleton components, including actin (**f**) and tubulin (**e**), also change when exposed to dilatational and deviatoric stress induced through different target densities and seeding protocols as well as exposure to fluid flow [46], used with permission. Please refer to online version of chapter for color version of the figure

gene expression of differentiation markers associated with mesenchymal condensation [46] (Fig. 7). Although correlation does not equal causation, it will be interesting to determine whether cell shape changes in response to mechanical signals are more of a response to minimize energy demands, e.g., by streamlining (laminar) flow to reduce losses associated with chaotic flow in boundary layers or as an active means for a cell to adapt to its environment (perhaps concomitant to reducing metabolic energy costs).

Computational methods can be applied to analyze cell deformations in near real time under shear flow; computational methods can predict deformation based on a given geometry and mechanical or chemical properties under external forces, which is passive behavior [7, 39]. On the other hand, active morphological changes of cells can occur through biological changes such as gene expression and reorientation of cytoskeletal filaments [10, 12, 31]. This active behavior can be also analyzed by computational models based on a given results under forces, which is an important approach to elucidate mechanisms of mechanoadaptation [17, 20, 31].

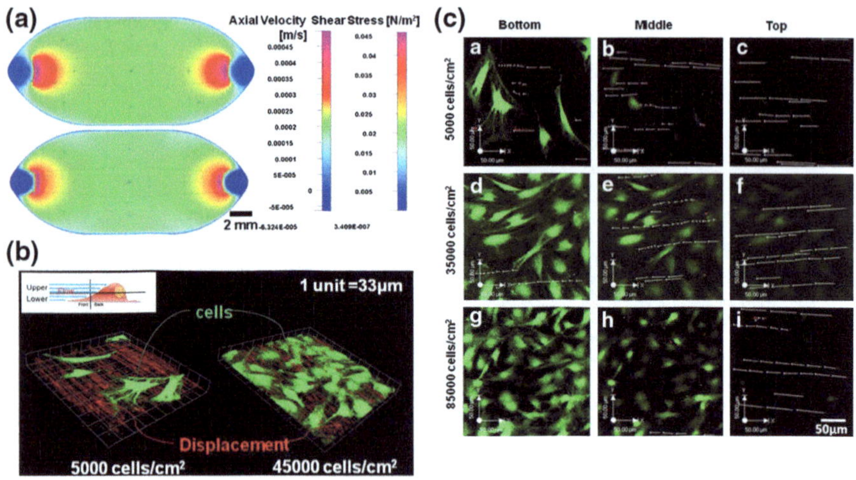

Fig. 8 CFD predictions and μ-PIV for flow regimes of interest. **a** CFD predictions of axial flow velocity (*upper*) and shear stress (*lower*) in the flow chamber at 10 μm from the bottom of the flow chamber. Flow moves from *right* (inlet) to *left* (outlet). **b** 3D images of flow fields (*red arrows* indicate microsphere displacements) around cells (*green*). Three dimensional confocal image stacks are analyzed to quantify the flow fields with respect to distance from the substrate and cell density. **d– l** Confocal images closest to the basal surface, approximately 2 μm from the substrate on which cells are seeded (**d**, **g**, **j**), 5 μm from the substrate (**e**, **h**, **k**), and 10 μm from the substrate (**f**, **i**, **l**). Cells are labeled with calcein *green*, microspheres exhibit *red fluorescence*, and white arrows indicated microsphere displacements in 990 ms. *Green*, *red*, and *white* indicate cells, microspheres, and displacements, respectively. *Adapted from* [40]

5 CFD Modeling for Precise Delivery of Mechanical Signals to MSCs to Direct Cell Fate

Based on CFD predictions, studies of parallel plate perfusion systems have underscored how important it is to translate signals delivered in two dimensions to three dimensions (3D), and at the subcellular length scale [1]. CFD can be used to predict flow regimes around cells (in 3D) in known flow chamber geometries and flow fields, to deliver controlled mechanical signals to cells seeded within. In situ microscopy allows for concomitant imaging of live cell mechano-adaptation while tracking the cell's dynamic mechanical environment [39, 41, 43]. Specifically, microscale particle image velocimetry (μ-PIV) allows not only for the validation of CFD predictions at the length scale of the coverslip (2D) and scaffolds (3D) onto which cells are seeded for mechanotransduction studies, but also for observing cell seeding effects on three dimensional flow field in the vicinity of cells. (Fig. 8, 9) Combined with rtPCR [34, 39, 41] and/or genetically modified stem cells that express green fluorescent protein upon differentiation.

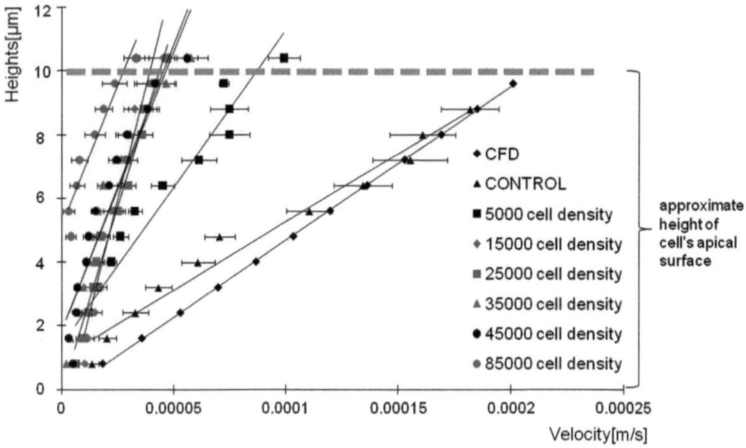

Fig. 9 CFD predictions compared to experimental measures (μ-PIV) of velocity in the flow chamber, in the absence of cells. Error bars report standard error (n = 5) at each data point. All linear regressions show $R^2 > 0.8$. The slopes of the linear regression lines represent the inverse of strain rate. *Adapted from* [40], used with permission

Flow is calculated from the continuity Eq. (1) and Navier–Stokes Eq. (2) using a second order upwind-discretization scheme in three dimensions. Wall shear stress is calculated from the wall strain rate (3). Hence,

$$\nabla \cdot v = 0, \tag{1}$$

$$\rho\ (v \cdot \nabla v)\Delta^2 v - \nabla P, \tag{2}$$

$$\tau_{cell} = \mu \frac{\partial v}{\partial x}\Big|_{cellheight}, \tag{3}$$

where v is the velocity vector, ρ is density, P is pressure, μ is viscosity, τ_{cell} is the shear stress at typical cell height, $\frac{\partial v}{\partial x}$ is the strain rate, and x is the height from bottom of chamber. (Fig. 8a).

CFD and microsphere displacement tracking enable the unprecedented prediction, validation and spatiotemporal delivery of mechanical signals on cell surfaces bordering other cells or the environment. This approach is rapidly translatable to the design of geometries and surfaces for tissue engineering scaffolds as delivery devices for mechanical and biochemical signals to steer the fate of cells seeded within.

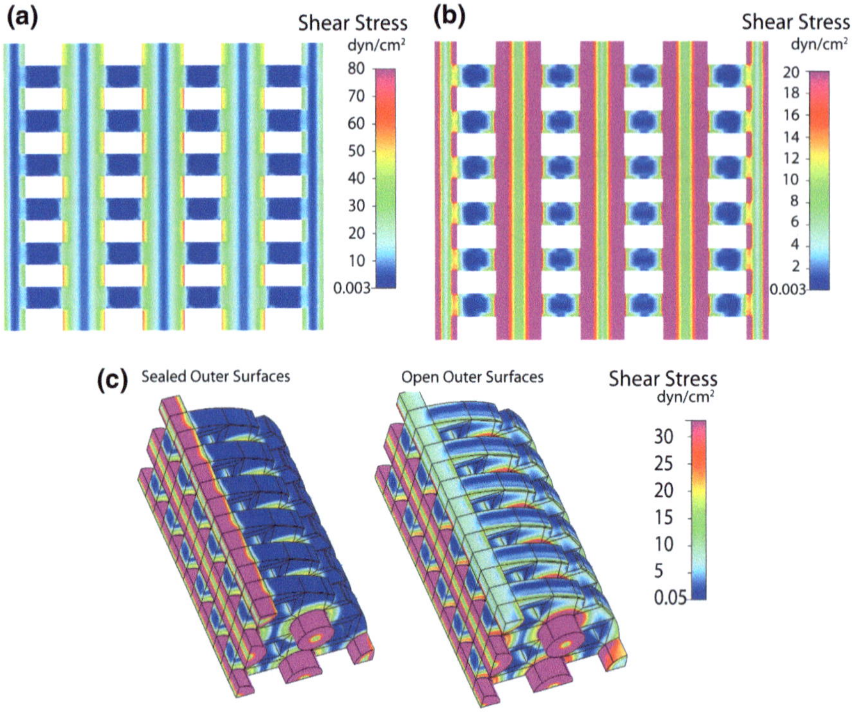

Fig. 10 CFD predicts shear stress to vary by an order of magnitude in longitudinal channels of a target tissue engineering scaffold geometry (80 dyn/cm^2, **a**, **c**) compared to transverse channels (10 dyn/cm^2, **b**, **c**). *Adapted from* [2], used with permission

6 CFD Modeling for Design of Tissue Engineering Scaffolds Geometry

Mechanically induced fluid movement through tissue engineering scaffolds provides a driving force for nutrient transport and waste removal. Furthermore, fluid flow can accelerate or activate bone development (osteogenesis and chondrogenesis) of MSCs [8, 11] through convective augmentation of anabolic biochemical factor transport or by delivery of mechanical cues at the interface between the fluid and the cell [2, 9, 23, 25].

A recently published CFD model explores effects of fluid flow in tissue engineering scaffolds designed to treat cranial defects (Fig. 2c). CFD predicts the different mechanical environments prevailing in longitudinal and transverse flow conduits, as well as on the conduit walls (inner surfaces of the scaffold) where cells are seeded (Fig. 10). Hence, by changing the size and relative geometry fluid channels in scaffolds, CFD can be used to optimize flow regimes within the scaffold to maintain nutrient and waste transport while achieving target mechanical stresses to guide stem cells toward target lineages using flow induced

Table 1 Effect of tissue engineering scaffold geometry (porosity and pore size) on fluid flow by CFD analysis

Reference	Scaffold type	Flow	Average shear surface rate (l/s) and stress (dyn/cm^2)
Anderson et al. [2]	Idealized design for critical sized critical defects Pore diameter: 800 μm Post diameter: 600 μm	Pressure gradient (Pa) 100	Longitudinal channels: 30–40 dyn/cm^2 Transverseflow: 0–10 dyn/crn^2
Lesman et al. [30]	PLLA/PLGA scaffold Porosity: 90 % Average pore diameter: 500 μm	Inflow velocity (cm/s) 0.5 1 1.5 2	Four different cell layer thicknesses (0, 50, 75, 125 μm) 2–18 dyn/cm^2
Voronov et al. [44]	Poly(L-Iacticacid) Porosity: 80–95% Average pore size: 215–402.5 μm	Pressure gradient (Pa) 10 1 0.1 0.01	 ~0.1 dyn/crn^2 ~0.01 dyn/cm^2 ~0.001 dyn/crn^2 ~0.0001 dyn/cm^2
Melchels et al. [36]	Photo-polymerisable poly(D/Llactide) (PDLLA) macromers Porosity: 62 ± 1 % Isotropic pore size: 412 ± 13 μm	Average fluid Flow velocity: 0.86 mm/s	15–24/s
	Porosity: 35–85 % Gradient in pore size: 250–500 μm		12–38/s
Y Yao et al. [45]	Idealized design with unit spherical cells porosity: 70 %	Flow rate: 0.1 ml/min Flow rate: 0.5 ml/min Flow rate: 1 ml/min	0.03 dyn/cm^2 0.15 dyn/cm^2 0.3 dyn/cm^2

deviatoric stresses and seeding density and protocol induced dilatational stresses [2, 12, 34, 35, 38, 46].

CFD has also been used to optimize geometries of tissue engineered scaffolds that are manufactured using rapid prototyping or other methods. Furthermore, material choice will have a large effect on target and actual scaffold geometries and signal delivery [38]. Comparing two types of scaffolds, which are made by Collagen-Glycosaminoglycan (CG) and Calcium Phosphate, CFD predicts that shear stress is 2.8 times higher in the CG as in the calcium-phosphate scaffold. This is mainly caused by pore size differences (CG pore size ~96 mm, calcium phosphate pore size ~350 mm). A scaffold with bigger pore sizes (215–402.5 μm average pore size), made of Poly(L-lactic acid), has been also analysed using CFD

[21], which shows that different pore sizes and porosities cause different wall shear stress on surface of the scaffolds (Table 1) [2, 30, 36, 44, 45].

Flow regimes within a rapid prototyped scaffold have been studied previously to predict the effect of geometric discrepancies due to imperfect manufacturing, e.g., differences between target and actual manufactured scaffold geometries [38]. CFD can also be used to analyze differences in flow that are attributable to differences in actual geometries compared to target ideal geometries; for example, CFD accurately predicts the average fluid velocity across 81 % of the sample volume and the wall shear stress across 73 % of the sample surface, when compared to experimental validation of actual values using μ-PIV [16].

7 Conclusion

This chapter has outlined to major advantages of using CFD to design and optimize tissue engineering scaffold geometry, material behavior (including biodegradation), and tissue ingrowth over time. One major advantage is that CFD predicts flow conditions for precise delivery of mechanical signals to stem cells to direct cell fate (cell scale). A second advantage is that CFD can be used to optimize tissue engineering scaffold geometries, including pore size, pore distribution and connectivity, as well as scaffold wall thickness, to modulate flow regimes within the scaffold (tissue to cell scale). As multiscale CFD and finite element modeling approaches converge, we anticipate that multiscale modeling will become an even more powerful tool to predict cell-organ scale system behavior, seamlessly, over multiple time scales that span the life of the patient or test subject.

References

1. Anderson, E.J., Falls, T.D., Sorkin, A.M., Knothe Tate, M.L.: The imperative for controlled mechanical stresses in unraveling cellular mechanisms of mechanotransduction. Biomed Eng Online **5**, 27 (2006). doi:10.1186/1475-925X-5-27. 1475-925X-5-27 [pii]
2. Anderson, E.J., Knothe Tate, M.L.: Design of tissue engineering scaffolds as delivery devices for mechanical and mechanically modulated signals. Tissue Eng **13**, 2525–2538 (2007). doi:10.1089/ten.2006.0443
3. Anderson, E.J., Knothe Tate, M.L.: Design of tissue engineering scaffolds as delivery devices for mechanical and mechanically modulated signals. Tissue Eng **13**, 2525–2538 (2007). doi:10.1089/ten.2006.0443
4. Anderson, E.J., Knothe Tate, M.L.: Open access to novel dual flow chamber technology for in vitro cell mechanotransduction, toxicity and pharamacokinetic studies. Biomed Eng Online **6**, 46 (2007). doi:10.1186/1475-925X-6-46. 1475-925X-6-46 [pii]
5. Anderson, E.J., Kreuzer, S.M., Small, O., Knothe Tate, M.L.: Pairing computational and scaled physical models to determine permeability as a measure of cellular communication in micro- and nano-scale pericellular spaces. Microflu Nanoflu **4**, 193–204 (2008). doi:10.1007/s10404-007-0156-5
6. Anderson, E.J., Knothe Tate, M.L.: Idealization of pericellular fluid space geometry and dimension results in a profound underprediction of nano-microscale stresses imparted by fluid

drag on osteocytes. J Biomech **41**, 1736–1746 (2008). doi:10.1016/j.jbiomech.2008.02.035. S0021-9290(08)00107-3 [pii]
7. Bagchi, P., Johnson, P.C., Popel, A.S.: Computational fluid dynamic simulation of aggregation of deformable cells in a shear flow. J Biomech Eng **127**, 1070–1080 (2005)
8. Baksh, D., Tuan, R.S.: Canonical and non-canonical Wnts differentially affect the development potential of primary isolate of human bone marrow mesenchymal stem cells. J Cell Physiol **212**, 817–826 (2007). doi:10.1002/jcp.21080
9. Bonewald, L.F.: Mechanosensation and Transduction in Osteocytes. Bonekey Osteovision **3**, 7–15 (2006). doi:10.1138/20060233
10. Boon, R.A., Leyen, T.A., Fontijn, R.D., Fledderus, J.O., Baggen, J.M., Volger, O.L., van Nieuw Amerongen, G.P., Horrevoets, A.J.: KLF2-induced actin shear fibers control both alignment to flow and JNK signaling in vascular endothelium. Blood **115**, 2533–2542 (2010). doi:10.1182/blood-2009-06-228726. blood-2009-06-228726 [pii]
11. Campbell, J.J., Lee, D.A., Bader, D.L.: Dynamic compressive strain influences chondrogenic gene expression in human mesenchymal stem cells. Biorheology **43**, 455–470 (2006)
12. Chang, H., Knothe Tate, M.L.: Structure-function relationships in the stem cell's mechanical world B: emergent anisotropy of the cytoskeleton correlates to volume and shape changing stress exposure. Mol Cell Biomech **8**, 297–318 (2011)
13. Cooke MN, Fisher JP, Dean D, Rimnac C, Mikos AG (2003) Use of stereolithography to manufacture critical-sized 3D biodegradable scaffolds for bone ingrowth. J Biomed Mater Res B Appl Biomater 64: 65–69.doi:10.1002/jbm.b.10485
14. Datta, N., Pham, Q.P., Sharma, U., Sikavitsas, V.I., Jansen, J.A., Mikos, A.G.: In vitro generated extracellular matrix and fluid shear stress synergistically enhance 3D osteoblastic differentiation. Proc Natl Acad Sci USA **103**, 2488–2493 (2006). doi:10.1073/pnas.0505661103. 0505661103 [pii]
15. David, V., Martin, A., Lafage-Proust, M.H., Malaval, L., Peyroche, S., Jones, D.B., Vico, L., Guignandon, A.: Mechanical loading down-regulates peroxisome proliferator-activated receptor gamma in bone marrow stromal cells and favors osteoblastogenesis at the expense of adipogenesis. Endocrinology **148**, 2553–2562 (2007). doi:10.1210/en.2006-1704. en.2006-1704 [pii]
16. De, B.S., Truscello, S., Ozcan, S.E., Leroy, T., Van, O.H., Berckmans, D., Schrooten, J.: Bimodular flow characterization in tissue engineering scaffolds using computational fluid dynamics and particle imaging velocimetry. Tissue Eng Part C Methods **16**, 1553–1564 (2010). doi:10.1089/ten.tec.2010.0107
17. Fletcher DA, Mullins RD (2010) Cell mechanics and the cytoskeleton. Nature **463**, 485–492. doi: 10.1038/nature08908, nature08908 [pii]
18. Gibson LJ, Ashby MF (1997) Cellular solids: structure and properties. In: Cambridge University Press, New York
19. Habal, M.B., Reddi, A.H.: Bone grafts and bone induction substitutes. Clin Plast Surg **21**, 525–542 (1994)
20. Inoue Y, Deji T, Shimada Y, Hojo M, Adachi T (2010) Simulations of dynamics of actin filaments by remodeling them in shear flows. Comput Biol Med 40: 876–882. doi: 10.1016/j.compbiomed.2010.09.008, S0010-4825(10)00137-X [pii]
21. Jungreuthmayer, C., Donahue, S.W., Jaasma, M.J., Al-Munajjed, A.A., Zanghellini, J., Kelly, D.J., O'Brien, F.J.: A comparative study of shear stresses in collagen-glycosaminoglycan and calcium phosphate scaffolds in bone tissue-engineering bioreactors. Tissue Eng Part A **15**, 1141–1149 (2009). doi:10.1089/ten.tea.2008.0204
22. Kearney, E.M., Farrell, E., Prendergast, P.J., Campbell, V.A.: Tensile strain as a regulator of mesenchymal stem cell osteogenesis. Ann Biomed Eng **38**, 1767–1779 (2010). doi:10.1007/s10439-010-9979-4
23. Knothe Tate ML (2003) Whither flows the fluid in bone? An osteocyte's perspective. J Biomech 36:1409–1424. S0021929003001234 [pii]
24. Knothe Tate ML (2007) Engineering of functional skeletal tissues, multiscale computational engineering of bones: state-of-the-art insights for the future, Springer, New York

25. Knothe Tate ML (2011) Top down and bottom up engineering of bone. J Biomech 44: 304–312. doi: 10.1016/j.jbiomech.2010.10.019, S0021-9290(10)00575-0 [pii]
26. Knothe Tate ML, Chang H, Moore SR, Knothe UR (2011) Surgical membranes as directional delivery devices to generate tissue: testing in an ovine critical sized defect model. PLoS One 6: e28702. doi: 10.1371/journal.pone.0028702, PONE-D-11-07884 [pii]
27. Knothe Tate ML, Falls TD, McBride SH, Atit R, Knothe UR (2008) Mechanical modulation of osteochondroprogenitor cell fate. Int J Biochem Cell Biol 40:2720–2738. doi: 10.1016/j.biocel.2008.05.011, S1357-2725(08)00198-2 [pii]
28. Knothe Tate ML, Ritzman TF, Schneider E, Knothe UR (2007) Testing of a new one-stage bone-transport surgical procedure exploiting the periosteum for the repair of long-bone defects. J Bone Joint Surg Am 89: 307–316. doi: 10.2106/JBJS.E.00512, 89/2/307 [pii]
29. Knothe UR, Dolejs S, Matthew MR, Knothe Tate ML (2010) Effects of mechanical loading patterns, bone graft, and proximity to periosteum on bone defect healing. J Biomech 43: 2728–2737. doi: 10.1016/j.jbiomech.2010.06.026, S0021-9290(10)00362-3 [pii]
30. Lesman, A., Blinder, Y., Levenberg, S.: Modeling of flow-induced shear stress applied on 3D cellular scaffolds: implications for vascular tissue engineering. Biotechnol Bioeng **105**, 645–654 (2010). doi:10.1002/bit.22555
31. Loufrani, L., Henrion, D.: Role of the cytoskeleton in flow (shear stress)-induced dilation and remodeling in resistance arteries. Med Biol Eng Comput **46**, 451–460 (2008). doi:10.1007/s11517-008-0306-2
32. McBeath R, Pirone DM, Nelson CM, Bhadriraju K, Chen CS (2004) Cell shape, cytoskeletal tension, and RhoA regulate stem cell lineage commitment. Dev Cell 6:483–495. S1534580704000759 [pii]
33. McBride, S.H., Dolejs, S., Brianza, S., Knothe, U., Tate, M.L.: Net change in periosteal strain during stance shift loading after surgery correlates to rapid de novo bone generation in critically sized defects. Ann Biomed Eng **39**, 1570–1581 (2011). doi:10.1007/s10439-010-0242-9
34. McBride, S.H., Falls, T., Knothe Tate, M.L.: Modulation of stem cell shape and fate B: mechanical modulation of cell shape and gene expression. Tissue Eng Part A **14**, 1573–1580 (2008). doi:10.1089/ten.tea.2008.0113
35. McBride, S.H., Knothe Tate, M.L.: Modulation of stem cell shape and fate A: the role of density and seeding protocol on nucleus shape and gene expression. Tissue Eng Part A **14**, 1561–1572 (2008). doi:10.1089/ten.tea.2008.0112
36. Melchels FP, Tonnarelli B, Olivares AL, Martin I, Lacroix D, Feijen J, Wendt DJ, Grijpma DW (2011) The influence of the scaffold design on the distribution of adhering cells after perfusion cell seeding. Biomaterials **32**, 2878–2884. doi: 10.1016/j.biomaterials.2011.01.023, S0142-9612(11)00036-6 [pii]
37. Pauwels, F.: A new theory on the influence of mechanical stimuli on the differentiation of supporting tissue. The tenth contribution to the functional anatomy and causal morphology of the supporting structure. Z Anat Entwicklungsgesch **121**, 478–515 (1960)
38. Sanan, A., Haines, S.J.: Repairing holes in the head: a history of cranioplasty. Neurosurgery **40**, 588–603 (1997)
39. Song, M., Brady-Kalnay, S., Dean, D. Knothe Tate, M.L.: Relating Stem Cell Shape to Fate Commitment by Mapping Cell Surface Strains In Situ. Trans ORS **37**, 0826 (2012)
40. Song MJ, Dean D, Knothe Tate ML (2010) In situ spatiotemporal mapping of flow fields around seeded stem cells at the subcellular length scale. PLoS One 5. doi: 10.1371/journal.pone.0012796
41. Song, M.J., Dean, D., Brady-Kalnay, S., Knothe Tate, M.L.: Optimization of Tissue Engineering Scaffold Geometry, Seeding & Flow Conditions to Steer Stem Cell Shape and Fate. TERMIS 1, 0243 (2011)
42. Sorkin AM, Dee KC, Knothe Tate ML (2004) "Culture shock" from the bone cell's perspective: emulating physiological conditions for mechanobiological investigations. Am J Physiol Cell Physiol 287:C1527–C1536. doi: 10.1152/ajpcell.00059.2004, 00059.2004 [pii]
43. Stops AJ, Heraty KB, Browne M, O'Brien FJ, McHugh PE (2010) A prediction of cell differentiation and proliferation within a collagen-glycosaminoglycan scaffold subjected to

mechanical strain and perfusive fluid flow. J Biomech 43:618–626. doi: 10.1016/j.jbiomech.2009.10.037, S0021-9290(09)00624-1 [pii]
44. Voronov R, Vangordon S, Sikavitsas VI, Papavassiliou DV (2010) Computational modeling of flow-induced shear stresses within 3D salt-leached porous scaffolds imaged via micro-CT. J Biomech 43: 1279–1286. doi: 10.1016/j.jbiomech.2010.01.007, S0021-9290(10)00039-4 [pii]
45. YaoY, Chen W , Jin W: The influence of pore structure on internal flow field shear stress within scaffold. Adv Mater Res **308–310**, 771–775 (2011). doi:10.4028/www.scientific.net/AMR.308-310.771
46. Zimmermann JA, Knothe Tate ML (2011) Structure-function relationships in the stem cell's mechanical world A: seeding protocols as a means to control shape and fate of live stem cells. Mol Cell Biomech 8:275–296

Modelling the Cryopreservation Process of a Suspension of Cells: The Effect of a Size-Distributed Cell Population

Alberto Cincotti and Sarah Fadda

Abstract Cryopreservation of biological material is a crucial step of tissue engineering, but biological material can be damaged by the cryopreservation process itself. Depending on some bio-physical properties that change from cell to cell lineages, an optimum cryopreservation protocol needs to be identified for any cell type to maximise post-thaw cell viability. Since a prohibitively large set of operating conditions has to be determined to avoid the principal origins of cell damage (i.e., ice formation and solution injuries), mathematical modelling represents a valuable alternative to experimental optimisation. The theoretical analysis traditionally adopted for the cryopreservation of a cell suspension addresses only a single, average cell size and ascribes the experimental evidence of different ice formation temperatures to statistical variations. In this chapter our efforts to develop a novel mathematical model based on the population balance approach that comprehensively takes into account the size distribution of a cell population are reviewed. According to this novel approach, a sound explanation for the experimental evidence of different ice formation temperatures may now be given by adopting a fully deterministic criterion based on the size distribution of the cell population. In this regard, the proposed model represents a clear novelty for the cryopreservation field and provides an original perspective to interpret system behaviour as experimentally measured so far. First our efforts to successfully validate the proposed model by comparison with suitable experimental data taken from the literature are reported. Then, in absence of suitable experimental data, the model is used to theoretically investigate system behaviour

A. Cincotti (✉) · S. Fadda
Dipartimento di Ingegneria Meccanica, Chimica e dei Materiali,
Università degli Studi di Cagliari, Cagliari, Italy
e-mail: cincotti@dicm.unica.it

S. Fadda
e-mail: s.fadda@dicm.unica.it

Stud Mechanobiol Tissue Eng Biomater (2013) 10: 145–181

DOI: 10.1007/8415_2012_134

Published Online: 20 March 2012

at various operating conditions. This is done both in absence or presence of a cryo-protectant agent, as well as when the extra-cellular ice is assumed to form under thermodynamic equilibrium or its dynamics is taken into account consistently by means of an additional population balance. More specifically, the effect of the cell size distribution on system behaviour when varying cooling rate and cryo-protectant content within practicable values for a standard cryopreservation protocol is investigated. It is demonstrated that, cell survival due to intra-cellular ice formation depends on the initial cell size distribution and its osmotic parameters. At practicable operating conditions in terms of cooling rate and cryo-protectant concentration, intra-cellular ice formation may be lethal for the fraction of larger size classes of the cell population whilst it may not reach a dangerous level for the intermediate size class cells and it will not even take place for the smaller ones.

1 Introduction

In the tissue engineering field, preservation is a core technology to bring cell-based products to market, as they have to be supplied on demand [18]. In general, preservation can be defined as the process of reversibly arresting the biochemical reactions, i.e. the metabolism of an organism, thus reaching the so-called state of suspended animation.

The principal preservation method consists of freezing the bio-specimens to cryogenic temperature in order to take advantage of the preservative power of the cold. Indeed, compared to the other preservation methods like maintaining the bio samples in continuous culture, cryopreservation has the benefits of affording long shelf lives, genetic stability, reduced microbial contamination risks, and cost effectiveness [18]. The other side of the coin is that cryopreserved biological material can be damaged by the cryopreservation process itself [25]. Cryopreservation consists of cooling to subzero temperatures with or without a Cryo-Protectant Agent (CPA), storage, thawing and return to physiological environment for specific usages. A part from the storage, all these stages are potentially able to damage the cells due to the physical and/or chemical phenomena involved. Limiting the analysis to the cryopreservation process of a cell suspension where heat transfer phenomena are less restrictive than for 2-D and 3-D tissue samples, in general during the cooling stage ice initially forms in the extra-cellular medium surrounding the cells. As the extra-cellular ice phase grows, the extra-cellular solute concentration increases, thus imposing a chemical potential difference between the cytoplasm and the unfrozen external solution which acts as the driving force for water diffusion out of the cell through the plasma membrane, i.e. the osmotic transport. The rate of the osmotic water transport is limited by the permeability of the plasma membrane to water, and the osmotic equilibrium can be maintained only if the rate of cooling is sufficiently slow. If the temperature is lowered too fast, a significant dehydration cannot take place and the cytoplasm becomes super-cooled when low temperatures are reached, indeed. In such a case, a driving force for Intra-cellular Ice Formation (IIF) is established and a severe damage of cells results that may even lead to lethality as a consequence of membrane rupture and mechanical deformation. Actually, during freezing cell injury occurs due not only to IIF taking place at high cooling rates, but also to the "solution effects" taking place at low cooling rates, when excessive, intolerable cell dehydration and electrolyte concentrations may be reached [27]. As a result, post thaw-viability of cryopreserved cell suspensions may be relatively low both at high and low cooling rates, depending on the specific operating conditions adopted and sample geometry and size [28].

In principle, an optimum cooling rate exists where the two mechanisms of damage (i.e. IIF and solution injuries) are balanced and the probability of cell survival reaches a maximum. The optimum cooling rate differs from cell to cell lineage, as the result of different bio-physical properties such as water and CPA permeabilities through plasma membrane which affect the dehydration rate and the Probability of IIF (PIIF). In this context, the advent of CPAs represents a sort of panacea in the field of cryopreservation technology, so much so that they are routinely used in any standard cryopreservation protocol experimentally adopted in a bio lab. This widespread use of CPAs is due to the fact that they allow one to significantly prevent cell damage from both the two major sources of risk. When CPA is added, on one hand the thermodynamic temperature of ice formation is naturally reduced according to the corresponding phase diagram (i.e. protection against IIF injury in favour of the eventual vitrification of water). On the other hand, a lower intra-cellular concentration of electrolytes which otherwise would increase due to water exo-osmosis during cooling may be obtained (i.e. protection against solution injury). Unfortunately, CPAs are known to be cytotoxic. This cytotoxic effect varies from cell to cell lineage and it is more dangerous at higher temperatures, depending on the contact time with the cells. Thus, in order to minimise cell damage one has to select not only the proper type of CPA but, also the best way to put a CPA in contact with the cells: the CPA concentration levels, the temperatures for pre-freezing addition and post-thawing removal of CPA (i.e. washing). Even the choice between one step or continuous modality when adding/removing CPA from the cell suspension has to be made [13, 14, 31].

It is apparent that, every cell lineage requires its own cryopreservation protocol to be developed and optimised through a rational design method. However, since the number of experiments required for a rigorous optimisation of a cryopreservation procedure is prohibitively large due to the high number of operating conditions to be set, shortcomings of experimental protocol optimisation are necessarily adopted. In this context, a sequential optimisation of individual operating conditions is typically performed, thus reaching only a pseudo-optimal solution, strongly dependent on the initial values used in the search. As a result, the empirical development of cryopreservation procedures has been limited to systems that are relatively robust, or those in which damage can be tolerated [28].

On the basis of these considerations, it is clear that modelling may represent a helpful tool. Indeed, a successful design of cryopreservation procedures may be obtained through the use of theoretical models and computer-aided optimisation methods, since computer simulations are less expensive and faster than experimental studies. They can be used to explore conditions that are impossible to achieve in the lab and they can account for the effect of more than one parameter at time. This is why mathematical models have been extremely powerful and have been proposed so far in the literature to analyse and interpret cryobiologically relevant phenomena [26]. Starting from the pioneering work by Mazur [24] some literature has been accumulated on cryopreservation modelling [16], but the main chronological steps where significant improvements were reached are relatively few. The bi-compartmental transport model proposed by Mazur addressed only the osmotic behaviour of cells during the cooling stage of a cryopreservation protocol. Thus, IIF was initially neglected whereas in the extra-cellular solution ice was assumed to form under thermodynamic equilibrium conditions. Later, the simulation of IIF dynamics was accounted for, whilst no further improvement for describing the behaviour of the extra-cellular compartment was proposed. More specifically, osmosis was always described on the basis of the bi-compartmental transport model first proposed by Mazur, but ice formation inside the cells was simulated according to the classical nucleation theory albeit considering only the ice nucleation phenomenon [38]. Ice crystal growth inside the cells showed up later, when Karlsson et al. [19] developed a complete physico–chemical theory of ice formation inside biological cells by coupling the water transport model with the theory of ice nucleation and crystal growth. More recently, Gao and co-workers [44] proposed a modified version of Karlsson's model, where the growing, time dependent ice volume inside a cell during cooling is accounted for into Mazur's equation, thus coupling more tightly osmosis and IIF phenomena. Intriguingly, a coupled description of water osmosis and CPA permeation followed by cooling was never performed, even if it is well known that these represent two fundamental steps taken in sequence from an experimental perspective. In this regard, addition of a permeant CPA (i.e. CPA equilibration between the two compartments) was always investigated separately from the cooling stage [1, 3, 7, 13, 31, 33, 41].

These models and other more sophisticated ones where non-ideal, multicomponent liquid aqueous solutions are taken into account, are able to predict IIF as a function of cooling rate, ice seeding temperature, initial CPA concentration, and CPA type. However, in all these modelling studies ice formation is either accounted for as taking place within a single representative (i.e. average) cell, or the PIIF is related to the nucleation rate by assuming sporadic nucleation of identical cells [40]. In particular, the PIIF adopted in all these models is defined and strictly valid only for the case of a crystal growth time negligible with respect to the nucleation time [38], i.e. it is assumed that a single ice nucleus per cell is formed and ice growth is extremely fast. All these represent relevant limitations for interpreting the behaviour of the system under practicable operating conditions during a standard cryopreservation process. Indeed, it is difficult to accept through reasoning based on physical grounds that, in the relatively high number of cell population actually suspended during a standard cryopreservation protocols, the cells are truly identical. Moreover, it is difficult to accept that different temperatures of IIF for a high number of identical cells subjected to the same freezing cycle may be ascribed to a nucleation process stochastic in nature. In our opinion a deterministic approach should be adopted, instead, and a more general definition of PIIF (taking into account not only ice nucleation but also ice growth and independently from the number of nuclei formed inside a cell) needs to be provided, given that the comparisons between theoretical results and experimental data are classically carried out in terms of PIIF. On the other hand, as already observed in the literature by Mazur almost 50 years ago [24], large cells are characterised by smaller surface-to-volume ratio than small ones and, therefore, are expected to loose less water during the cooling stage. As a consequence, at any given cooling rate, larger cells retain a higher percentage of internal water than smaller ones, so that super-cooled conditions are reached earlier, i.e. at higher temperatures. Following these considerations, we have recently proposed a novel theoretical interpretation of the PIIF for a suspension of cells by addressing the effect of its size distribution during the cooling stage of a standard cryopreservation protocol [9, 10, 11]. The present Chapter reviews this new modelling approach and its main implications in the cryopreservation field.

First, model equations are provided in the Model Section. For better highlighting our contribution to the field, the model equations are reported by starting from the traditional description of a single cell response when cryopreserved in the presence or absence of a CPA. Then, the single cell model is embedded in the Population Balance Modelling (PBM) approach adopted to properly take into account the effect of a size-distributed cell population. This naturally leads to a novel formulation of the PIIF in line with the proposed modelling framework where statistical variations are neglected. The mathematical description of Extracellular Ice Formation (EIF) dynamics through PBM in order to remove the classical simplifying assumption of thermodynamic equilibrium conditions for the extra-cellular compartment is also reported. Model equations are described by highlighting the coupling between intra- and extra-cellular compartments and the mutual effects between all the physico–chemical phenomena that might be progressively taken into account (i.e. water osmosis, CPA permeation, IIF and EIF dynamics through ice nucleation and crystal growth).

Later, the experimental validation of the novel modelling approach is provided in the Results and Discussion Section, by directly comparing the theoretical results to suitable experimental data taken from the literature, where CPA was not used and EIF dynamics may be neglected. A two-step experimental validation is performed in order to test model reliability more extensively. First the adjustable parameters of the proposed model are tuned through fitting procedures (i.e. regression analysis). Then, system behaviour measured under operating conditions different from those used during the fitting is predicted by keeping constant the values of the model parameters adjusted before. The novel formulation of PIIF appears to be far more consistent with the experimental measurements than the previous theoretical interpretations. As expected, it is found that IIF temperature depends on the cell size, i.e. it is higher for larger cells and smaller cells are less prone to IIF injury. Correspondingly, the PIIF results to be dependent on the initial size distribution of the cell population and should not be related to statistical variations. On the basis of this successful two-step experimental validation we felt safe to carry out model investigations on system behaviour when CPA is used and EIF dynamics is taken into account. More specifically, the effect of the cell size distribution when varying cooling rate and CPA concentration within practicable values for a standard cryopreservation protocol is investigated. On the other hand, experimental data related to these specific cases and suitable for a proper comparison with our novel modelling approach are not available in the literature. Thus, even if only from a theoretical perspective, the effect of removing the limiting assumption of thermodynamic equilibrium in the extra-cellular compartment may be carefully analysed along with the sequence of CPA addition and cooling stages experimentally adopted. In this regard, we bridged the gap between the modelling of the two subsequent stages (i.e. CPA equilibration and cooling) that were independently studied so far in the literature, thus reaching a more comprehensive description of system behaviour. The possibility to predict the effect of specific combinations of operating conditions (i.e. cooling rate and CPA content) on cell survival confirmed that IIF may be lethal for a fraction of the cell population (i.e. larger size classes) whilst it may not reach a dangerous level for the intermediate size class cells and it will not even take place for the smaller ones. EIF dynamics may play a role in this context, depending on the used cooling rate. In addition, an original and physically sound explanation for several, well-known experimental evidences, which appeared in a number of articles available in the literature of cell cryopreservation in presence of CPA, was comprehensively reached.

Finally, in the concluding section, the general outputs of the novel modelling approach are summarised and the future directions are outlined by addressing the limitations still present in the proposed model and the corresponding potential improvements.

2 Model Section

2.1 Cell Model

In the framework of the mathematical modelling of cell response to cryopreservation, the behaviour of a single cell is usually described through the "saltwater sack" model [45], whose schematic representation is given in Fig. 1.

Basically, the cell is seen as a spherical drop of a salty (NaCl) aqueous solution representing the cytoplasm, where proteins, organelles, and other macromolecules are suspended. For cryopreservation all these materials suspended in the cytoplasm are actually inactive. As such, they are lumped together in the *inactive* cell fraction which remains constant during cryopreservation, depending only on the specific level of maturation achieved by the cell along its mitotic life-cycle. In this regard,

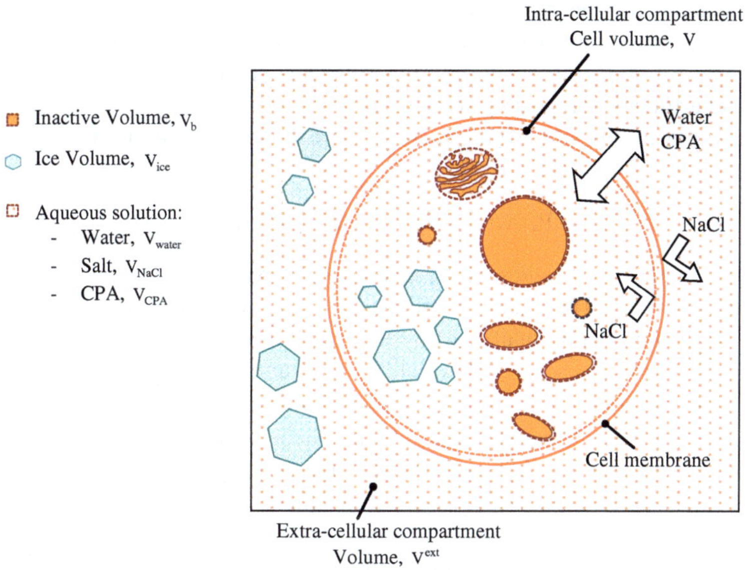

Fig. 1 Schematic representation of the cell model

differently sized cells belonging to the same lineage are assumed to be constituted by the same inactive cell fraction, which is then considered a characteristic feature of the cellular lineage.

This cell structure is separated from the exterior environment by an oil semi-permeable membrane that rules the movement of any molecule between the intra- and the extra-cellular compartments composing the system. Cell membrane is characterised by a low resistance to water transport, whilst the permeability to a specific solute depends on its size, electric charge and the features of hydrogen bonding [29]. Traditionally, in cryopreservation the sodium chloride is assumed to be not permeant through the membrane. Thus, the content of NaCl does not change in both the intra- and the extra-cellular compartments. On the contrary, a permeant CPA like glycerol or DMSO may be transferred from the external medium to the interior of the cells and vice versa, so that its content in the two compartments may change. The water and solutes permeabilities to membrane are considered characteristic features of a cellular lineage just like the inactive cell fraction.

Apart from osmosis, during the cooling stage of cryopreservation liquid water may turn into ice. This phase change takes place inside both the intra- and the extra-cellular compartment, separately. Thus, the water content in any compartment does not change due to ice formation but simply splits between the liquid and solid fractions therein.

For simplicity, a negligible difference between ice and liquid water densities is typically assumed. In addition, the densities of water, salt, CPA and inactive materials are considered constant with respect to temperature variations, so that a mass balance may be replaced by a volume balance. Finally, an ideal perfect mixing is assumed for both the intra- and the extra-cellular compartment, so that a single concentration value may be defined for any chemical species present in the two compartments composing the system.

According to this simplifying picture, the following equation may be adopted for the model of a single cell:

$$V(t) = V_{water}(t) + V_b + V_{\mathrm{NaCl}} + V_{ice}(t) + V_{CPA}(t) \tag{1}$$

where V_b and $V_{water}(t)$ represent the inactive cell volume and the volume of liquid water, respectively. The total cell volume $V(t)$ may vary due to water osmosis and CPA permeation, being unaffected by the water phase change from liquid to ice (i.e. $\Delta V_{water} = \Delta V_{water}|_{osmosis} + \Delta V_{water}|_{phase\ change}$ and $\Delta V_{water}|_{phase\ change} = -\Delta V_{ice}$).

Typically, during cryopreservation one million of cells is suspended in 1 ml of aqueous solution inside suitable vials. Assuming an average diameter of 10 μm for each cell, a maximum total cytoplasmic volume of 5.2×10^{-4} ml is obtained. It is apparent that the capacity of the suspending aqueous solution is much higher than the cumulative cytoplasm capacity of the suspended population of cells. Thus, the mass transfer with the cells through cellular membrane (i.e. water osmosis and CPA permeation) may affect only negligibly the volume and the concentrations of the extra-cellular compartment. Then, by contrast with the cell model given in Eq. 1, the behaviour of the extra-cellular compartment may be described as follows:

$$V^{ext} = V^{ext}_{water}(t) + V^{ext}_{\mathrm{NaCl}} + V^{ext}_{ice}(t) + V^{ext}_{CPA} \tag{2}$$

where the total extra-cellular volume does not change with time given that $\Delta V^{ext}_{water} = \Delta V^{ext}_{water}\big|_{phase\ change}$ and $\Delta V^{ext}_{water}\big|_{phase\ change} = -\Delta V^{ext}_{ice}$.

2.2 Temperature Dynamics

Traditionally in cryopreservation, temperature is assumed homogenous throughout the entire system, i.e. intra- and extra-cellular spatial temperature gradients are neglected. This is a relevant simplification from modelling perspective, since it permits to drop the simulation of the thermal behaviour of the system thus avoiding to couple a thermal balance to the mass one represented by Eqs. 1 and 2. Actually, the validity of this assumption depends on the geometries, sizes and operating conditions specifically adopted. Certainly, the standard use from experimental perspective of thin vials in large capacity cooling chambers justifies this simplifying assumption of the theoretical description.

According to these considerations, the modelling of cryopreservation processes typically assigns a specific profile to the decreasing temperature of the entire system, regardless of any thermal accumulation, latent heat and heat transfer resistance. Actually, in any standard lab the biological material is cryopreserved by simply storing it inside a freezer operating at a fixed low temperature. In such a case, system temperature is left free to vary downwards starting from an initial value (T_{init}), where the cell suspension was prepared. However, in order to optimise the cryopreservation of cells by limiting intra-cellular ice formation, system temperature needs to be controlled. This is one of the results reached by theoretical cryobiologists in the last 50 years as briefly described in the Introduction Section. Thus, the modelling of cryopreservation processes typically adopts a linear profile obtained with a constant cooling rate B as follows:

$$T(t) = \begin{cases} T_{init} & 0 \leq t < t_{eq} \\ T_{init} - B \cdot \left(t - t_{eq}\right) & t \geq t_{eq} \end{cases} \quad (3)$$

In Eq. 3, the general case of using a permeant CPA is accounted for. Indeed, the equilibration stage of CPA entering the cells from the extra-cellular medium is performed before the system is cooled down. Due to its relatively slow permeability, CPA equilibration between intra- and extra-cellular compartments requires a temporal interval, t_{eq}, and it is typically performed at a fixed temperature in a one step fashion.

2.3 Cell Volume Variation: Water Osmosis and CPA Permeation

According to Eq. 1, when CPA is not used, cell volume changes only in response to water osmosis. The rate at which water leaves or enters a cell is proportional to a driving force given by the difference between internal and external osmotic pressures Π [15, 24].

For simplicity, in cryopreservation the solutions of the intra- and extra-cellular compartments are traditionally assumed ideal and dilute, so that the osmotic pressure is proportional to the concentration of the dissolved salt expressed in terms of osmolality m_{NaCl}:

$$\Pi = \Re \cdot T \cdot m_{\mathrm{NaCl}} \quad (4)$$

According to this simplifying assumption, cell volume varies as follows

$$\frac{dV}{dt} = \left.\frac{dV_{water}}{dt}\right|_{osmosis} = L_p(t) \cdot A(t) \cdot \Re \cdot T \cdot \left(m_{NaCl}^{\mathrm{int}}(t) - m_{NaCl}^{ext}(t)\right) \quad (5)$$

where $L_p(t)$ represents the effective membrane permeability to water, whilst $A(t)$ is the cellular membrane surface area that varies with time as the cell shrinks or swells. The higher the temperature, the higher the hydraulic permeability, according to the classic Arrhenius-like dependence [23]:

$$L_p(t) = L_{p,ref} \cdot \exp\left(-\frac{E}{\Re}\left(\frac{1}{T(t)} - \frac{1}{T_{ref}}\right)\right), \tag{6}$$

where E is the apparent activation energy, and the pre-exponential factor $L_{p,ref}$ represents the value of hydraulic permeability at a given, reference temperature T_{ref}.

When a permeant CPA is used, cell volume changes in response not only to water osmosis but also to CPA permeation. In this case, a rather more complex description results, since water osmosis and CPA permeation are coupled phenomena from the material transport perspective. More specifically, if water and CPA are co-transported (i.e. water and CPA molecules are transferred through cell membrane by travelling along the same transport channels) the Kedem and Katchalsky formalism [21] may be adopted to simulate cell volume variation:

$$\begin{aligned}\frac{dV}{dt} &= \left.\frac{dV_{water}}{dt}\right|_{osmosis} + \frac{dV_{CPA}}{dt} \\ &= L_p(t) \cdot A(t) \cdot \Re \cdot T \cdot \left[\left(m_{\mathrm{NaCl}}^{\mathrm{int}}(t) - m_{\mathrm{NaCl}}^{ext}(t)\right) + \sigma \cdot \left(m_{CPA}^{\mathrm{int}}(t) - m_{CPA}^{ext}(t)\right)\right]\end{aligned} \tag{7}$$

Here σ (called the *reflection coefficient*) is the adjustable parameter that takes into account the interactions between water and CPA transport across cellular membrane.

Equation 5 or 7 are used to determine cell volume $V(t)$ in presence or absence of CPA, respectively. To this aim, the concentrations of NaCl and CPA in the intra- and the extra-cellular terms of the driving forces appearing in Eq. 5 or 7 have to be determined. These concentrations are calculated as inversely proportional to the liquid water volumes, i.e. $V_{water}(t)$ and $V_{water}^{ext}(t)$, correspondingly, which are determined from Eqs. 1 and 2. Of course, NaCl and CPA concentrations are also proportional to their corresponding contents (i.e. osmoles) in the intra- and the extra-cellular compartments. In this regard, whilst the content in both compartments of the not-permeant solute NaCl is constantly equal to a given initial value (i.e. the so-called isotonic condition), the intra-cellular CPA content may change due to permeation and, according to Kedem and Katchalsky formalism, it is calculated as:

$$\begin{aligned}\frac{dn_{CPA}}{dt} &= (1-\sigma) \cdot \left(\frac{m_{CPA}^{\mathrm{int}}(t) + m_{CPA}^{ext}(t)}{2}\right) \cdot \frac{dV}{dt} \\ &\quad + P_{CPA}(t) \cdot A(t) \cdot \left(m_{CPA}^{ext}(t) - m_{CPA}^{\mathrm{int}}(t)\right)\end{aligned} \tag{8}$$

where n_{CPA} represents the number of CPA moles inside the cell, whilst P_{CPA} is the permeability of CPA that shows an Arrhenius-like temperature dependence as the hydraulic permeability L_p reported in Eq. 6.

At this point it is worth noting that, some assumptions detailed above cannot be made and should be actually avoided, even though they are widely used in the technical literature of cryopreservation modelling. Among these, the van't Hoff equation that describe the osmotic pressure in analogy with the ideal-gas equation (see Eq. 4) is not valid when cooling at low temperatures, where very concentrated solutions may be obtained due to liquid water sequestering for ice formation. In this latter case, the concentrations should be replaced by activities to express the osmotic driving forces of Eqs. 5–7, thus accounting for non-ideal and/or concentrated liquid mixtures through the Debye–Huckel theory [36] or the osmotic virial equation [8].

An additional assumption that may not be justified is represented by water and CPA co-transport across cellular membrane, i.e. the Kedem and Katchalsky formalism given in Eqs. 7 and 8. Indeed, in natural biological membranes co-transport is often unlikely and/or negligible, since transport channels are highly selective and stereo-specific [20, 22, 26]. In such a case, the two-parameter formalism reported by Kleinhans [22] may be adopted thus making the experimental validation of the model easier by reducing the number of adjustable parameters (the reflection coefficient σ is abandoned).

2.4 Intra-Cellular Ice Formation

In order to evaluate the volume of liquid water in a cell $V_{water}(t)$ through Eq. 1, an independent equation is required to determine the intra-cellular volume of ice $V_{ice}(t)$. This task is accomplished by modelling IIF dynamics. To this aim, a mechanistic scenario of nucleation and diffusion-limited growth of ice crystals is typically adopted.

In this context, if cryopreservation is carried out following standard protocols (i.e. slow cooling and low CPA concentrations), ice nucleation may take place only at relatively high temperatures, when crystal growth is relatively fast. Therefore, the rapid ice crystal growth in association with a small control volume (i.e. cell volume) generate only a very limited number of ice crystals inside a cell (typically no more than one). On the other hand, when cooling down at higher rates IIF is hindered kinetically, so that the intra-cellular liquid solution eventually reaches vitrification at low temperatures. This aspect, clearly prevents the application of a continuum modelling approach (i.e. PBM) to simulate ice formation inside a cell.

As a consequence, the intra-cellular ice volume $V_{ice}(t)$ appearing in Eq. 1 has to be determined necessarily through a discrete modelling approach, by following the fate of any single ice nucleus that is first formed, and then grows as time increases [19]:

$$V_{ice}(t) = \begin{cases} 0 & \text{if } N_{ice}(t) = 0 \\ \sum_{i=1}^{N_{ice}(t)} \frac{4\pi}{3}[r_i(t)]^3 & \text{if } N_{ice}(t) \geq 1 \end{cases}, \tag{9}$$

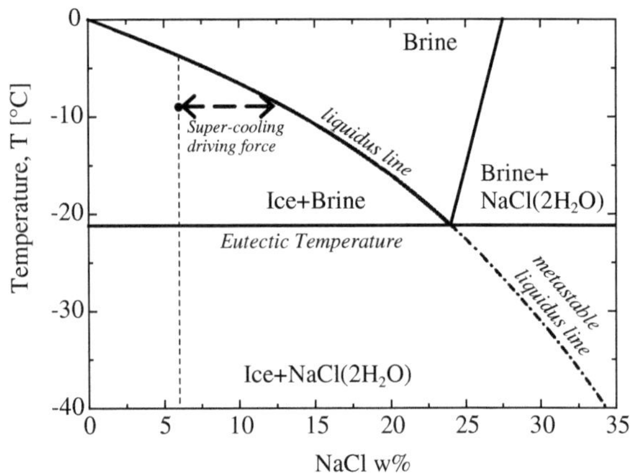

Fig. 2 Binary NaCl/water phase diagram

where the number of spherical ice crystals $N_{ice}(t)$ is determined by the nucleation rate, B_0, as

$$\frac{dN_{ice}}{dt} = B_0(t); \quad N_{ice} = 0 \quad \text{at} \quad t = 0. \tag{10}$$

Following the classical nucleation theory [6] B_0 is proportional to the liquid volume inside a cell as it follows [9]:

$$B_0(t) = J(t) \cdot (V(t) - V_b - V_{ice}(t)), \tag{11}$$

where the specific nucleation rate J depends on the temperature and the level of super-cooling in the liquid phase (driving force).

The temporal evolution of the radius of the ith ice crystal, $r_i(t)$ appearing in Eq. 9 is evaluated through the classical diffusion-controlled growth law:

$$\begin{cases} r_i(t) = 0 \quad \text{at} \quad t < t_i \quad \text{when} \quad N_{ice}(t) = 0 \\ \dfrac{dr_i}{dt} = \bar{D}(t) \cdot \dfrac{\Omega_g(t)}{r_i}; \quad r_i(t) = r^*(t) \quad \text{at} \quad t = t_i \quad \forall i \in [1; N_{ice}(t)] \end{cases}, \tag{12}$$

where t_i is the birth instant of the i-th ice nucleus which starts growing from the actual initial critical radius $r^*(t_i)$, whilst $\Omega_g(t)$ is the driving force of ice crystals growth, and $\bar{D}$ represents the effective diffusivity of water.

Clearly, the driving forces of ice nucleation and growth are determined as departures from the thermodynamic equilibrium conditions evaluated on the *liquidus line* of the corresponding phase diagram. In this context it is worth noting that, the liquidus line needs to be extended under the eutectic point (thus considering a metastable liquidus line) where ice nucleation and growth may still take place under-non-equilibrium conditions (see Fig. 2).

The model of IIF just described is valid both for binary and multicomponent solutions, i.e. when CPA is absent or not, respectively. The only difference is represented by the phase diagram adopted to evaluate the driving forces of nucleation and diffusion-limited growth of ice crystals (i.e. a binary phase diagram for the water/NaCl system, and a ternary phase diagram for the water/NaCl/CPA system).

2.5 Extra-Cellular Ice Formation

In analogy to the intra-cellular compartment, even for the extra-cellular one an independent equation is required to determine the extra-cellular volume of ice $V_{ice}^{ext}(t)$ when evaluating $V_{water}^{ext}(t)$ from Eq. 2. This task is accomplished by modelling EIF.

In contrast with the intra-cellular compartment, EIF was traditionally described by assuming that liquid solution and ice are constantly in thermodynamic equilibrium conditions, thus neglecting the dynamics of EIF. The different assumptions between intra- and extra-cellular compartments to describe the same physico-chemical phenomenon of ice formation may be ascribed to the difference in the volume capacities of the two compartments used in a standard cryopreservation protocol. This reasoning has been already adopted above to justify that water osmosis and CPA permeation cannot significantly change the extra-cellular solution composition and volume. Thus, the relatively higher capacity of the extra-cellular solution is also responsible for a faster ice formation with respect to the intra-cellular one [5, 43]. In fact, according to the classical nucleation theory, ice nucleation rate increases if water volume increases (see Eq. 11, due to the statistically more frequent nucleation impact between water molecules. Therefore, for the sake of model simplicity, traditionally in cryopreservation it was implicitly assumed that the time scale of nucleation is much smaller in the extra-cellular solution than inside the cells, and, accordingly, the corresponding ice formation dynamics is so fast to practically reach equilibrium condition instantaneously. In this case, Eq. 2 may not be used since all the variables necessary to describe water osmosis and CPA permeation through Eqs. 5 and 7 (i.e. NaCl and/or CPA concentrations in the extra-cellular compartment) may be directly obtained from the corresponding phase diagram.

Actually, the assumption of thermodynamic equilibrium conditions should be regarded as a model simplification, since any natural phenomenon is characterised by a dynamics. If the time scales of the different phenomena involved in a process are relatively different, this model simplification is reasonable. However, this is not always the case, since the operating conditions may be changed, and the time scales change accordingly. Thus, the assumption of equilibrium conditions needs to be verified on a case by case basis.

On the other hand, there is experimental evidence that the temperature of EIF plays an important role in optimising post-thaw viability of the cells [30]. For example, different kinetics of IIF at slow cooling rates was found depending on whether or not ice-crystals were seeded in the external solution [42]. Moreover, the kinetics of IIF was found to strongly depend on the specific temperature at which the ice was seeded [32]. As explained by Petersen et al. [32], if the temperature at which EIF sets in is reduced, cell dehydration decreases and IIF increases similar to the case of using higher cooling rates. Needless to say that, generally speaking, EIF does not take place at a single temperature but instead it involves a range of temperatures, starting from an initial value. The latter one along with the size of the temperature range involved actually depend on EIF dynamics and the adopted cooling rate, unless EIF is triggered by seeding ice on purpose at a specific temperature as commonly done in the cryo-microscopic analysis addressed in the papers cited just above. Thus, not only the prediction of a single temperature, but the more general description of the EIF dynamics is necessary when aiming to develop a reliable and comprehensive modelling of the cryopreservation process.

When removing the restrictive assumption of thermodynamic equilibrium thus taking into account EIF dynamics, the classic nucleation theory and diffusion-controlled growth may be adopted, thus following the same constitutive laws used above to simulate IIF. The only difference between the two compartments is that the extra-cellular control volume is large enough to reach a sufficient number of ice crystals, i.e. a population of ice crystals, thus justifying the continuum-modelling approach through a PBM that we adopted [11]. On the basis of these considerations, $V_{ice}^{ext}(t)$ is calculated as being proportional to the third-order moment of the extra-cellular ice crystal size distribution $n_{ice}^{ext}(r;t)$:

$$V_{ice}^{ext}(t) = \int_{0}^{+\infty} \frac{4}{3} \cdot \pi \cdot r^3 \cdot n_{ice}^{ext}(r;t)dr. \tag{13}$$

The distribution $n_{ice}^{ext}(r;t)$ represents a number density distribution (i.e. $n_{ice}^{ext}(r;t) \cdot dr$ is the number of extra-cellular ice crystals that at time t have a radius between r and $r + dr$). The dynamics of $n_{ice}^{ext}(r;t)$ is described by the following 1-D PB equation and the corresponding initial and boundary conditions:

$$\begin{cases} \dfrac{\partial n_{ice}^{ext}(r;t)}{\partial t} + \dfrac{\partial}{\partial r}\left[G^{ext}(r) \cdot n_{ice}^{ext}(r;t)\right] = 0 \\ n_{ice}^{ext}(r;t) = 0 \quad \text{at } t = 0 \quad \forall r \in \left[r^{*,ext}; +\infty\right). \\ n_{ice}^{ext}(r^{*,ext};t) = \dfrac{B_0^{ext}(t)}{G^{ext}(r^{*,ext})} \quad \text{at } r = r^{*,ext} \quad \forall t > 0 \end{cases} \tag{14}$$

The quantities $B_0^{ext}(t)$ and $G^{ext}(r)$ are the ice crystal's nucleation and growth rates, respectively. In the initial condition reported above it is assumed that ice crystals are not present in the extra-cellular compartment when cooling starts at $t = 0$. Later, they are formed and increase their size starting from the critical radius $r^{*,ext}$. The value of the boundary condition at the lower end of the domain $n_{ice}^{ext}(r^{*,ext};t)$ depends on both the nucleation and growth rates [35].

2.6 The Dynamics of a Size-Distributed Population of Cells

The equations reported above for cell volume variation, IIF dynamics and EIF at thermodynamic equilibrium conditions have been proposed in the literature during the past decades. This set of equations allowed researchers to predict the ice formation inside the intra-cellular compartment as a function of different operating conditions as cooling rate, seeding temperature, initial CPA concentration and CPA type, but always taking into account a single, average cell size, i.e. addressing only a mono-sized population of cells. In this context, the striking experimental evidence that inside a system at homogeneous temperature identical cells show different temperatures of IIF even if subjected to the same freezing cycle, was explained by assuming that the nucleation process is stochastic in nature. Accordingly, the PIIF was related to the nucleation rate by assuming a sporadic nucleation of identical cells [38, 40].

In his pioneering work on osmotic behaviour of cells during cryopreservation, Mazur observed that, large cells are characterised by a smaller surface-to-volume ratio than small ones, and, therefore, are expected to loose less water during the cooling stage [24]. Consequently, at any given cooling rate, larger cells retain an higher percentage of internal water than smaller ones, so that super-cooled conditions are reached earlier, i.e. at higher temperatures. Even though this aspect was already highlighted almost 50 years ago, it has never been accounted for when modelling cryopreservation until the authors recognised that a population of cells of the same lineage is always distributed in size [9]. This is mainly due to the mitotic cycle progression rather than the biological diversity. According to this picture, a population of differently sized cells subjected to the same freezing protocol may exhibit different IIF temperatures. Specifically, one would expect that the IIF temperature of large cells is higher with respect to that one of small cells. In other words, PIIF should be related to the initial size distribution of a cell population, and the sporadic nucleation of identical cells should be discarded in favour of a deterministic criterion.

To this aim, a population of cells characterised by a size distribution needs to be considered and the dynamics of this distribution during the cryopreservation protocol needs to be quantitatively described. In the framework of a continuum modelling approach, this task may be accomplished by means of the following PBM:

$$\begin{cases} \frac{\partial n(V;t)}{\partial t} + \frac{\partial [G_v(V) \cdot n(V;t)]}{\partial V} = 0 & \\ n(V;t) = n^0(V) & \text{at } t = 0;\ \forall V \in [0, +\infty] \\ n(V;t) = n(\infty;t) = 0 & \forall t \end{cases}, \tag{15}$$

where $n(V; t)$ represents the cell number density distribution (i.e. $n(V; t)\ dV$ is the number of cells that at time t possess a volume between V and $V + dV$), whilst $n^0(V)$ refers to the initial size distribution of the cell population at isotonic conditions. The PBM formulation inherently ensures a constant number of cells N_{tot}

throughout the process (i.e. $\int_0^{+\infty} n^0(V) \cdot dV = \int_0^{+\infty} n(V;t) \cdot dV = N_{tot}$). $G_v(V)$ appearing in Eq. 15 is the rate of change of the cell volume due to water osmosis and CPA permeation. A negative growth rate $G_v(V)$ shifts the cell number density distribution $n(V;t)$ towards smaller volumes, thus simulating cells shrinkage in a hypertonic environment. In reverse, cell volume expansion in a hypotonic extra-cellular solution is reflected by a positive $G_v(V)$.

The rate of cell volume variation $G_v(V)$ may be evaluated by means of Eq. 5 for the case of CPA absence or Eq. 7 when a CPA is used. Actually, all the equations reported above for the case of a single, average cell may be adopted when addressing a population of cells characterised by a size distribution, paying attention to make proper distinctions. Basically, any single sized class of cells composing the distribution of the cell population behaves independently by following the same equations given above for the simulation of water osmosis, CPA permeation and the discrete IIF approach. As such, any single sized class of cells is characterised at any time during cryopreservation by its own intra-cellular water content, NaCl and CPA concentrations. Even the IIF and the extent of water phase change may differ from class to class of cells, whilst all of them communicate through water osmosis and CPA permeation with the same extra-cellular compartment which is characterised by a single, homogenous value of chemical species concentrations and EIF extent.

2.7 Probability of Intra-Cellular Ice Formation

Experimentally, the temperature at which IIF occurs at given cooling conditions is commonly determined looking at cells under a microscope from the darkening of the cell due to light scattering of the small ice crystals that form inside. The number of iced-up cells is then evaluated using cryomicroscopy, and the corresponding PIIF is computed at each temperature level as the fraction of cells (with respect to the total initial ones) cumulatively frozen at that temperature.

When accounting for a size-distributed population of cells, the model of sporadic nucleation of identical cells used so far in the literature to define the PIIF has to be discarded in favour of a deterministic criterion that takes into account the different extent of IIF from class to class of cells, i.e. the distribution of IIF [9, 10]. To this aim, it is assumed that cells are iced-up when their corresponding internal ice volume fraction $\eta_{ice}(t) = \frac{V_{ice}(t)}{V(t)-V_b}$ reaches the value of 0.5. Accordingly, the PIIF may be defined as:

$$PIIF(t) = \frac{\int_0^{+\infty} n(V;t)|_{\eta_{ice} \geq 0.5}\, dV}{N_{tot}^0}. \tag{16}$$

Besides, a detectable value for the probability of internal ice formation may be defined as:

$$PIIF_{detect}(t) = \frac{\int_0^{+\infty} n(V;t)|_{\substack{\eta_{ice} \geq 50\%; \\ \eta_{water} \geq 10\%}} dV}{N_{tot}^0}, \tag{17}$$

where $\eta_{water}(t) = \frac{V_{ice}(t)+V_{water}(t)}{V(t)}$ is the fraction of water (solid plus liquid) inside any single size class of the cell population. This way, iced-up cells are assumed to be experimentally detectable by human eye looking at a microscope if the ice volume fraction reaches the value of 0.5 (i.e. $\eta_{ice} \geq 0.5$), whilst a relatively small content of intra-cellular liquid water equal to 10% vol. is maintained (i.e. $\eta_{water} \geq 0.1$). Accordingly, the $PIIF_{detect}$ is evaluated at each temperature level as the fraction of frozen cells at that temperature containing a minimum amount of water (threshold). The choice of a specific threshold value for η_{water} in order to detect ice formation is quite reasonable. Indeed, the volumetric content of total water ($V_{water} + V_{ice}$) inside the cells may be relatively small with respect to the inactive volume V_b and/or CPA volume V_{CPA} in Eq. 1. Therefore, ice formation may not be experimentally detectable through the darkening/flashing of the cell commonly used in cryomicroscopy, even if the liquid water contained in the cell is completely turned into ice (i.e. $\eta_{ice} \to 1$ and $V_{water} \to 0$). On the basis of the chosen thresholds for η_{ice} and η_{water}, in general the size distribution of the cell population may be divided in three adjacent portions, for which different fates are assumed: unfrozen cells characterised by $\eta_{ice} < 0.5$; iced-up cells with a small, innocuous amount of water when $\eta_{ice} \geq 0.5$ and $\eta_{water} < 0.1$; iced-up cells with a large, lethal amount of water when $\eta_{ice} \geq 0.5$ and $\eta_{water} \geq 0.1$.

Alternatively, as very recently proposed in the literature [37], IIF may be measured by means of a high-speed (8000 fps) video cryomicroscopy system. This way, cells at the initial stage of ice formation (before flashing takes place) may be numbered and a cumulative incidence of initial IIF ($PIIF_{init}$) may be experimentally determined. This new cumulative fraction of cells with an initial ice formation (i.e. when η_{ice} becomes greater than zero, $\eta_{ice} > 0$) is defined as:

$$PIIF_{init}(t) = \frac{\int_0^{+\infty} n(V;t)|_{\eta_{ice} > 0} dV}{N_{tot}^0} \tag{18}$$

3 Results and Discussion

3.1 Experimental Validation of the Model

The most relevant aspect of the mathematical modelling of cryopreservation based on PBM is that the different IIF temperatures experimentally evidenced for a pool of a given cell lineage is not ascribed to statistical variations but is explained through a deterministic criterion: the size distribution of the cell population.

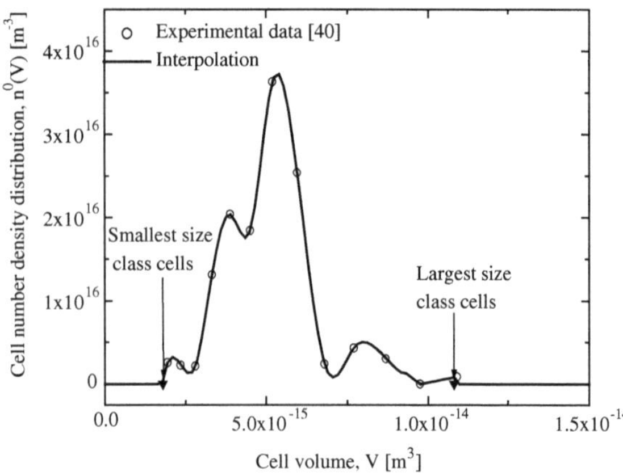

Fig. 3 Initial condition for Eq. 15: experimental and interpolated cell number density distribution as a function of cell volume (adapted from [9])

This hypothesis represents a clear novelty in the scientific field of cryopreservation modelling which has been addressing a population of identically sized cells in the last 50 years.

The first step taken towards the experimental validation of the new theory is the direct comparison between theoretical results and measured data available in the literature, thus tuning some adjustable parameters through a regression analysis, i.e. fitting procedure [9]. Actually, this revealed to be not a straightforward step, since the experimental data to be used have to be referred necessarily to a population of cells. This means that, the behaviour of a relatively high number of cells needs to be followed simultaneously under a cryomicroscope. In addition, the corresponding size distribution of the cell population under investigation needs to be measured and accurately reported. On the contrary, the experimental PIIF is typically measured through cryomicroscopic analysis by observing the behaviour of a restricted number of cells. Moreover, the size distribution is typically not reported in the literature since the model of sporadic nucleation traditionally adopted to interpret the system behaviour accounts only for the average cell size considered as representative of the entire population. On the other hand, at least the measure of the initial, isotonic size distribution of the cell population subjected to cryopreservation is needed to validate the novel theory since it represents the initial condition of the PBM ($n^0(V)$ in Eq. 15).

Fortunately, Toner et al. [40] published their experimental measurements of the PIIF of a suspension (i.e. relatively high number) of isolated rat hepatocytes subjected to cryopreservation in absence of CPA at different cooling rates. In this paper, the corresponding size distribution in terms of number frequency vs cell diameter at isotonic conditions was also reported. Then, the initial, isotonic condition $n^0(V)$ of the PBM was derived from the experimental histogram distribution of the isolated rat hepatocytes available in the literature. It is reported in Fig. 3 in terms of cell number density distribution as a function of cell volume. As it can be seen, all cells fall in the range of volumes between a smallest size class and a

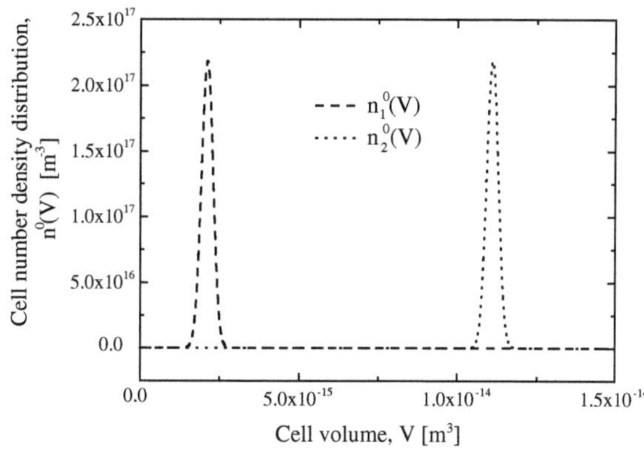

Fig. 4 Initial condition for Eq. 15: cell number density distributions of fictitious populations of small and large cells (adapted from [9])

largest size class. In order to better highlight the effect of a cell size distribution, two additional fictitious, narrow, initial cell number density distributions, $n_1^0(V)$ and $n_2^0(V)$, are considered. These distributions, reported in Fig. 4, are characterised by mean values which correspond to the smallest and the largest size class cells of the experimental distribution given in Fig. 3, respectively.

The complete model describing trans-membrane transport phenomena along with IIF in a cell population involves a large number of parameters. Many of them may be taken from the literature, whilst only five of them related to osmosis and ice nucleation have to be determined through a fitting procedure. Ideally, they should be fitted independently, using specifically designed experimental runs, to reduce ambiguity whilst preserving model consistency: first the osmotic behaviour should be determined whilst neglecting ice nucleation, then IIF may be safely analysed. Unfortunately, the set of experimental runs necessary for this ideal fitting procedure is not available. Therefore, all the five parameters have been determined simultaneously by direct comparison between model results and experimental data in terms of PIIF of 60 isolated rat hepatocytes cooled at −400°C/min in a cryo-microscopy system [40] as reported in Fig. 5. As it can be seen, the PBM approach is able to simulate the completion of IIF at −400°C/min in the temperature range [-15 °C to -5 °C] when the initial cell size distribution reported in Fig. 3 is taken into account.

On the contrary, by adopting the two fictitious initial distributions reported in Fig. 4, whilst keeping constant all the other parameters and operative conditions of the experimental run, the PIIF vs system temperature simulated by the proposed model is completely different. Indeed, the model predicts no IIF down to −16°C for the case of a population of large cells, $n_1^0(V)$, whilst an abrupt increase of PIIF takes place around −5.5°C, for the case of a population of small cells, $n_2^0(V)$.

The different behaviour of cells according to their size is shown from a different perspective in Fig. 6. Here the path of the intra-cellular sodium chloride concentrations for the smallest and the largest size class cells of the initial population

Fig. 5 Comparison between experimental and model results in terms of PIIF when CPA is absent at a cooling rate of −400°C/min (adapted from [9])

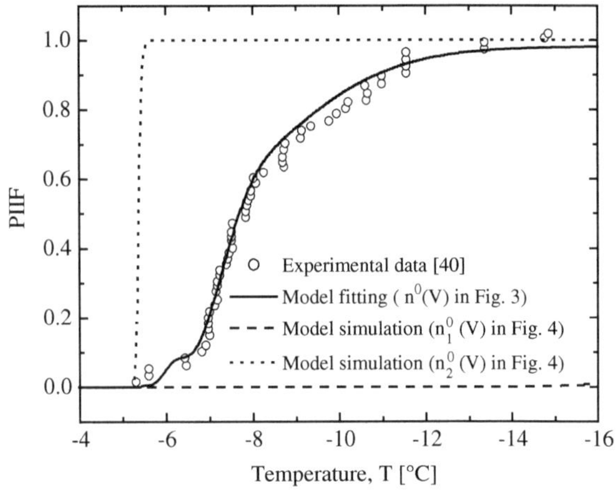

Fig. 6 NaCl concentration path of the smallest and the largest classes of the cell population on the phase diagram when CPA is absent at a cooling rate of −400°C/min (adapted from [10]

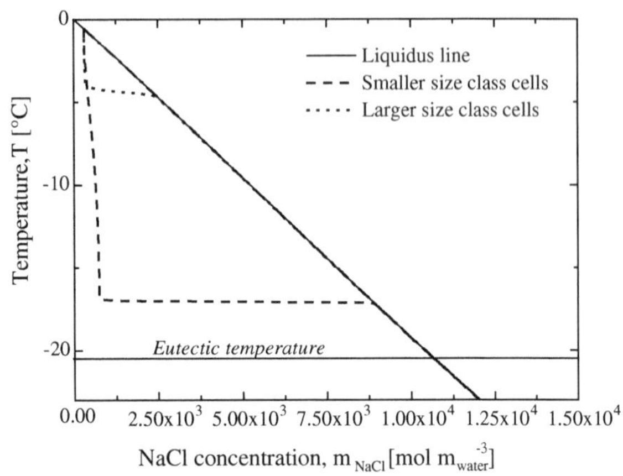

$n^0(V)$ is depicted on the relevant phase diagram as a function of the system temperature, during the cooling stage at −400°C/min. It is worth noting that, the occurrence of IIF may be recognised from the abrupt slope change of the path, towards higher salt concentration values. It may be seen that, largest and smallest size class cells follow different paths. Ice forms inside the largest size class cells at about −5°C, then the intra-cellular salt concentration rapidly increases, thus reaching a final intra-cellular salt concentration higher than that one of the smallest size class cells. Once IIF takes place, the computed sodium chloride concentration profile significantly overlaps with the liquidus line. The smallest size class cells undergo the same fate, but at quite lower temperature (≈ -17°C).

Thus, differently sized cells in a population exhibit a different IIF temperature under the same operative conditions. Specifically, the temperature of IIF is higher for the larger cells and, correspondingly, the smaller cells are less prone to IIF.

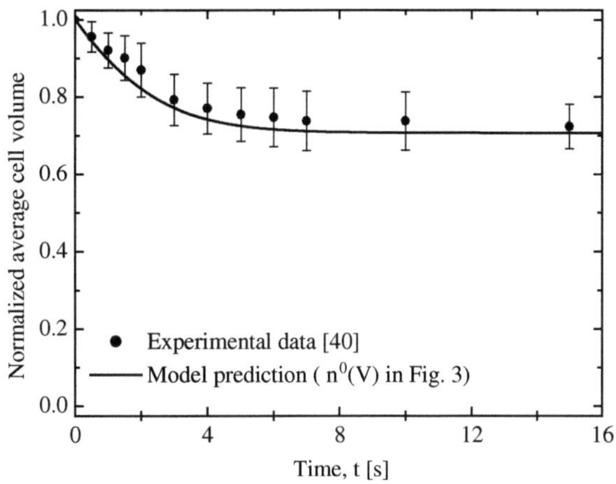

Fig. 7 Comparison between experimental data and model results in terms of temporal behaviour of the normalised average cell volume of isolated rat hepatocytes kept at $-1.1°C$ when CPA is absent (adapted from [9])

Although model output reasonably matches with the experimental data shown in Fig. 5 (the average percentage error is about 10%), this result has been obtained through a non-ideal fitting procedure involving five adjustable parameters. Since this represents a relatively large set of parameters, the experimental validation of the new approach needs to be completed by testing model reliability also through prediction runs. More specifically, system behaviour measured at operating conditions different from those adopted during the experimental runs used for the regression analysis is predicted by keeping constant the values of the adjustable model parameters as tuned in the fitting procedure. For example, the temporal behaviour of the normalised average cell volume of rat hepatocytes measured at a constant holding temperature of $-1.1°C$ in absence of CPA is reported in Fig. 7 along with model predictions ($\int_0^{+\infty} V \cdot n(V;t)\, dV \Big/ \int_0^{+\infty} V \cdot n^0(V)\, dV$).

Actually, due to the relatively high value of system temperature, IIF does not really take place during this experimental run. Accordingly, the proposed model does not predict IIF in any size class of the cell population (not shown). Thus the good matching shown in Fig. 7 demonstrates the prediction capability of the proposed PBM to describe the osmosis induced by the external ice formation when IIF does not take place. In other words, in Fig. 7 only the reliability of the values assigned to osmotic parameters during the fitting procedure may be evaluated. Then, model reliability has been further tested by predicting system behaviour when IIF occurs, in order to weigh up also the values fitted for the parameters related to ice nucleation rate. In Fig. 8, experimental data are compared to model results in terms of PIIF measured at $-40°C$ as a function of the cooling rate.

As can be seen, the cumulative fractional number of cells that underwent IIF is predicted with a reasonable accuracy when the cell size distribution reported in Fig. 3 is adopted. The matching is definitely better than the one obtained by means of the sporadic nucleation model as reported by Toner et al. [40]. In particular, an abrupt increment of $PIIF|_{-40\,°C}$ at about $-100°C/min$ is predicted by Toner et al. [40].

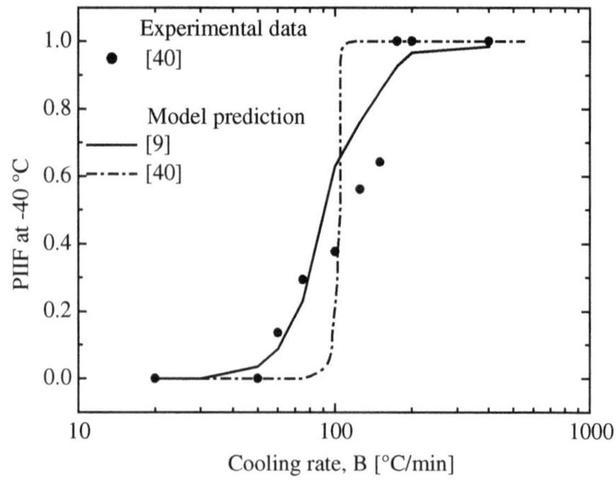

Fig. 8 Comparison between experimental data and model predictions in terms of PIIF at −40°C for a population of rat hepatocytes at different cooling rates when CPA is absent (adapted from [9])

Interestingly, this behaviour of an abrupt increment of PIIF at a certain temperature could be obtained through our model if a population of identically sized cells was taken into account. The successful comparisons reported in Figs. 7 and 8 demonstrate that taking into account the cell size distribution allows one to better predict the response of a cells suspension to a cryo-preservation process.

For sake of brevity a comprehensive list of all parameters values necessary for the model is not given here. The interested reader is invited to refer to the original papers summarised in the present one for collecting all the parameters values used in the simulations [9–11].

3.2 The Effect of Cooling Rate

Starting from the successful experimental validation, a series of model simulations have been performed. In particular, the PBM approach has been used to investigate theoretically the effect of the cooling rate on the fate of the size-distributed cell population reported in Fig. 3. Initially, the case of extra-cellular ice formed under the assumption of thermodynamic equilibrium conditions and without using a CPA has been analysed at different cooling rates. Then, the dynamics of EIF has been accounted for and the simulations have been repeated. This way, the effect of cooling rate in absence of CPA may be investigated also by evaluating if the classic assumption of thermodynamic equilibrium conditions for EIF may play a role.

Three different cooling rates have been adopted, namely −1, −50 and −400°C/min. These represent practicable values for the current technology on standard cryopreservation of a cell suspension [10]. The model results corresponding to the three cooling rates in absence of CPA are compared on Fig. 9 in terms of size distribution at different times and PIIF as a function of temperature.

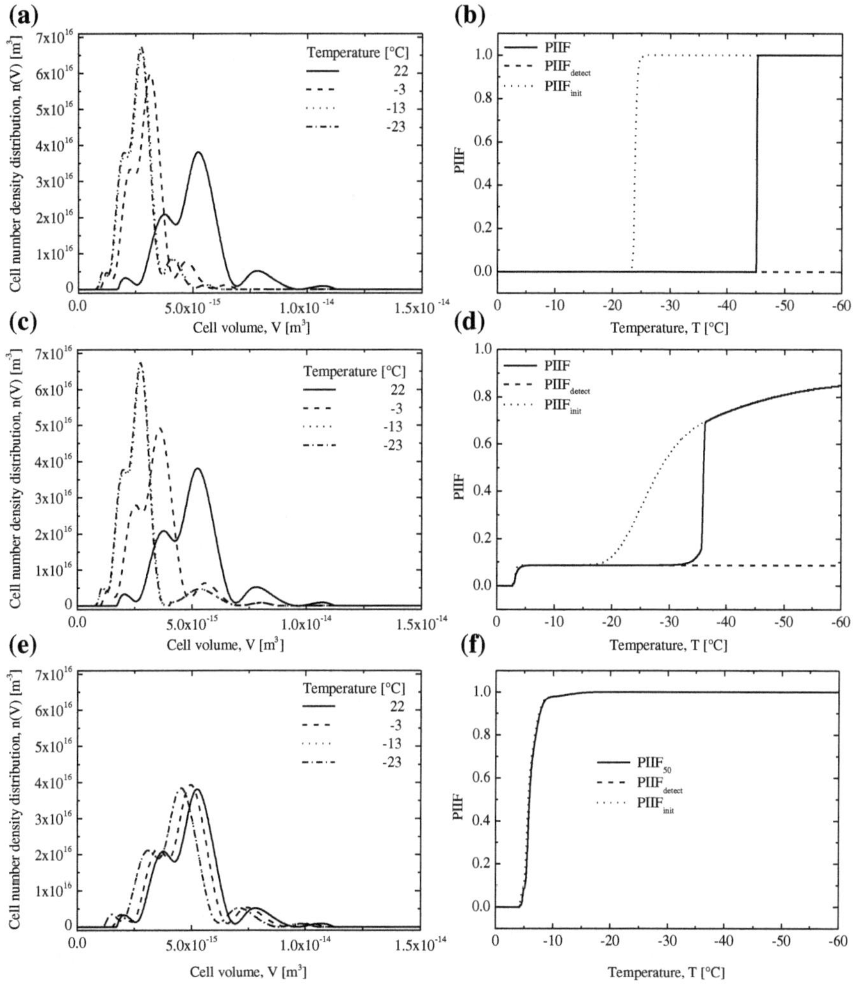

Fig. 9 Size distribution and PIIF computed in absence of CPA at different cooling rates: (**a–b**) −1°C/min; (**c–d**) −50°C/min; (**e–f**) −400°C/min (adapted from [10])

It is shown that, at the lowest cooling rate investigated (i.e. −1°C/min) the behaviour of the cells of any size class of the population is very similar: significant dehydration (see Fig. 9a, where cell shrinkage is reflected by a cell size distribution moving significantly towards smaller volumes), slow and only partial icing (see in Fig. 9b, PIIF very different from $PIIF_{init}$ and negligible $PIIF_{detect}$). As a consequence, at this low cooling rate a non-lethal IIF may result for any size class of the cell population, but *solution injury* may play a role due to the extended dehydration.

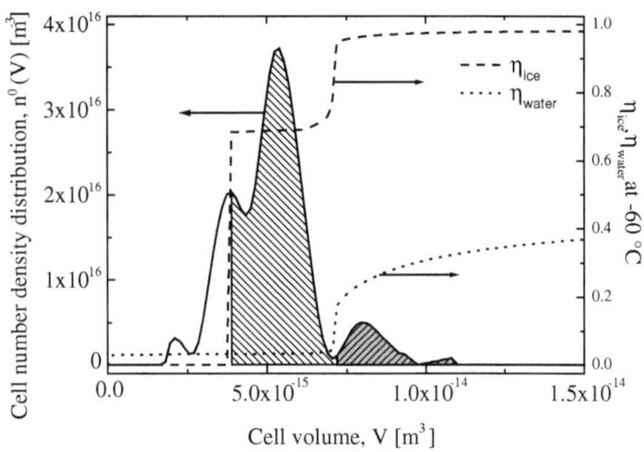

Fig. 10 Partitioning of the initial cell size distribution at −60°C based on the thresholds of internal ice volume percentage η_{ice} and total water content η_{water} when cooling at −50°C/min in absence of CPA (adapted from [10])

The picture changes dramatically when the cooling rate is increased to −50°C/min. Now, the fate of the cells, as a consequence of the different interplay between osmosis and IIF, strongly depends on their initial size. Indeed, the small size class cells dehydrate significantly (see Fig. 9c), so that IIF does not take place (see Fig. 9d, where PIIF < 0.81). On the other hand, for the largest size class cells, osmosis is relatively slower, so that undercooling and the consequent IIF takes place above the eutectic temperature. In particular, these cells are responsible of the first step of PIIF reported in Fig. 9d, where it is seen that a certain portion of the cell population ices up at about −4°C.

At this relatively high temperature, ice growth is rapid (see Fig. 9d, where PIIF and $PIIF_{init}$ overlap significantly around −4°C) and the entrapped and iced-up intra-cellular water volume represents a relevant portion of these cells (see Fig. 9d, where PIIF and $PIIF_{detect}$ overlap significantly around −4°C).

For the medium size class cells, an intermediate behaviour between these two extremes occurs. The latter ones, similarly to the small size class cells, dehydrate significantly without forming internal ice until the eutectic temperature is reached when osmosis phenomenon stops, and IIF inevitably occurs. These cells are responsible of the second step of PIIF reported in Fig. 9d, where it is seen that a portion of the cell population ices up at about −35°C. However, at this low temperature and high saline concentration ice growth is a relatively slow process (in Fig. 9d, PIIF very different from $PIIF_{init}$ at intermediate temperatures). Moreover, due to the significant osmosis, the entrapped and iced-up intra-cellular water volume does not represent a relevant portion of these cells (see Fig. 9d where an additional, second step for $PIIF_{detect}$ at −35°C is absent).

In summary, at a cooling rate of −50°C/min the behaviour of the cells strongly depends on their initial size, as reported in Fig. 10. Here the initial cell size distribution is shown along with the η_{ice} and η_{water} distributions in order to distinguish the fractions of cells that result to be unfrozen (i.e. small size class cells represented by the white area, i.e. $\eta_{ice} < 0.5$), iced-up with a small, innocuous amount of water (i.e. intermediate size class cells represented by the light shadowed area, i.e. $\eta_{ice} \geq 0.5$ and $\eta_{water} < 0.1$), or iced-up with a large, lethal amount of water (i.e. large size class cells represented by the dark shadowed area, i.e. $\eta_{ice} \geq 0.5$ and $\eta_{water} \geq 0.1$).

The results of the simulations related to −400°C/min are reported in Fig. 9e–f. At this relatively high cooling rate, ice forms inside cells of any size class at temperatures above the eutectic one, in the range [-15 °C to -5 °C] (see Fig. 9f).

Therefore, in Fig. 9 it is shown at a glance that, the different size classes of cells behave similarly at the lowest and the highest cooling rates, albeit for different reasons. Specifically, at low cooling rates a significant osmosis takes place thus limiting IIF for all the different size classes of cells. On the contrary, at high cooling rates osmosis is very limited, IIF is favoured and takes place rapidly and lethally for all the different size classes of cells. Different fates among the size classes of cells are obtained only at intermediate cooling rates, when osmosis and IIF do not dominate each other and their interplay leads to completely unfrozen small cells, lethally iced-up larger cells, and iced-up for intermediate class cells but with an innocuous amount of ice (see Fig. 10).

On passing, it is worth noting that the two-step profile of PIIF versus temperature shown in Fig. 9d as obtained at the intermediate cooling rate of −50°C/min is qualitatively similar to the experimental behaviour reported by Toner et al. [39] and Berrada et al. [4] even if with different cell lineages and operating conditions. The corresponding interpretation was performed by invoking two different ice nucleation mechanisms (namely surface and volumic catalysed heterogeneous nucleation), acting at different temperatures on the averaged size cells, in the framework of the sporadic nucleation modelling approach. Instead, in this work such behaviour is simulated by considering only one heterogeneous nucleation mechanism of ice formation taking place in a population of differently sized cells.

When the assumption of thermodynamic equilibrium conditions is removed thus accounting for a dynamic EIF, very similar results are obtained for the extreme cooling rates (not reported for the sake of brevity). Indeed, at relatively low and high cooling rates osmosis or IIF dominates each other, independently to what occurs in the extra-cellular environment. More specifically, at a low cooling rate of −1°C/min osmosis is so slow that a dynamic EIF results to be as rapid as the instantaneous ice formation obtained under the assumption of thermodynamic equilibrium conditions. As such no difference is obtained between the two cases. On the other hand, at relatively high cooling rates as −400°C/min, osmosis is so slow with respect to IIF that there is no interplay between any size class of cells and the surrounding environment: cells of any size are basically isolated from the surrounding medium. As such, considering a dynamic modelling of EIF or thermodynamic equilibrium conditions is not relevant either.

It is then apparent that, at intermediate cooling rates, even when a dynamic EIF is accounted for, the interplay between osmosis and IIF generates different fates among the differently sized classes of cells just like it was obtained assuming thermodynamic equilibrium conditions in the suspending medium. This is shown in Fig. 11, where the comparison between the simulation results obtained for the

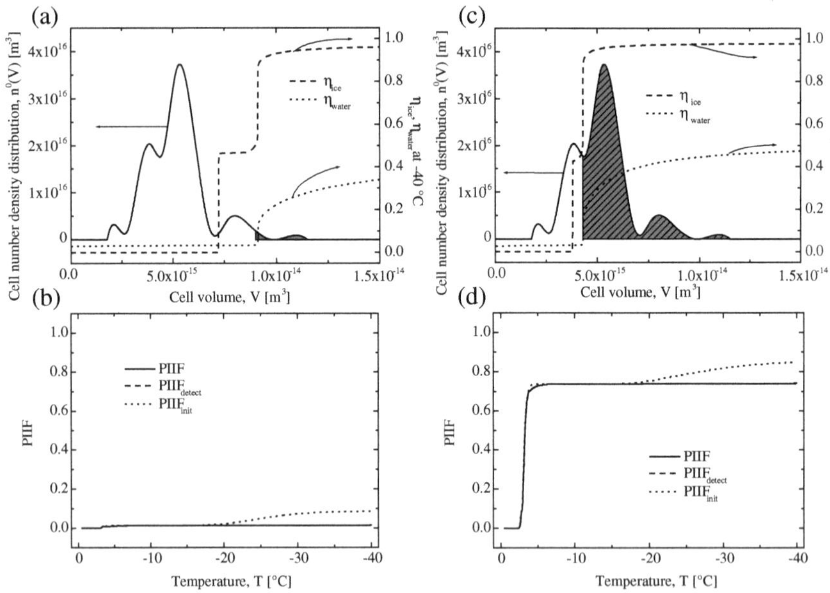

Fig. 11 Simulation results for the cooling stage carried out at −50°C/min in absence of CPA under the assumption of extra-cellular thermodynamic equilibrium in *left* column (**a–b**) and dynamics for EIF in *right* column (**c–d**) (adapted from [11])

case of −50°C/min when considering thermodynamic equilibrium conditions and the dynamics of EIF are compared to better highlight the differences. Specifically, the plots in the left column of Fig. 11 (i.e. panels (a–b)) show the results obtained when considering thermodynamic equilibrium conditions, whilst in the right column the results obtained for dynamic EIF are reported (i.e. panels (c–d)). The initial cell size distribution, the internal ice volume fraction η_{ice}, and the total water fraction η_{water} at −40°C as a functions of the initial cell volume are compared in top plots 11a and c, whilst in bottom plots 11b and d the profiles of PIIF, $\text{PIIF}_{\text{init}}$, and $\text{PIIF}_{\text{detect}}$ vs. temperature are shown.

As can be seen, considering thermodynamic equilibrium conditions or dynamic EIF for the suspending solution does have an effect and different model results are obtained using either one or the other approach. In fact, whilst differently sized cells in a single population exhibit different IIF temperatures under the same operating conditions in both cases, when using a dynamic EIF a smaller ice percentage is generally obtained at −40°C in comparison to the case of equilibrium conditions (i.e. Fig. 11c–d vs Fig. 11a–b). This is because, at this cooling rate osmosis may actually take place for a relatively short period of time and the dynamic EIF is slow enough to limit further the time window available for the water efflux out of the cells, thus facilitating IIF.

Therefore it can be concluded that, depending on the particular system of interest (cell lineage) and the employed cooling rates, the assumption of extra-cellular thermodynamic equilibrium is not generally valid. A dynamic description of the extra-cellular ice formation should be used to increase the predictive capabilities of the model for realistic operating conditions.

3.3 The Effect of CPA

When CPA comes into play, system behaviour is more complex. For this reason, our model investigations in presence of CPA have been performed only under the classic assumption of thermodynamic equilibrium conditions for EIF. This way only one factor at time with respect to our previous simulations have been changed, thus helping the discussion and interpretation of the obtained theoretical results.

Before beginning the analysis of the effect of CPA in our modelling approach, it is necessary to recognise that the investigations performed have been limited to practicable operating conditions. This is true from two different viewpoints. First, in order to resemble the experimental procedure actually adopted in a standard cryopreservation protocol, our model takes into account the simulation of the initial equilibration stage as coupled to the subsequent cooling stage. This actually represents a novelty in the field of cryopreservation modelling. As such, in our investigations the simulation of the cooling stage always starts at the end of the simulation of the preceding equilibration stage, especially in terms of the size distribution of the cell population which we guess may play a role in our main hypothesis. Secondly, the simulation of the two stages is always performed under practicable conditions, i.e. cells initially under isotonic conditions are equilibrated with CPA in a one-step addition at ambient temperature and then cooling starts with a constant rate ("linear" profile of system temperature). These represent standard operations, even though different modalities have been proposed so far in the literature especially for the equilibration stage (i.e. drop-by-drop or step-wise addition and/or at low temperature). Moreover, only the region of relatively slow cooling rates (i.e. 1–400°C/min) and relatively low CPA concentrations (i.e. 0–5000 $\mathrm{osmol_{CPA}/m^3_{water}}$) reported in the literature [19] is considered in our investigations. Indeed, higher values of the cooling rates, i.e. 10^6–10^{13}°C/min investigated by Karlsson et al. [19], are not usually achieved, regardless the CPA concentration levels. These extreme conditions do not resemble the standard cryopreservation protocols adopted for biological samples of suspended cells, and may not be obtained easily from a technological point of view. As a consequence, the simulation of system behaviour under these relatively high cooling rates where IIF is basically replaced by vitrification has not been performed. On the other hand, due to its well known cytotoxic character, the cases of high CPA concentrations are not considered either. This allows us to avoid drawing erroneous conclusions

Fig. 12 Temporal evolution of the cell number density distribution during the equilibration at ambient temperature of cells in a solution of 5000 $osmol_{CPA}/m^3_{water}$ glycerol solution (adapted from [10])

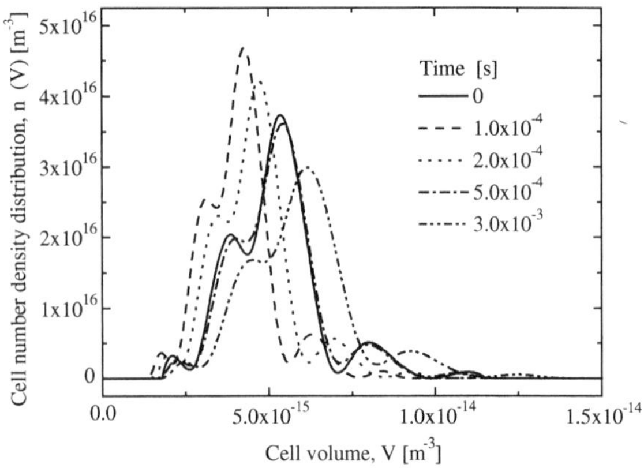

from the theoretical point of view since our model does not even take into account this mechanism of cell lethality. In addition, by keeping CPA content at low levels cell volumic excursions during the equilibration stage are also limited, thus avoiding this source of risk for cell damage.

Starting with the size distribution of the cell population depicted in Fig. 3, the dynamic evolution of the size distribution of the cell population initially at isotonic conditions and suspended in an aqueous solution of glycerol at 5000 $osmol_{CPA}/m^3_{water}$ at ambient temperature is simulated as shown in Fig. 12. As well known in the literature, the response of the cells consists of shrinkage followed by swelling, so that the corresponding size distribution moves back and forth (i.e. initially towards smaller volumes, and then towards larger ones), as water and CPA move in and out through cell membrane, until the driving forces are cancelled out when intra- and extra-cellular concentrations of solutes become equal.

However, depending on the CPA level considered, the cell size distribution at the end of equilibration may not coincide with the initial, isotonic one, as clearly reported in Fig. 12.

As a consequence, when cooling starts the size distribution of the cell population may be quite different from the initial, isotonic one. This aspect is highlighted in Fig. 13a, where the temporal evolution of the average cell volume of the cell population with respect to the initial one is reported for the cases of glycerol addition at different concentrations, namely 1000, 2000 and 5000 $osmol_{CPA}/m^3_{water}$. This difference between initial, isotonic size distribution of the cell population and the corresponding one at the end of the equilibration stage for different CPA concentration values may be further observed from Fig. 13b, where the temporal profile during equilibration of the average cell volume of the size-distributed cell population is reported at different CPA concentrations. Clearly, the equilibration time depends on the specific values adopted for water permeability through cell membrane. The latter one is relatively high for the rat hepatocytes considered in our simulations (typically 30–150 s for other cell lineages), due to their peculiar ability to modulate water permeability through aquaporins, necessary for carrying out their physiological task of bile secretion.

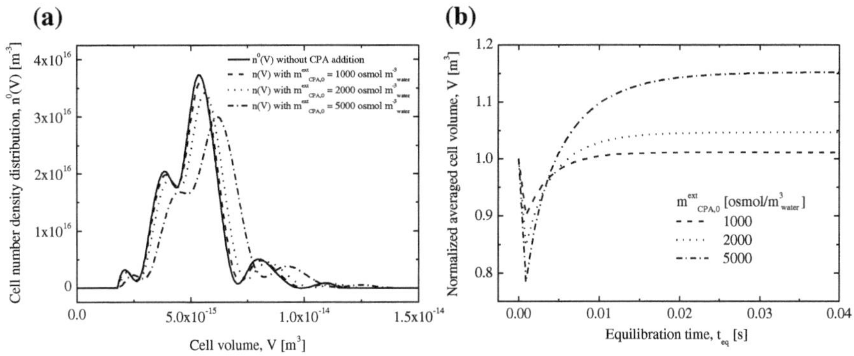

Fig. 13 Model results of the equilibration stage at different CPA concentrations: cell number density distribution at start and end of equilibration (**a**), and average cell volume of the size-distributed population as a function of time during equilibration (**b**) (adapted from [10])

Fig. 14 Histogram representing the volume composition of largest size class cells after equilibration with different CPA concentrations (adapted from [10])

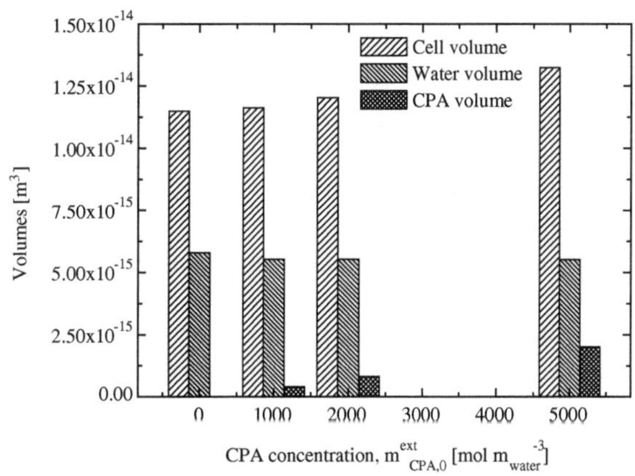

The start of the simulation for the cooling stage from the modelling results obtained at the end of the equilibration stage represents a novelty that permits to overcome the apparent limitation of previous IIF models. There, the modelling of equilibration and cooling stages are not coupled, and at the beginning of cooling stage it was assumed that intra-cellular water of the cells with isotonic volume is simply replaced by a proper amount of CPA [19]. However, during the equilibration stage the CPA does not replace water, but it is basically added to the cell. This is actually well known by all the investigators addressing the osmotic response of cells and it may be clearly seen from Fig. 14. Here, the behaviour of the largest size class cells is considered as representative of the entire population. Cell volume, as well as those ones of intracellular water and CPA, are respectively shown after equilibration with different extra-cellular CPA concentration levels. It is seen that whilst water volume remains constant, cell volume increases due to

CPA permeation through the membrane. As a consequence, at the end of equilibration stage (i.e. at the beginning of the cooling one) the actual intracellular water volumic content is almost similar to the initial, isotonic one, so that it remarkably affects IIF kinetics (i.e. Eqs. 1 and 11) and its interplay with osmosis rate. This result is in contrast with the typical consideration that IIF inhibition by CPA is partially related to the reduction of intra-cellular water volume through CPA entrapment during equilibration [16, 19].

Turning our attention to the cooling stage of the modelled cryopreservation process, it is noteworthy that when using an increasing CPA content two main effects may generally result: the osmosis phenomenon is hindered and the viscosity of the intracellular liquid solution (cytoplasm) increases. These have contrasting effects on IIF and may prevail each other depending on the adopted cooling rate and the volume class of cells. More specifically, according to the phase diagram as CPA increases the thermodynamic temperature for phase change from liquid solution to ice is lowered, so that EIF may take place only at lower temperatures. As a consequence, during cooling at a given cooling rate the cells start to loose their internal water through osmosis later, at lower temperatures. Then, more water eventually remains entrapped inside the cells, thus favoring IIF. In conclusion, if CPA increases IIF is thermodynamically favored, because exo-osmosis is thermodynamically limited. On the other hand, if CPA increases the intra-cellular solution viscosity increases as well. This represents a kinetic limitation for IIF, since the liquid diffusion that limits the nucleation and growth of ice crystals is reduced, accordingly.

These are the general considerations that help to understand and rationalise the results of the simulations performed in presence of CPA as grouped in Fig. 15. Here the PIIF is reported as a function of temperature for different cooling rates, at various CPA concentrations. It is shown that, the sigmoidal plot of PIIF as a function of temperature dramatically changes when varying the cooling rate. If CPA is absent, the PIIF sigmoid moves towards higher temperatures when increasing the cooling rate (the solid lines of Fig. 15a–f).

This is expected and it is a well-known behaviour of the system, i.e. cells are iced up at higher temperatures (IIF is favoured) when increasing the cooling rate, if CPA is absent.

On the other hand, IIF may be favored even in presence of CPA, which is not consistent with the common thought that the antifreeze should always decrease the IIF temperature [2, 17]. This is shown in Fig. 15a–b where at the lowest cooling rates considered (i.e. -1°C/min and -10°C/min) the paradox of increasing IIF temperature when increasing CPA concentration is now obtained. Actually, also this undesired effect is well known in the literature [17, 19]. It is due to the thermodynamic limitation of water exo-osmosis that enhances IIF, if CPA content increases at low cooling rates. On the contrary, at relatively higher cooling rates (i.e. -400°C/min, in Fig. 15f) IIF is inhibited and it occurs at lower temperatures when increasing CPA content, as commonly expected. This is due to the kinetic limitations resulting from an higher viscosity and lower diffusivity of the cytoplasmic solution that, at these operating conditions, overcomes the thermodynamic

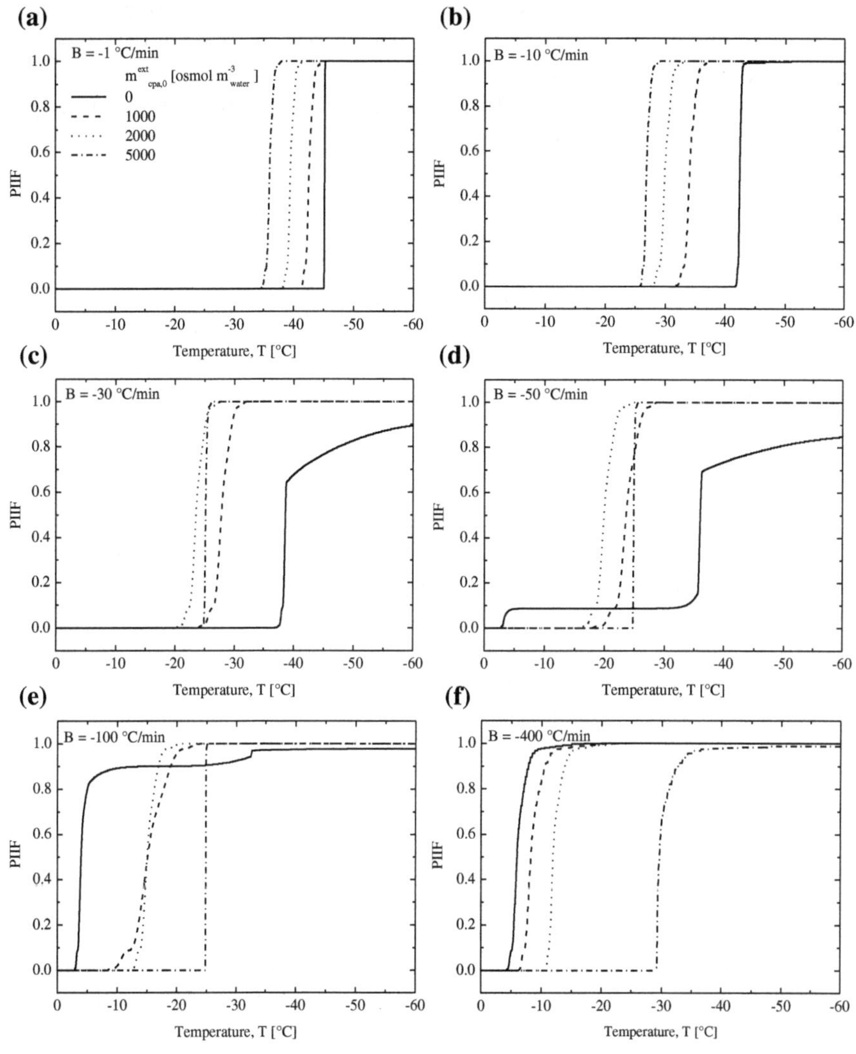

Fig. 15 Model results in terms of PIIF vs temperature at various cooling rates and CPA concentrations (adapted from [10]). In all panels CPA content has been varied from 0 to 5000 $osmol/m^3_{water}$ as indicated in the legend of panel (**a**); cooling rates equal to −1°C/min (**a**), −10°C/min (**b**), −30°C/min (**c**), −50°C/min (**d**), −100°C/min (**e**), −400°C/min (**f**) has been considered

limitation of osmosis. Of course, at the intermediate cooling rates considered in this work (i.e. −30, −50 and −100°C/min, in Fig. 15c–e), a continuous variation between these two extreme system behaviours is obtained, so that the effects of a reduced time for osmosis or an increased solution viscosity contrast each other until one of them eventually prevails, depending on the CPA content and the size class of the cell population. This generates the PIIF curve intersection reported in Fig. 15e. In this regard, it is worth noting that some experimental data reporting this bizarre behaviour of intersecting PIIF curves are available in the literature [12, 34]. However, the corresponding theoretical interpretation based on the previous modelling approach addressing the sporadic nucleation of the single sized, average cell was not provided. As such, these data were supposed to be the consequence of scattered measurements. It is presumable that, the previous modelling approach may be adopted to simulate this behaviour only by resorting to different mechanisms of ice nucleation as it was done in order to justify the two-step profiles reported in Fig. 9d that are found also in Fig. 15d–e. These different mechanisms of ice nucleation should be assumed to act differently and eventually prevail each other at different temperatures, as a function of the operating conditions due to unknown (statistical) reasons. On the contrary, all these behaviours may now be comprehensively taken into account through our novel modelling approach, i.e. by addressing the effect of the size distribution of the cell population with a single ice nucleation mechanism. This way differently sized cells behave differently and their fate during the cooling stage depends on the initial cells size distribution of the cell population. Moreover, differently sized cells behave differently at the intermediate cooling rates considered in our simulations -1 to -100 °C/min. This is noteworthy from a practical perspective since only the intermediate cooling rates can actually be attained nowadays for a suspension of cells inside a standard programmable cooling chambers commercially available in the market.

As a representative example, the case of -50°C/min is addressed in Fig. 16. Here, by keeping constant the cooling rate, the system behaviour at several, increasing CPA contents (from 0 to 12 kosmol/$m^3{}_{water}$) is shown in terms of η_{ice} and η_{water} distributions at -60°C reported as functions of the initial cell volume in order to easily identify the classes of cells with different fates. It is apparent that, the size distribution may affect the cooling stage of a cell population not only in the absence (Fig. 16a) but even in the presence of CPA (Fig. 16b–e), respectively.

In general, the larger cells are more prone to form a lethal amount of intracellular ice, whilst the smaller ones may even remain unfrozen and will eventually vitrify. Depending on the specific CPA content and the adopted thresholds of η_{ice} and η_{water}, it may also happen that only a small amount of ice is formed inside medium sized cells (Fig. 16a–b). On the other hand, all the differently sized cells of the population may behave similarly, as it may be seen from Fig. 16c and 16f, where all size classes result to be lethally iced-up at -60°C or unfrozen, correspondingly.

Specifically, by progressively increasing the CPA content, the lethally frozen portion of the cell population indicated by the dark shadowed area initially increases and then decreases, till it completely vanishes, thus showing a maximum. Correspondingly, the PIIF sigmoid first moves towards higher temperatures reaching its maximum value equal to 1 whilst displaying a more abrupt, sharp variation, and then returns back to the lower ones (not shown here for brevity). Clearly, as already discussed previously, the CPA content of 12 kosmol/m^3 considered in Fig. 16f may be too high so that cytotoxicity and excessive volumic

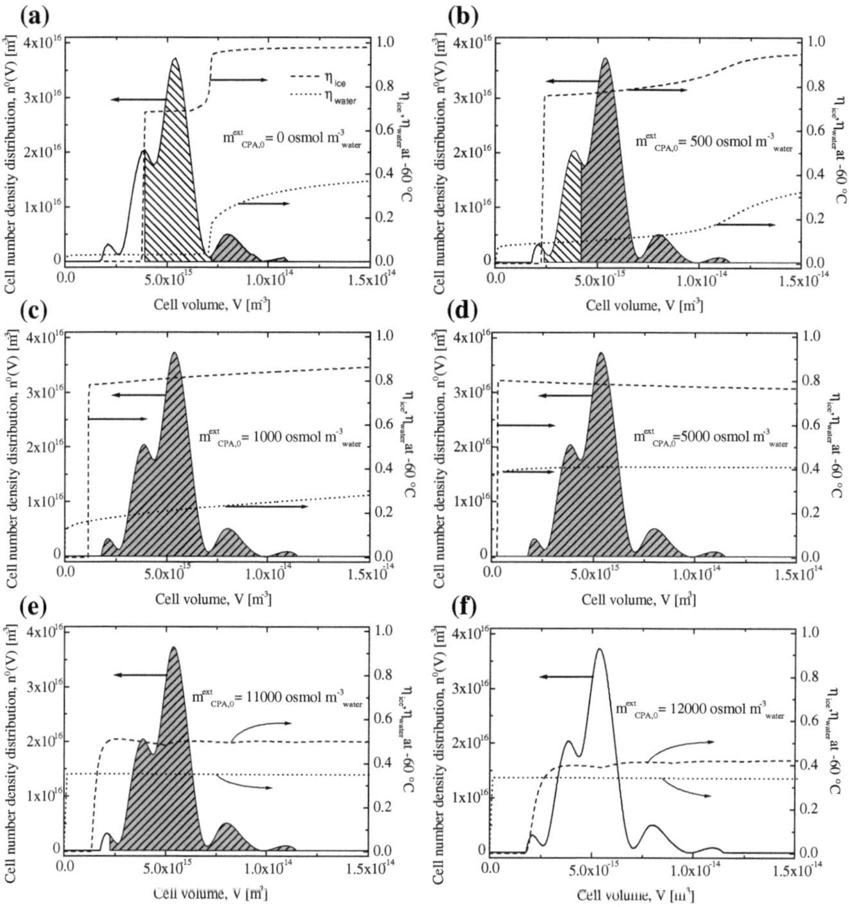

Fig. 16 Model results of cooling stage of a cell suspension carried out at −50°C/min with increasing CPA content in terms of the initial cell-size, internal ice volume percentage η_{ice}, and total water content η_{water} profiles at −60°C as a function of initial cell volume (adapted from [10]): $m^{ext}_{CPA,0}$= 0 osmol/m^3_{water} (**a**), 500 osmol/m^3_{water} (**b**), 1000 osmol/m^3_{water} (**c**), 5000 osmol/m^3_{water} (**d**), 11000 osmol/m^3_{water} (**e**) 12000 osmol/m^3_{water}(**f**) are considered

range excursions during equilibration stage may not be avoided as mechanisms of cell mortality, even if all the classes of the size-distributed cell population will be unfrozen at −60°C and eventually vitrified. However, it is worth noting that, the behaviour of IIF described above, i.e. the occurrence of a maximum value of PIIF, is actually observed at all cooling rates considered in this work. More specifically, to reach a completely unfrozen size-distributed cell population at −60°C, a lower CPA content is needed when increasing the cooling rate. Thus, a specific combination of operative conditions (i.e. cooling rate and CPA content) may be found to avoid cytotoxicity and excessive volumic range excursions during equilibration stage, whilst obtaining a completely unfrozen size-distributed cell population.

Basically, a system behaviour at varying CPA content similar to the one reported in Fig. 16 for the case of −50°C/min may be found also for the other cooling rates considered in this work. Specifically, at −1°C/min a very high CPA concentration needs to be used to obtain a completely unfrozen size-distributed cell population, whilst, if CPA is completely absent, an innocuous level of IIF is reached in every size class of the cell population. Conversely, at −400°C/min a significant, lethal IIF for any cell is obtained without using CPA, whilst, with a CPA content as low as 7 kosmol/m^3, a completely unfrozen size-distributed cell population results. Along these lines, high cooling rates need to be combined to high CPA contents for reaching vitrification, whilst at low cooling rates cell viability may be increased by adopting relatively low CPA concentrations. These conclusions perfectly match the well known rule of thumb for a successful cryopreservation protocol. Clearly, the values of the operative conditions needed to optimise the process in terms of final cell viability strongly depend on the osmotic characteristics and initial cell size distribution of the specific cell lineage at hand.

4 Concluding Remarks and Future Directions

A novel theoretical interpretation of the freezing process for a suspension of cells has been proposed. It is based on the PBM approach addressing the size distribution of the cell population. The latter one is invoked to interpret and rationalise the experimental evidence measured so far in a more comprehensive fashion than the other theoretical analysis developed before.

After the initial experimental validation by direct comparison with suitable experimental data taken from the literature, model investigations on system behaviour at various operating conditions have been performed, i.e. different CPA content and cooling rates. It is demonstrated that, cell survival due to intracellular ice formation depends on the initial cell size distribution and its osmotic parameters. At practicable operating conditions in terms of cooling rate and cryo-protectant concentration, IIF may be lethal for the fraction of larger size classes of the cell population whilst it may not reach a dangerous level for the intermediate size class cells and it will not even take place for the smaller ones. This has been also tested by removing the classic assumption of EIF under thermodynamic equilibrium thus accounting for its dynamics. As such it may be concluded that, when cryopreserving a cell suspension under practicable operating conditions the cell size distribution of the cell population and the dynamics of EIF should be carefully taken into account, contrarily to what has been classically assumed.

In the next future, the model approach developed so far will be adopted to simulate in a comprehensive way a complete cyopreservation cycle by accounting for the cooling, storage and thawing stages. To reach this goal, the model capability to quantitatively describe system behaviour needs to be improved by overcoming several limitations. Specifically, some physico-chemical phenomena previously omitted that take place during the cooling phase down to cryogenic temperatures (−196°C) followed by the thawing up to 37°C, will be taken into account: a double mechanism (heterogeneous and homogeneous) of nucleation for intra-cellular ice, vitrification–devitrification of intra- and extra-cellular water, solution injury). This way, the whole cryopreservation cycle will be simulated instead of limiting the description of system behaviour to only those physico-chemical phenomena involved in the cooling phase down to a given temperature (−40 or −60°C). Since a very complex mathematical model would result if all the physical–chemical phenomena involved during the complete cryopreservation cycle are considered, a specific strategy will be followed to avoid ineffective model improvements that only make harder the numerical solution. Specifically, a sequential removal of the simplifying assumptions will be adopted, thus increasing the complexity of the mathematical model, i.e. by continuously adding the modelling improvements that will result to be effective on influencing system behaviour through the analysis of its numerical solution.

Acknowledgments The Fondazione Banco di Sardegna, Italy, and The Regional Government of Sardegna are gratefully acknowledged for the finantial support of the project "Crioconservazione di cellule staminali da cordone ombelicale" (2010–2011) and the fellowship within the "Master and Back" program (2010–2012), respectively.

References

1. Agca, Y., Gilmore, J., Byers, M., Woods, E.J., Liu, J., Critser, J.K.: Osmotic characteristics of mouse spermatozoa in the presence of extenders and sugars. Biol. Reprod. **67**, 1493–1501 (2002)
2. Aksan, A., Toner, M.: Roles of thermodynamic state and molecularmobility in biopreservation. In: Bronzino, J.D. (ed.) Tissue Engineering and Artificial Organs. Taylor & Francis, Boca Raton (2006)
3. Benson, C.T., Liu, C., Gao, D.Y., Critser, E.S., Benson, J.D., Critser, J.K.: Hydraulic conductivity (L_p) and its activation energy (E_a), cryoprotectant agent permeability (P_s) and its E_a, and reflection coefficients (σ) for golden hamster individual pancreatic islet cell membrane. Cryobiology **37**, 290–299 (1998)
4. Berrada, M.S., Bischof, J.C.: Evaluation of freezing effects on human microvascular-endothelial cells (HMEC). CryoLetters **22**, 353–366 (2001)
5. Bigg, E.K.: The supercooling of water. Proc. Phys. Soc. B **66**, 688–694 (1953)
6. Dirksen, J.A., Ring, T.A.: Fundamentals of crystallization: kinetic effects on particle size distributions and morphology. Chem. Eng. Sci. **46**, 2389–2427 (1991)
7. Ebertz, S.L., McGann, L.E.: Cryoprotectant permeability parameters for cells used in a bioengineered human corneal equivalent and applications for cryopreservation. Cryobiology **49**, 169–180 (2004)
8. Elmoazzen, H.Y., Elliott, J.A.W., McGann, L.E.: Osmotic transport across cell membranes in nondilute solutions: a new non dilute solute transport equations. Biophys. J. **96**, 2559–2571 (2009)

9. Fadda, S., Cincotti, A., Cao, G.: The effect of cell size distribution during the cooling stage of cryopreservation without CPA. AIChE. J. **56**, 2173–2185 (2010)
10. Fadda, S., Cincotti. A., Cao, G.: Rationalizing the equilibration and cooling stages of cryopreservation: the effect of cell size distribution. AIChE. J. **54**, 1075–1095 (2011)
11. Fadda, S., Briesen, H., Cincotti, A.: The effect of EIF dynamics on the cryopreservation process of a size distributed cell population. Cryobiology **62**, 218–231 (2011)
12. Harris, C.L., Toner, M., Hubel, A., Cravalho, E.G., Yarmush, M.L., Tompkins, R.G.: Cryopreservation of isolated hepatocytes: intracellular ice formation under various chemical and physical conditions. Cryobiology **28**, 436–444 (1991)
13. Hunt, C.J., Armitage, S.E., Pegg, D.E.: Cryopreservation of umbilical cord blood: 1. osmotically inactive volume, hydraulic conductivity and permeability of $CD34^+$ cells to dimethyl sulphoxide. Cryobiology **46**, 61–75 (2003)
14. Hunt, C.J., Armitage, S.E., Pegg, D.E. Cryopreservation of umbilical cord blood: 2. Tolerance of $CD34^+$ cells to multimolar dimethyl sulphoxide and effect of cooling rate on recovery after freezing and thawing. Cryobiology **46**, 76–87 (2003)
15. Jacobs, M.H.: The simultaneous measurement of cell permeability to water and to dissolved substances. J. Cell. Comp. Physiol. **2**, 427–444 (1933)
16. Karlsson, J.O.M.: Effect of solution composition on the theoretical prediction of the ice nucleation kinetics and thermodynamics. Cryobiology **60**, 43–51 (2010)
17. Karlsson, J.O.M., Toner, M.: Long-term storage of tissues by cryopreservation: critical issues. Biomaterials **17**, 243–256 (1996)
18. Karlsson, J.O.M., Toner, M.: Cryopreservation. In: Lanza, R., Langer, R., Vacanti, J. (eds.) Principles of Tissue Engineering, 2nd edn. Academic press, San Diego (2000)
19. Karlsson, J.O.M., Cravalho, E.G., Toner, M.: A model diffusion—limited ice growth inside biological cells during freezing. J. Appl. Phys. **75**, 4442–4445 (1994)
20. Katkov, I.I.: *Amicus Plato, sed magis amica veritas*: plots must obey the laws they refer to and models shall describe biophysical reality! Cryobiology **62**, 242–244 (2011)
21. Kedem, O., Katchalsky, A.: Thermodynamics analysis of the permeability of biological membranes to non-electrolytes. Biochim. Biophys. Acta. **27**, 229–246 (1958)
22. Kleinhans, F.W.: Membrane permeability modeling: Kedem-Katchalsky vs a two-parameters formalism. Cryobiology **37**, 271–289 (1998)
23. Levin, R.L., Cravalho, E.G., Huggins, C.E.: A membrane model describing the effect of temperature on the water conductivity of erythrocyte membranes at subzero temperatures. Cryobiology **13**, 415–429 (1976)
24. Mazur, P.: Kinetics of water loss from cells at subzero temperatures and likelihood of intracellular freezing. J. Gen. Physiol. **47**, 347–369 (1963)
25. Mazur, P.: Principles of cryobiology. In: Fuller, B., Lane, N., Benson, E.E. (eds.) Life in Frozen State. CRC Press, London (2004)
26. Mazur, P.: A biologist's view of the relevance of thermodynamics and physical chemistry to cryobiology. Cryobiology **60**, 4–10 (2010)
27. Mazur, P., Leibo, S.P., Chu, E.H.Y.: A two-factor hypothesis of freezing injury—evidence from Chinese hamster tissue culture cells. Exp. Cell. Res. **71**, 345–355 (1972)
28. McGrath, J.J.: Preservation of biological material by freezing and thawing. In: Shitzer, A., Eberhart, R.C. (eds.) Heat Transfer in Medicine and Biology, Analysis and Application vol. 2, pp. 185–238. Plenum Press, New York (1985)
29. McGrath, J.J.: Membrane transport properties. In: McGrath, J.J., Diller, K.R. (eds.) Low Temperature Biotechnology: Emerging Applications and Engineering Contributions. ASME, New York (1988)
30. Miyamoto, H., Ishibashi, T.: Effects of the temperature of ice-seeding on survival of frozen-and-thawed mouse morulae. Cell. Mol. Life Sci. **37**, 187–188 (1981)
31. Mukherjee, I.N., Song, Y.C., Sambanis, A.: Cryoprotectant delivery and removal from murine insulinomas at vitrification-relevant concentrations. Cryobiology **55**, 10–18 (2007)
32. Petersen, A., Schneider, H., Rau, G., Glasmacher, B.: A new approach for freezing of aqueous solutions under active control of the nucleation temperature. Cryobiology **53**, 248–257 (2006)

33. Phelps, M.J., Liu, J., Benson, J.D., Willoughby, C.E., Gilmore, J.A., Critser, J.K.: Effect of percoll separation, cryoprotective agents, and temperature on plasma membrane permeability characteristics of murine spermatozoa and their relevance to cryopreservation. Biol. Reprod. **61**, 1031–1041 (1999)
34. Rall, W.F., Mazur, P., McGrath, J.J.: Depression of the ice-nucleation temperature of rapidly cooled mouse embryos by glycerol and dimethyl sulphoxide. Biophys. J. **41**, 1–12 (1983)
35. Randolph, A.D., Larson, M.A.: Theory of particulate processes, analysis and techniques of continuous crystallization, 2nd edn. Academic Press, San Diego (1988)
36. Saenz, J., Toner, M., Risco, R.: Comparison between ideal and nonideal solution models for single-cell cryopreservation protocols. J. Phys. Chem. B **113**, 4853–4864 (2009)
37. Stott, S.L., Karlsson, J.O.M.: Visualization of intracellular ice formation using high-speed video cryomicroscopy. Cryobiology **58**, 84–95 (2009)
38. Toner, M., Cravalho, E.G., Karel, M.: Thermodynamics and kinetics of intracellular ice formation during freezing of biological cells. J. Appl. Phys. **67**, 1582–1593 (1990)
39. Toner, M., Cravalho, E.G., Karel, M., Armant, D.R.: Cryomicroscopic analysis of intracellular ice formation during freezing of mouse oocytes without cryoadditives. Cryobiology **28**, 55–71 (1991)
40. Toner, M., Tompkins, R.G., Cravalho, E.G., Yarmush, M.L.: Transport phenomena during freezing of isolated hepatocytes. AIChE. J. **38**, 1512–1522 (1992)
41. Woods, E.J., Liu, J., Gilmore, J.A., Reid, T.J., Gao, D.Y., Crister, J.K.: Determination of human platelet membrane permeability coefficients using the Kedem-Katchalsky formalism: estimates from two-vs three-parameter fits. Cryobiology **38**, 200–208 (1999)
42. Yang, G., Veres, M., Szalai, G., Zhang, A., Xu, L.X., He, X.: Biotransport Phenomena in Freezing Mammalian Oocytes. Ann. Biomed. Eng. **39**, 580–591 (2011)
43. Zachariassen, K.E., Kristiansen, E., Andre Pedersen, S., Hammel, H.T.: Ice nucleation in solutions and freeze-avoiding insects—homogeneous or heterogenous. Cryobiology **48**, 309–321 (2004)
44. Zhao, G., Luo, D., Gao, D.: Universal model for intracellular ice formation and its growth. AIChE. J. **52**, 2596–2606 (2006)
45. Zhmakin, A.I.: Physical aspects of cryobiology. Phys. Usp. **51**, 231–252 (2008)

Mesenchymal Stem Cell Heterogeneity and Ageing In Vitro: A Model Approach

Jörg Galle, Martin Hoffmann and Axel Krinner

Abstract Mesenchymal Stem Cell (MSC)-based therapies have been suggested as a particular promising strategy in tissue regeneration. These cells can easily be obtained from the patient and are able to produce a large number of progeny that can be induced to form connective tissue. However, rapid amplification of the isolated cells is required for their therapeutic application. While already the isolated populations are heterogeneous regarding various functional and molecular aspects, this heterogeneity further evolves during amplification. Understanding the origin and development of MSC heterogeneity will help to improve MSC culture conditions and thus facilitate their clinical use. We here review recent results on MSC heterogeneity and introduce a mathematical framework that approaches MSC heterogeneity on the single cell level. This approach bases on the concept of noise-driven MSC differentiation and allows describing MSC heterogeneity with respect to their differentiation state and age. It is capable of describing the impact of MSC heterogeneity on in vitro expansion and differentiation. We present new results on the formation of an age structure in MSC populations in vitro and the age-dependent differentiation structure of MSC populations. Moreover, we discuss open questions regarding MSC adaptation to changing environments and the cell intrinsic control of state fluctuations.

J. Galle (✉) · A. Krinner
Interdisciplinary Centre for Bioinformatics (IZBI),
University of Leipzig, Leipzig, Germany
e-mail: galle@izbi.uni-leipzig.de

M. Hoffmann
Biomathematics and Bioinformatics Group,
Department of Knowledge Engineering,
Maastricht University, Maastricht, The Netherlands

Stud Mechanobiol Tissue Eng Biomater (2013) 10: 183–205
DOI: 10.1007/8415_2012_116

Published Online: 14 February 2012

1 Introduction

Mesenchymal stem cells (MSCs, also known as multipotent mesenchymal stromal cells) have been characterised as a heterogeneous cell population of adherent spindle-shaped cells capable of differentiating into bone-marrow stromal cells, osteoblasts, chondrocytes, adipocytes and myocytes. Related lineage priming of MSCs has been nicely demonstrated in monoclonal culture [23]. In some tissue types, e.g. bone-marrow stroma, adipose, skeletal muscle and synovium, MSCs persist in adult life without loosing their capacity to proliferate and differentiate [41, 76]. Accordingly, MSCs have been proposed as innovative therapeutic tools in tissue regeneration. Their therapeutic deployment comprises treatment of various diseases, including osteoarthritis [12] and myocardial infarction [40]. MSC application in asthma, radiation exposure, and neurological disorders has been explored as well [9]. Moreover, beside of having high regenerative potential these cells have been shown to carry immunosuppressive capacities, to improve angiogenesis and to prevent fibrosis [24].

Therapeutic applications of MSCs require massive in vitro expansion of the isolated cells [7, 21]. The populations typically can be expanded for up to 20 population doublings (PDs) until they enter a senescent state. Figure 1a shows a single cell-derived, expanding clone of MSCs. As shown in Fig. 1b the growth properties of such clones can vary largely during expansion. Various culture protocols have been suggested in order to isolate MSCs with high regenerative potential [18, 88, 92]. However, independent of the culture condition applied, massive replication of MSCs was found to be associated with continuous decline of the cell's functional competence, which was called "MSC ageing" [91]. During expansion the MSCs show a decreasing proliferation potential [84, 88]. Moreover, the efficiency of differentiating into local tissue after transplantation was found to severely decrease during expansion [75]. During prolonged in vitro culture MSCs frequently undergo spontaneous malignant transformation which represents a biohazard in long-term expansion [77]. Both ageing and transformation appear to be stochastic in nature and render the MSC populations an additional layer of heterogeneity.

In the following we discuss different hypotheses on the origin and development of MSC heterogeneity and introduce a model framework which allows addressing related questions. We show that a noise-driven approach to MSC differentiation combined with a model of intrinsic MSC ageing can explain MSC heterogeneity on both the individual and population level. We highlight open questions on MSC heterogeneity and suggest future investigations.

2 MSC Heterogeneity on Different Scales

Due to a growing body of evidence it is has been generally accepted that tissue stem cells are heterogeneous with regard to their function. This heterogeneity has been suggested to involve properties like their cycling activity, engraftment

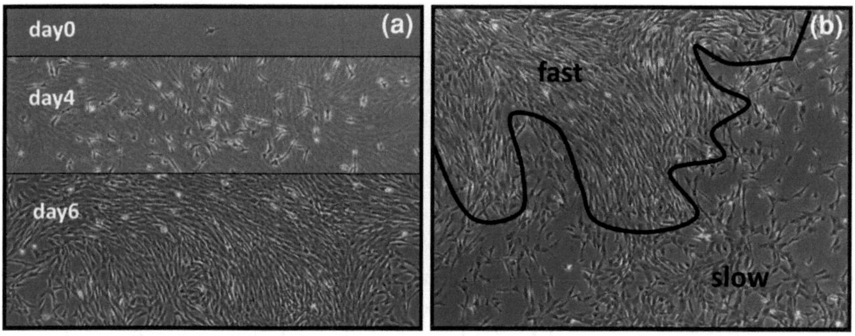

Fig. 1 Expansion of MSCs in vitro. **a** Massive expansion of an ovine MSC clone within 6 days of in vitro culture at standard culture conditions [101]. **b** Spatial competition of ovine MSC clones in vitro. A fast growing clone spreads into the area already occupied by a slow growing clone

potential or differentiation status as well as their expression of adhesion molecules or cell surface antigens [54]. Accordingly, a functional definition of tissue stem cells has been given by Loeffler and Roeder [53] that includes heterogeneity as a characteristic feature: 'Tissue stem cells are a potentially heterogeneous population of undifferentiated cells that are capable of proliferation and production of a large number of differentiated tissue cells for replenishing tissue continuously and after injury, while self-maintaining their population.'

However, isolation of stem cells for subsequent applications by applying these functional criteria is impossible, simply because a cell cannot be differentiated in different lineages while simultaneously maintaining its 'stem cell' state. As an alternative method marker systems have been studied in order to identify individual cells that carry stem cell properties. But, while well defined marker systems are available for other stem cells, e.g. for haematopoietic stem cells (HSCs, [27]), for MSCs such marker systems are still a matter of debate [42]. First effective protocols have been suggested more than ten years ago [67]. Currently, standard isolation protocols for human MSCs use combinations of up to ten surface markers. There is common agreement that more than 95% of the cells must express CD73, CD90 and CD105 and less than 2% CD35, CD45 together with CD11b or CD14 [24, 25, 48]. However, the MSCs isolated applying related protocols show still a large variance in the expression of these markers. Subpopulations of them with more defined expression have been demonstrated to carry different functional potential (e.g. Sca-1, [13]). Moreover, gene expression analysis of individual MSCs [82, 83] demonstrated MSC heterogeneity even on the single cell level. Despite a common molecular signature of potential multilineage differentiation capacity the analysed cells show considerable variance in expression.

During expansion MSCs adapt to the in vitro conditions and change their expression profile until they reach a senescent state as described in detail by Wagner et al. [94]. How this adaptation process affects MSC heterogeneity will be discussed below. The related changes in the expression profile are accompanied by morphological changes of the cells. Initially many of the cells are round and small, so-called RS cells, [14, 15]. But relatively fast most of the cells develop a fibroblast-like morphology. With increasing culture time they spread more and more and their morphology becomes irregular. This suggests using morphological cell properties, as e.g. cell size, as alternative sorting targets. Actually, characterisation of high proliferative subpopulations by cell size has been demonstrated [31]. Unfortunately, cell size does also change according to spontaneous transformation events during long term culture. Such transformations can occur frequently and can start early. They result in the emergence of a cell type characterised by an elevated proliferation, reduced plasticity and round and small morphology [2, 77]. Thus, transformed cells may be hardly distinguishable from high potential untransformed MSCs by morphological characterization only.

Interestingly, the time point of spontaneous transformation seems to be stochastic in nature like that of acquiring senescence [77]. Experimental results suggest that the transformed fate is not correlated to a defined regulatory state. So, transformed clones have been observed to differ in their expansion rate [77] as well as in the number of chromosomes present in the cells [2]. In the following we will focus on non-transformed MSCs and will neglect transformation as a particular source of population heterogeneity.

In summary, experimental findings demonstrate that MSC heterogeneity is a multi-scale phenomenon and is subject to significant changes during in vitro cultivation.

3 On the Origin of MSC Heterogeneity

While heterogeneity is accepted to be a characteristic property of stem cell populations, its origin is still not well understood. Heterogeneity of non-transformed functional stem cells has been discussed as a consequence (i) of cell adaptation to dynamic environments and (ii) of the flexibility and reversibility of stem cell fate decisions. Additional variance in cell fates may be associated with cell ageing.

3.1 Heterogeneity as a Consequence of MSC Environmental Adaptation (Extrinsic)

It has been demonstrated that MSCs from different tissues, including bone-marrow stroma, adipose, skeletal muscle, synovium and umbilical cord differ in both molecular and functional properties [39, 58, 63, 99]. Typically a large number of genes and proteins have been found to be differentially expressed (see e.g. [61, 62]) and, although the MSCs were expanded over many PDs in vitro, these differences in expression appeared to be conserved. Moreover, they also manifest in functional differences of the MSCs including their expansion and differentiation potential. For example experiments on human bone marrow MSCs revealed that about one-third of the clones are able to acquire phenotypes of pre-adipocytes, osteocytes and chondrocytes [67]. In contrast, only 1.4% of single MSCs isolated from adipose-derived adult stem cell populations were tri-potent, the others being bi-potent or uni-potent [102]. Dental MSCs were found to have less potential to differentiate into adipogenic and chondrogenic, but more potential for neural and odontogenic differentiation compared to bone marrow MSCs [39].

In agreement with these in vivo results in vitro culture conditions strongly impact MSC phenotypes. A large number of environmental factors such as oxygen [19, 41, 52, 101], glucose [91] and growth factors [5, 36] have been demonstrated to affect MSC expansion and differentiation. These processes are also affected by substrate stiffness [29], geometry [66], micro/nano-structure [20, 60] as well as surface chemistry [17]. Thus, one may expect that population heterogeneity observed in a specific environment (compare Fig. 2) is a consequence of adaptation of the MSCs to different microenvironments. Thereby, stem cells may either reside in microenvironments that support stem cell maintenance, so-called stem cell niches, or in activating environments that enforce their expansion. While e.g. HSC-niches are well described [27], MSC-niches are discussed controversial. Several studies suggest that MSCs reside in a perivascular niche in almost all adult tissues, where they associate with blood vessels [16]. However, they are also found in non-vascularised cartilage tissue where they show a well defined spatial distribution [68]. In fact, until now it has not been shown directly that heterogeneous microenvironments are responsible for the experimentally observed MSC population heterogeneity.

3.2 *Heterogeneity as a Consequence of MSC Self-Organisation (Flexibility)*

MSC differentiation and lineage specification have been found to be at least partially reversible, demonstrating a limited but significant plasticity of MSCs [87]. Moreover, plasticity has been described as a fundamental feature of these cells [100]. This is supported by recent experimental findings that they represent an excellent source for generating induced pluripotent stem cells [98] which appeared to be the closest equivalent to embryonic stem cells as demonstrated by DNA microarray gene profile and germline-transmission efficiency [64]. Thus, the question raises whether individual MSCs exhibit a stable tissue specific phenotype or whether they permanently change their regulatory states exhibiting a tissue specific phenotype only on population level.

Evidence that fate decisions of stem cells are reversible and that these cell populations self-organise permanently emerges from so-called regeneration experiments [10, 11]. In particular it has been shown that for a haematopoietic progenitor cell line the expression profile of the stem cell marker Sca-1 regenerates from different subpopulations. These experiments suggest fluctuations in the expression of such markers for each individual cell in a fixed environment. For primary MSCs such a regeneration of the tissue specific distribution of regulatory

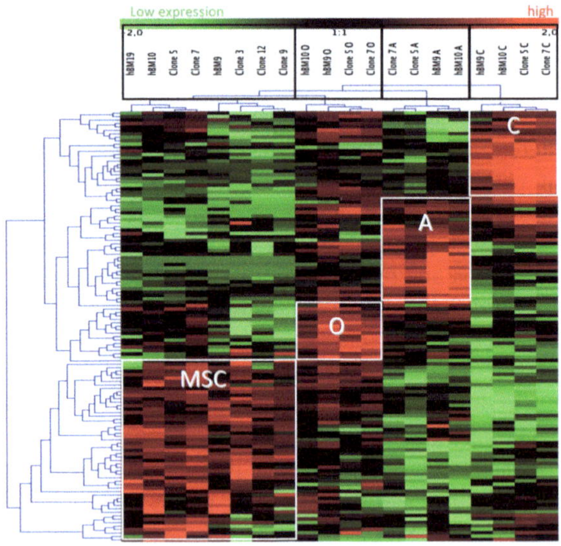

Fig. 2 Clonal heterogeneity of bone marrow MSCs. Hierarchical clustering using individual samples and genes down regulated after adipogenic (A), osteogenic (O), and chondrogenic (C) differentiation and genes implicated in A, O, and C differentiation. Individual samples are three primary MSC layers and five MSC clones before differentiation and two primary MSC layers and two MSC clones after A, O, and C differentiation. The individual samples show considerable differences in expression before and after differentiation. From: Delorme et al. [23]

states is expected within a few days [46]. However, such experiments have not been carried out so far. While such observations do not rule out the impact of microenvironments on the formation of heterogeneous populations, they suggest that stem cells populations can develop heterogeneity also independent of the presence of complex in vivo environments. However, a direct proof of the assumption of (stem cell) state fluctuations would require demonstrating them in individual cells in a defined environment. A prerequisite of such studies is long term single cell tracking, which has been successfully demonstrated for HSCs [72], and for which excellent new technology has been established [43].

In summary, there is evidence that MSCs are very flexible in their fate decisions and can adapt to a large variety of different environments. The experimental findings can be understood assuming that individual cells underlie permanent fluctuations of their regulatory states and that these fluctuations are modulated by the cell environment.

3.3 Heterogeneity as Consequence of MSC Ageing (Intrinsic)

The functional potential of MSCs has been shown to decrease throughout life. Under homeostatic conditions, there are limited demands on the self-renewing stem cells and so these cells divide infrequently, sparing stem cells the perils of

DNA-replication and mitosis. However, under stress the metabolic activity of the stem cells increases. The stem cells are exposed to higher levels of DNA-damage-inducing metabolic side products such as reactive oxygen species [85]. It has been suggested that the damages induced thereby impact not only the stem cells but the whole organism. In fact, excess replicative demands alone can induce progeroid phenotypes [78].

Recent experiments demonstrated that artificial in vitro ageing and in vivo ageing of MSCs induce related changes on the cellular as well as on the molecular scale [45, 94]. Thereby, pre-mature senescence has been implicated as a major cause of the in vitro decline in MSC function [90]. The accumulation of this phenotype, also called replicative senescence, has been demonstrated to be a continuous process in MSCs [95]. Interestingly, expansion at low oxygen pressure and low glucose culture decreases the number of accumulating senescent cells compared to high oxygen pressure and high glucose culture, respectively [89, 101]. Whether the accumulation of damage is actually pre-requisite for MSC ageing is still unknown. However, DNA-damage rarely affects the 2% of protein-coding sequences. Instead, it is expected to alter regulatory regions [69] and expression and function of non-coding RNAs that are involved in chromatin regulation [4]. Accordingly, an increasing number of ageing studies investigate the impact of epigenetic changes, such as DNA hyper/hypo-methylation, and histone modifications [65, 86]. The results suggest that decreasing accessibility of certain regulatory states of MSCs due to epigenetic remodelling may represent an alternative or at least a complementary explanation of ageing.

Without any question, the distribution of MSC states changes over time, i.e. with age and in a tissue specific manner. Whether this is due to changing environments or due to epigenetic remodelling and accumulation of damage is currently not fully understood.

4 General Model Approaches

A large number of theoretical approaches to MSC culture aim at quantitatively describing culture conditions and their impact on processes such as matrix formation [81]. Theoretical models of MSC expansion and differentiation are rather rare [22, 51] and do not include single cell-level population heterogeneity. In order to provide reliable predictions on the dynamics of such systems, theoretical approaches are required that account for: (i) composition and structure of the cell environment and (ii) particular stem cell properties such as functional differentiation and self-renewal in individual cells. Currently there are different concepts to approach these problems in general. In particular there are different concepts of modelling stem cell organisation. The most prominent are the ‘pedigree concept’ and the ‘plasticity concept’.

The pedigree concept treats ‘stemness’ as a property that if once lost is lost forever. In the models that obey this concept this loss can be a deterministic or a stochastic process. Accordingly the development of individual progenitors may more or less differ. However, over time they all will approach a defined state (Fig. 3a). Stem cells inherit ‘stemness’ by performing asymmetric cell divisions. This concept was very successful in describing the hierarchical organisation of tissues. However, experimental results on HSCs have led to the development of a concept that allows for more flexibility, the plasticity concept. According to this concept cells can loose and gain stem cell properties. If this applies to all possible states of a population, the states approach a stationary distribution which will be in general a broad distribution (Fig. 3b). In the following we will focus on models based on the plasticity concept.

The reversibility and stochasticity of cellular fate decisions has been studied by Loeffler and Roeder [54]. In their models [34, 73, 74] individual cells gain and loose stem cell properties depending on whether they localise inside or outside a specific niche environment, respectively. Thus, the environment directs the cellular fate and the reversibility of cell fate decisions is enabled by probabilistic switches between different micro-environments. The models were successfully applied to in vivo organisation of normal and malignant HSC populations.

However, MSC populations have been shown to expand while maintaining stem cell properties also in a homogenous environment. For modelling these systems we have expanded the ideas of Loeffler and Roeder by assuming that cells gain and loose stem cell properties according to a probabilistic process whose state-specific amplitudes are set by the environment. Within this approach cell fate decisions are basically reversible. The assumed fluctuations are hypothesised to be generated by intra- and extracellular noise triggering random transitions between different regulatory network activation patterns. This assumption is supported by experimental findings demonstrating that epigenetic gene silencing has a strong stochastic component [70, 96]. In the following we will give a brief description of a MSC population model that is based on this approach.

5 The Concept of Noise-Driven Differentiation

A growing body of evidence indicates that noise is not generally detrimental to biological systems but can be employed to generate genotypic, phenotypic, and behavioural diversity [44, 79, 80]. In particular, noise-driven solutions are expected to prevail in cellular adaptation to variable environments. Moreover, it has been proposed that biological systems have built-in molecular devices for noise control [1, 3, 28]. Together with the experimental results on MSCs, reviewed above, this has led us to suggest a model of noise-driven differentiation [37, 47], which will be introduced in the following.

5.1 General Assumption

In the model of noise-driven differentiation, cell differentiation is defined as loss of stem cell properties. It is quantified by a continuous state variable α that can adopt values between 0 (full stem cell competency) and 1 (fully differentiated cell).

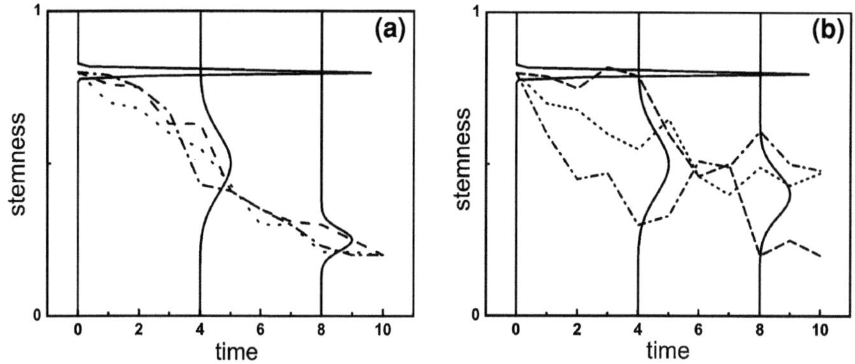

Fig. 3 Cell fate trajectories of individual cells according to different stem cell concepts. Cells of a defined differentiation state were selected (t = 0) and cultivated. **a** Pedigree concept: The cells loose stem cell properties due to a random process. Over time they accumulate again in a defined but more differentiated state. **b** Plasticity concept: The cells loose and gain stem cell properties in any state. The population approaches a stationary distribution over time. *Solid lines* denote sketches of the distribution of cell states at different time points. *Dashed lines* are trajectories of individual cells

Each value of α may represent a set of regulatory network activation patterns. From the molecular point of view, α may depend on abundance and subcellular localization of proteins and RNAs, as well as other types of signalling and metabolic molecules [50]. In general, cell differentiation is assumed to be reversible.

Each cell's α -value fluctuates randomly with a state-dependent noise amplitude $\sigma(\alpha)$ (Fig. 4a). From its current α-value a cell adopts a new value α' with a randomization rate R which may in general depend on α. We assume R to be constant. α' is drawn from a Gaussian distribution $p(\alpha'|\alpha)$, centred around α with standard deviation $\sigma(\alpha)$.

$$p(\alpha'|\alpha) \propto \exp(-(\alpha' - \alpha)^2/2\sigma^2(\alpha)) \tag{1}$$

The state dependence of $\sigma(\alpha)$ is assumed to be determined by the environment. We describe this dependency by:

$$\sigma(\alpha) = \sigma_0[1 - \alpha f(E)] \geq 0 \tag{2}$$

Here, σ_0 denotes the noise amplitude for stem cells, i.e. $\sigma(\alpha = 0)$. f(E) is a function describing the environmental impact. Positive fluctuation amplitudes require that $f(E) < 1$. In a simple approach it can be a constant $f(E) = f_0$. In this case $f_0 < 0$ describes stem cell supporting environments and $1 > f_0 > 0$ describes differentiating environments [37]. Differentiation is assumed to occur independently of cell proliferation as found in progenitor systems [8].

In contrast, cell proliferation is assumed to be differentiation state dependent. We assume that only cells in intermediate differentiation states with $\alpha_s < \alpha < \alpha_d$ proliferate (see Fig. 4b). For these states we assume an identical growth time τ.

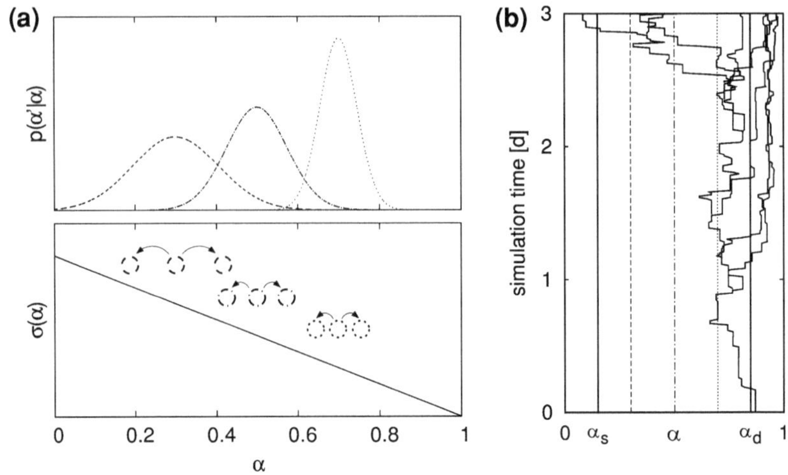

Fig. 4 Mechanisms of noise driven MSC differentiation. **a** Modelling fluctuations of the differentiation state α. *Upper panel*: Gaussian conditional probability function $p(\alpha'|\alpha)$ for the transition $\alpha \to \alpha'$ for $\alpha = 0.3$, 0.5 and 0.7. *Lower panel*: A decrease of the noise amplitude $\sigma(\alpha)$ with α results in an accumulation of cells at higher values of α. **b** A pedigree of a differentiated, quiescent cell ($\alpha > \alpha_d$) illustrating the model concept. After some time the cell regains proliferative capacity ($\alpha_s < \alpha < \alpha_d$) and generates a number of progeny. Moreover, there is a non vanishing probability of even regaining stem cell properties ($\alpha < \alpha_s$)

Stem cells ($\alpha < \alpha_s$) and differentiated cells ($\alpha > \alpha_d$) do not proliferate. During the growth process cells may frequently switch between proliferative and non-proliferative states. This will result in an effective cell growth time larger than τ.

The rate of randomization R is a model parameter that can be used for fitting experimental data. For MSCs it has been set to about 1 update per hour for a stem cell noise-amplitude of $\sigma_0 = 0.15$. Growth times are given by experimental observed minimal cell cycles times, which is about 11 h for MSC. The choice of α_s and α_d is somewhat arbitrary, as there is no clear phenotype related to it. We have chosen $\alpha_s = 0.15$ and $\alpha_d = 0.85$. Details can be found in Krinner et al. [47].

A straightforward method for quantifying the population heterogeneity is to calculate the normalised Shannon entropy from the distribution of the α states (here for a histogram of n bins):

$$H = \frac{-1}{\ln(n)} \sum_n p(\alpha_n) \ln(p(\alpha_n)) \qquad (3)$$

Figure 5 shows the dependence of H on f(E) assuming a constant $f(E) = f_0$ between 0 and 1. H decreases with increasing f(E). However, the distribution remains heterogeneous (close to maximum entropy) as long as f(E) gets close to 1. These changes are much more pronounced in quiescent populations $1/\tau = 0$ compared to proliferating populations $1/\tau > 0$. Thus, very efficient noise control mechanisms are required in order to generate a homogeneous proliferating population. It appears that this is given under artificial in vitro conditions, as e.g. high oxygen tension, leading to fast accumulation of cells in senescent states [101].

5.2 *Exemplifying the Environmental Impact*

A factor that strongly impacts MSC proliferation and differentiation is oxygen. It has been shown that MSC-derived cell populations show higher proliferative activity when cultured under low oxygen tension (2–5% pO_2) compared to high oxygen tension (20–21% pO_2) [19, 52, 71]. Moreover, cell populations expanded at low oxygen tension show a faster and more directed differentiation into osteoblasts, adipocytes [35, 52] and chondrocytes [59, 97]. As shown in Fig. 5d, our model predicts that increased proliferation increases the population heterogeneity. This would suggest describing the oxygen dependence in terms of the growth rate, i.e. by assuming a higher growth rate $1/\tau$ at low oxygen tension. However, in case of a proliferation stop, e.g. according to contact inhibition of growth (see below), cells at 5% pO_2 would acquire a senescent state as fast as at 20% pO_2, which has not been observed. Thus, we decided to model the observed effects of oxygen tension by changes of the noise-profile exclusively.

Actually, we assumed that cell adaptation to non-physiological high oxygen tension results in decreased state fluctuations which subsequently lead to an accumulation of cells in differentiated states. In contrast, low oxygen tension conserves stem cell and progenitor states by enabling high amplitudes of the state fluctuation. We described this dependency by a Hill function $f(pO_2/pO_2^{max})$ approaching 0 and 1 at low and high pO_2, respectively:

$$\sigma(\alpha) = \sigma_0\left[1 - \alpha f\left(pO_2/pO_2^{max}\right)\right] \text{ with } f(X) = \frac{X^n}{X^n + K^n} \tag{4}$$

The related 'noise landscape' is shown in Fig. 6 together with the differentiation state profiles for 5% and 20% pO_2. The parameters of the profiles were obtained by reproducing the clonal growth properties at these conditions [47]. Using these distributions as input we were able to simulate the experimentally observed impact of low oxygen expansion on subsequent differentiation in pellet culture which appears to be a result of limited lineage plasticity of MSC populations.

5.3 *Impact of Cell–Cell Interactions (Contact Inhibition of Growth)*

The above assumptions exclusively consider intrinsic regulation of the MSC state. The impact of cell-cell interaction has been neglected so far. However, it is well known that MSC expansion, lineage specification and terminal differentiation

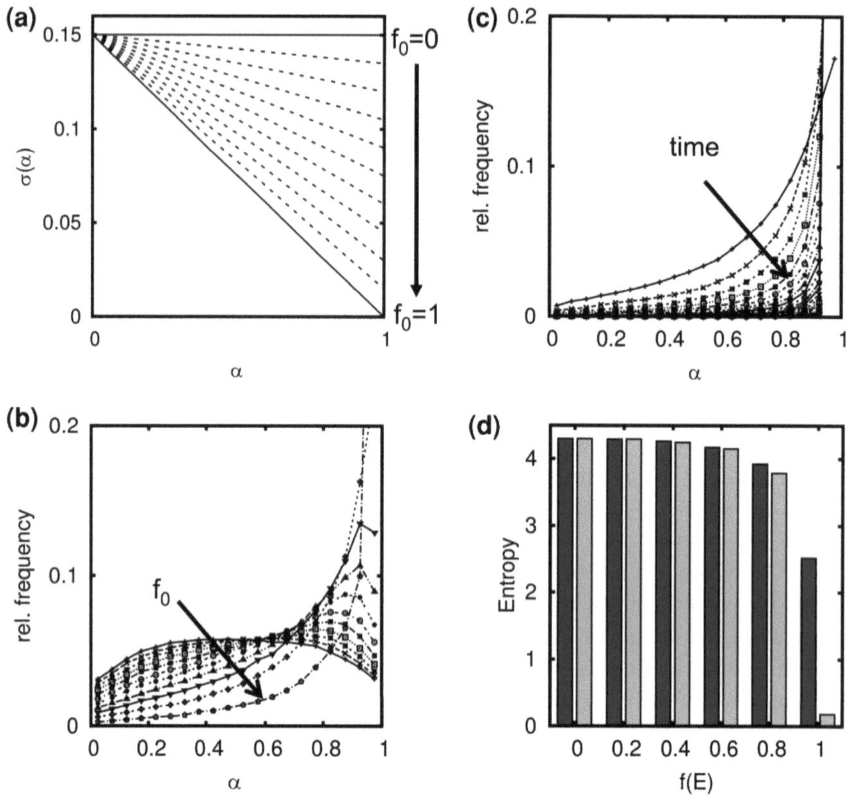

Fig. 5 Simulation results on noise driven MSC differentiation. **a** Noise-profiles with monotonously increasing f(E): $0 < f(E) = f_0 < 1$ were applied. **b** The calculated differentiation profiles of the MSCs demonstrate accumulation of the cells in differentiated states $\alpha > \alpha_d$. The amount of differentiated cells increases with f_0. **c** Equilibration of the profiles occurs on the scale of a few days. Shown are snapshots of a fast equilibrating system with $1/\tau = 0$ after switching f(E) from 0 to 1 (time steps $\Delta t = 3$ h). Eventually all cells accumulate in the fully differentiated state (bin: $0.95 < \alpha < 1.00$). Except for the first profile at 3h (+) its fraction exceeds 0.2. **d** The normalised entropy of the α-distributions strongly decreases if f(E) gets close to one. For $1/\tau = 11$ h (*dark grey*) the systems equilibrate as shown in **b**. For $1/\tau = 0$ (*light grey*) the tendency to differentiate is more pronounced and more homogeneous populations are formed

depend or even require such interactions [93]. A well known control mechanism of proliferation based on cell-cell interaction is contact inhibition of growth. According to this mechanism cells that form close contacts stop proliferation. Preventing this phenomenon by applying sophisticated culture conditions can significantly enlarge cell culture harvest [57].

In order to consider such regulation individual cell-based models of cell populations have been established [26, 32, 33]. In these models each individual cell is described by a physical object that can deform, adhere to other cells or a

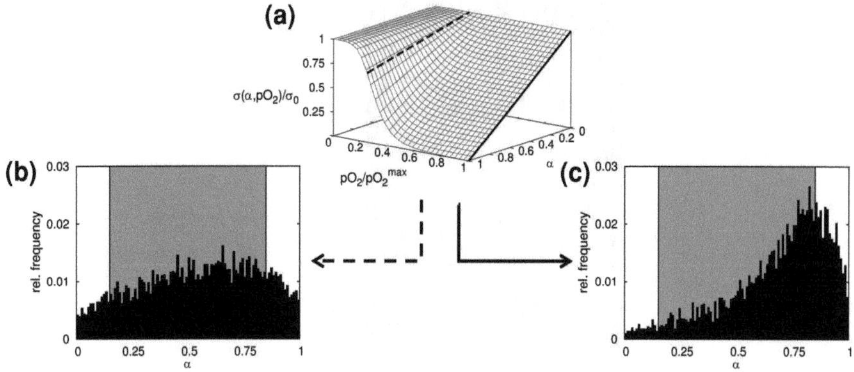

Fig. 6 Noise-landscapes of MSCs in dependence of the oxygen tension. **a** A low oxygen environment (5% pO_2) is characterised by high noise amplitudes in all states (*dashed line*), while a high oxygen environment (20% pO_2) is characterised by low amplitudes in high α-states (*solid line*). **b**, **c** The α-profiles of the low and high oxygen environments for $\tau = 11$ h

substrate, can move, grow and divide. Cell proliferation is modelled assuming a two phase cell cycle where during the interphase, a cell doubles its volume by stochastic increments and during the mitotic phase, a cell divides into two daughter cells of equal properties. Thereby, cell shape is often approximated by a sphere (see Fig. 7a) and the elastic deformation of a cell subject to compression by other cells or substrate is modelled by the Hertz-Model [49]. Contact inhibition of growth is considered by these models assuming that a cell stops growth if the sum of contact forces on it exceeds a critical threshold value.

Simulating monoclonal expansion by applying such an individual cell-based model one observes a specific distribution of proliferating cells within the growing clone [26, 32], with quiescent cells located in the core and proliferating cells at the periphery of the clone. In such simulations one can simply record the pedigree of the clone and can calculate the spatial distribution of the different generations throughout it. Thereby, the generation numbers are uniquely defined for all cells by the recursion $m_l = m_m = m_k + 1$ for the daughter cells l and m of mother cell k; i.e. m counts for the number of divisions that has been carried out until the cell was born. The result of such a simulation is shown in Fig. 7c. One observes a heterogeneous distribution with 'younger' cells in the core and 'older' cells at the periphery. Within about 12 days differences in 'age' of more than ten generation have been established.

Obviously, MSC shape is very different from being spherical-like. Considering that the cell–cell interactions resulting in contact inhibition of growth depend on the cell shape, the reliability of simulation results, as shown in the top row of Fig. 7, seems to be questionable. We therefore developed a more sophisticated model of MSCs that explicitly accounts for podia formation [38]. This approach builds on the model assuming spherical cell bodies. In addition, cells feature podia that generate forces for cell spreading and movement. Podia of model cells retract

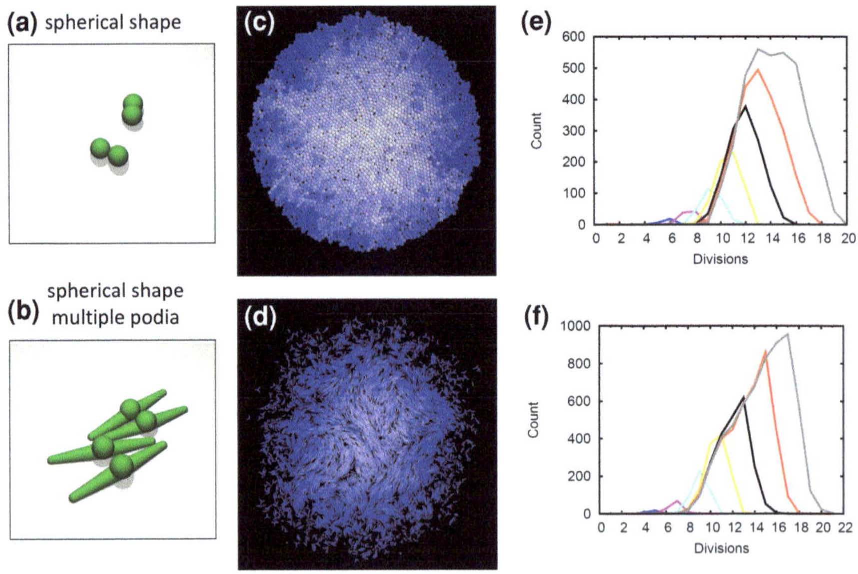

Fig. 7 Formation of an age distribution within expanding MSC clones. Comparison of a simple model assuming spherical cell shape (**a**, **c**, **e**) with an advanced model assuming that cells do form multiple podia (**b**, **d**, **f**). **a**, **b** Cell shape of the respective model. **c**, **d** Due to active contact inhibition the growing cells in the centre of the populations stop proliferation. Accordingly, a gradient in generation number is formed, here indicated by colour saturation. *White* cells are of generation 8, and *dark blue* cells of generation 22. **e**, **f** Development of the distribution of generation numbers in a growing colony over 8 days (*blue*: d2, *magenta*: d3, *cyan*: d4, *yellow*: d5, *black*: d6, *orange*: d7, *grey*: d8). While the formation of an age gradient is a generic phenomenon, the details of the distribution of generation numbers depend on the biophysical model.

prior to cell division and align to each other as is also experimentally observed in proliferating MSCs in vitro (see Fig. 1b, [15]). In this model cell migration is accomplished by protrusion and traction forces exerted by model podia. The number of podia is dynamically controlled by adaptation of the probabilities of podium generation and inactivation. The migration phenotype largely differs between cells with only one active podium (mostly ballistic movement with random turns) and cells with multiple active podia (mostly stretched out and resting with random reorientation moves). Details can be found in Hoffmann et al. [38].

Simulating monoclonal growth applying this podia model one observes results comparable to those observed in the simple model. In particular this regards the spatial distribution of cell generations (see Fig. 7d). However, quantitative differences can be observed depending on the choice of the model parameters, which include parameters determining cell friction forces, podia lengths and activation/deactivation rates as well as the parameters defining the sensitivity of the cells to contact inhibition. Regardless of these differences the age structure appears to be a generic feature of growing cell populations. Nevertheless, it has been not considered in MSC model approaches so far.

6 The Case of Stable, Inherited Heterogeneity: MSC Ageing

In order to define age dependent properties of the MSC populations we extended our former approach and assumed that stem cell states become de-stabilised with 'cellular age' resulting in an increased tendency for spontaneous differentiation. We model this scenario assuming that the noise amplitude of a cell $\sigma(\alpha)$ depends on its generation number m. Accordingly, each individual cell is characterised by its noise amplitude, which depends on the cell's differentiation state α and age m:

$$\sigma(\alpha, m) = \sigma_0[1 - \alpha f(E)] + m\, r_D[1 - 2\alpha] \geq 0. \tag{5}$$

The first term on the right hand side of Eq. 5 defines the extrinsic, environmentally determined noise amplitude (see Eq. 2). Here, σ_0 denotes the noise amplitude for initial stem cells, i.e. $\sigma(\alpha = 0,\ m = 0)$. The second term on the right hand side defines the effect of ageing, which is most obvious considering the noise amplitudes in stem cell states $\sigma(\alpha = 0)$:

$$\sigma(\alpha = 0) = \sigma_0 + m r_D. \tag{6a}$$

This stem cell noise amplitude increases with each generation by the rate r_D. This assumption allows us to quantify the 'age' of a cell by its stem cell noise amplitude $\sigma(\alpha = 0)$. The noise amplitude of differentiated cells is given by:

$$\sigma(\alpha = 1) = \sigma_0[1 - f(E)] - m r_D > 0 \tag{6b}$$

This amplitude decreases with m. For $m_{RE} = \sigma_0[1 - f(E)]/r_D$ it is equal to zero. In order to ensure that $\sigma(\alpha = 1) > 0$, we assume that σ becomes independent of m for $m > m_{RE}$.

According to these assumptions the heterogeneity of a population can be described by the probability distribution to find a cell of age m and in state α. During expansion the average number of cell divisions grows, thus noise amplitudes in differentiated states decrease and, consequently, cells accumulate in these states. This effect is independent of the environment. It is determined by the ageing rate r_D. This rate determines also the age m_{RE} of replicative senescence. Assuming $r_D \sim 2.5 \times 10^{-3}$ one obtains an upper limit of m_{RE} of about 60. This can be seen as an upper bound to the experimental findings so far (<30 at 20% pO_2, [95] and <40 at 3% pO_2, [30]).

In simulations of the model we assumed for the environmental term of the noise amplitude:

$$f(E) = 2\ (1 - \sigma_E/\sigma_0), \tag{7}$$

where σ_E is the mean noise amplitude defined by the environment.

Figure 8 shows results on the simulated age-structure and population heterogeneity applying the age model and assuming $\sigma_e = \sigma_0 = 0.075$. After 20 days of

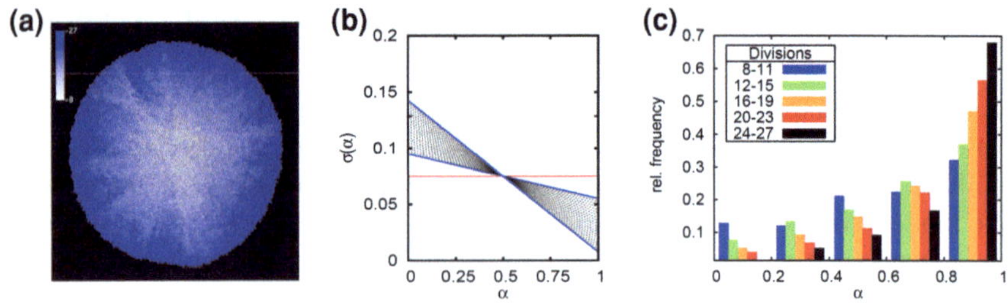

Fig. 8 Simulation results applying the ageing-model. **a** The observed distribution of the generations is comparable to that observed without ageing. **b** Age-dependent noise profiles as set before (*red*) and observed after expansion (*grey*). A considerable heterogeneity regarding age has developed. **c** Distribution of differentiation states depending on the age. Young cells (*blue*, m = 8–11) include a relevant fraction of undifferentiated cells with $\alpha < 0.2$, while most of the old cells (*black*, m = 24–27) reside in a differentiation state of $\alpha > 0.8$

expansion a few cells of the population have nearly reached $\sigma(\alpha = 0) = 0.15$ as assumed in the other applications. However, a large heterogeneity regarding age has been established (7 < m < 28). With increasing age the cells tend to accumulate more and more in differentiated, senescent states (here, $\alpha > 0.8$). This is in nice agreement with experimental observations on ageing MSCs [45, 95].

As found in our simulation studies comparing MSCs that were expanded at high and low oxygen tension (Sect. 5.2, [47]), these differences between young and old cells in the distribution of the differentiation states will also manifest in differences in their lineage specification and functional differentiation potential. Accordingly, aged cells are characterised by a lower regenerative potential compared to young cells. This is again in agreement with experimental findings [45, 95].

In summary the model is capable of describing experimental findings on the environmental dependent MSC ageing. The underlying concept is general in the sense that in vitro and in vivo ageing can be assumed to base on the same principles. This is supported by recent experimental findings demonstrating that ageing and replicative senescence have related effects on stem cells [94] .

7 Discussion

Grafting of MSCs is an emerging technology to repair tissues and organs of mesenchymal origin. A prerequisite of these applications is rapid and massive MSC expansion. This becomes obvious considering the isolation process of these cells. Bone marrow is an important source of MSCs from where they can be easily isolated and purified. However, only about 0.01% of nucleated bone marrow cells carry MSC properties. Accordingly from a 20 ml aspirate usually only up to 50.000 MSCs are obtained [6]. Consequently, at least 5PDs are required in order to obtain the 10^6–10^7 cells that are required in a typical application.

While even the isolated populations are heterogeneous regarding different functional aspects, this heterogeneity further evolves during amplification. Thereby the actual changes depend on the details of the culture conditions which will essentially impact the cell harvest. Thus, to understand the origin and development of MSC heterogeneity will help to improve their cell culture conditions. Moreover, it represents a basic step towards individualised therapies [31].

We here presented a mathematical approach to MSC populations which allows considering each cell individually. Thus, the approach enables us to simulate the changes in the population behaviour based on changes in single cells which are assumed to depend on their particular regulatory state. Our model allows simulations to closely follow standard in vitro expansion and differentiation protocols of MSCs such that they may be viewed as experiments 'in silico' (on the computer) and the results can directly be compared to those found from experiments.

The approach presented here assumes that noise is predominant in most cellular states. Its essence is that MSC population structures are determined by state-specific noise [37] forming a 'noise landscape', where low noise states represent the attractor states. Cells subjected to an environment not matching their internal state are assumed to be destabilised by a high noise amplitude. They subsequently adapt to this environment by travelling towards low noise states. Extensions of the proposed noise-driven approach to lineage specification and functional differentiation have been described [47]. Thereby, decision making in individual cells during these processes was linked to particular differentiation states and cell–cell interactions. In particular, lineage specification was assumed to require sufficiently high stemness, i.e. α values below a certain threshold. In this application MSC ageing was not considered. However, while changing the distribution of the differentiation states, obviously ageing will also affect the lineage specification dynamics, with aged cells rarely or never able to switch lineage. Accordingly, we expect also functional differentiation to be hampered.

A question not raised so far is whether there is a functional relevance of MSC heterogeneity regarding expansion and differentiation. The observed heterogeneity of MSCs could actually represent a functional aspect of their organisation. For example high proliferation potential of a subpopulation could originate from secretory properties of another subpopulation in the close neighbourhood. In a recent study Krinner et al. analysed the expansion and differentiation properties of single cell derived clones and compared them with the properties of their mother clone (unpublished results). As the high expansion potential of individual clones outcompetes that of their mother population, they were not able to provide any evidence for a specification of subclones into fast expanding clones and expansion supporting secretory clones.

In order to get deeper insights into MSC organisation a detailed characterisation of the microenvironments in which MSCs reside in vivo is required. Effective methods for screening MSC properties in complex in vitro environments have been already developed [55, 56]. Moreover, regeneration experiments on primary MSCs as suggested in Krinner et al. [46] will help to understand MSC adaptation to changing environments and to identify properties that are inherited. This would also help to solve the question whether proliferation alone (as assumed in the models presented) or/and adaptation to changing environments is a source of persistent clonal differences. For this purpose FACS analysis of selected markers should be combined with gene expression analysis as demonstrated by Chang and co-workers [10]. Moreover, time-lapse online analysis of the expression of stemness and differentiation markers by individual MSCs could not only provide direct evidence for the assumptions made in the noise-driven differentiation model, it would also delineate the emerging structure of noise-landscapes.

The proposed model is capable of explaining an entire panel of experimental observations regarding MSC heterogeneity. However, the molecular basis of the assumed noise-profiles and their dependence on the differentiation states, the environment and age remain speculative. Models have been suggested that involve e.g. Wnt-pathway activity [3]. A general model framework linking noise-landscapes to the dynamics of regulatory networks is missing. Thus, building up multi-scale models that bridge the current gap between the increasing amount of molecular data and observed cellular phenotypes represents a current need in order to improve our understanding of the heterogeneity of MSCs in vitro. Their validation will require sophisticated experimental studies on the single cell level.

References

1. Acar, M., Becskei, A., van Oudenaarden, A.: Enhancement of cellular memory by reducing stochastic transitions. Nature **435**(7039), 228–232 (2005)
2. Ahmadbeigi, N., Shafiee, A., Seyedjafari, E., Gheisari, Y., Vassei, M., Amanpour, S., Amini, S., Bagherizadeh, I., Soleimani, M.: Early spontaneous immortalization and loss of plasticity of rabbit bone marrow mesenchymal stem cells. Cell Prolif. **44**(1), 67–74 (2011)
3. Arias, A.M., Hayward, P.: Filtering transcriptional noise during development: concepts and mechanisms. Nat. Rev. Genet. **7**(1), 34–44 (2006)
4. Backofen, R., Bernhart, S.H., Flamm, C., Fried, C., Fritzsch, G., Hackermuller, J., Hertel, J., Hofacker, I.L., Missal, K., Mosig, A., Prohaska, S.J., Rose, D., Stadler, P.F., Tanzer, A., Washietl, S., Will, S.: RNAs everywhere: genome-wide annotation of structured RNAs. J. Exp. Zool. B. Mol. Dev. Evol. **308**(1), 1–25 (2007)
5. Ball, S.G., Shuttleworth, C.A., Kielty, C.M.: Mesenchymal stem cells and neovascularization: role of platelet-derived growth factor receptors. J. Cell. Mol. Med. **11**(5), 1012–1030 (2007)
6. Braccini, A., Wendt, D., Jaquiery, C., Jakob, M., Heberer, M., Kenins, L., Wodnar-Filipowicz, A., Quarto, R., Martin, I.: Three-dimensional perfusion culture of human bone marrow cells and generation of osteoinductive grafts. Stem Cells **23**(8), 1066–1072 (2005)
7. Brooke, G., Cook, M., Blair, C., Han, R., Heazlewood, C., Jones, B., Kambouris, M., Kollar, K., McTaggart, S., Pelekanos, R., Rice, A., Rossetti, T., Atkinson, K.: Therapeutic applications of mesenchymal stromal cells. Semin. Cell. Dev. Biol. **18**(6), 846–858 (2007)
8. Brown, G., Drayson, M.T., Durham, J., Toellner, K.M., Hughes, P.J., Choudhry, M.A., Taylor, D.R., Bird, R., Michell, R.H.: HL60 cells halted in G1 or S phase differentiate normally. Exp. Cell Res. **281**(1), 28–38 (2002)
9. Caplan, A.I.: Why are MSCs therapeutic? New data: new insight. J. Pathol. **217**(2), 318–324 (2009)

10. Chang, H.H., Hemberg, M., Barahona, M., Ingber, D.E., Huang, S.: Transcriptome-wide noise controls lineage choice in mammalian progenitor cells. Nature **453**(7194), 544–547 (2008)
11. Chang, H.H., Oh, P.Y., Ingber, D.E., Huang, S.: Multistable and multistep dynamics in neutrophil differentiation. BMC. Cell Biol. **7**, 11 (2006)
12. Chen, F.H., Tuan, R.S.: Mesenchymal stem cells in arthritic diseases. Arthritis Res. Ther. **10**(5), 223 (2008)
13. Choi, J., Curtis, S.J., Roy, D.M., Flesken-Nikitin, A., Nikitin, A.Y.: Local mesenchymal stem/progenitor cells are a preferential target for initiation of adult soft tissue sarcomas associated with p53 and Rb deficiency. Am. J. Pathol. **177**(5), 2645–2658 (2010)
14. Colter, D.C., Class, R., DiGirolamo, C.M., Prockop, D.J.: Rapid expansion of recycling stem cells in cultures of plastic-adherent cells from human bone marrow. Proc. Natl. Acad. Sci. U S A **97**(7), 3213–3218 (2000)
15. Colter, D.C., Sekiya, I., Prockop, D.J.: Identification of a subpopulation of rapidly self-renewing and multipotential adult stem cells in colonies of human marrow stromal cells. Proc. Natl. Acad. Sci. U S A **98**(14), 7841–7845 (2001)
16. Crisan, M., Yap, S., Casteilla, L., Chen, C.W., Corselli, M., Park, T.S., Andriolo, G., Sun, B., Zheng, B., Zhang, L., Norotte, C., Teng, P.N., Traas, J., Schugar, R., Deasy, B.M., Badylak, S., Buhring, H.J., Giacobino, J.P., Lazzari, L., Huard, J., Peault, B.: A perivascular origin for mesenchymal stem cells in multiple human organs. Cell Stem Cell **3**(3), 301–313 (2008)
17. Curran, J.M., Chen, R., Hunt, J.A.: The guidance of human mesenchymal stem cell differentiation in vitro by controlled modifications to the cell substrate. Biomaterials **27**(27), 4783–4793 (2006)
18. D'Ippolito, G., Diabira, S., Howard, G.A., Menei, P., Roos, B.A., Schiller, P.C.: Marrow-isolated adult multilineage inducible (MIAMI) cells, a unique population of postnatal young and old human cells with extensive expansion and differentiation potential. J. Cell Sci. **117**(Pt 14), 2971–2981 (2004)
19. D'Ippolito, G., Diabira, S., Howard, G.A., Roos, B.A., Schiller, P.C.: Low oxygen tension inhibits osteogenic differentiation and enhances stemness of human MIAMI cells. Bone **39**(3), 513–522 (2006)
20. Dalby, M.J., Gadegaard, N., Tare, R., Andar, A., Riehle, M.O., Herzyk, P., Wilkinson, C.D.W., Oreffo, R.O.C.: The control of human mesenchymal cell differentiation using nanoscale symmetry and disorder. Nat. Mater. **6**(12), 997–1003 (2007)
21. Dazzi, F., Horwood, N.J.: Potential of mesenchymal stem cell therapy. Curr. Opin. Oncol. **19**(6), 650–655 (2007)
22. Deasy, B.M., Jankowski, R.J., Payne, T.R., Cao, B., Goff, J.P., Greenberger, J.S., Huard, J.: Modeling stem cell population growth: incorporating terms for proliferative heterogeneity. Stem Cells **21**(5), 536–545 (2003)
23. Delorme, B., Ringe, J., Pontikoglou, C., Gaillard, J., Langonne, A., Sensebe, L., Noel, D., Jorgensen, C., Haupl, T., Charbord, P.: Specific lineage-priming of bone marrow mesenchymal stem cells provides the molecular framework for their plasticity. Stem Cells **27**(5), 1142–1151 (2009)
24. Djouad, F., Bouffi, C., Ghannam, S., Noel, D., Jorgensen, C.: Mesenchymal stem cells: innovative therapeutic tools for rheumatic diseases. Nat. Rev. Rheumatol. **5**(7), 392–399 (2009)
25. Dominici, M., Le Blanc, K., Mueller, I., Slaper-Cortenbach, I., Marini, F., Krause, D., Deans, R., Keating, A., Prockop, D., Horwitz, E.: Minimal criteria for defining multipotent mesenchymal stromal cells. The International Society for Cellular Therapy position statement. Cytotherapy **8**(4), 315–317 (2006)
26. Drasdo, D., Hoehme, S., Block, M.: On the Role of Physics in the Growth and Pattern Formation of Multi-Cellular Systems: What can we Learn from Individual-Cell Based Models? J. Stat. Phys. **128**, 287–345 (2007)

27. Ehninger, A., Trumpp, A.: The bone marrow stem cell niche grows up: mesenchymal stem cells and macrophages move in. J. Exp. Med. **208**(3), 421–428 (2011)
28. El-Samad, H., Khammash, M.: Regulated degradation is a mechanism for suppressing stochastic fluctuations in gene regulatory networks. Biophys. J. **90**(10), 3749–3761 (2006)
29. Engler, A.J., Sen, S., Sweeney, H.L., Discher, D.E.: Matrix elasticity directs stem cell lineage specification. Cell **126**(4), 677–689 (2006)
30. Fehrer, C., Brunauer, R., Laschober, G., Unterluggauer, H., Reitinger, S., Kloss, F., Gülly, C., Gassner, R., Lepperdinger, G.: Reduced oxygen tension attenuates differentiation capacity of human mesenchymal stem cells and prolongs their lifespan. Aging Cell **6**(6), 745–757 (2007)
31. Galle, J., Bader, A., Hepp, P., Grill, W., Fuchs, B., Kas, J.A., Krinner, A., Marquass, B., Muller, K., Schiller, J., Schulz, R.M., von Buttlar, M., von der Burg, E., Zscharnack, M., Loffler, M.: Mesenchymal stem cells in cartilage repair: state of the art and methods to monitor cell growth, differentiation and cartilage regeneration. Curr. Med. Chem. **17**(21), 2274–2291 (2010)
32. Galle, J., Hoffmann, M., Aust, G.: From single cells to tissue architecture-a bottom-up approach to modelling the spatio-temporal organisation of complex multi-cellular systems. J. Math. Biol. **58**(1–2), 261–283 (2009)
33. Galle, J., Loeffler, M., Drasdo, D.: Modeling the effect of deregulated proliferation and apoptosis on the growth dynamics of epithelial cell populations in vitro. Biophys. J. **88**(1), 62–75 (2005)
34. Glauche, I., Cross, M., Loeffler, M., Roeder, I.: Lineage specification of hematopoietic stem cells: mathematical modeling and biological implications. Stem Cells **25**(7), 1791–1799 (2007)
35. Grayson, W.L., Zhao, F., Bunnell, B., Ma, T.: Hypoxia enhances proliferation and tissue formation of human mesenchymal stem cells. Biochem. Biophys. Res. Commun. **358**(3), 948–953 (2007)
36. Heng, B.C., Cao, T., Lee, E.H.: Directing stem cell differentiation into the chondrogenic lineage in vitro. Stem Cells **22**(7), 1152–1167 (2004)
37. Hoffmann, M., Chang, H.H., Huang, S., Ingber, D.E., Loeffler, M., Galle, J.: Noise-driven stem cell and progenitor population dynamics. PLoS One **3**(8), e2922 (2008)
38. Hoffmann, M., Kuska, J.P., Zscharnack, M., Loeffler, M., Galle, J.: Spatial organization of mesenchymal stem cells in vitro–results from a new individual cell-based model with podia. PLoS One **6**(7), e21960 (2011) (Unpublished)
39. Huang, G.T., Gronthos, S., Shi, S.: Mesenchymal stem cells derived from dental tissues vs. those from other sources: their biology and role in regenerative medicine. J. Dent. Res. **88**(9), 792–806 (2009)
40. Huang, N.F., Li, S.: Mesenchymal stem cells for vascular regeneration. Regen. Med. **3**(6), 877–892 (2008)
41. Hunziker, E.B.: Articular cartilage repair: basic science and clinical progress. A review of the current status and prospects. Osteoarthritis Cartilage **10**(6), 432–463 (2002)
42. Kaltz, N., Ringe, J., Holzwarth, C., Charbord, P., Niemeyer, M., Jacobs, V.R., Peschel, C., Haupl, T., Oostendorp, R.A.: Novel markers of mesenchymal stem cells defined by genome-wide gene expression analysis of stromal cells from different sources. Exp. Cell. Res. **316**(16), 2609–2617 (2010)
43. Kobel, S., Valero, A., Latt, J., Renaud, P., Lutolf, M.: Optimization of microfluidic single cell trapping for long-term on-chip culture. Lab. Chip **10**(7), 857–863 (2010)
44. Korobkova, E., Emonet, T., Vilar, J.M., Shimizu, T.S., Cluzel, P.: From molecular noise to behavioural variability in a single bacterium. Nature **428**(6982), 574–578 (2004)
45. Kretlow, J.D., Jin, Y.-Q., Liu, W., Zhang, W.J., Hong, T.-H., Zhou, G., Baggett, L.S., Mikos, A.G., Cao, Y.: Donor age and cell passage affects differentiation potential of murine bone marrow-derived stem cells. BMC Cell Biol. **9**, 60 (2008)
46. Krinner, A., Hoffmann, M., Loeffler, M., Drasdo, D., Galle, J.: Individual fates of mesenchymal stem cells in vitro. BMC Syst. Biol. **4**, 73 (2010)

47. Krinner, A., Zscharnack, M., Bader, A., Drasdo, D., Galle, J.: Impact of oxygen environment on mesenchymal stem cell expansion and chondrogenic differentiation. Cell Prolif. **42**(4), 471–484 (2009)
48. Kuhn, N.Z., Tuan, R.S.: Regulation of stemness and stem cell niche of mesenchymal stem cells: implications in tumorigenesis and metastasis. J. Cell Physiol. **222**(2), 268–277 (2010)
49. Landau, L.D., Lifschitz, E.M.: Theory of Elasticity. 3rd rev. edn. Pergamon Press, Oxford (1986)
50. Lécuyer, E., Yoshida, H., Parthasarathy, N., Alm, C., Babak, T., Cerovina, T., Hughes, T.R., Tomancak, P., Krause, H.M.: Global analysis of mRNA localization reveals a prominent role in organizing cellular architecture and function. Cell **131**(1), 174–187 (2007)
51. Lemon, G., Waters, S.L., Rose, F.R.A.J., King, J.R.: Mathematical modelling of human mesenchymal stem cell proliferation and differentiation inside artificial porous scaffolds. J. Theor. Biol. **249**, 543–553 (2007)
52. Lennon, D.P., Edmison, J.M., Caplan, A.I.: Cultivation of rat marrow-derived mesenchymal stem cells in reduced oxygen tension: effects on in vitro and in vivo osteochondrogenesis. J. Cell Physiol. **187**(3), 345–355 (2001)
53. Loeffler, M., Roeder, I.: Tissue stem cells: definition, plasticity, heterogeneity, self-organization and models—a conceptual approach. Cells Tissues Organs **171**(1), 8–26 (2002)
54. Loeffler, M., Roeder, I.: Conceptual models to understand tissue stem cell organization. Curr. Opin. Hematol. **11**(2), 81–87 (2004)
55. Lutolf, M.P., Blau, H.M.: Artificial stem cell niches. Adv. Mater. **21**(32–33), 3255–3268 (2009)
56. Lutolf, M.P., Gilbert, P.M., Blau, H.M.: Designing materials to direct stem-cell fate. Nature **462**(7272), 433–441 (2009)
57. Majd, H., Wipff, P.J., Buscemi, L., Bueno, M., Vonwil, D., Quinn, T.M., Hinz, B.: A novel method of dynamic culture surface expansion improves mesenchymal stem cell proliferation and phenotype. Stem Cells **27**(1), 200–209 (2009)
58. Manini, I., Gulino, L., Gava, B., Pierantozzi, E., Curina, C., Rossi, D., Brafa, A., D'Aniello, C., Sorrentino, V.: Multi-potent progenitors in freshly isolated and cultured human mesenchymal stem cells: a comparison between adipose and dermal tissue. Cell Tissue Res. **344**(1), 85–95 (2011)
59. Martin-Rendon, E., Hale, S.J.M., Ryan, D., Baban, D., Forde, S.P., Roubelakis, M., Sweeney, D., Moukayed, M., Harris, A.L., Davies, K., Watt, S.: Transcriptional profiling of human cord blood CD133+ and cultured bone marrow mesenchymal stem cells in response to hypoxia. Stem Cells **25**(4), 1003–1012 (2007)
60. Mata, A., Boehm, C., Fleischman, A.J., Muschler, G.F., Roy, S.: Connective tissue progenitor cell growth characteristics on textured substrates. Int. J. Nanomedicine **2**(3), 389–406 (2007)
61. Menicanin, D., Bartold, P.M., Zannettino, A.C., Gronthos, S.: Identification of a common gene expression signature associated with immature clonal mesenchymal cell populations derived from bone marrow and dental tissues. Stem Cells Dev. **19**(10), 1501–1510 (2010)
62. Mrozik, K.M., Zilm, P.S., Bagley, C.J., Hack, S., Hoffmann, P., Gronthos, S., Bartold, P.M.: Proteomic characterization of mesenchymal stem cell-like populations derived from ovine periodontal ligament, dental pulp, and bone marrow: analysis of differentially expressed proteins. Stem Cells Dev. **19**(10), 1485–1499 (2010)
63. Nekanti, U., Rao, V.B., Bahirvani, A.G., Jan, M., Totey, S., Ta, M.: Long-term expansion and pluripotent marker array analysis of Wharton's jelly-derived mesenchymal stem cells. Stem Cells Dev. **19**(1), 117–130 (2010)
64. Niibe, K., Kawamura, Y., Araki, D., Morikawa, S., Miura, K., Suzuki, S., Shimmura, S., Sunabori, T., Mabuchi, Y., Nagai, Y., Nakagawa, T., Okano, H., Matsuzaki, Y.: Purified mesenchymal stem cells are an efficient source for iPS cell induction. PLoS One **6**(3), e17610 (2011)
65. Oakley, E.J., Zant, G.V.: Unraveling the complex regulation of stem cells: implications for aging and cancer. Leukemia **21**(4), 612–621 (2007)

66. Oh S., Brammer K.S., Li Y.S., Teng D., Engler A.J., Chien S., Jin S.: Stem cell fate dictated solely by altered nanotube dimension. Proc. Natl. Acad. Sci. U S A **106**(7):2130–2135 (2009)
67. Pittenger M.F., Mackay A.M., Beck S.C., Jaiswal R.K., Douglas R., Mosca J.D., Moorman M.A., Simonetti D.W., Craig S., Marshak D.R.: Multilineage potential of adult human mesenchymal stem cells. Science **284**(5411):143–147 (1998)
68. Pretzel D., Linss S., Rochler S., Endres M., Kaps C., Alsalameh S., Kinne R.W.: Relative percentage and zonal distribution of mesenchymal progenitor cells in human osteoarthritic and normal cartilage. Arthritis. Res. Ther. **13**(2):R64 (2011)
70. Rando, O.J., Paulsson, J.: Noisy silencing of chromatin. Dev. Cell **11**(2), 134–136 (2006)
71. Ren, H., Cao, Y., Zhao, Q., Li, J., Zhou, C., Liao, L., Cai, H., Han, Z.C., Yang, R., Chen, G., Zhao, R.C.: Proliferation and differentiation of bone marrow stromal cells under hypoxic conditions. Biochem. Biophys. Res. Commun. **347**(1), 12–21 (2006)
72. Rieger, M.A., Hoppe, P.S., Smejkal, B.M., Eitelhuber, A.C., Schroeder, T.: Hematopoietic cytokines can instruct lineage choice. Science **325**(5937), 217–218 (2009)
73. Roeder, I., Braesel, K., Lorenz, R.: Loeffler, M.: Stem cell fate analysis revisited: interpretation of individual clone dynamics in the light of a new paradigm of stem cell organization. J. Biomed. Biotechnol. **3**, 84656 (2007)
74. Roeder, I., Horn, M., Glauche, I., Hochhaus, A., Mueller, M.C., Loeffler, M.: Dynamic modeling of imatinib-treated chronic myeloid leukemia: functional insights and clinical implications. Nat. Med. **12**(10), 1181–1184 (2006)
75. Rombouts, W.J.C., Ploemacher, R.E.: Primary murine MSC show highly efficient homing to the bone marrow but lose homing ability following culture. Leukemia **17**(1), 160–170 (2003)
76. Rosenbaum, A.J., Grande, D.A., Dines, J.S.: The use of mesenchymal stem cells in tissue engineering: a global assessment. Organogenesis **4**(1), 23–27 (2008)
77. Rosland, G.V., Svendsen, A., Torsvik, A., Sobala, E., McCormack, E., Immervoll, H., Mysliwietz, J., Tonn, J.C., Goldbrunner, R., Lonning, P.E., Bjerkvig, R., Schichor, C.: Long-term cultures of bone marrow-derived human mesenchymal stem cells frequently undergo spontaneous malignant transformation. Cancer Res. **69**(13), 5331–5339 (2009)
78. Ruzankina, Y., Brown, E.J.: Relationships between stem cell exhaustion, tumour suppression and ageing. Br. J. Cancer **97**(9), 1189–1193 (2007)
79. Samoilov, M., Plyasunov, S., Arkin, A.P.: Stochastic amplification and signaling in enzymatic futile cycles through noise-induced bistability with oscillations. Proc. Natl. Acad. Sci. U S A **102**(7), 2310–2315 (2005)
80. Samoilov, M.S., Price, G., Arkin, A.P.: From fluctuations to phenotypes: the physiology of noise. Sci. STKE 2006 (366):re17 (2006)
81. Sengers, B.G., Taylor, M., Please, C.P., Oreffo, R.O.C.: Computational modelling of cell spreading and tissue regeneration in porous scaffolds. Biomaterials **28**(10), 1926–1940 (2007)
82. Seshi, B.: Gene expression analysis at the single cell level using the human bone marrow stromal cell as a model: sample preparation methods. Methods Mol. Biol. **449**, 117–132 (2008)
83. Seshi, B., Kumar, S., King, D.: Multilineage gene expression in human bone marrow stromal cells as evidenced by single-cell microarray analysis. Blood Cells Mol. Dis. **31**(2), 268–285 (2003)
84. Sethe, S., Scutt, A., Stolzing, A.: Aging of mesenchymal stem cells. Ageing Res. Rev. **5**(1), 91–116 (2006)
85. Sharpless, N.E., DePinho, R.A.: How stem cells age and why this makes us grow old. Nat. Rev. Mol. Cell Biol. **8**(9), 703–713 (2007)
86. Sinclair, D.A., Oberdoerffer, P.: The ageing epigenome: damaged beyond repair? Ageing Res. Rev. **8**(3), 189–198 (2009)

87. Song, L., Webb, N.E., Song, Y., Tuan, R.S.: Identification and functional analysis of candidate genes regulating mesenchymal stem cell self-renewal and multipotency. Stem Cells **24**(7), 1707–1718 (2006)
88. Sorrentino, A., Ferracin, M., Castelli, G., Biffoni, M., Tomaselli, G., Baiocchi, M., Fatica, A., Negrini, M., Peschle, C., Valtieri, M.: Isolation and characterization of CD146+ multipotent mesenchymal stromal cells. Exp. Hematol. **36**(8), 1035–1046 (2008)
89. Stolzing, A., Coleman, N., Scutt, A.: Glucose-induced replicative senescence in mesenchymal stem cells. Rejuvenation Res. **9**(1), 31–35 (2006)
90. Stolzing, A., Jones, E., McGonagle, D., Scutt, A.: Age-related changes in human bone marrow-derived mesenchymal stem cells: consequences for cell therapies. Mech. Ageing Dev. **129**(3), 163–173 (2008)
91. Stolzing, A., Scutt, A.: Age-related impairment of mesenchymal progenitor cell function. Aging Cell **5**, 213–224 (2006)
92. Suva, D., Garavaglia, G., Menetrey, J., Chapuis, B., Hoffmeyer, P., Bernheim, L., Kindler, V.: Non-hematopoietic human bone marrow contains long-lasting, pluripotential mesenchymal stem cells. J. Cell Physiol. **198**(1), 110–118 (2004)
93. Tuli, R., Tuli, S., Nandi, S., Huang, X., Manner, P.A., Hozack, W.J., Danielson, K.G., Hall, D.J., Tuan, R.S.: Transforming growth factor-beta-mediated chondrogenesis of human mesenchymal progenitor cells involves N-cadherin and mitogen-activated protein kinase and Wnt signaling cross-talk. J. Biol. Chem. **278**(42), 41227–41236 (2003)
94. Wagner, W., Bork, S., Horn, P., Krunic, D., Walenda, T., Diehlmann, A., Benes, V., Blake, J., Huber, F.-X., Eckstein, V., Boukamp, P., Ho, A.D.: Aging and replicative senescence have related effects on human stem and progenitor cells. PLoS One **4**(6), e5846 (2009)
95. Wagner, W., Horn, P., Castoldi, M., Diehlmann, A., Bork, S., Saffrich, R., Benes, V., Blake, J., Pfister, S., Eckstein, V., Ho, A.D.: Replicative senescence of mesenchymal stem cells: a continuous and organized process. PLoS One **3**(5), e2213 (2008)
96. Xu, B.L., Tao, Y.: External noise and feedback regulation: steady-state statistics of auto-regulatory genetic network. J. Theor. Biol. **243**(2), 214–221 (2006)
97. Xu, Y., Malladi, P., Chiou, M., Bekerman, E., Giaccia, A.J., Longaker, M.T.: In vitro expansion of adipose-derived adult stromal cells in hypoxia enhances early chondrogenesis. Tissue Eng. **13**(12), 2981–2993 (2007)
98. Yan, X., Qin, H., Qu, C., Tuan, R.S., Shi, S., Huang, G.T.: iPS cells reprogrammed from human mesenchymal-like stem/progenitor cells of dental tissue origin. Stem Cells Dev. **19**(4), 469–480 (2010)
99. Yoshimura, H., Muneta, T., Nimura, A., Yokoyama, A., Koga, H., Sekiya, I.: Comparison of rat mesenchymal stem cells derived from bone marrow, synovium, periosteum, adipose tissue, and muscle. Cell Tissue Res. **327**(3), 449–462 (2007)
100. Zipori, D.: The stem state: mesenchymal plasticity as a paradigm. Curr. Stem Cell Res. Ther. **1**(1), 95–102 (2006)
101. Zscharnack, M., Poesel, C., Galle, J., Bader, A.: Low Oxygen Expansion Improves Subsequent Chondrogenesis of Ovine Bone-Marrow-Derived Mesenchymal Stem Cells in Collagen Type I Hydrogel. Cells Tissues Organs **190**(2), 81–93 (2009)
102. Zuk, P.A., Zhu, M., Ashjian, P., De Ugarte, D.A., Huang, J.I., Mizuno, H., Alfonso, Z.C., Fraser, J.K., Benhaim, P., Hedrick, M.H.: Human adipose tissue is a source of multipotent stem cells. Mol. Biol. Cell **13**(12), 4279–4295 (2002)

Image-Based Cell Quality Assessment: Modeling of Cell Morphology and Quality for Clinical Cell Therapy

Hiroto Sasaki, Fumiko Matsuoka, Wakana Yamamoto, Kenji Kojima, Hiroyuki Honda and Ryuji Kato

Abstract In clinical tissue engineering, both safety and effectiveness are definite requirements that should be satisfied. Conventional cell biology techniques are facing limitations in the quality assurance step of cell production for clinical therapy. Image-based cell quality assessment offers a great potential, because it is the only way to non-destructively and repeatedly assess cellular phenotypes and irregularities. To effectively assess cell quality using the multiple parameters derived from time course cell imaging, machine learning models, which have been effectively used to connect biological phenomena with biological measurements in the field of bioinformatics, are promising approaches for achieving high accuracy. Here, we present the recent results of our successful cell quality modeling and discuss its possibility and considerations on further application in clinical cell therapy.

1 Introduction

In clinical tissue engineering and cell therapy, although the cell is a "live material" with great variety and a highly sensitive nature, its production should be strictly controlled for safe and effective therapy.

If cellular irregularity is overlooked, it could cause serious side-effects such as tumorigenesis [14]. If cell yields do not fulfill the criteria on the day of the operation, the operation has to be cancelled or the cells be used with less activity.

H. Sasaki · F. Matsuoka · W. Yamamoto
K. Kojima · H. Honda · R. Kato (✉)
Department of Biotechnology, Graduate School of Engineering,
Nagoya University, Furocho, Chikusaku, Nagoya 464-8603, Japan
e-mail: kato-r@nubio.nagoya-u.ac.jp

Stud Mechanobiol Tissue Eng Biomater (2013) 10: 207–226
DOI: 10.1007/8415_2012_132

Published Online: 14 February 2012

If cell growth is unexpectedly slow during expansion culture, clinicians have to waste precious time waiting for its recovery, and cannot schedule the operation. During this process, costly consumables are wasted, and future cell behavior is most likely unpredictable. If cellular activity (e.g., differentiation rate, cellular growth, protein production, etc.) is disturbed by unresolved technical errors, the therapeutic effects of the cells will not be as expected. Hence, the many unsolved problems associated with quality assurance in clinical cell therapy should be conquered by technological achievements.

Despite the existence of such ambiguous problems in clinical cell therapy, conventional assay technologies in biology have not yet conquered any of these problems. Furthermore, conventional molecular biology assay techniques are basically incompatible with the production of cellular products, because they lack the 4 major characteristics listed below.

The first and the most fundamental criterion is non-invasiveness. Cells for clinical application should be as intact as possible, because the artificial manipulation process itself could trigger cellular abnormalities. In addition, cells derived from patients are usually limited, considering the limited source of cells and reduction in the patient's load during collection of source cells. For greater safety, fluorescent staining or gene transfer should be avoided.

The second characteristic is exhaustiveness. In clinical cell therapy, "sampling check" is a commonly used method for cell assessment. However, compared to chemical compound production, human cells exhibit huge variances; hence, partial sampling will not assure the quality of the cell population. Therefore, assessment of cells used for clinical purposes should shift to "total cell assessment" with technological advances.

The third property is synchronism. Given that cultured cells differ drastically with respect to their individual mobility, duplication, senescence, differentiation, and production activity, end-point assays are associated with high error rates. Kinetic examination is expected to detect small irregularities in cells for monitoring cellular status and optimizing cell culture conditions. High error rates together with the lack of speed limits the use of end-point assays. Most cellular contamination checks require hours, days, or weeks. This problem forces the patients to wait to be informed of problems in the implanted cells until after the operation. For practical cellular assessment, "on-time" evaluation right before the operation would strongly assist the doctors.

The fourth characteristic is correlation with future status. However accurate the on-time monitoring results of conventional assays may be, they do not quantitatively predict the future state. Therefore, in the cell production process, planning effective protocols to revise the current cell culture process for better production is extremely difficult. To enable the smooth operation of clinical cell therapy, future state prediction would facilitate the development of new clinical protocols. Because cells are not uniform and change with time, the culturing process should be optimized in an ongoing manner using effective "feedback information" with the future results in mind. Such a feedback information loop of cell quality information will improve the quality of the final cellular product, resulting in greater therapeutic effects. Furthermore, from the clinical doctors' perspective, proper scheduling of surgery is essential for treating more patients in a given facility.

To satisfy all the above-listed criteria, "image-based cell quality assessment" offers great potential for quality assurance in cell therapy. Image-based cell quality assessment enables non-invasive, fully exhaustive, timely, and predictive evaluation of cells.

Technologies that evaluate and assess cellular activities using cell image measurement methods are being reported [3, 4, 6, 8–11, 13, 16–24, 26]. Takagi has widely reviewed especially the non-invasive cell-imaging technologies, and has introduced novel technologies. The reviews strongly indicate that cellular morphologies significantly correlate with cellular activities. These findings underscore the importance of cellular morphology monitoring in traditional cell culture methods. Many textbooks have indicated that cellular morphology is an indicator of cell quality, and therefore, it should be carefully monitored. In many facilities that offer clinical tissue engineering therapy, cells are maintained and controlled during the culture process by the experts' experienced "visual expertise." Considering the successes of image-based cell assessments and the strong requirements in clinical cell therapy, analysis of datasets of morphological features and biological phenomena are attractive for machine learning researchers. In recent years, machine learning algorithms that have been widely applied in the field of bioinformatics and systems biology (gene analysis, mRNA profile analysis, protein data analysis, etc.) have been effectively used in image-based cellular analysis studies [1, 5, 15]. However, as far as we recognize, studies using non-labeled cellular images are still limited. Furthermore, applications of machine learning algorithms to assist with cell quality assessments, especially focusing on the requirements of cell therapy, are rare.

In this book chapter, we present some of our successful modeling results that support the effectiveness of machine learning applications in clinical tissue engineering and cell therapy.

2 Strategy of Image-Based Cell Assessment Model Construction

The construction of a model for image-based cell quality assessment comprises 4 major steps: (1) Image data collection, (2) Image processing, (3) Experimental data collection, and (4) Modeling (Fig. 1).

In our studies, image data comprise phase contrast microscopic images, because they represent the type of images most frequently used by cell biologists therefore considered to contain historically-proven indicative information. For image data collection, we use BioStation CT (Nikon Corporation, Tokyo, Japan), the fully automated cell incubation and monitoring system for stable, periodical, and mass time-lapse image data. For experimental data collection, we assayed the "observed cells" using conventional cell biological techniques. From carefully selected

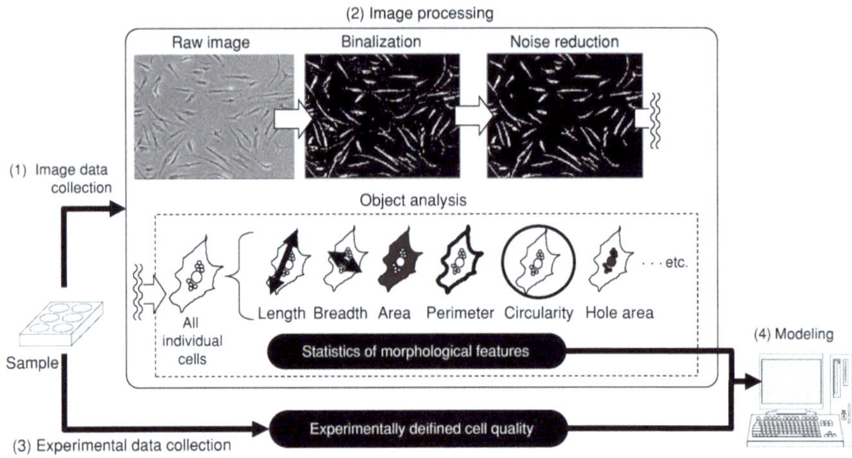

Fig. 1 Schematic illustration of image-based cell quality assessments

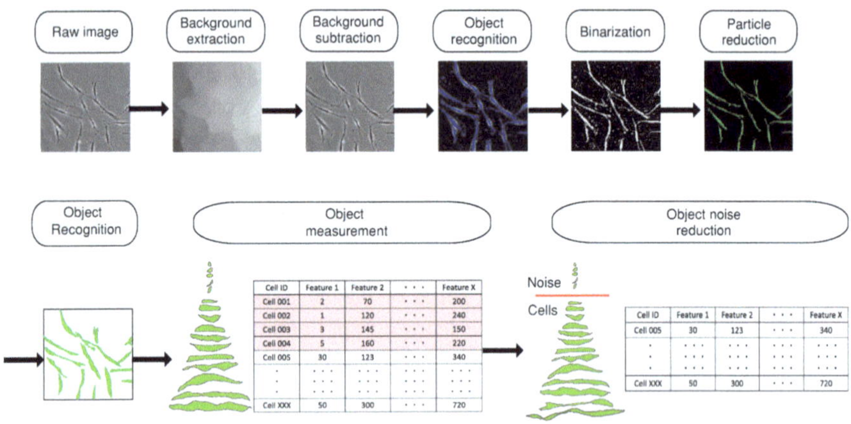

Fig. 2 Schematic illustration of image processing for assessing morphological features

assays, teacher signals can be obtained. There are cases where the culturing condition itself can be used as a teacher signal. For image processing, we customized our original image processing filters combining image analysis software and original programs in C and R languages (Fig. 2). Briefly, the raw image was processed to have the minimum error compared to manual cell counts after binarization. In this process, we applied an original combination of filters that were optimized using 26 types of human cells, including tumor cell lines and primary cells. From the objects extracted after the binarization, we measured 9–30 morphological features based on the characteristics of cells, together with multicollinearity examinations and interviews with cell culture experts. Statistical measures from each morphological feature were tagged with the teacher signal (the target prediction value determined by the experimental data collection) and were applied as a dataset. For the modeling, we chose regression analysis models, discriminant analysis models, clustering analysis models, or neural network-based models, according to the complexity and quality of the teacher signal measurements. Further in this chapter, we present some of our successful models.

3 Regression Analysis Model for Image-Based Cell Quality Assessment

Regression is a modeling approach to understand the quantitative relationship between multiple independent variables (input features) and a dependent variable (target prediction value). During the process of constructing the regression function, users can estimate the combinational importance of various features such as "morphological features" in the case of image-based assessment. In other words, users can quantitatively understand the characteristic morphological parameter combinations that cell-culture experts unconsciously recognize and apply for their judgements.

For the target prediction value, there are various biological parameters that require prediction in clinical tissue engineering, such as cellular activity, cellular proliferation rate, cellular lineage, cellular differentiation rate, cellular production rate. Among the many candidate parameters, we chose one of the essential parameters in the cell production process, the future cell yield. The cell yield greatly affects the scheduling of operations, because most medical facilities assure the quality of cell therapy by defining the "cell number for injection." Therefore, cell culture experts are required to predict the operation date based on their expertise.

To replace such ambiguous cell production procedures, we sought to quantitatively predict the future cell yield (14 days later) of clinically obtained primary human dermal fibroblasts with multiple regression models using early cellular images (images from 1 to 3 days culture period). Fibroblasts are the cell source for skin defect and wrinkle medications already used in clinical cell therapy [2, 25]. The concept of the cell yield prediction model using cellular images is illustrated in Fig. 3.

To obtain input features from cell culture images, we collected a total of 270 phase contrast microscopic images (20 ×) of cultured primary dermal fibroblasts obtained from ten healthy volunteers (3 males, 7 females, 29–72 years old). Informed consent was obtained according to a protocol approved by the Ethics Committee of Nagoya University Hospital. Cells derived from passages 3 and 4 of the primary expansion were maintained in modified Eagle's medium (DMEM, Life technologies, Carlsbad, CA, USA) containing 10% fetal bovine serum (FBS, Life technologies) at 37°C in the presence of 5% CO_2. All images (.bmp files), manually obtained by three operators in this work, were processed using MetaMorph software (Molecular Devices, LLC., Sunnyvale, CA, USA) with

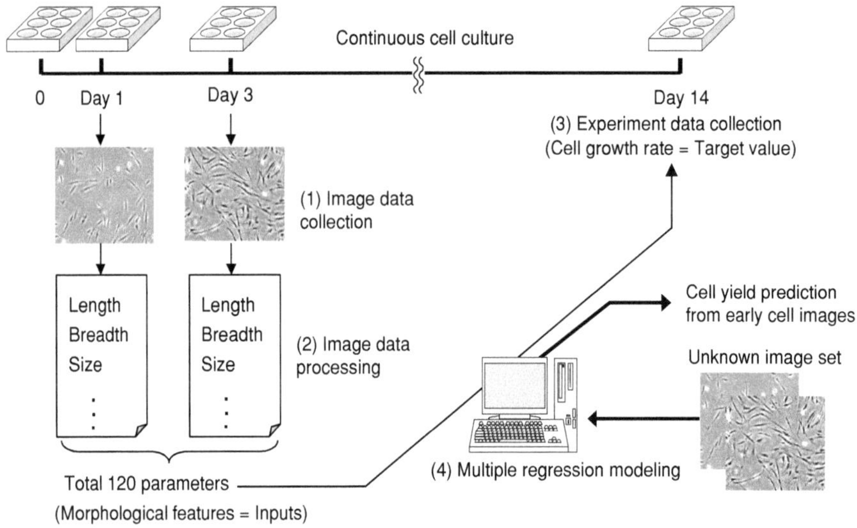

Fig. 3 Schematic illustration of the cell yield prediction model and its construction

original filter sets. Briefly, the raw images were pre-processed using open-close filters and binarized using the optimized threshold, and all objects in the image were measured with the integrated morphometry analysis process. Image data were pre-processed using the universal threshold optimized with 20 randomly picked image samples. Using integrated morphometry analysis according to the manufactuer's manual (http://mdc.custhelp.com/app/answers/list/c/110), the number of objects was measured together with 22 individual morphological features in MetaMorph, such as total area, standard area count, perimeter, width, height, orientation, length, breadth, fiber length, fiber breadth, shape factor, elliptical form factor, inner radius, outer radius, mean radius, equivalent radius, equivalent sphere surface area, equivalent sphere volume, equivalent prolate volume, equivalent oblate volume, hole area, and relative hole area. Prior to statistical analysis, morphological data from noise objects (non-cellular objects) were automatically removed using the original noise-reduction algorithm (image auto-wash method). Statistics were calculated with each measured feature in five view fields from the same well. For each feature, (a) average of day 1, (b) standard deviation of day 1, (c) average of day 3, (d) standard deviation of day 3, (e) average ratio of the day 1, and day 3 averages, (f) average ratio of the day 1 and day 3 standard deviations, were calculated as input features. From a total of 184 input features, 120 features with a CV (coefficient of variation) of <20 were used as inputs in the modeling. Input features were selected by stepwise parameter selection in the multiple regression model analysis using SPSS software (for Windows 11.5.1, IBM, Armonk, NY, USA).

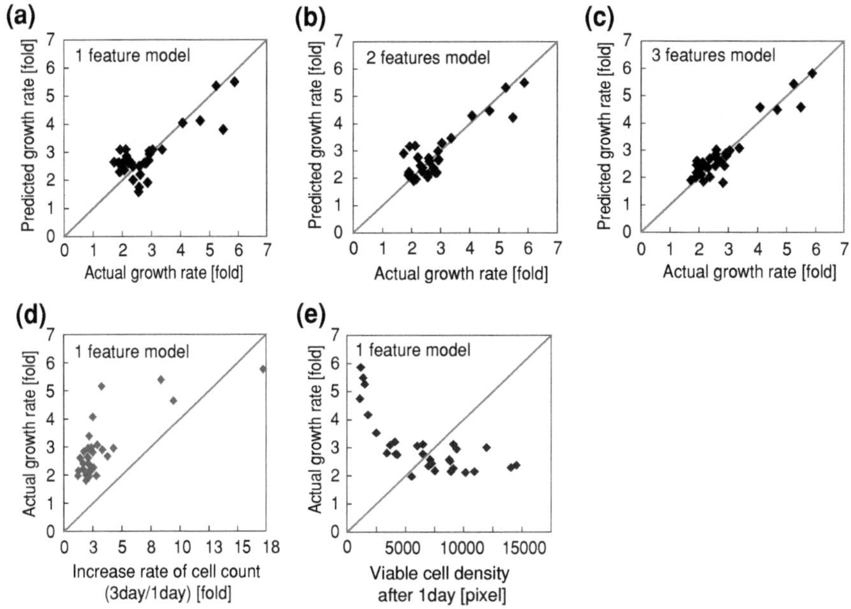

Fig. 4 Results of prediction using cell yield prediction models. **a** Prediction model using 1 feature input. **b** Prediction model using 2 feature inputs. **c** Prediction model using 3 feature inputs (the best prediction model). **d** Correlation plot of growth rate and 1 selected feature (cell density on day 1). **e** Correlation plot of growth rate and 1 selected feature (cell growth rate (day 3/day 1). Each plot represents 1 image dataset. Morphological features in models **a b**, and **c** were selected by a stepwise parameter selection process during model construction. The selected parameters were (P1) change rate of the standard variation of elliptical form factor (day 1–3) (P2) size of inner radius on day 3, and (P3) cell number on day 1. Morphological features in models **d** and **e** were selected on the basis of the "feeling" of a cell culture expert

To obtain the teacher signal or the target prediction value, which is based on the experimental data, three volunteers counted the partial sample of the total cell suspension from three different wells to estimate the average cell yield after 14 days of culture, and calculated the cell growth rate as the ratio of cell yield to the seeding cell number. The cell growth rate was used as the quantitative teacher signal in the modeling.

Among the 120 parameters extracted from the image data, three parameters, change rate of the standard variation of elliptical form factor (day 1–3) (P1), size of inner radius on day 3 (P2), and cell number on day 1 (P3), were found to be the best combination of cell morphology information for predicting future cell yield (average squared error = 0.14). If ±0.5 error can be accepted by the medical facility, the predict performance of this model is 87%.

The rise in prediction accuracy of models using different numbers of input features strongly indicated that multiple feature combinations provide higher prediction accuracy (Fig. 4a–c). As a comparison, it is interesting that other single parameters, intentionally selected by cell culture experts, correlated poorly with cell yield. The correlation ratio of the model using increase rate of cell count was 0.50 (Fig. 4d), and the model using viable cell density after 1 day was 0.59 (Fig. 4e). These results strongly suggest that morphological feature selection combined with not only "morphologies" but also "changes of morphologies" is effective in image-based analysis. Since many previous image-based cell analysis works depend on "intentionally selected" morphological features, our work points out the importance of introduction of machine learning approach to construct better models for practical usage. Cell count, one of the easiest parameter that could be raised for cell yield prediction, was selected in the best prediction model. However, it should be noted that cell count was the "last parameter" selected for constructing the model, indicating that it only works in combination with morphological information. This is also clear from the above-compared prediction results in the model using a single parameter, "cell growth rate." Together with such combinational effects of parameters, another important finding using this model is that the "exact period of culture" was quantitatively pointed out. For example, cell count (nearly equal to cell growth) is important in the first 24 h, and not very informative in the latter period. Such a timing definition is extremely important for setting the image acquisition schedule, and also for defining the prediction date in the early cell culture process.

Fibroblasts change their morphology from the sharp spindle shape to the flat and polygonal shape when their growth activity decreases. The automatically determined combination of morphological parameters directly correlated with this known morphological change, indicating that the expert's feeling could be effectively modeled with this technique.

From the data, we arrived at three conclusions: (1) morphological cell information is informative for cell growth prediction, (2) objectively selected parameters are more effective in cell growth prediction than the ones selected on the basis of feeling, and (3) a combination of multiple parameters is more effective in the prediction than a single parameter. It should also be noted that such quantitative cell quality prediction can be further extended to cell differentiation rate prediction [7]. Kagami et al. [7] have shown that osteogenic differentiation status of human mesenchymal stem cells after 2 weeks of cell culture could be predicted by the morphological features priorily.

4 Discriminant Function Model for Image-Based Cell Quality Assessment

Discriminant function analysis is a statistical analysis to predict a categorical dependent variable using one or more continuous or binary independent variables. Compared to the regression analysis model, the discriminant function model satisfies clinical cell therapy requirements for assessing "binary categorical events." This is because many events in the cell production process cannot be quantitatively measured or measured data is usually categorized even if they could be measured quantitatively. For example, bacterial contamination is the most well known risk in the cell culture process. The contamination can be measured by colony-forming assays or quantitative polymerase chain reaction (PCR) of bacterial markers. However, the quantity of contamination is meaningless in a model, because any amount of contamination should be eliminated from the process. In such cases, a decision should be made as "yes, OK to continue" or "no, discard the sample," and simple discriminant function models are effective in incorporating combinational features in the binary decision with high accuracy.

For discriminant function analysis, we designed a model to detect "human processing error", because from the aspect of establishing more safe cell production process, detection of "processing error" is essential. Practically, we attempted to model various types of human error that could be involved in the culture process and considered to affect the cellular damage rate. In various studies, we succeeded in constructing a model to detect the "error in trypsin treatment." Trypsin treatment is an essential step in subculture to digest the cellular adhesion molecule and collect cells from the culture plate, although known as a physical stress that damage cells. Some delicate cells are known to be very sensitive to trypsin concentration, and in such cases, a very low trypsin concentration is recommended. Commonly, protocol of trypsin treatment is fixed in the medical facilities' standard protocols. However, if the human error in diluting such a damageable solution can be non-destructively monitored, the model can assist the rigid protocol in the cell production process, and can reduce costly error-monitoring process of additional manual checks. The concept of the human error detection model using cellular images is illustrated in Fig. 5.

To obtain input features from cell culture images, we collected a total of 450 phase contrast microscopic images (10×) [3 wells × 5 fields × 15 time points (24–136 h, every 8 h) × 2 treatment conditions (0.25%, 0.025%)] of primary human gingival fibroblasts obtained from healthy volunteers (22 and 24 years old, female and male) and cultured in DMEM (Life technologies) containing 10% FBS (Life technologies) at 37°C in the presence of 5% CO_2 by BioStaion CT(Nikon Corporation). All images (.bmp files) were processed using MetaMorph software (Molecular Devices) and our original programs with original filter sets, as described for the regression analysis model. Using integrated morphometry analysis, the number of objects was measured together with seven individual morphological features in MetaMorph, such as total area, breadth, fiber length, fiber breadth, shape factor, elliptical form factor, and inner radius. Prior to statistical analysis, all morphological data of the noise objects (non-cellular objects) were automatically removed by the original noise-reduction algorithm (image auto-wash method). The features were averaged in 5 view fields from the same well. For each feature, average and average change ratio were calculated for and between each time point (24, 48, 72, and 96 h), and a total of 203 features were used as input features. In the modeling process, input features were selected by stepwise parameter selection in the linear discriminant analysis using SPSS software (for Windows 11.5.1).

For the target event, we intentionally designed two conditions of trypsin treatment during cell passage: 0.25% trypsin treatment (too dense = error) and 0.025% trypsin

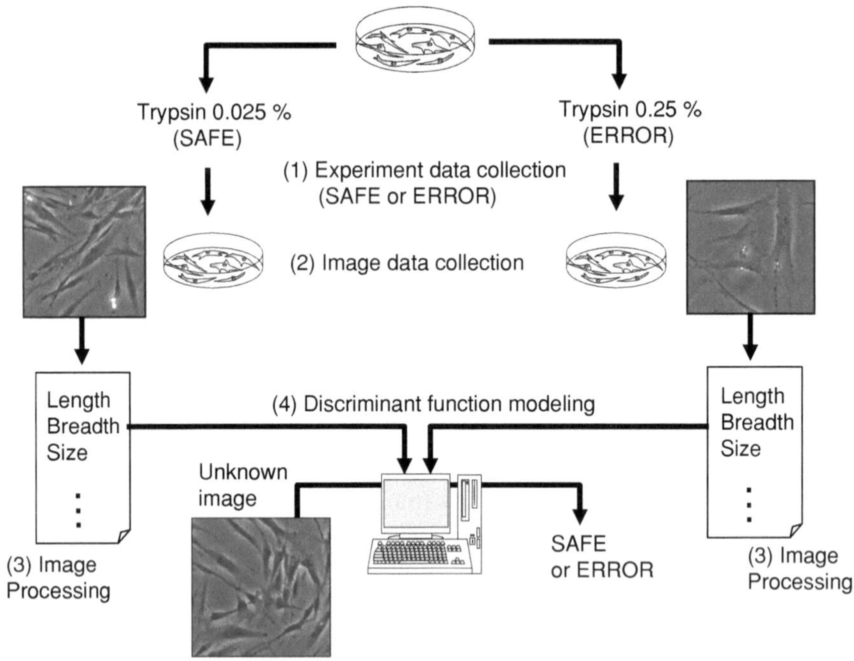

Fig. 5 Schematic illustration of the human error detection model and its construction

treatment (normal density). When cell images were collected using the former condition, the operation category was labeled as ERROR, and when cell images were from the latter condition, the operation category was labeled as SAFE.

From the 203 parameters, elliptical form factor at 24 and 48 h, breadth change between 72 h and 96 h, and fiber breadth change from 72 to 96 h were selected (Fig. 6). Damaged fibroblasts are known to lose their sharp spindle shape morphology and convert to a flat and polygonal shape; both morphological features were considered to indicate the same "morphological pattern." Interestingly, the "period of morphological change" was determined by this modeling. The "narrowness" of the first 2 days was important, but the damage from the error handling also affected the rate of morphological change during the later period (72–96 h). The prediction accuracy was 92.9%; such "operation error" is impossible to detect with conventional biological measurements.

5 Clustering Model for Image-Based Cell Quality Assessment

Clustering is a modeling approach to understand the similarity between objects by assigning them to groups. One of the features that make clustering a useful modeling method is that it enables "multi-class prediction." Biological quality,

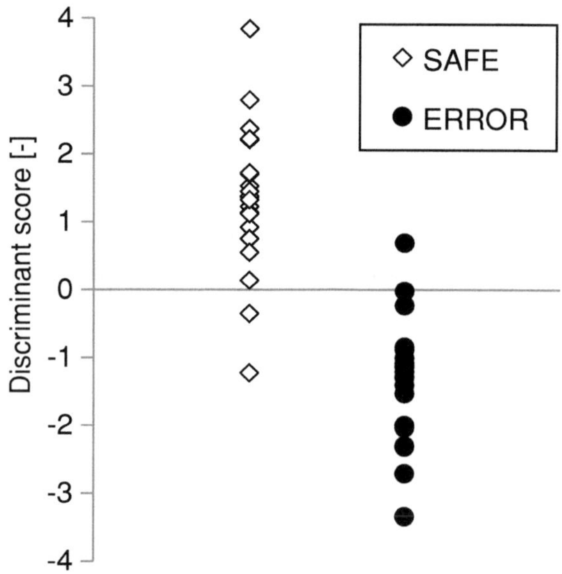

Fig. 6 Prediction results from the human error detection model. Each plot represents 1 image dataset. Morphological features in the model were selected by a stepwise parameter selection process during model construction

which relates to cells in clinical cell therapy, is often very complicated and ambiguous. Machine learning researchers should be aware that most cellular qualities are too complex to define using a single value. For example, although various biomarkers to define "stemness" of cells are measured, there exist various profiles because not all the markers are completely the same in cells. Therefore, even with the detailed profiling of biomarkers with flow cytometry analysis, cell qualities are commonly discussed as "levels" or "categories." There are many more examples of such categorical definitions of cell qualities, such as cell lineage or cell damage rate. For such cellular qualities, there is still no good single marker to define the phenomenon quantitatively.

Among such complex quality conditions of cells, "cell-type classification" is a strongly required solution in clinical cell therapy. In clinical cell therapy, a cell source is obtained from the tissues of patients. Cells are isolated from tissues by a process of primary culture, including enzyme digestion and explant culture process. Because a tissue is a complex of different cell types, there is a high possibility of incorporating "unwanted cells," which are designated as "cell contamination," in the primary cell population. For example, in the dermal defect therapy using fibroblasts, keratinocyte contamination increases the risk of dermal cyst formation. Similarly, in skin burn treatment, fibroblast contamination in keratinocytes causes a risk of skin contraction. Although such cell types are clearly known, there are still cells without good markers, such as "fibroblasts." Without a clear marker protein, fibroblasts cannot be stained or detected in the keratinocyte population. Furthermore, if clinical doctors require detection of keratinocytes among fibroblasts, staining of cells for keratinocyte-specific protein markers depletes precious patients' cell source. Therefore, "cell type classification" is an important challenge for image

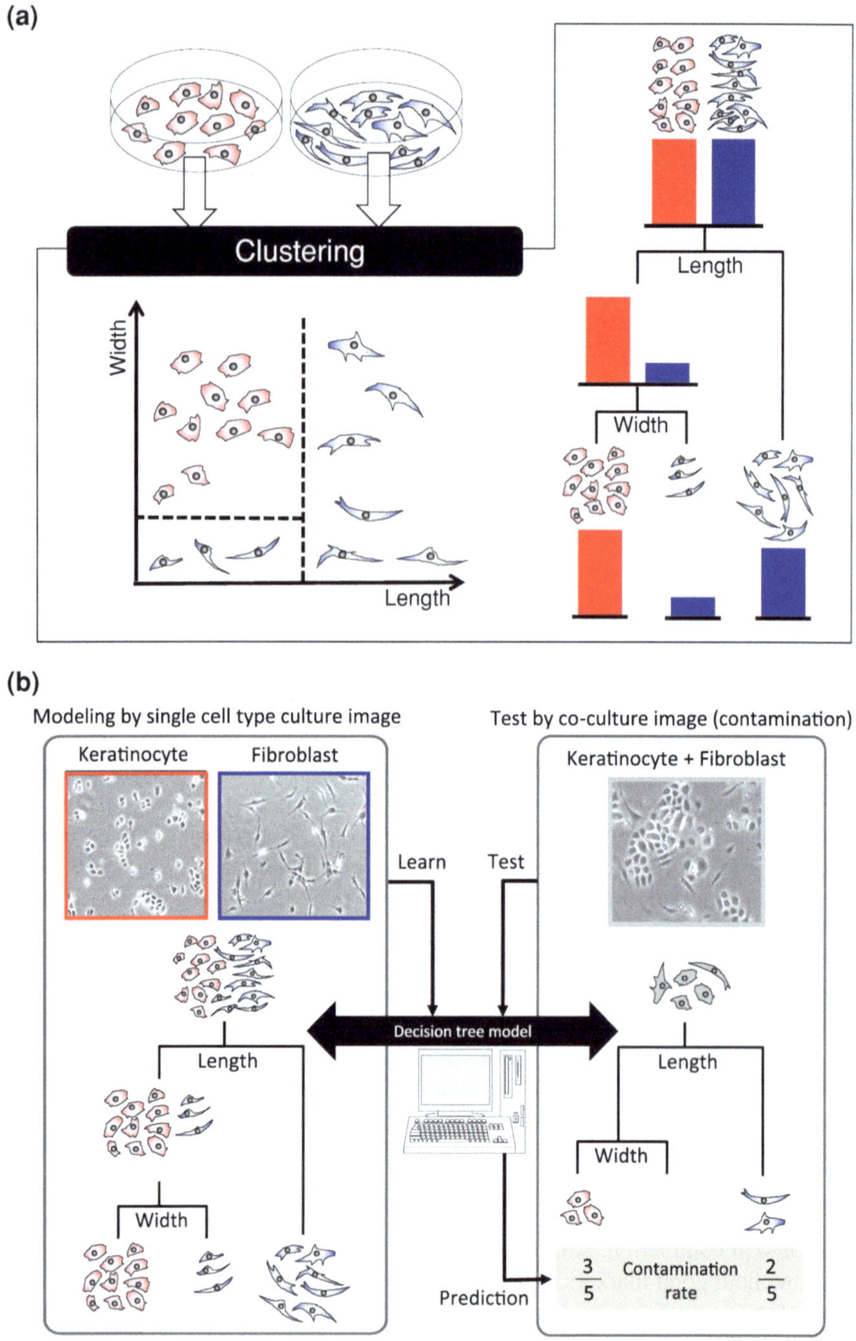
(a)
Clustering
Width
Length
Length
Width
(b)
Modeling by single cell type culture image
Test by co-culture image (contamination)
Keratinocyte
Fibroblast
Keratinocyte + Fibroblast
Learn
Test
Decision tree model
Length
Width
Length
Width
Prediction
3/5
Contamination rate
2/5

◀ **Fig. 7** Schematic illustration of the cell contamination evaluation model and its construction. **a** Schematic illustration of decision tree model for cell morphology modeling. Combination of morphological parameters (*left*) could be determined using decision tree (*right*). **b** Practical scheme for constructing cell contamination evaluation model. Model can be constructed by single cell type culture images if all cell objects could be effectively extracted from the image data. From the unknown image of co-cultured cells, fibroblasts mixed in keratinocytes in this case, the constructed model can predict the contamination rate

analysis. Considering the existence of "sub-populations" in both types of cells, a multiple-grouping method, decision tree modeling, was applied. The concept of the cell contamination evaluation model using cellular images is illustrated in Fig. 7.

To obtain input features from cell culture images, we collected a total of 270 phase contrast microscopic images (10 ×) of cultured normal human dermal fibroblasts (NHDF; KURABO, Osaka, Japan) and normal human epithelial keratinocytes (NHEK; KURABO) grown at 37°C in the presence of 5% CO_2. NHDFs were maintained in modified Eagle's medium (MEM, Life technologies) with 10% FBS, and NHEKs were maintained in EpiLife-KG2 (KURABO). For image acquisition, NHDFs were mixed with NHEK (0%/1%–30%/100%) in an NHDF contamination model of NEHKs. A total of 1912 phase contrast microscopic images (4×) were acquired from five view fields per well of a six well plate over 8 h periods for 5 days using BioStation CT (Nikon Corporation). All images (.bmp files) were processed using MetaMorph software (Molecular Devices) and our original program with original filter sets, as described in the prior section. Using integrated morphometry analysis, the number of objects was measured together with 19 individual morphological features in MetaMorph, such as total area, hole area, relative hole area, perimeter, width, height, length, breadth, fiber length, fiber breadth, shape factor, elliptical form factor, inner radius, outer radius, mean radius, equivalent radius, pixel area, area, orientation. From both cell types, totally 2,792,527 objects were measured with these morphological parameters. In the modeling process, input features were selected by recursive partitioning and regression tree (rpart) modeling using R statistics.

For the modeling of teacher signals in rpart, we carefully classified the cellular morphology types of both cells, and identified nine types of objects in the image data: (1) F_n: objects with the most typical fibroblast morphology (2) F_s: objects with small fibroblast morphology in the process of expansion (3) F_o: objects that represent a fusion of several overlapping fibroblasts (4) K_n: objects with the most typical keratinocyte morphology (5) K_s: objects with small keratinocyte morphology in the process of expansion (6) K_o: objects with holes after binarization, indicating the halo in the middle of cell (commonly caused by hill-top structure of cells) (7) K_c: objects with a "C" shape, indicating that the halo is relatively large in cell area (8) n_s; tiny non-cellular objects (9) n_l: long, but tiny non-cellular objects (Fig. 8). Groups 8 and 9 represented the common noise found in both cell types. From the modeling, we chose 100 objects from the total cells which fit to each cluster from 100% fibroblast and 100% keratinocyte images, constructed a decision tree model by tenfold cross validation, and examined the accuracy of the

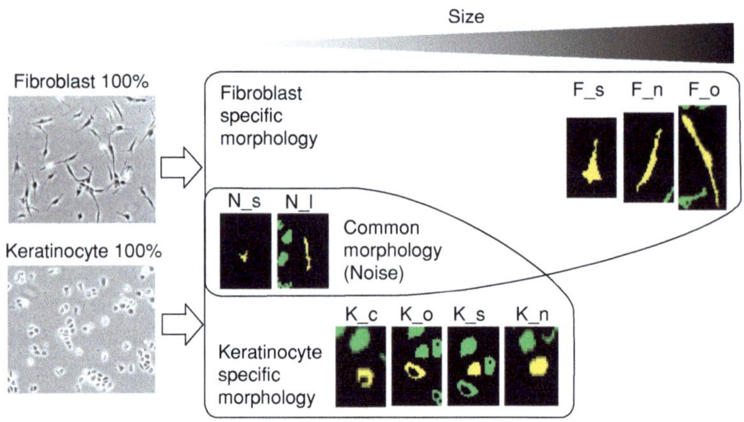

Fig. 8 Object grouping for effective decision tree construction

model for predicting the percentage of contamination of fibroblasts in keratinocytes by using only the analysis of objects from the mixed culture images.

From the clustering, nine groups of morphologically different objects found in the two cell types could be assigned to 9-clusters with high accuracy (Fig. 9). And surprisingly, although the nine-cluster model was originally constructed from the objects in 100% single cell-type images, this model succeeded in predicting cellular objects in images of co-culture (13 samples with different mixed percentages ranging from 1 to 30% fibroblast contamination in keratinocytes) with an accuracy of 93% (±5%). This prediction performance of non-labeling image cytometry is extremely high, and it should be noted that this cytometry was performed with non-stained cellular images.

The process of the decision tree modeling, i.e., the clustering with teacher signals, should provide important insights into the problem of cell classification. For cell classification, one should realize that cells are swinging between two phases of different morphologies. One is the characteristic morphology of the cell type, and the other is the round and small morphology, which is identical between all the cell types. This round and small morphology may indicate "cells before division" or "cells beginning to apoptose" or "cells in migration." With this common morphology, the classification of "individual cells" is extremely difficult. Because with the case of cells, a certain common population of objects will share exact same morphologies, and become the noise in the modeling process. The existence of such a population was very clear when we first attempted to classify the two cell types by using discrimination analysis (Fig. 10). Figure 10 shows that there is an "indiscriminative" population within each cell type. We found that such an "overlapping" population can be reduced by defining the "true-negative noise objects common to both cell types," which is represented by groups 8 and 9 in the above-mentioned criteria (Fig. 9). However, such noise reduction alone could not completely improve the prediction accuracy, because of the presence of

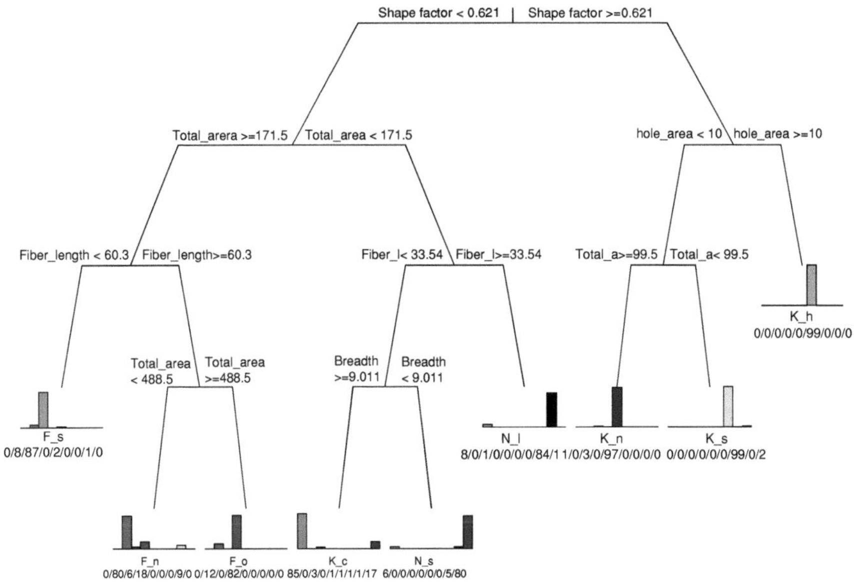

Fig. 9 Structure of the cell contamination evaluation model. The selected parameters are described in each node. Under the terminus of branches, *bars* indicate the percentage of all cell objects in the terminal node. The *numbers separated by slashes* indicates the object numbers clustered at the end branch

another "sub-population of objects" which has similar morphology and disrupts the classification accuracy. Such sub-populations are individually assigned in the decision tree model; therefore, it provided the highest accuracy in classifying the two cell types at the single cell level.

6 Discussion

Although some reports have described a connection between "morphology" and "cell quality," few reports focus on the practical tasks and problems associated with clinical cell therapy or demonstrate the effectiveness of various machine learning model approaches. Therefore, in this chapter, we have reviewed some of our successful studies on image-based bioinformatics modeling in clinical cell therapy.

Image-based cell assessment is gaining popularity along with the advances in "imaging equipment" and "high content analysis software." Indeed, machine learning approaches are important for the analysis of "multi-variant data measurements" together with "biological complex phenomena." Bioinformatics, widely applied in the research fields of DNA microarray, SNP typing, and proteome data mining, has been the leading application in the field of machine learning algorithms. Image-based data analysis may be the next frontier.

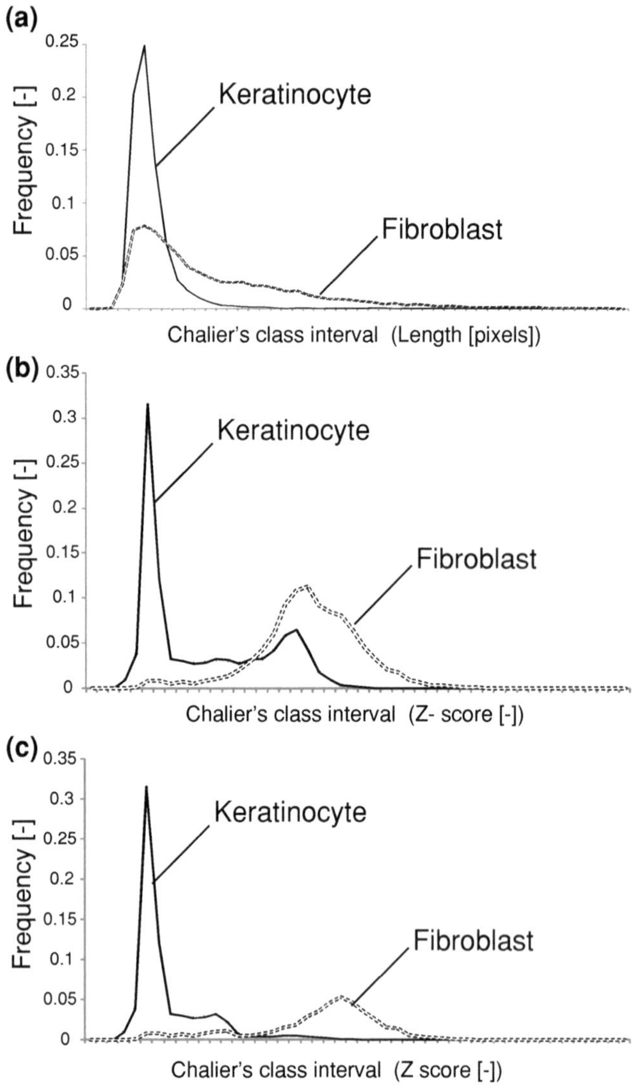

Fig. 10 Parameter distribution comparing keratinocytes and fibroblasts. **a** Distribution of length between two types of cells. There is huge overlapping area in one parameter distribution. **b** Distribution of Z-score from the discriminant analysis using 19 morphological features between two types of cells. There is still overlapping area even after the discriminant modeling. **c** Distribution of Z-score from the discriminant analysis using 19 morphological features using "noise reduced data". Noise objects were excluded from the discriminant analysis by eliminating the group N_s and N_l objects from the data. Such noise data **c** reduction greatly reduces overlapping area from cell distributions

However, many important points have been neglected in the recent image-based biological analyses. Although machine learning approaches offer strong prediction performances, analysts should be aware of the problems arising from the combination of "nature of image data" and "nature of cells."

First, cellular image data has a much higher bias compared to mRNA and protein data. This is because mRNA and proteins are total molecules summed within all the cells in one vessel, whereas image data biases depend on where and by whom the cells are observed. Typically, cells grow locally in culture dishes, and migrate toward empty spaces. Thus, if cells are not uniformly seeded, the cell number and migration rates differ greatly between images. Furthermore, analysts should be aware that most cellular images from cell biologists are taken from the researchers' favorite "view field" with favorite "focus and lighting." Therefore, if images are not acquired randomly or scheduled by automated machinery, random selection of images from the "researcher's image library" already has a huge bias.

Second, due to the nature of cells, cellular images always contain a certain percentage of "common features." This is because every cell type exhibits the same round and small morphology when the cells are "dead," "proliferating," or "rapidly migrating." In addition, cell-derived debris increases on the surface of the culture plate during the culture process. Such a "common sub-population," which contributes to fatal noise in machine learning algorithms, drastically lowers model accuracy. Therefore, the objects in cellular images have to be effectively filtered or classified by detailed observation and statistical analysis of the cell populations before model construction. We also have to consider the fact that primary cells sometimes contain "different cell types," although they are usually overlooked.

Third, the cell variation is huge between cell lines, cell passages, and cell origins. Therefore, the variation arising from the source of cellular images is also extremely important. In our experience, such variation is extremely large; therefore, the quantity of images should be large enough to provide enough cell numbers to minimize the standard deviation. Such quantity and variation in cellular images are commonly neglected, mostly because of the cost and labor involved in the experiment. For effective machine learning, sufficient data for cross-validation is required; therefore, the experimental design for image acquisition is extremely important.

Fourth, cellular image processing is commonly completely dependent on the researcher's feeling. In most cases, binarization is processed with "a threshold." However, this threshold is commonly decided by some value "considered OK by the researcher after evaluating fewer than 20 images." Such a threshold is rarely "thoroughly scanned," because such a function is lacking in most cellular image analysis software. Therefore, most cellular images are processed individually and differently with a "feeling-based threshold," or processed by a single threshold that is "roughly decided." Therefore, a high bias is inherent in image processing when image data are processed into numerical data.

To reduce these four major biases, we applied original solutions before analysis. (i) Image bias: we used an original seeding device for equal cell seeding, and acquired more than 400 images per condition, including different view fields, wells, and time-points. We also used a fully auto-scheduled and auto-focusing image acquisition system, BioStation CT (Nikon corporation) to minimize the image acquisition bias. (ii) Population bias: we have shown that population bias can be classified using decision tree models, too. We also remove the "common population" before analysis by using our original noise-reduction algorithm. (iii) Cell and culture bias: we routinely used three or more lots or passages of cells in every experiment. We also set image acquisition periods of less than 8 h to obtain more information during the average doubling time. (iv) Processing bias: we optimized the universal threshold by examining the threshold ranges of randomly picked images from 5 to 10 cell lines at different time-points for the best threshold that resulted in minimum error in the cell count. We also strongly recommend using pattern matching algorithms to recognize objects in the image, such as the software "CL-Quant (Nikon Corporation).

To assist clinical cell therapy with such machine learning models combined with image analysis, "what to predict" should be carefully researched. There are strong expectations and requirements in clinical cell therapy toward non-invasive cell quality assessments. A large number of "qualities" and "events" are expected to be predicted in the clinical setting. However, to build a prediction model, the teacher signal should be carefully selected and defined. For example, beta galactosidase activity is a biological marker of cellular senescence [12]. However, the efficiency of beta galactosidase detection in a single stained image is actually very low. It may correlate with senescence when the data are averaged in "1 well," but not in "images that do not cover all the wells." In such cases, differences in detection power exist between biological assays and image detection. Without knowing the detection power of the target event, inconsiderate modeling and data acquisition will result in no harvest. Moreover, there are some cases where a "1 marker measurement" is not sufficient for definition. For example, in the case of defining stemness of cells, the single marker staining result would not produce the expected stemness prediction model.

Additionally, the "acceptable accruracy" of model should be also carefully discussed by the users of these machine-learning models for cell quality evaluations. The accuracy should also be carefully checked with the aspect of consistency and reproducibility. In many cases, the non-labeling and real-time estimation of cell quality are difficult to compare its performance with conventional methods, because the image-based cell quality assessment has uncomparable advantage features. However, such comparison difficulty never means that conventional assay can be eliminated. The acceptable accuracy of models should be defined by each medical facility with their specific verification data, and should be used only to reduce the excessiveness and unstableness in the conventional cell production process.

7 Conclusion

Machine learning modeling algorithms have great potential in image-based cell quality assessments. Based on the selection of appropriate models with sufficient quantities of images for predicting target events, our results suggest a potential of non-labeled and real-time assessment of cellular quality strongly required in the present industrialization era of tissue engineering and regenerative medicine. Our results, most of them are newly presented in this chapter, indicate that quantitative prediction, categorical prediction, and multi-categorical classification can be achieved with high accuracy.

Like the other machine learning models in different research fields, robustness should be repeatedly examined to build universally effective cell quality prediction models. We are conducting further investigations to determine the extent of cellular variations to build a practically useful model.

We strongly expect that the feedback loop of advances and improvements in biology and computational technology will advance this field of cell assessment with bioinformatics machine learning models.

Acknowledgments We are grateful to the New Energy and Industrial Technology Development Organization (NEDO) for the Grant for Industrial Technology Research (Financial Support to Young Researchers, 09C46036a) for the support. We also thank to the Nikon Corporation for their research collaboration and financial support. We also thank Mai Okada and Yurika Nonogaki for supporting the experiments and data storage. Finally we greatly thank Yoshihide Nagura, Kazuhiro Mukaiyama, and Asuka Miwa for establishing basal techniques for our image analysis procedure.

References

1. Carpenter, A.E., Jones, T.R., Lamprecht, M.R., Clarke, C., Kang, I.H., Friman, O., Guertin, D.A., Chang, J.H., Lindquist, R.A., Moffat, J., Golland, P., Sabatini, D.M.: CellProfiler: image analysis software for identifying and quantifying cell phenotypes. Genome Biol. **7**, R100 (2006)
2. Ebisawa, K., Kato, R., Okada, M., Sugimura, T., Latif, M.A., Hori, Y., Narita, Y., Ueda, M., Honda, H., Kagami, H.: Gingival and dermal fibroblasts: their similarities and differences revealed from gene expression. J. Biosci. Bioeng. **111**, 255–258 (2011)
3. Galon, J., Costes, A., Sanchez-Cabo, F., Kirilovsky, A., Mlecnik, B., Lagorce-Pages, C., Tosolini, M., Camus, M., Berger, A., Wind, P., Zinzindohoue, F., Bruneval, P., Cugnenc, P.H., Trajanoski, Z., Fridman, W.H., Pages, F.: Type, density, and location of immune cells within human colorectal tumors predict clinical outcome. **313**, 1960–1964 (2006)
4. Harder, N., Mora-Bermudez, F., Godinez, W.J., Wunsche, A., Eils, R., Ellenberg, J., Rohr, K.: Automatic analysis of dividing cells in live cell movies to detect mitotic delays and correlate phenotypes in time. Genome Res. **19**, 2113–2124 (2009)
5. Held, M., Schmitz, M.H., Fischer, B., Walter, T., Neumann, B., Olma, M.H., Peter, M., Ellenberg, J., Gerlich, D.W.: CellCognition: time-resolved phenotype annotation in high-throughput live cell imaging. Nat. Methods **7**, 747–754 (2010)
6. Jones, T.R., Carpenter, A.E., Lamprecht, M.R., Moffat, J., Silver, S.J., Grenier, J.K., Castoreno, A.B., Eggert, U.S., Root, D.E., Golland, P., Sabatini, D.M.: Scoring diverse cellular morphologies in image-based screens with iterative feedback and machine learning. Proc. Natl. Acad. Sci. U S A. **106**, 1826–1831 (2009)
7. Kagami, H., Agata, H., Kato, R., Matsuoka, F., Tojo, A.: Fundamental technological developments Required for increased availability of tissue engineering. InTech (2010)
8. Kamentsky, L., Jones, T.R., Fraser, A., Bray, M.A., Logan, D.J., Madden, K.L., Ljosa, V., Rueden, C., Eliceiri, K.W., Carpenter, A.E.: Improved structure, function and compatibility

for CellProfiler: modular high-throughput image analysis software. Bioinformatics **15**, 1179–1180 (2011)
9. Kino-oka, M., Agatahama, Y., Haga, Y., Inoie, M., Taya, M.: Long-term subculture of human keratinocytes under an anoxic condition. J. Biosci. Bioeng. **100**, 119–122 (2005)
10. Kirillov, V., Stebenyaeva, E., Paplevka, A., Demidchik, E.: A rapid method for diagnosing regional metastases of papillary thyroid cancer with morphometry. Microsc. Res. Tech. **69**, 721–728 (2006)
11. Kittler, R., Pelletier, L., Heninger, A.K., Slabicki, M., Theis, M., Miroslaw, L., Poser, I., Lawo, S., Grabner, H., Kozak, K., Wagner, J., Surendranath, V., Richter, C., Bowen, W., Jackson, A.L., Habermann, B., Hyman, A.A., Buchholz, F.: Genome-scale RNAi profiling of cell division in human tissue culture cells. Nat. Cell. Biol. **9**, 1401–1412 (2007)
12. Kurz, D.J., Decary, S., Hong, Y., Erusalimsky, J.D.: Senescence-associated (beta)-galactosidase reflects an increase in lysosomal mass during replicative ageing of human endothelial cells. J. Cell Sci. **113** (Pt 20), 3613–3622 (2000)
13. Logan. D.J., Carpenter. A.E.: Screening cellular feature measurements for image-based assay development. J. Biomol. Screen. **15**, 840–846 (2010)
14. Maekawa, M., Yamanaka, S.: Glis1, unique pro-reprogramming factor, may facilitate clinical applications of iPSC technology. Nature **474**, 225–229 (2011)
15. Misselwitz, B., Strittmatter, G., Periaswamy, B., Schlumberger, M.C., Rout, S., Horvath, P., Kozak, K., Hardt, W.D.: Enhanced CellClassifier: a multi-class classification tool for microscopy images. BMC Bioinformatics. **11**, 30 (2010)
16. Nafe, R., Franz, K., Schlote, W., Schneider, B.: Morphology of tumor cell nuclei is significantly related with survival time of patients with glioblastomas. Clin. Cancer Res. **11**, 2141–2148 (2005)
17. Narkilahti, S., Rajala, K., Pihlajamaki, H., Suuronen, R., Hovatta, O., Skottman, H.: Monitoring and analysis of dynamic growth of human embryonic stem cells: comparison of automated instrumentation and conventional culturing methods. Biomed. Eng. Online **6**, 11 (2007)
18. Neumann, B., Held, M., Liebel, U., Erfle, H., Rogers, P., Pepperkok, R., Ellenberg, J.: High-throughput RNAi screening by time-lapse imaging of live human cells. Nat. Methods **3**, 385–390 (2006)
19. Sekiya, I., Larson, B.L., Smith, J.R., Pochampally, R., Cui, J.G., Prockop, D.J.: Expansion of human adult stem cells from bone marrow stroma: conditions that maximize the yields of early progenitors and evaluate their quality. Stem Cells **20**, 530–541 (2002)
20. Simpson, K.J., Selfors, L.M., Bui, J., Reynolds, A., Leake, D., Khvorova, A., Brugge, J.S.: Identification of genes that regulate epithelial cell migration using an siRNA screening approach. Nat. Cell Biol. **10**, 1027–1038 (2008)
21. Takagi, M.: Noninvasive quality estimation of adherent mammalian cells for transplantation. Biotech. Bioproc. Eng. **15**, 54–60 (2010)
22. Tokumitsu, A., Wakitani, S., Takagi, M.: Noninvasive discrimination of human normal cells and malignant tumor cells by phase-shifting laser microscopy. J. Biosci. Bioeng. **109**, 499–503 (2010)
23. Treiser, M.D., Yang, E.H., Gordonov, S., Cohen, D.M., Androulakis, I.P., Kohn, J., Chen, C.S., Moghe, P.V.: Cytoskeleton-based forecasting of stem cell lineage fates. Proc. Natl. Acad. Sci. U S A **107**, 610–615 (2010)
24. Umegaki, R., Murai, K., Kino-Oka, M., Taya, M.: Correlation of cellular life span with growth parameters observed in successive cultures of human keratinocytes. J. Biosci. Bioeng. **94**, 231–236 (2002)
25. Watson, D., Keller, G.S., Lacombe, V., Fodor, P.B., Rawnsley, J., Lask, G.P.: Autologous fibroblasts for treatment of facial rhytids and dermal depressions. A pilot study. Arch. Facial Plast. Surg. **1**, 165–170 (1999)
26. Zhou, X., Cao, X., Perlman, Z., Wong, S.T.: A computerized cellular imaging system for high content analysis in Monastrol suppressor screens. J. Biomed. Inform. **39**, 115–125 (2006)

Part II
Computational Tools For Process Design

Continuum Modelling of In Vitro Tissue Engineering: A Review

RD O'Dea, HM Byrne and SL Waters

Abstract By providing replacements for damaged tissues and organs, in vitro tissue engineering has the potential to become a viable alternative to donor-provided organ transplant, which is increasingly hampered by a shortage of available tissue. The complexity of the myriad biophysical and biochemical processes that together regulate tissue growth renders almost impossible understanding by experimental investigation alone. Mathematical modelling applied to tissue engineering represents a powerful tool with which to investigate how the different underlying processes interact to produce functional tissues for implantation. The aim of this review is to demonstrate how a combination of mathematical modelling, analysis and in silico computation, undertaken in collaboration with experimental studies, may lead to significant advances in our understanding of the fundamental processes that regulate biological tissue growth and the optimal design of in vitro methods for generating replacement tissues that are fully functional. With this in mind, we review the state-of-the-art in theoretical research in the field of in vitro tissue engineering, concentrating on continuum modelling of cell culture in bioreactor systems and with particular emphasis on the generation of new tissues from cells seeded on porous scaffolds. We highlight the advantages and limitations of different mathematical modelling approaches that can be used to study aspects of cell

R. O'Dea (✉)
School of Science and Technology, Nottingham Trent University Clifton Campus, Nottingham, NG11 8NS, UK
e-mail: reuben.odea@ntu.ac.uk

H. Byrne
Oxford Centre for Collaborative Applied Mathematics, Mathematical Institute, Gibson Building, Oxford, OX2 6HA, UK

S. Waters
Oxford Centre for Industrial and Applied Mathematics, Mathematical Institute, 24-29 St Giles', Oxford, OX1 3LB, UK

Stud Mechanobiol Tissue Eng Biomater (2013) 10: 229–266
DOI: 10.1007/8415_2012_140

Published Online: 7 July 2012

population growth. We also discuss future mathematical and computational challenges and interesting open questions.

1 Introduction

The interdisciplinary field of tissue engineering is emerging as a valuable tool in the quest for viable clinical solutions to health problems associated with tissue damage, degeneration and failure. Currently, the most successful surgical approaches involve the implantation of tissue grafts, or entire replacement organs, taken from suitable donors. Due in part to greater longevity in society (Palferman 2003), there is a chronic shortage of donor tissue. For example, in the UK during 2009/2010 552 patients died while awaiting transplants, and at 31 March 2010, there were approximately 8000 NHS patients registered for organ transplant (Johnson 2010). Furthermore, engineered tissues with the correct in vivo properties have applications in toxicology screening and drug testing. While therapeutic tissue engineering has the potential to address the issue of donor tissue shortage, this field remains in its infancy, since tissue growth is exceedingly complex, being regulated by an enormous variety of processes, from intracellular transduction pathways to tissue-level mechanics. Understanding these mechanisms is crucial to the development of reliable methods for engineering viable replacement tissue; mathematical analyses can provide fundamental insight into these mechanisms. In addition, collaboration between experimental and theoretical researchers enables in silico testing of experimental protocols (thereby reducing experimental cost) and stimulates the generation of model-driven experimentally-testable hypotheses. In this way, mathematical modelling can provide a key scientific tool with which to improve tissue engineering approaches.

In this chapter, we review the contribution of mathematical modelling to the understanding of tissue growth processes. We focus on continuum approaches which consider the influence of cells' biochemical and biophysical environment on tissue growth. Before surveying such theoretical analyses (Sects. 3.1–3.3), it is instructive to consider the key biological considerations in more detail.

1.1 Tissue Engineering

Broadly speaking, there are two distinct approaches to tissue engineering: in vivo and in vitro tissue engineering. The latter involves growing replacement tissue in the laboratory for implantation into patients. A common in vitro method involves seeding porous scaffolds with cells of the desired tissue type. After a period of incubation, during which the cells proliferate and colonise the scaffold, the resulting tissue construct is implanted into a patient. In contrast, in vivo tissue engineering involves implanting (e.g.) degradable scaffolds or gels loaded with cells directly into the body. In each case, degradation of these artificial supporting structures and their replacement by extracellular materials—such as collagen and proteoglycans (Freed et al. 1994; Hutmacher 2000)—leads to eventual tissue repair. In vivo tissue engineering uses the human body as a natural bioreactor, the perceived advantage being that the human body offers the correct physical and biochemical cues to enable creation of functional, viable tissue. However, the mechanisms by which these cues are employed by the cells are not well understood; a thorough review of in vivo tissue engineering considerations is given by Zdrahala and Zdrahala (1999). In what follows, we concentrate on in vitro tissue engineering, where tissue growth occurs under closely monitored and controlled environmental and operating conditions.

1.1.1 Cell Populations

The tissue engineering concepts outlined above are conceptually straightforward; however, in practice many barriers remain to be overcome. Fundamental problems include stimulating sufficient cellular proliferation to colonise the scaffold and preventing dedifferentiation of the seeded cell population. An approach mitigating the former problem involves using tissue precursor cells or multipotent stem cells, which have high proliferative capacity and can be induced to differentiate to a number of different cell types (Risbud and Sittinger 2002). The literature regarding the use of stem cells in tissue engineering and, for instance, the methods by which they can be induced to differentiate along different cell lines is extensive; a good introduction is given by Salgado et al. (2004) and references therein. We choose not to review this literature here, preferring to focus on continuum modelling of biochemical (and biophysical) aspects of tissue growth; implicit in the mathematical models that we analyse are the assumptions that, on the timescale of interest, the cell population has sufficient proliferative capacity to colonise the scaffold.

The mechanical forces that cells experience affect their differentiation, proliferation, orientation, gene activity and a host of other activities; indeed, as we indicate below, culturing cells in an in vivo-like mechanical environment can maintain differentiated function of the seeded cells. The stimuli are integrated into the cellular response via a process known as mechanotransduction. The mechanical environment of the tissue comprises both internally generated and externally applied forces. Internally generated forces may include active forces generated by cells during movement and adhesion, or residual stresses brought about by tissue growth and remodelling; the presence of such stresses has been observed in a variety of soft tissues (examples include arterial and venous tissue, myocardium and the trachea) and, in many cases, is crucial to their correct function. For example, residual stresses act to minimise the peak stress across the depth of the arterial wall (Chuong and Fung 1986) and are involved in wall remodelling (Fung 1991). Externally applied loads that act in vivo include macroscale forces due to movement and muscle contraction, and shear stress induced by fluid flow. Such forces are known to be important in the correct functioning of various tissues; placing patients with broken limbs under traction to prevent incorrect bone repair or misshapen bones is a simple example of this phenomenon that has been practised in hospitals for many years. Furthermore, flow-mediated shear stress has been shown to effect the culture of a variety of mechanosensitive cell types such as bone, cartilage, muscle, liver and blood vessels. For example, many studies have shown how stimulation via fluid shear stress enhances extracellular matrix formation (Bakker et al. 2004a; Klein-Nulend et al. 1995b; You et al. 2000).

Many studies have investigated the influence of the mechanical environment on cells' phenotype. For example, the response of osteocytes to mechanical loading has been investigated by Klein-Nulend et al. (1995b) and Bakker et al. (2004a). It is generally accepted that these terminally-differentiated human cells are the most mechanosensitive in bone and that they direct the formation and resorption of bone tissue at the microscopic level (Noble and Reeve 2000). Osteocytes have several thin processes which extend into the porous structure of bone and respond to interstitial fluid flows which exist in bone under loading. In this way, bone remodelling can be directed by physiological loading, despite the small strains allowed by the stiff calcified matrix.

Due in part to its avascular nature, cartilaginous tissue is notorious for its poor capacity to self repair and much experimental work has concentrated on developing suitable implants. Experimental studies (reviewed in Urban 1994) indicate that, under physiological conditions, moderate levels of mechanical stress regulate cartilage cell (chondrocyte) metabolism and ensure maintenance of extra-cellular matrix (ECM) integrity. Further, these processes are profoundly influenced by mechanical compression and hydrostatic pressure, such stimulation leading to accelerated chondrocyte growth and ECM synthesis, depending upon its regime of application (e.g. loading magnitude or, in the case of cyclic loading, frequency).

Another important area of tissue engineering is the culture of sheets of keratinocytes, which are used as replacement epithelium in a host of clinical settings (notably wound closure and/or skin grafts for severe burns). Mechanical strain is known to influence the proliferation rate of keratinocytes and activate them to express keratin, the constituent of intermediate filaments expressed specifically in keratinocytes (Yano et al. 2004).

1.1.2 Artificial Scaffolds

As indicated above, in vitro tissue engineering often involves seeding a porous scaffold with cells, to create a 'tissue construct'. The properties of the scaffold are therefore of central importance to the success of this approach.

As the ultimate aim is the in vivo implantation of the scaffold, the first requirement is that it is compatible with the host tissue, and does not elicit an immune response (Salgado et al. 2004). Furthermore, the scaffold acts as a surrogate for the significant amount of acellular material that is present in living tissue and defines its mechanical properties (for instance the collagen and elastin fibres present in ECM, or the calcium hydroxyapatite which lends bone tissue its rigidity). The scaffold material must, therefore, be chosen so that its mechanical integrity is maintained under physiological conditions, a factor of especial importance when the construct is load-bearing (as is the case for bone or cartilage implants). Additionally, since tissue repair is effected by the replacement of artificial scaffolds by extracellular materials, ensuring that the rates of nascent tissue growth and scaffold degradation (e.g. due to hydrolysis) are appropriately matched is crucial in maintaining the mechanical integrity of the construct. Lastly, porous scaffolds with a highly porous, interconnected structure are required to encourage cell penetration, vascularisation of the construct from surrounding tissue (in vivo) and efficient mass transfer of nutrients and waste products. There is, therefore, a trade-off between mechanical loading and mass transfer considerations in scaffold design.

Topographical and biochemical characteristics of the supporting scaffold can be used to enhance cell-scaffold adherence (crucial in most cells' growth), and to direct cell movement or differentiation (Weiss 1945; Salgado et al. 2004); for example, scaffolds may be engineered to deliver growth factors or DNA (Sipe 2002), or to contain specific cellular recognition molecules (Freed and Vunjak-Novakovic 1998). The suitability of a wide range of materials for tissue engineering has been investigated; examples include hydroxyapatite, poly(α-hydroxyesters), and natural polymers such as collagen and chitin. Poly(α-hydroxyesters) comprise a range of materials, including widely used polymers such as poly(lactide-co-glycolide)(PLGA), poly(glycolic acid) and poly(L-lactic acid). A comprehensive discussion of scaffolds for bone and cartilage tissue engineering applications is given by Hutmacher (2000); more general treatments are given by Atala et al. (1997) and Hollister (2005).

1.1.3 Bioreactor Systems

In a tissue engineering context, a bioreactor may be viewed as a cell culture system, in which biochemical and/or biophysical processes are closely monitored and controlled. For the reasons described above, many in vitro culture systems aim to mimic the in vivo environment; bespoke bioreactors are therefore required to provide the appropriate biochemical and biophysical conditions for individual tissue engineering applications. Specific examples of bioreactor types, which we investigate in more detail in later sections, are described below; reviews of bioreactor designs for specific applications are given by Martin et al. (2004), Cartmell and El Haj (2005), Eibl and Eibl (2009).

For the purposes of this review, it is convenient to classify cell culture techniques into two distinct groups, termed static and dynamic culture, respectively. Static culture refers to the simplest scenario in which mass transfer of (e.g.) nutrients or waste products through the culture is effected by diffusion; dynamic culture encompasses a range of bioreactor systems, designed to improve mass transfer to the cell population and/or to provide mechanical stimulation. Typically, enhanced mass transfer is effected via advection of nutrients and waste products by an imposed flow of culture medium (two strategies for achieving this are discussed below); mechanical stimulation of cells may be restricted to the influence of such a flow, or by additional cell stimulation strategies. We pause to remark that the operating parameters (e.g. flow rate) of the majority of dynamic culture systems do not change during the cell culture period, or are manually adjusted. Butler et al. (2009) highlight the need for the development of bioreactor systems which regulate automatically the biochemical and biophysical environment to accommodate the evolving requirements of the developing tissue.

The simplest example of a static culture bioreactor is a petri dish. Here, cells are grown in a monolayer to confluence. A weakness of this approach is rapid cell de-differentiation and loss of phenotype due to the absence of a three-dimensional architecture (Lin and Chang 2008). Cell populations cultured as three-dimensional multicellular spheroids (via culture in a non-adhesive environment) aim to address this by maximising cell-cell contacts, and have been found to positively affect tissue functionality (Riccalton-Banks et al. 2003). However, since the transport of nutrients and waste products is by diffusion only, scale-up to produce tissue of a size appropriate for implant results in the formation of constructs with a viable, proliferating periphery but a necrotic core (Cartmell and El Haj 2005). Bioreactors which employ dynamic culture conditions aim to mitigate the effects of diffusion-limited transport. Two such systems are described below.

Rotating wall bioreactors comprise a cylindrical vessel of circular cross-section rotating about its longitudinal axis with constant angular speed (see Fig. 1a). While specific designs vary, key parameters with which to vary the culture environment include: (i) rotation speed, (ii) bioreactor aspect ratio and (iii) bioreactor orientation. See Hammond and Hammond (2001) for more details of rotating-wall bioreactors. The system can be used as a suspension culture method, in which an initial cell suspension spontaneously self-assembles to form 3D spheroids (Lappa 2003); alternatively, a porous scaffold can be added, either pre-seeded, or to which the suspended cells adhere. A wide range of cell types have been cultured in this bioreactor system. Representative examples include human breast carcinoma cells cultured in suspension to produce in vivo-like nodules suitable for in vitro drug screening [experimental investigation discussed in Sawyer et al. (2007) and Waters et al. (2006)], and osteoblasts seeded on porous scaffolds (Yu et al. 2004).

Commonly, the longitudinal axis of the vessel is oriented horizontally and the rotation speed is chosen such that the upward hydrodynamic force balances the downward gravitational force and the spheroid or scaffolds contained within exist in a state of perpetual free-fall. This is thought to provide an optimal environment for tissue engineering applications, since the relative motion between the culture medium and tissue construct ensures that the culture medium remains mixed, aiding mass transport, while the shear stresses experienced by the cells can be controlled (Martin et al. 2004; Waters et al. 2006).

An alternative approach is to directly perfuse a cell-seeded scaffold with culture medium via a pump. By confining the scaffold such that culture medium is driven through its interconnected pores (perfusion bioreactor), rather than around the periphery (perfusion chamber), mass transfer limitations throughout the tissue

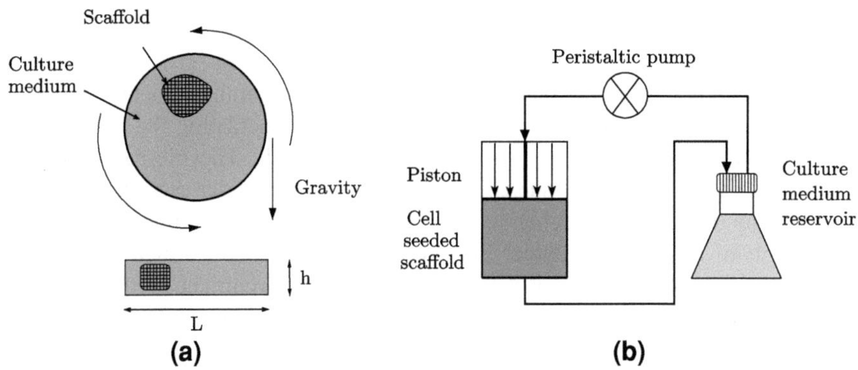

Fig. 1 **a** A rotating bioreactor system comprising a cylindrical vessel of diameter L and depth h rotating about its longitudinal axis, which contains a porous cell-seeded scaffold immersed in culture medium. *Upper figure* is a face-on view of the bioreactor. *Lower figure* is a side view showing the small gaps between the scaffold and the bioreactor walls. **b** The perfusion/compression bioreactor system of El-Haj et al. (1990), comprising a cell-seeded scaffold contained within a cylindrical vessel and constrained by a perforated piston, through which culture medium is perfused via a peristaltic pump. The perforations in the face of the piston allows simultaneous perfusion and compression of the scaffold

construct are reduced. In the rotating wall bioreactor, the rotation speed determines the mechanical environment of the cells (and the mass transfer characteristics of the system); here, the fluid shear stress experienced by the cells may be controlled by varying the perfusion rate. Indeed this type of bioreactor is frequently employed in the generation of bone tissue constructs, due to the sensitivity of such cells to stimulation by fluid shear stress (McCoy and O'Brien 2010). A modified perfusion bioreactor system is employed by El-Haj et al. (1990), in which cells seeded in a porous scaffold are subjected to perfusion with media and direct compression using a piston. This system is illustrated in Fig. 1b.

The above bioreactors provide convenient model systems with which to investigate aspects of tissue growth modelling relevant to tissue engineering applications. Outstanding problems that need to be addressed in order to optimise the culture environment in these bioreactors include the following

1. How can diffusion-limited mass-transfer limitations be overcome?
2. How do the culture medium, cells and scaffold and other extracellular materials interact to produce a viable tissue construct?
3. What influence does the mechanical stimulation and biochemical environment provided by the bioreactor have on tissue growth?

The models that we review in the remainder of this chapter illustrate how mathematical models can be used to address these questions and, in so doing, improve the efficiency of cell culture systems.

2 Mathematical Modelling Approaches

As outlined above, it is clear that a wide variety of factors influences the formation of tissue, and considerable effort has been invested in elucidating the mechanisms by which cells experience and respond to these stimuli. The bioreactor system employed is specific to the tissue engineering application under consideration, and generates its own unique biochemical and biomechanical environment, tailored to the growth of a particular tissue. Such variety necessitates a range of versatile mathematical modelling approaches, reflecting the environment of the cells and the experimental questions being posed. Such theoretical models can be used to predict the flow and nutrient transport characteristics within a specific bioreactor system, and in particular to determine local information about nutrient and shear stress fields that is not straightforward (or even possible) to obtain experimentally. The resulting models may be validated against measurable experimental data, such as perfusion flow rate, outlet nutrient concentration, and then exploited to reveal details of the mechanical and nutrient fields within the bioreactor system. Once validated, the model can then be used to predict the outcome of a particular experimental scenario (limiting the need for numerous and expensive bioreactor experiments, potentially saving time and resources) and to optimise the bioreactor operating conditions.

We focus our review on the use of multiphase modelling to describe biological tissues, and present a brief overview in Sect. 2.1 below. We conclude this section with a short description of alternative modelling approaches, before describing in detail models specific for bioreactor systems in Sect. 3.

2.1 Multiphase Modelling

Biological tissue is a composite material, comprising a wide variety of interacting constituents, including, for example, a large number of different cell types, their associated ECM and other deposited materials, and interstitial fluid. Multiphase, or mixture theory, models provide a natural continuum framework within which to investigate such interactions. These models are based upon the idea that tissues may be represented by a mixture of continua, which are able to occupy the same region of space; interactions between the different 'phases' are specified via mass and force balance equations, together with appropriate constitutive relations, the choice of which allows a wide variety of physical systems to be modelled. This methodology also reflects the idea that, as composite materials, tissue properties are reflected by the relative volume fractions (and properties) of their constituents (Trelstad and Silver 1981); changes in tissue composition occur via processes such as mitosis, apoptosis, necrosis, (de-)differentiation and ECM production.

Since the study of (Treusdell and Noll 1960), an enormous number of studies have been dedicated to formulating a rigorous framework of conservation laws for mixtures of interacting continua (Bowen 1976; Marle 1982; Passman and Nunziato 1984; Whitaker 2000; Kolev 2002; Ateshian 2007). Such approaches have been used widely to model biological tissue mechanics (Mow et al. 1980; Lai et al. 1991) and, more recently, tissue growth and remodelling (Please and McElwain 1998; Please et al. 1999; Landman and Please 2001; Preziosi and Tosin 2009; Ambrosi et al. 2010). Since the aim of this review is to highlight modelling approaches and challenges, we do not provide a historical perspective of multiphase modelling approaches, nor do we present a comprehensive derivation of relevant multiphase equations; it is markworthy, however, that although by definition such a multiphase continuum approach does not account for the precise microscopic detail of (for example) cell–cell interactions, the averaging process involved in deriving models of this type ensures that terms present in the model equations arise from appropriate microscopic considerations (see, e.g. Drew 1983). In Sect. 3, we provide some example studies which highlight modelling issues in tissue engineering and mixture theory approaches with which to investigate them. We remark that in many cases, modelling tumour growth was the original focus of these studies; the formulations are, nevertheless, relevant to tissue engineering applications.

As indicated above, the basis of these multiphase models is a set of conservation equations, governing mass and momentum transfer between the constituent phases. For a system comprising n incompressible phases, if inertial effects and body forces are negligible, then the equations governing the ith phase (with density ρ_i, volume fraction θ_i, velocity $\boldsymbol{u}_i$ and stress tensor σ_i) may be expressed:

$$\rho_i \left\{ \frac{\partial \theta_i}{\partial t} + \nabla \cdot (\theta_i \boldsymbol{u}_i) \right\} = S_i \,, \tag{1}$$

$$\nabla \cdot (\theta_i \boldsymbol{\sigma_i}) + \sum_{j \neq i} \mathbf{F}_{ij} = 0 \,, \tag{2}$$

where t denotes time, ∇ is the spatial gradient operator, S_i is the net rate of mass transfer into the ith phase and $\mathbf{F}_{ij}$ denotes the force acting on phase i as a result of interactions with phase j. We note that in some cases, flow in bioreactors is modelled by the full Navier–Stokes equation, in which case, inertial terms are retained in Eq. (2).

Within the multiphase modelling context, assumptions can be made that simplify the resulting systems of equations. For example, one approach is to assume that the cells occupy no volume, and therefore have no effect on the fluid flow. When determining the mechanical load that the flow exerts on the cells, it is assumed that the shear stress exerted on the substrate will be that experienced by the adherent cells; nutrient transport is incorporated by considering cells to be sinks or sources of metabolites or waste products. In some cases, such flow and transport problems require equations for the flow of nutrient-rich culture medium surrounding a porous scaffold to be coupled to those describing the flow within the scaffold. These equations are linked by specifying boundary conditions at the interface between the single-phase flow domain (the surrounding fluid) and the multiphase flow domain (the porous scaffold in which both scaffold and fluid volume fractions must be considered). For example, Navier–Stokes equations may be used to describe the surrounding fluid via

$$\nabla \cdot \boldsymbol{u} = 0, \frac{\partial \boldsymbol{u}}{\partial t} + \boldsymbol{u} \cdot \nabla \boldsymbol{u} = -\frac{1}{\rho} \nabla p + v \nabla^2 \boldsymbol{u} \tag{3}$$

where $\boldsymbol{u}$ is fluid velocity, p is fluid pressure, ρ is the fluid density and v is the kinematic viscosity. The flow within the porous scaffold may be described by Darcy's law

$$\nabla \cdot \boldsymbol{u} = 0, \boldsymbol{u} = \frac{k}{\mu} \nabla p, \tag{4}$$

where k is the permeability of the porous medium and μ is the dynamic viscosity. Alternatively, the Brinkman equations may be used:

$$\nabla \cdot \boldsymbol{u} = 0, \boldsymbol{u} = \frac{k}{\mu} \nabla p - k \nabla^2 \boldsymbol{u}. \tag{5}$$

In Eqs. (4) and (5) the Darcy flux $\boldsymbol{u}$ is the fluid velocity weighted by the scaffold porosity.

Both the Darcy and Brinkman equations are examples of multiphase models that may be obtained from Eqs. (1) and (2) via appropriate choices of constitutive laws. Models of this type are reviewed in Sect. 3.2. A key aspect of the macroscale Darcy and Brinkman models is that the microscale properties are captured via a parameter at the macroscale, for example, the scaffold permeability, without the need for detailed knowledge of the pore geometry, information that is expensive to obtain, and moreover, changes from one scaffold to another. Furthermore, if details of the pore geometry *are* known, they can be incorporated into an expressions for the permeability, for example via homogenisation techniques (Shipley et al. 2009).

An alternative to the above macroscale approach when considering flow and nutrient transport problems is to consider a *microscale* one in which the fluid flow within the interconnected pores is solved using Navier–Stokes equations. Such a method, especially when complemented with imaging techniques such as μ-CT that provide detailed information about the geometry and the microstructure of the porous scaffold, is a powerful tool for the full characterisation of the 3D flow fields and stresses within dynamic culture systems. However, this can be computationally intensive, and requires a simulation to be run for every scaffold architecture. Examples of such studies are reviewed in Sect. 3.2.

As an alternative to assuming that the cells occupy no volume, they may be assumed to occupy volume, but that their interaction with the flow can be neglected (so that the role of any culture medium in the system is simply to provide a supply of nutrients to the cells). Such models, which are particularly appropriate for static culture systems where the transport of nutrients and metabolites is by diffusion, are reviewed in Sect. 3.1. Finally, in some applications, detailed information regarding culture medium flow and cell population dynamics is required; here it is necessary to consider explicitly the cells as a separate phase (with their own volume fraction, local velocity, and so on), and to consider their interaction with the surrounding fluid flow. Examples of models using this approach are given in Sect. 3.3.

2.1.1 Mathematical Reduction

When adopting a multiphase approach (and the various modelling simplifications) the resulting continuum models comprise coupled partial differential equations (PDEs). However, their solution may be computationally intensive and may not reveal details of the mechanisms underlying observed tissue growth phenomena. An alternative approach is to exploit the typically disparate length- and time-scales inherent in these systems: for example, the bioreactor may be long and thin, or the timescale for cell growth may be long compared to that for diffusion of solutes. In the former case, the spatial dependence of the problem may be reduced, in the latter, quasi-steady assumptions may be made. Alternatively, the magnitude of problem specific parameters (such as the Reynolds number or Peclet number) may be exploited to simplify the equations. For example, the Reynolds number characterises the ratio of inertia to viscous effects, and the Peclet number the ratio of advection to diffusion timescales, and by considering these parameters to be either large or small, the governing equations may be simplified to retain only the key aspects of the underlying physics of the system under consideration. In all cases, sophisticated asymptotic methods can be used rationally to simplify the governing equations, leading to reduced models that are tractable, yet remain physically well-grounded. The reduced models can then be attacked using analytical techniques and simpler numerical methods. By allowing maximum analytical progress to be made (enabling, for instance, the influence of various governing parameters to be relatively easily investigated), the analysis of reduced models can provide significantly more physical insight into the underlying mechanisms than would be obtained by computational simulations of the full system alone. Such approaches are reviewed in Sect. 3, and contrasted with more computational studies.

2.2 *Alternative Modelling Approaches*

When modelling cell population growth, many authors have employed a discrete, rather than a continuum, approach and considered individual cells explicitly. Such models provide a natural framework within which to accommodate, for example, cell signalling interactions, movement and proliferation, thereby providing comprehensive and detailed information about the dynamics of the cell population. Representative studies include Ouchi et al. (2003), in which a cellular Potts

(Graner-Glazier-Hogeweg) model is employed and formal rules for cellular interactions are postulated; such a modelling approach has been widely used, capturing cell sorting phenomena as well as a range of morphogenetic processes in, for example, the development of the slime mould *Dictyostelium discoideum*. To accommodate the physics of cell–cell or cell–environment interactions, Meineke et al. (2001) incorporate linear over-damped springs connecting cell centres within the modelling framework. This approach was included in a multiscale computational model developed to investigate the role of Wnt signalling in regulating tissue renewal in the intestinal crypt by Van Leeuwen et al. (2009). The preceding models employ a deterministic approach to model cell behaviour; however, at the single-cell scale, biological processes are inherently stochastic. Biological systems are profoundly affected by such stochastic noise; for example, genetic selection may be influenced by the stochastic nature of mRNA transcription (see Wilkinson 2009 and references therein). The importance of such stochasticity to a variety of tissue growth processes has been widely investigated. Common approaches include the use of master and Langevin equations (Othmer et al. 1988; Hadeler et al. 2004).

The analysis of discrete models typically necessitates a computational approach, since realistic simulations demand large numbers of cells, for which analytic results may be difficult or impossible to obtain. In an attempt to circumvent this problem, multiscale (homogenisation) techniques have been employed to derive continuum models directly from underlying discrete systems, enabling some of these discrete effects to be incorporated into tissue-scale models in a mathematically precise way. Representative examples include Turner et al. (2004); Murray et al. (2009) and Fozard et al. (2010), who employ such techniques to represent collective movement of adherent cells within a continuum framework, while (O'Dea and King 2011a, b) analyse pattern formation due to cell signalling processes. Shipley et al. (2009) determine the macroscale flow and transport properties of a specific type of tissue engineering scaffold, in order to specify criteria for effective tissue growth. In all cases, the resulting continuum models are formulated as small systems of PDEs which may be amenable to analysis and/or numerical investigation. For these reasons, much research has concentrated on continuum representations of tissue growth, and it is models of this type for tissue engineering bioreactors that we now consider in detail below.

3 Mathematical Modelling of Bioreactor Systems

The biochemical and biomechanical cues which lead to optimum growth are specific to the tissue under consideration; correspondingly, the mathematical modelling approach required is determined by both the bioreactor system and the particular biological processes under investigation. In the following we review theoretical studies of such systems, concentrating on the following broad themes: (i) models of static culture systems which focus on biochemical effects such as nutrient-limited growth, (ii) dynamic culture models in which the cells are viewed as point sources and sinks of metabolites and nutrients respectively, and the cells occupy no volume; and (iii) dynamic culture models in which the volume fraction of cells is finite and biochemical and biomechanical interactions with the environment are incorporated.

3.1 Modelling Static Tissue Culture

In this section we focus on mathematical models that have been developed to describe tissue growth in static culture, when the delivery of vital nutrients and removal of waste products are diffusion-limited. The motivation for many of these models, particularly those of Galban and Locke (1997, 1999a, b), stems from experiments performed by Freed and coworkers in which highly porous polymer scaffolds of varying thickness were seeded with chondrocytes and immersed in fluid containing (diffusible) nutrients such as oxygen and glucose (Freed et al. 1993, 1994). The average cell density achieved within the scaffolds was found to decrease as the thickness of the scaffold increased, leading the authors to conclude that limitations in nutrient transport, caused by an increase in cell mass, could be responsible for inhibiting cell growth. As we explain below, the continuum models of tissue growth in static culture range from phenomenological ones (Galban and Locke 1997) to multiphase models that distinguish between different components of the developing tissue (Galban and Locke, 1999a, b). Some models are cast as moving boundary problems, the free boundary typically delineating regions of the scaffold that have been colonized by cells and tissue matrix from regions that are devoid of cells and tissue matrix (Galban and Locke 1997; Wilson et al. 2007). A key and unifying feature of these models is the assumption that nutrient availability is growth-rate limiting.

3.1.1 Quasi-Steady Nutrient Distribution

In an attempt to determine whether Freed et al.'s mechanistic explanation was consistent with their experimental observations, Galban and Locke developed a series of mathematical models for the growth of chondrocytes seeded in a porous polymer matrix (Galban and Locke 1997, 1999a, b). In their initial work, two moving boundary problems were developed to study the influence of nutrient transport and consumption on cell growth (Galban and Locke 1997). They focused on a single, growth-rate-limiting nutrient (e.g. oxygen or glucose) whose consumption rate was assumed to be highest in regions containing cells. By exploiting the difference in time-scales for nutrient diffusion within a scaffold ($\sim$minutes) and cell growth ($\sim$hours), they were able to make a quasi-steady approximation, neglecting time-derivatives in the reaction–diffusion equation that defines the nutrient distribution within the porous matrix. The position of the moving

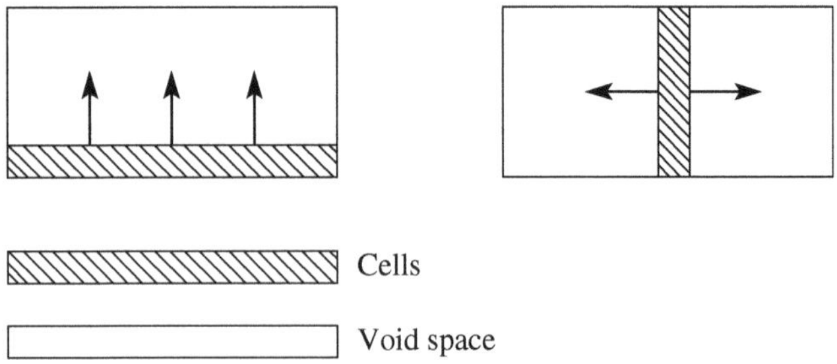

Fig. 2 In Galban and Locke (1997), moving boundary problems were developed to study the influence of nutrient transport and consumption on cell growth. The moving boundaries separate regions devoid of cells from those containing cells. Two seeding protocols were considered: in one case, cells were seeded on the scaffold periphery (where nutrient levels are highest), and in the second case the cells were seeded on a plane passing through the centre of the scaffold. The direction of cell growth is indicated by the arrows in the figure

boundary separating regions containing cells from those devoid of cells was determined by assuming that cell growth is localised at the moving boundary where it occurs at a rate which is proportional to the local nutrient concentration. This phenomenological model was used to compare two different seeding protocols: in one case, cells were seeded on the periphery of the scaffold, where nutrient levels are highest, and in the second case the cells were seeded in a plane passing through the centre of the scaffold. The direction of cell growth in each case is indicated in Fig. 2.

Numerical simulations revealed that the second case, with cells initially migrating outwards from the centre of the scaffold and in the direction of increasing nutrient levels, gave results which were in better agreement with Freed et al.'s experimental data than those obtained for the case in which the cell migrate towards the centre of the scaffold and into regions with lower nutrient levels. Good agreement was, however, only achieved for scaffolds of intermediate thickness (i.e. neither excessively thick nor overly thin). This deficiency motivated Galban and Locke to consider more detailed, multiphase models, in which the scaffold was viewed as a two-phase mixture of (effectively) cells and extracellular fluid, and cell growth was no longer localised on a moving boundary (Galban and Locke 1999a, b). In Galban and Locke (1999a), a one-dimensional geometry was employed, and the nutrient concentration assumed to vary parallel to the axis of the cylindrical scaffold only. The cell and fluid volume fractions were taken to be spatially uniform and to evolve over time, the cell volume fraction increasing at a rate which depends on the nutrient distribution within the scaffold. As in Galban and Locke (1997), the nutrient distribution was governed by a quasi-steady, reaction–diffusion equation. However the diffusion coefficient in Galban and Locke (1999a) was no longer taken to be a constant: it was assumed to decrease as the cell volume fraction increased. In this way, cell growth was found to be self-limiting: as cells proliferate, their volume fraction increases but this hinders nutrient transport through the scaffold and, hence, slows the rate of cell growth until, eventually, nutrient levels are too low to support further growth. This behaviour was evident in their numerical results which showed the cell volume fraction increasing towards a fixed value at long times, this value depending on the system parameters and, for example, decreasing as the scaffold thickness was increased. In this respect, the one-dimensional model yielded results which were in good agreement with the experimental observations reported in Freed et al. (1993, 1994). However, the predicted changes were not large enough to match those observed in the experimental data, possibly because the cell volume fraction was assumed to be uniform throughout the scaffold. Therefore, in Galban and Locke (1999b), the two-phase model was generalised by extending it to two spatial dimensions and allowing the cell and fluid volume fractions, and, hence, the effective nutrient diffusion coefficient, to vary with position and time. For typical simulations, spatial variation in the cell volume fraction and the effective diffusion coefficient became more pronounced over time, with the cell density being maximal on the periphery of the scaffold, where nutrient levels were highest, and declining towards the centre, where nutrient levels were lowest (the diffusion coefficient exhibited the opposite behaviour, with transport inhibited where the cell volume fraction was high). The agreement between the simulations obtained from this model and Freed et al.'s experimental data was much better than that obtained for the earlier, simpler models: the model was able to capture the dynamic trends that had been observed in the spatial distribution of the cell volume fraction and nutrient concentration when the scaffold thickness was varied.

The predictions generated from Galban and Locke's work, particularly the results in Galban and Locke (1999b), highlight some of the problems associated with culturing cells in scaffolds under static conditions. For example, as the thickness of the desired tissue construct (and, hence, the scaffold) increases, limitations associated with transporting nutrient to cells at the centre of the scaffold can lead to the formation of nutrient-deprived regions which are characterised by cell quiescence and even necrotic cell death. As we explain in Sects. 3.2 and 3.3, perfusion bioreactors, in which nutrient-rich culture medium is driven through the scaffold, offer scope for increasing nutrient supply to the central regions of the scaffolds and, thereby, improving the integrity of the engineered tissue.

3.1.2 Reaction and Diffusion of Nutrients

Galban and Locke's work has inspired the development of a number of similar mathematical models. For example, Lewis et al. (2005) develop a simple, one-dimensional model to investigate interactions between the evolving nutrient profiles and cell distributions within cartilaginous tissue constructs. The nutrient profile is modelled with a reaction–diffusion equation in which the diffusion coefficient is assumed to be constant and the local rate of nutrient consumption taken to be proportional to both the nutrient concentration and the cell density. The cells are assumed to be immobile and to proliferate at a rate proportional to that at which they consume nutrients. By comparing their simulations with experimental data for the spatio-temporal evolution of the oxygen and cell distributions, they found that their model represents a good description of cartilaginous tissue growth for the first two weeks after seeding. At later times, mechanisms not included in the model, such as contact inhibition of cell proliferation and reduced nutrient transport, are likely to become significant and may explain the poor agreement with the data after two weeks in culture.

Dunn et al. (2006) extended Lewis et al.'s model to account for contact inhibition of growth, using a logistic growth law to describe cell growth, the growth rate being proportional to the nutrient concentration, and by considering a two-dimensional cylindrical geometry. Their model was found to give good agreement with data from experiments in which pre-osteoblasts were initially seeded uniformly throughout a porous scaffold, these results being qualitatively similar to those reported by Freed et al. for chondrocytes. Dunn et al.'s model was, however, unable to reproduce experiments in which the scaffolds initially consisted of a series of thinner scaffolds that alternated between cell-seeded and unseeded: when the initial conditions were altered to mimic the layered structure, the experimentally observed peak in cell density at the interface between seeded and unseeded regions was not evident in the numerical simulations. Two possible explanations for this discrepancy were proposed. First, convective flow, which is not included in the mathematical model, may enhance nutrient transport at the boundaries between the seeded and unseeded regions. Alternatively, cell migration may not be negligible: cells may migrate away from regions of low oxygen and accumulate at the interface between the seeded and unseeded regions, where oxygen levels are higher.

More recently, Jeong et al. (2011) have focussed on developing an efficient numerical method for solving an extension of Lewis et al.'s model for the growth of cartilaginous tissue constructs in two-dimensional cylindrical geometry, in which the cells are allowed to move by random motion. Jeong and coworkers propose an operator-splitting algorithm to solve the resulting pair of coupled reaction–diffusion equations. Their approach involves alternating finite-difference approximations and analytical solutions in order to update the cell and nutrient profiles on successive time-steps. We remark that the model that Jeong et al. solve is similar in form to an earlier model by Obradovic et al. (2000) that was used to investigate how the local oxygen concentration within a cartilaginous construct influences the rate at which cells seeded within the scaffold produce glycosaminoglycan, an important component of cartilage.

The work of Chung et al. (2006) places the phenomenological reaction–diffusion models developed in Obradovic et al. (2000) and Jeong et al. (2011) on a stronger theoretical foundation. Following Galban and Locke (1999a, b), Chung and coworkers use a volume-averaging approach to derive an extension to Galban and Locke's two-phase model that accounts for a small degree of random cell movement. Their model simulations suggest that cell motility leads to more uniform cell distributions within the scaffold and higher overall rates of cell growth than when cell movement is neglected. Chung et al. also use their model to demonstrate that uniform cell seeding is likely to be a better strategy for tissue engineering (in static culture) than non-uniform seeding, arguing that concentrated cell seeding results in competition for nutrients which, when combined with limitations in nutrient transport, will reduce the overall cell growth rate.

3.1.3 The Cells' Mechanical Environment

While nutrient availability undoubtedly plays a crucial role in regulating cell growth within tissue constructs, mechanical effects are also important. In Wilson et al. (2007), Wilson and coworkers develop a simple deterministic model of tissue growth in which the cells are initially seeded around the periphery of a porous scaffold. They assume that nutrient is freely available, and use Darcy's law to model cell movement towards the centre of the scaffold. As in Galban and Locke (1997), a moving boundary problem is introduced to delineate regions of the scaffold that have been colonised by cells from regions which are devoid of cells. The Baiocchi transformation is used to transform the model to a linear complementarity problem for which one-dimensional analytical solutions and two-dimensional numerical ones are presented. Attention focusses on the behaviour of the moving boundary as the cells reach the centre of the scaffold and the colony approaches confluence: asymptotic techniques are used to derive approximate expressions for the time to confluence and to show that, near closure, the moving boundary evolves to an ellipse and that the ratio of its semi-major and semi-minor axes is identical to that of the two-dimensional, rectangular scaffold. The pressure within the scaffold is found to increase considerably shortly before it is filled, highlighting the potential problem for tissue engineers of a "slit" persisting within the tissue construct and compromising its mechanical integrity (see Fig. 3).

3.2 Modelling Dynamic Tissue Culture: Cell Volume Neglected

As discussed in Sect. 1.1.3, since static culture systems rely on diffusive transport of solutes, it is not possible to engineer constructs of a size suitable for implantation. Dynamic culture systems exploit the flow of culture medium to enhance nutrient and waste product transport by advection. Furthermore, such systems can provide mechanical load to mechanosensitive tissues, e.g. via the application of fluid shear stress to the cells. We review here models in which the cells occupy no volume, and have no affect on the fluid flow. We start by describing computational approaches and then consider approaches in which simplifying assumptions are used to derive reduced models that may be solved using analytical or simpler numerical techniques.

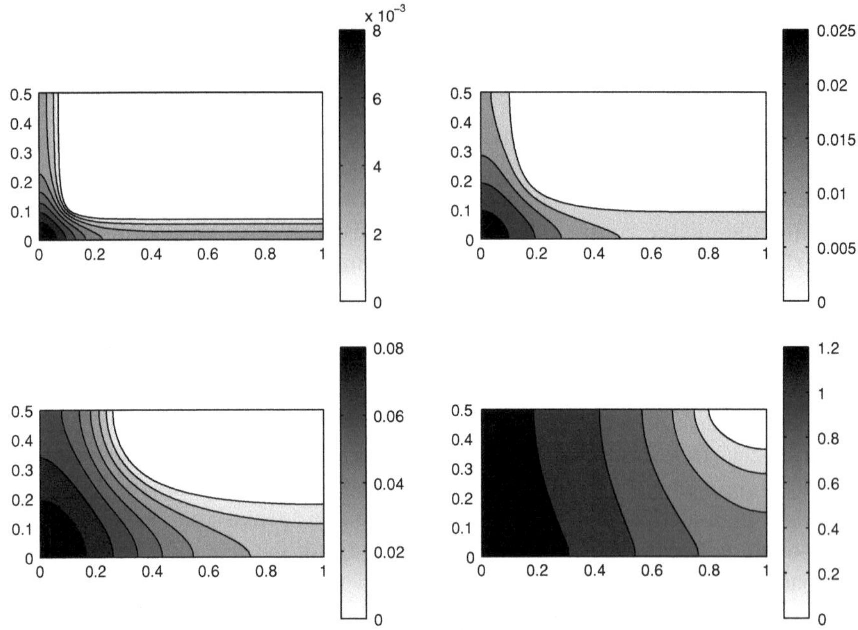

Fig. 3 Simulation results from the model developed in Wilson et al. (2007) showing how cells seeded around the periphery of a two-dimensional, rectangular porous scaffold eventually colonise the scaffold. The pressure profiles for the quadrant are plotted at dimensionless times $t = 0.5$, (*top left panel*), 1.0, 1.5 and 1.66 (*bottom right panel*), and show how the maximum pressure within the scaffold increases as the cells colonise more of the scaffold. The origin $(x, y) = (0, 0)$ is in the far corner so that the profiles may be easily viewed

3.2.1 Computational Approaches

When using computational fluid dynamics (CFD) approaches, the geometry of the system is retained (i.e. no simplifying assumptions are made e.g. due to the bioreactor being long and thin), and the reactor geometries can be generated using CAD packages. Such an approach was taken in Lawrence et al. (2008) who considered the effect of bioreactor geometry on the fluid flow through porous constructs. Rectangular and circular bioreactors were considered, with and without porous scaffolds, and with different inlet and outlet patterns. The governing equations (in this case, Navier–Stokes for the single-phase fluid, and Brinkman equations for flow through the porous scaffold) were solved using computational fluid dynamics software (COMSOL Multiphysics and ANSYS CFX). The authors identified non-ideal fluid distribution profiles by looking for "channeling" (where the fluid finds a short cut and leaves the reactor without dispersing through it), and "dead zones" which reduce the effective volume of the reactor and prolong the residence time of some fluid elements. Such non-uniform flow patterns can lead to poor nutrient distribution, and non-uniform shear stress distribution. The authors focus on fluid flow only, and highlight nutrient transfer as an open problem. Many authors have considered similar approaches, using either Brinkman (Lawrence et al. 2008; Devarapalli et al. 2009) or Darcy constitutive laws Cioffi et al. (2008) to model flow through the porous scaffold.

In addition to the above macroscale approach, there are two alternative ways to incorporate scaffold topology into the models. In the first, imaging techniques are used to provide detailed information about the pore scale geometry, which in turn provides a realistic computational domain within which to simulate the governing model equations. As an example, Cioffi et al. (2006) reconstructed the scaffold micro-geometry from μ-CT images acquired from a sample of the actual scaffold. The steady-state Navier–Stokes equations through the domain were solved using the commercial CFD software ANSYS FLUENT, which enabled quantification of the flow field and fluid shear stresses acting on the internal walls of the matrix in a tissue engineered construct subject to direct perfusion of culture medium within a bioreactor. A second approach is to idealise the micro-scale geometry. Boschetti et al. (2006) used a simplification of the geometry of a polymeric scaffold obtained by particulate leaching. A micro-domain of the scaffold was idealised as 27 sub-units arranged in a honey-comb pattern. Each sub-unit was obtained by subtracting a solid sphere from a concentric solid cube. The aim was to predict how the shear stress experienced by the cells depends on physical quantities such as scaffold porosity, pore size, and the imposed flow rate. By solving the Navier–Stokes equations in this complex domain, subject to appropriate boundary conditions (such as no-slip), it was possible to determine the exact nature of the flow properties. In Cioffi et al. (2006), the results of the simulations using the exact detailed scaffold architecture were compared with that from a simplified micro-scale model (Raimondi et al. 2004). A key finding is that, within the parameter regime considered (low Reynolds number flow and interconnected pores), the micro-geometry of the scaffold did not affect significantly the median shear stress acting on the inner scaffold walls. Hence, for the scaffolds considered in Cioffi et al. (2006), it is not necessary to build a detailed model for each new scaffold geometry, when a simple estimation of the median, mode and mean shear stress is required. The use of 3D geometrical models simplifies the scaffold geometry and reduces significantly the cost of the computation. Cioffi et al. (2008) considered a combined macro-scale/microscale computational approach to quantify oxygen transport and flow-mediated shear stress experienced by cells cultured in three-dimensional scaffolds in perfusion bioreactor systems. The macro-scale model consisted of Darcy equations for flow in the porous scaffold and an advection–reaction–diffusion equation for nutrient transport. The micro-scale approach was based on μ-CT reconstruction of the scaffold geometry. While both modelling approaches predicted similar average oxygen levels at different depths, the μ-CT model captured the micro-scale variations associated with the scaffold architecture. Thus, the choice of modelling approach for the scaffold geometry should be motivated by the required model outputs—if estimations of average shear stress and oxygen concentration are of interest, a simplified modelling approach can be taken, whereas if precise details of the spatial distribution of shear stress and oxygen concentration are required, then a more detailed modelling approach, where the scaffold geometry is reconstructed using μ-CT images is more appropriate.

3.2.2 Asymptotic Model Simplification

As outlined in Sect. 2.1.1, asymptotic methods can be used to simplify the governing equations, leading to reduced models that provide insights into the key underlying mechanisms involved in tissue growth within bioreactors. This approach is exemplified in Cummings and Waters (2006) (for complementary computational approaches see e.g. Lappa 2003) where a model for a rotating bioreactor is developed, in which a cylindrical vessel of circular cross-section rotates about its longitudinal axis. The bioreactor is filled with nutrient-rich culture medium and contains a growing tissue construct, which is modelled as a cylindrical solid object of circular cross-section. The axial length of the bioreactor is small relative to its radius. This fact, together with the observation that the reduced Reynolds number for the system is small, leads to a simplified model, in which the dependent variables are averaged across the axial direction, reducing the problem to two space dimensions. The rotation of the system introduces Coriolis and centrifugal force terms in the governing equations. The resulting fluid-dynamical problem is similar to that for flow around a moving circular obstacle (the tissue construct) within a rotating Hele-Shaw cell (see, e.g. Schwartz 1989; Waters and Cummings 2005). Additionally, the Peclet number is typically large in these biological applications, which simplifies the nutrient transport problem to one in which the nutrient concentration is constant along the two-dimensional fluid flow streamlines, with diffusion taking place perpendicular to the streamlines.

The authors exploit the fact that fluid flow, nutrient transport, and construct growth occur on very different time-scales, and decompose the problem into four distinct stages. Firstly, the fluid flow around the tissue construct is determined, assuming its location within the bioreactor is known. Then the position of the construct is determined by considering the balance of forces acting on the construct. Two classes of periodic motions are found, distinguished by whether or not the centre of the construct orbits the bioreactor centre (a special subcase of the latter being where the tissue construct remains stationary in the laboratory frame of reference). The fluid flow and construct trajectory model were validated in Cummings et al. (2009) by comparing the theoretical results with an experimental study of the trajectory of a large solid cylindrical disc suspended within a fluid-filled rotating vessel. All three flow regimes described above were observed experimentally, and good agreement between experiment and theory was found.

Having successfully solved the fluid flow problem, the nutrient distribution around the construct was then found, assuming a fresh supply of nutrient is maintained at the bioreactor's circular boundary and that at the tissue construct boundary nutrient is taken up at a rate proportional to how much is available in the surrounding culture medium. Using the nutrient concentration solution, and assuming the cells grow in response to the nutrient field, the growth rate of the construct was determined.

The theoretical study led to the important biological insight that if the construct position is stationary when viewed from the laboratory frame, then a nutrient depletion zone will form in the neighbourhood of the construct. Thus it is more desirable to operate the bioreactor in a regime in which the construct undergoes a time-dependent trajectory when viewed from the laboratory frame, mixing the nutrient concentration as it goes, and eliminating the development of nutrient depletion zones.

Still motivated by tissue growth in rotating bioreactors, Waters et al. (2006) considered a two-region, single-phase model to determine the morphology of a tissue construct formed from a single-cell suspension in culture media, within a rotating bioreactor. Experimental results indicated that at rotation rates below a critical value, the cells 'self-assemble' to form smooth nodules which are approximately elliptical in cross-section. However, at higher rotation rates, an amorphous construct forms with a highly irregular boundary. Histological studies show that the construct consists of a fluid interior which is a mix of apoptotic cells and culture media, surrounded by a membrane of proliferating cells and collagen. Motivated by these observations, Waters et al. (2006) consider a more dense viscous fluid drop surrounded by an extensible membrane of constant membrane tension in a less dense immiscible viscous fluid within a rotating bioreactor system. Both thin-disc and slender-pipe bioreactors (for which the bioreactor aspect ratio is small or large respectively) were considered so that, similarly to Cummings and Waters (2006), the fluid-dynamical equations may be simplified leading to a series of spatially 2D problems. The authors consider the interfacial stability of the initially circular fluid-fluid interface to small-amplitude, oscillatory perturbations; the instability arises as a result of the competition between the destabilising effects of centrifugal forces, and the stabilising effects of surface tension. The theoretical results indicate that culture within thin-disc bioreactors is more likely to result in irregular shape constructs than culture within slender-pipe bioreactors, and that in the thin-disc regime the wave number of the most unstable mode increased as the rotation rate increases, in line with experimental observations. Of course, the instability mechanism examined in Waters et al. (2006) is not the only one by which irregular constructs might be expected to arise, and the modelling highlighted here does not include any biochemical features, e.g. growth in response to nutrient fields, or cell proliferation, in the model.

In Cummings and Waters (2006) bioactive processes that modify the porous scaffold are neglected—rather the construct is modelled as an impermeable solid object. However, a number of studies have explicitly accounted for the properties of the porous scaffold and their effect on the fluid flow and nutrient transport. In Whittaker et al. (2009) a simple mathematical model is developed for forced flow of culture medium through a porous scaffold. Flow is forced through the scaffold via inlet and outlet pipes, and also through identical porous-walled fibres inserted through the scaffold: fluid is pumped into one end of each fibre and the other end is sealed so that the fluid is forced to travel through the porous fibre walls into the scaffold. There are therefore two flow domains—the fibres and the space occupied by the scaffold around them—and the flows are coupled via continuity conditions at the interfaces. The scaffold is modelled as a uniform isotropic porous medium, with a point source and point sink modelling the inlet and outlet pipes respectively, and line sources representing flow from the fibres. A separate fluid dynamical problem is solved within the fibres to compute the strength of the source terms in the line sources representing flux of fluid from the fibres into the scaffold. Again, geometric features are exploited to simplify the governing equations: the slender geometry of the fibres allows lubrication theory to be used to simplify the full Navier–Stokes equations. A Poisson problem is obtained for flow in the scaffold, which can be solved numerically, and is a much easier to solve than direct numerical simulation of the full Navier–Stokes equations within the complex scaffold geometry. Having solved for the fluid flow, the authors were then able to determine the distributions of shear stress, nutrients and waste products, and asses the implications for cell proliferation.

3.2.3 Bioreactor Design

In a series of papers (Shipley et al. 2010, 2011; Shipley and Waters 2011), Shipley and co-workers consider flow and nutrient transport problems in hollow fibre membrane bioreactors (HFMB). HFMBs are designed to enhance nutrient delivery to, and metabolite removal from, cells by using fluid flow to provide advective transport in addition to diffusion. A single hollow fibre consists of a central lumen, surrounded by a porous fibre wall or membrane, which separates the lumen from the extra-capillary space (ECS). Cells are seeded in a single layer on the fibre walls, or throughout a matrix surrounding the fibre. Fluid is driven through the fibre lumen under the action of an applied pressure gradient. The porous fibre wall allows the passage of nutrients, metabolites and growth factors both to and from the cells, and acts as a membrane to protect the cells from the direct effect of fluid shear due to flow through the lumen, thus enabling relatively high flow rates to be used without cell damage. A bundle of such fibres is then housed within a bioreactor: in addition to flow through the fibre lumen, the bioreactor has entry and exit ports for ECS flow. By developing a series of theoretical models, Shipley et al. were able to specify a set of operating conditions which the end user can use to prescribe the bioreactor geometry (e.g. fibre length and ECS depth) and operating parameters (e.g. pressures, flow rates, nutrient inlet concentrations, cell-seeding density) to obtain the optimum cell culture environment for the tissue under consideration.

In Shipley et al. (2010) theoretical predictions for the fluid retentate (lumen outlet flowrate) and permeate (ECS outlet flowrate) are derived, and compared against experimental data to determine the membrane permeability and slip. Fluid transport in the lumen and ECS is described using Navier–Stokes equations, and flow through the porous fibre walls using Darcy equations. The model is simplified by exploiting the slender geometry of bioreactor system, so that lubrication theory may be used to describe flow in the lumen and ECS. Analysis of the width of the boundary layer region where slip effects are important, together with the sensitivity of the retentate and permeate equations to the slip parameter, showed that slip was not significant for these membranes. The model was then further reduced by assuming no-slip conditions at the membrane surface, and comparison of the theoretical predictions for the retentate and permeate with the experimental data enabled the membrane permeability to be determined. The validated and parameterized model was then used to determine the operating conditions that enable the lumen inlet flowrate and pressure at the lumen outlet to be controlled to provide a specific permeate to lumen inlet flowrate ratio. In Shipley et al. (2011) the complementary nutrient transport problem was studied: although the modelling was for a generic nutrient, the parameters were specialised for oxygen. Here fluid flow in the lumen was modelled by Poiseuille's law, while membrane and ECS flow were neglected. Nutrient concentration was governed by an advection–diffusion equation in the lumen, a diffusion equation in the membrane, and an advection–reaction–diffusion equation in the ECS, with Michaelis–Menten kinetics for the reaction term. These equations were coupled via specification of continuity of concentration and flux at the lumen-membrane and membrane-ECS boundaries. Analytical progress was made in the limit in which the ECS oxygen concentration is much larger than the half-maximal concentration of oxygen, so that the Michaelis–Menten reaction term can be approximated by a constant. These solutions were complemented by computations of the full system of equations, using COMSOL. The study enabled operating conditions to be defined to enable the user to determine the medium flow rate, lumen length and ECS depth that ensures a minimum oxygen concentration throughout the HFMB. In Shipley and Waters (2011), this work was extended to consider flow throughout the HFMB and the transport of lactate, in addition to oxygen. The study indicated that oxygen (as opposed to lactate) is the limiting solute with respect to bioreactor design, and that opening the ECS port to promote radial flow through the bioreactor provides significant benefit, enabling a greater volume of cells to be cultured within the desired nutrient environment.

All the models described above consider timescales appropriate to transport processes rather than cell proliferation, differentiation etc, and neglect neo-tissue formation, so that, for example, the effect of cell proliferation on the scaffold permeability and hence the resulting fluid flow is not considered. In Sect. 3.3 we consider models which account for the evolution of cell volume fraction, and the full coupling between the cell dynamics and their environment.

3.3 Modelling Dynamic Tissue Culture: Multiphase Phase Flow Models

As discussed in Sect. 3.2, a simple study of perfusion within porous tissue constructs was presented by Whittaker et al. (2009); here, the scaffold geometry is fixed, and the effects of cell proliferation on scaffold porosity, and hence fluid flow and nutrient transport fields, are neglected. Multiphase models which consider these features include Coletti et al. (2006) and Causin and Sacco (2011). In these studies, the dynamics of a cell seeded porous scaffold incubated in a perfusion bioreactor were considered. The scaffold was modelled as a rigid porous material, saturated with a viscous fluid. To account for cell proliferation, the porosity of the rigid scaffold is defined to be the sum of a time-invariant component (the porosity of the decellularised porous scaffold) and a time-dependent component (due to the proliferation of the cells); that is, the cell phase is assumed to be immobile, and with identical material properties to the scaffold phase. By appealing to the separation of timescales between flow and cell proliferation, the momentum balance equation for a viscous culture medium is simplified by assuming that fluid flow may be modelled by the Brinkman equations. Occlusion of scaffold pores due to cell proliferation and its influence on the flow field is accommodated via a cell density-dependent Darcy permeability of Carman-Kozeny type. The influence of this flow field on the distribution of a passive nutrient is modelled by reaction–diffusion–advection equations; nutrient consumption by the cells is modelled by Michaelis–Menton kinetics. Cell proliferation in response to this nutrient field is described by a Contois equation.

A more comprehensive study, investigating such ideas is presented by Chung et al. (2007). Here, cell movement is accommodated via linear diffusion, the fluid flow and nutrient fields being modelled in a similar manner to that described above. The cell density and nutrient profiles were calculated by numerical simulation, to demonstrate that perfusion of cell cultures can lead to enhanced proliferation and a more spatially-uniform cell distribution. Shakeel et al. (2011) extend this approach by employing nonlinear diffusion to represent cell movement, and by accommodating the influence of the mechanical environment on cell proliferation. Such a change in cell phenotype may be accommodated by suitable specification of the mass transfer rates S_i; Shakeel et al. (2011) capture the dependence of cell proliferation and nutrient consumption on fluid shear stress by adapting functional forms proposed by O'Dea and coworkers (reviewed below) to describe enhanced proliferation and nutrient consumption when cells are exposed to physiologically-relevant levels of fluid shear stress. By employing the commercial finite element software COMSOL Multiphysics, the influence on eventual tissue construct composition of different scaffold porosity distributions, and cell seeding protocols was investigated. The authors conclude that an effective means to ensure nutrient delivery to large tissue constructs in such bioreactors is through the use of high-porosity channels which span the construct.

3.3.1 Cell-Environment Interactions: Momentum Transfer

The above authors dramatically simplify the momentum transfer between phases by employing the Darcy or Brinkman equations to relate the fluid flow to the fluid pressure. A comprehensive multiphase formulation relevant to tissue growth processes within a perfusion bioreactor system, which addresses in detail the interactions between phases was presented by Lemon et al. (2006). The modelling framework accommodates an arbitrary number of fluid phases (representing, for instance, cells and culture medium), contained within a porous medium (such as an artificial scaffold and/or ECM). The model differs from other fluid-based models (such as Breward et al. (2002), Franks and King (2003) and references therein) by the explicit consideration of a tissue's solid constituents (here, interpreted as a porous scaffold and/or ECM), and the general nature of the interphase interaction terms, which extends the range of tissue engineering applications that can be addressed. In general, the formulation allows for deformation of the porous medium; however, in Lemon et al. (2006), attention is restricted to a rigid porous scaffold and as such, the stimulation provided by mechanical bioreactors (e.g. the bioreactor system of El-Haj et al. 1990) may not be accommodated. The adoption of fluid-like constitutive assumptions for the phases contained within the porous scaffold, which is appropriate on the timescale of tissue growth, simplifies significantly the modelling of tissue growth processes since mass transfer between phases (representing tissue growth) does not generate stress, as is the case when elastic constitutive modelling assumptions are made.

Referring to Eq. (2) in Sect. 2.1, the interphase force terms $\mathbf{F}_{ij}$ are assumed to comprise contributions from viscous drag, accommodated via terms proportional to the difference in phase velocity, and active forces (such as those which arise from interphase interactions), these being modelled via additional pressure contributions (Lemon et al. 2006):

$$\mathbf{F}_{ij} = p_{ij}\left(\theta_j \nabla \theta_i - \theta_i \nabla \theta_j\right) + k\theta_i\theta_j(\boldsymbol{u}_j - \boldsymbol{u}_i); p_{ij} = p + \psi_{ij}; j \neq i \tag{6}$$

$$p_i = p + \phi_i + \sum_{j \neq i} \theta_j \psi_{ij} \tag{7}$$

where p_{ij} denote the 'interphase pressures' which are assumed to comprise a contact-independent pressure, p, common to all phases in the mixture, and an additional term ψ_{ij} which accounts for the effect of tractions between phases (e.g. cell–ECM interactions). The pressure p_i of each phase accommodates an additional internally-generated pressure ϕ_i (due to cell–cell interactions, for example).

The multiphase approach embodied by Eqs. (1), (2), (6) and (7), together with suitable constitutive assumptions about σ_i and S_i, have been employed by many authors to investigate a range of tissue engineering problems. In each case, the relevance to the biological system under consideration is ensured by restriction to an appropriate numbers of phases, and by specification of σ_i, S_i, and the extra pressures ϕ_i and ψ_{ij}.

For example, Lemon and coworkers specialise the modelling framework to consider two fluid phases (cells and culture medium being modelled by a viscous fluid and an inviscid fluid, respectively) contained within a rigid porous scaffold, employing the resulting model to investigate proliferation, aggregation, dispersal and travelling wave behaviour of a population motile cells within an artificial scaffold. In Lemon et al. (2006), linear stability analysis was employed to determine how aggregative or dispersive behaviour depends upon the model parameters; in Lemon and King (2007b), by considering travelling wave solutions, taking distinguished limits of certain model parameters (such as the viscous drag between the cell and scaffold phases) and obtaining numerical solutions in one spatial dimension it was shown that this model displays a wide range of travelling wave behaviour. Notably, both backward and forward-travelling waves may exist, an unusual feature of models for tissue growth. By introducing a generic passive solute whose spatio-temporal dynamics are governed by an advection–reaction–diffusion equation, the influence of nutrient limitations on tissue growth was considered in Lemon and King (2007a). In addition, different initial cell seeding strategies were studied, and spatial heterogeneity of the scaffold was accommodated in a limited sense by prescribing a scaffold density distribution which is constant except near the periphery where it tapers to zero. Numerical simulation (in one spatial dimension), as well as simplifying asymptotic limits, indicate preferential tissue growth near the scaffold's periphery due to nutrient depletion by the cells; reduction of cell-scaffold drag ameliorates this feature.

These models reveal that scaffold properties (drag and porosity) are key determining factors affecting the expansion of the cell population to colonise the scaffold, and provides a methodology with which to determine model parameter regimes in which the cell population is able to colonise uniformly the scaffold.

In the above studies by Lemon and coworkers, attention was restricted to a single cell phase, which was assumed to proliferate at a constant rate (or at a rate proportional to the local nutrient concentration). O'Dea et al. (2008, 2010) employed the general multiphase formulation (1), (2), (6) and (7) to investigate the ability of different mechanical stimuli to influence tissue construct growth. By suitable specification of the mass transfer rates S_i (and employing viscous constitutive assumptions for both the cell and culture medium phases), progression between three different cell phenotypes was considered: (i) proliferative, (ii) ECM-depositing and (iii) apoptotic. Guided by experimental studies (Roelofsen et al. 1995; Klein-Nulend et al. 1995a; You et al. 2000; Bakker et al. 2004b; Han et al. 2004; Yourek et al. 2004), simple functional forms were proposed to model phenotypic changes in response to cell density (reflecting contact inhibition), and culture medium pressure and shear stress (reflecting the accepted influences of these stimuli on bone tissue growth). This approach was also exploited in the study of Shakeel et al. (2011), described above. Numerical simulations presented in O'Dea et al. (2010) indicated that the mechanical stimuli to which the cells respond can alter significantly the composition and morphology of the resulting tissue construct. This suggests that, in real applications, the histology of engineered tissue constructs may be employed to infer the dominant regulatory mechanisms at play in a given cell line.

3.3.2 Tissue Engineering Scaffolds

While the studies of Lemon, O'Dea and coworkers accommodate explicitly the momentum transfer between phases (unlike that of Shakeel et al. 2011) the models are significantly simplified by assuming that the tissue construct's solid components (scaffold, ECM) are rigid, spatially-homogeneous and constant in time, despite experimental evidence to the contrary (Lemon and King (2007a) consider spatial variation in scaffold density only near the scaffold periphery). By including additional mass conservation equations of the form (1) to govern the scaffold and ECM density distributions and employing experimental data to initialise the scaffold distribution, O'Dea et al. (2012) have shown that, due to cell-scaffold adhesion, spatial variations in scaffold density may lead to significant heterogeneity in cell and ECM distributions, with implications for their mechanical suitability for implant. In addition, it was highlighted that the model simulation results can be employed to demonstrate how rates of scaffold degradation and ECM deposition may be matched in order to maintain mechanical integrity of tissue constructs.

Despite the general nature of the multiphase formulation (1), (2), (6) and (7), all of the foregoing investigations employ significant modelling assumptions, or exploit suitable asymptotic limits, to simplify the governing equations, leading to reduced models that are analytically tractable or more straightforward to solve numerically. Lemon and co-workers study linear stability, travelling waves and numerical solution in one spatial dimension; Coletti et al. (2006) and Causin and Sacco (2011) consider an immobile cell phase; Chung et al. (2007) and Shakeel et al. (2011) simplify the momentum transfer between phases; and O'Dea and co-workers consider the simplifying limit of asymptotically-small tissue construct aspect ratio to obtain a one-dimensional model and calculate travelling wave and numerical solutions. To determine the applicability of such geometric simplifications, Osborne et al. (2010) use finite element methods to construct two dimensional solutions for bioreactors of varying aspect ratios. Comparison with results obtained by O'Dea et al. (2010) in the long-wavelength limit, demonstrated that the simplified model captures the majority of the qualitative behaviour; however, to capture accurately mechanotransduction-affected tissue growth, spatial effects in (at least) two dimensions are required.

The aforementioned studies employ the general multiphase model (1), (2), (6) and (7) in various incarnations to investigate in detail cell–cell and cell–substrate interactions. However, in all cases, the scaffold phase is modelled as a rigid porous medium. Deformation of the tissue construct forms a key feature of many mechanical bioreactor systems (in the specific example outlined in Sect. 1.1, this is provided by macroscale compression of the scaffold by a piston). This can be accommodated within the multiphase framework by appropriate choice of the constitutive law for the solid component(s) of the tissue. Furthermore, if the tissue stresses induced by volumetric growth are of interest, as is certainly the case for the engineering of soft tissues, fluid-based constitutive choices for the cell phase are inappropriate, since such stresses will dissipate (Araujo and McElwain 2005b).

Among the first multiphase formulation to accommodate such ideas, Roose et al. (2003) presented a poroelastic model for the growth of an avascular tumour spheroid in which the tissue comprises a porous solid phase, saturated by a fluid whose flow is governed by Darcy's law. The stress tensor for poroelastic materials (Biot, 1941) was modified by the addition of a term accommodating volumetric tissue growth. Byrne and Preziosi (2003) provide a more thorough derivation of equations appropriate to model tissue as a deformable porous media (comprising a solid and a liquid phase), as well as considering explicitly environmental influences on tissue growth.

A comprehensive derivation for a general n-phase mixture is provided by Araujo and McElwain (2005a, b), who state conservation equations for mass, linear and angular momentum, and energy; detailed application to a spherically-symmetric tissue represented as a biphasic mixture of linear-elastic solid (cell) phase, saturated with an inviscid fluid was given. Such a constitutive assumption leads to singular behaviour in stress evolution (Jones et al. 2000), typically regularised by the addition of viscosity to the elastic constitutive law (so that the tissue is modelled as a poroviscoelastic material); Araujo and McElwain show that anisotropic growth in the direction of least compressive stress is an alternative method for regularising growth-induced stress. Preziosi and Tosin (2009) present a similar multiphase formulation accounting for two different cell types, ECM and tissue fluid.

Many other authors have given extensive consideration to the derivation of multiphase models of tissue growth which can accommodate deformation and flow in a physiologically-accurate way (see e.g. Loret and Simões (2005); Ricken and Bluhm (2010) and references therein); however, the complexity of such models means that most studies are restricted to biphasic mixtures, or employ simplifying constitutive laws and/or geometries. An alternative approach is to exploit a CFD approach (see Sect. 3.2) to determine 3D characterisation of the flow and nutrient transport within cell culture systems. Examples include Kwon et al. (2008) and Consolo et al. (2011), in which the optimisation of conditions for embryonic stem cells encapsulated within hydrogel beads was considered, when cultured in a rotating bioreactor (see Sect. 1.1.3). Advection and diffusion of oxygen were coupled to comprehensive multiphase fluid dynamics calculations to investigate how the rotation speed ensures proper oxygen supply, while maintaining a low-stress condition. Theoretical predictions of optimal rotation speed were obtained, ensuring oxygen delivery to the cells while avoiding excessively dense bead packing and collision with the bioreactor walls. These were employed in in vitro experiments showing that the bead motion adheres to the in silico analysis and that such a dynamic culture strategy shows distinct benefits over static culture in terms of cell number and viability.

4 Discussion

In vitro tissue engineering is an emerging field of enormous importance, with the potential to alleviate the current shortage of tissues and organs for transplant, which are required for successful clinical intervention in problems associated with tissue damage, degeneration or failure. In addition, engineered tissues have applications in toxicology testing and drug screening.

Tissue engineering is an intrinsically interdisciplinary field; the complexity of the biological and biochemical processes involved obfuscates investigation by experimental biologists alone. Theoretical contributions from applied mathematics can provide important insight into tissue growth and clarify how the different processes interact. In this review, we have provided an exposition of the contribution of continuum mathematical modelling to the generation of new tissues from cells seeded on porous scaffolds. The mathematical models generate insight into nutrient-limited growth in static culture, and the dual role of fluid flow in enhancing nutrient transfer to the growing tissue construct, and in providing a source of mechanical stimulus to the cells e.g. via fluid shear stress.

It is clear that significant advances in tissue engineering have been made in recent years and that mathematical modelling is starting to play a central role in the design of bioreactors and the interpretation of the resulting experimental data (see, e.g. Shipley et al. (2011), Shipley and Waters (2011), Shipley et al. (2009), Cinar et al. (2003), Julien and Whitford (2007) and references therein). In spite of this progress, many open problems remain to be addressed. We outline below some of the theoretical, computational and modelling challenges that lie ahead.

- Appropriateness of continuum limit

As stated above, the mathematical models that we have reviewed typically treat the developing tissue and its constituents as continua and, as a result, do not distinguish between individual cells. When studying the initial growth phase of cells seeded within a tissue construct, it is natural to question the validity of adopting a continuum approach, particularly if the initial cell seeding density is low. In such cases, it may be more appropriate to use discrete, cell-based models of the type developed by Chung et al. (2010) and Cheng et al. (2009). In Cheng et al. (2009), the model has three components: a reaction–diffusion equation for the nutrient concentration, a cellular automata model describing cell migration, proliferation, and collision, and rules relating cell division rates and migration speeds to nutrient concentration. The hybrid discrete-continuous model was solved to study how transport limitations affect the tissue regeneration rates under conditions encountered in typical bioreactors. In Chung et al. (2010) a similar hybrid cellular automata approach is used to investigate the effect of nutrient-limitation on cell construct development for cartilage tissue engineering. The model was used to identify seeding strategies that result in enhanced cell number and a uniform cell distribution for the tissue engineered construct.

As a further example, The EU funded FP6 project Complex Automata Simulation Techniques (COAST) aims to develop computational methods to solve multi-scale models where a number of single-scale models interact on different length and timescales and pass information between themselves. The approach involves coupling of single-scale cellular automata or agent based models. The success of the approach has been demonstrated in the modelling of in-stent restenosis within blood vessels.

Discrete, cell-based descriptions, such as cellular automata models, are well suited to problems requiring the incorporation of cell signalling processes or subcellular phenomena, including the cell cycle and signal transduction pathways: see, for example, Van Leeuwen et al. (2009) and Alarcon et al. (2010). However, it is not at present clear how to incorporate the influence of mechanical forces into such a formulation; in this case, treating the cells as deformable objects is more appropriate. For example, by following Drasdo and coworkers (Drasdo and Hohme 2005; Byrne and Drasdo 2009) and viewing the cells as homogeneous, isotropic, elastic objects (or ellipsoids in 3D), it should be possible to compute the stresses and strains experienced by individual cells and relate them to the subcellular response that such mechanical stimuli elicit (Mullender et al. 2004).

- Hybrid models

In practice, as cells proliferate, the computational effort needed to track individual cells quickly becomes prohibitive. In such situations, when cell numbers are large, it may be more appropriate to resort once again to a continuum approach. While there are now a number of alternative approaches for modelling at the cell- and continuum scales, it is less clear how to transition between the two approaches or how to relate parameters appearing in the continuum models to those that appear in cell-level ones. A possible resolution to the latter problem is outlined in Byrne and Drasdo (2009): simulations of an individual-based model of avascular tumour growth, parameterised by measurable biophysical quantities, are compared with simulations from a continuum mechanical model and, in this way, parameters in the continuum model related to measurable quantities. The problem of deciding when it is appropriate to switch from a continuum to a discrete description (or vice versa) is considered in Kim et al. (2007). The authors propose a hybrid model for the growth of an avascular tumour embedded within a deformable gel. A cell-based approach is used in the outer annulus of the tumour, where nutrient levels are high and the cells are proliferating, while continuum descriptions are used for the gel surrounding the tumour and the central core of the tumour, where nutrient levels are low and the cells undergo necrosis. Careful consideration is given to the appropriate coupling of these representations at the boundary between the continuum and discrete domains.

- Computational challenges

The systems of governing equations arising from multiphase descriptions of biological tissues are complex: to solve these equations in the complex 2D and 3D geometries typically encountered within bioreactor systems requires the development of sophisticated numerical schemes. To this end, Osborne and Whiteley (2010) consider a numerical technique specifically for the solution of the multiphase flow equations. They demonstrate that these equations can be written as a mixed system of PDEs, consisting of first-order hyperbolic equations for the volume fraction of each phase, generalised Stokes equations for the velocity of each phase, and elliptic PDEs for the concentration of nutrients and messengers. This complex system of coupled nonlinear PDEs is solved via the development of finite element techniques.

As the mathematical and computational models being used to simulate tissue growth become more detailed and complex, and as they are extended by different researchers, it becomes increasingly important that the underlying software is robust, reliable and fully tested. Several groups are developing software to tackle such problems. For example, the software Cancer Heart and Soft Tissue Environment (CHASTE) has been developed and is continually being advanced to solve multiscale and multiphase problems in areas of physiology that encompass cardiac, cancer and tissue engineering applications (Pitt-Francis et al. 2000).

- Modelling challenges

Within the context of multiphase modelling, a key challenge is the specification of appropriate constitutive laws for the material properties of the constituent phases, and their interactions via interphase mass and momentum transfer. Current modelling approaches, such as O'Dea et al. (2010), have proposed simple candidate constitutive laws and have shown how the characteristic tissue morphologies that arise depend on the mechanical stimuli: by comparing model predictions with experimental data, it should be possible to determine the dominant regulatory stimuli for a given cell line. However, in order to account accurately for the material properties of the tissue construct in question (highlighted as a key consideration in in vitro tissue engineering in Sect. 1.1.2), simplified constitutive assumptions such as those employed by O'Dea et al. (2010) and Osborne et al. (2010) are inappropriate. Detailed modelling of growth-induced stresses and tissue construct deformation is required. Such approaches necessitate a dramatic increase in model complexity (see e.g. Loret and Simões 2005; Ricken and Bluhm 2010) and are therefore heavily reliant on numerical simulation, or model simplification in order to make analytic progress.

Once appropriate models have been developed for the bioreactor culture system, the models must be parameterised for the particular tissue type under consideration e.g. bone, cartilage or cardiac tissue. Such model parameterisation requires ongoing and close collaboration between modellers and experimentalists, in order that the parameters of interest can be identified and successfully measured. A significant barrier to the development of biologically realistic and powerful models is the generation of suitable experimental data against which they can be validated. Since the engineering of a tissue construct is a dynamic process, and it is often technically difficult (or infeasible) to track a single experiment, the collection of reliable quantitative data is an extremely difficult problem. An important output of theoretical models is detailed information regarding the spatio-temporal distributions of (e.g.) stress, cell density, or nutrient levels within evolving tissue constructs: since this data is difficult, or impossible, to obtain experimentally, it is an extremely challenging problem to validate the developed theoretical models against experimental data. However, by validating key features of the model for which suitable data may be procured, we may employ such models to provide detailed spatio-temporal information with confidence. Such an approach was demonstrated in Cummings et al. (2009) (see Sect. 3.2): here the authors validated the trajectory of a tissue construct within a rotating bioreactor by comparing theoretical and experimental results, and, having successfully done this, were then able to use the model to provide detailed spatial information about the nutrient field surrounding the tissue construct.

- In vitro to in vivo—integration and vascularisation

The models reviewed thus far have focussed on the in vitro generation of tissue constructs rather than their integration with normal tissue when they are implanted into patients. Recently, Lutianov and coworkers have formulated a mathematical model, comprising a system of nonlinear reaction–diffusion equations, to describe the in vivo regeneration of cartilage by isolated chondrocytes and/or mesenchymal stem cells that are seeded into a defect in the knee (Lutianov et al. 2011). Model simulations suggest that it takes around eighteen months for chondrocytes to regenerate a typical defect, of length 10–20 mm and thickness 2–3 mm. The authors also use their model to demonstrate that mesenchymal stem cells are no better at regenerating cartilage than chondrocytes. Landman and Cai (2007) have also used mathematical modelling to study the integration of tissue constructs into host tissues. A key aspect of integration is the development of an appropriate vascular supply to the engineered tissue (Novosel et al. 2011). We do not thoroughly review theoretical models for the vascularisation of engineered tissues here, but refer the interested reader to Landman and Cai (2007) and Lemon et al. (2009). In contrast to Lutianov et al. who consider avascular tissues, Landman and Cai develop a model to investigate the feasibility of stimulating the formation of new blood vessels within the tissue construct to prevent the development of nutrient-starved regions developing within the tissue construct. Lemon et al. (2009) described the evolution of different tissue constituents within an artificial scaffold, including vasculature, via a set of coupled non-linear ordinary differential equations. Bifurcation analysis was used to determine the extent of scaffold vascularisation as a function of the parameter values. The development of models which accommodate the formation and establishment of a vascular supply is a challenging open problem in the field: sophisticated models already exist describing angiogenesis in, e.g. wound healing and cancer, and it will be instructive to draw upon these models when considering vascularisation of tissue engineered constructs (Owen et al. 2009; Anderson et al. 2012).

In conclusion, despite, or perhaps because of, the many challenges outlined above, it is clear that tissue engineering is an exciting multidisciplinary field which is raising many demanding biological questions that can serve as the basis for fascinating, and equally demanding, mathematical problems for many years to come. Continued dialogue between mathematical modellers and tissue engineers, biologists, clinicians and a variety of other experimental scientists will be crucial to the resolution of these problems and should lead to wider recognition that mathematical modelling can be used as a powerful tool to advance understanding of tissue engineering.

Acknowledgments SLW is grateful to the EPSRC for funding in the form of an Advanced Research Fellowship and HMB thanks the Oxford Centre for Collaborative Applied Mathematics. This publication is based on work supported by Award No. KUK-C1-013-04, made by King Abdullah University of Science and Technology (KAUST).

References

Alarcon, T., Byrne, H.M., Maini, P.K.: A multiple scale model for tumour growth. Multiscale Mod. Sim. **3**, 440–475 (2010)

Ambrosi, D., Preziosi, L., Vitale, G.: The insight of mixtures theory for growth and remodelling. Z. Angew. Math. Phys. **61**, 177–191 (2010)

Anderson, A.R.A, Chaplain, M.A.J., McDougall, S.: A hybrid discrete-continuum model of tumour induced angiogenesis. In: Jackson Trachette, L. (ed.) Modeling Tumor Vasculature, pp. 105–133. Springer, New York (2012). ISBN:978-1-4614-0052-3

Araujo, R.P., McElwain, D.L.S.: A mixture theory for the genesis of residual stresses in growing tissues i: a general formulation. SIAM J. Appl. Math. **65**, 1261–1284 (2005a)

Araujo, R.P., McElwain, D.L.S.: A mixture theory for the genesis of residual stresses in growing tissues ii: solutions to the biphasic equations for a multicell spheroid. SIAM. J. Appl. Math. **66**, 447–467 (2005b)

Atala, A., Mooney, D.J., Vacanti, J.P, Langer, R.: Synthetic biodegradable polymer scaffolds. Birkhèauser, Boston (1997)

Ateshian, G.A.: On the theory of reactive mixtures for modeling biological growth. Biomech. Model. Mechanobiol. **6**(6), 423–445 (2007)

Bakker, A., Klein-Nulend, J., Burger, E.: Shear stress inhibits while disuse promotes osteocyte apoptosis. Biochem. Biophys. Res. Commun. **320**(4), 1163–1168 (2004a)

Bakker, A., Klein-Nulend, J., Burger, E.: Shear stress inhibits while disuse promotes osteocyte apoptosis. Biochem. Biophys. Res. Comm. **320**, 1163–1168 (2004b)

Biot, M.A.: General theory of three-dimensional consolidation. J. Appl. Phys. **12**(2), 155–164 (1941)

Boschetti, F., Raimondi, M.T., Migliavacca, F., Dubini, G.: Prediction of the micro-fluid dynamics environment imposed to three-dimensional engineered cell systems in bioreactors. J. Biomech. **39**, 418–425 (2006)

Bowen, R.M.: Theory of mixtures. In: Eringen, A.C. (ed.) Continuum Physics, vol. 3. Academic Press, New York (1976)

Breward, C.J.W., Byrne, H.M., Lewis, C.E.: The role of cell–cell interactions in a two-phase model for avascular tumour growth. J. Math. Biol. **45**(2), 125–152 (2002)

Butler, D.L., Hunter, S.A., Chokalingam, K., Cordray, M.J., Shearn, J., Juncosa-Melvin, N., Nirmalanandhan, S., Jain, A.: Using functional tissue engineering and bioreactors to mechanically stimulate tissue-engineered constructs. Tissue Eng. Part A **15**(4), 741–749 (2009)

Byrne, H., Drasdo, D.: Individual-based and continuum models of growing cell populations: a comparison. J. Math. Biol. **58**, 657–687 (2009)

Byrne, H.M, Preziosi, L.: Modelling solid tumour growth using the theory of mixtures. Math. Med. Biol. **20**(4), 341–366 (2003)

Cartmell, S.H., El Haj, A.J.: Mechanical bioreactors for tissue engineering. In: Chaudhuri, J., Al-Rubeai, M. (eds.) Bioreactors for Tissue Engineering: Principles Design and Operation, Chap. 8, pp. 193–209. Springer, Dordrecht (2005)
Causin, P., Sacco, R.: A computational model for biomass growth simulation in tissue engineering. Comm. Appl. Ind. Math. **2**(1), 1–20 doi:10.1685/journal.caim.370 (2011)
Cheng, G., Markenscoff, P., Zygourakis, K.: A 3D hybrid model for tissue growth: the interplay between cell population and mass transport dynamics. Biophys. J. **97**(2), 401–414 (2009)
Chung, C.A., Yang, C.W., Chen, C.W.: Analysis of cell growth and diffusion in a scaffold for cartilage tissue engineering. Biotech. Bioeng. **94**(6), 1138–1146 (2006)
Chung, C.A., Chen, C.W., Chen, C.P., Tseng, C.S.: Enhancement of cell growth in tissue-engineering constructs under direct perfusion: modeling and simulation. Biotech. Bioeng. **97**(6), 1603–1616 (2007)
Chung, C.A., Lin, T.-H., Chen, S.-D., Huang, H.-I: Hybrid cellular automaton modeling of nutrient modulated cell growth in tissue engineering constructs. J. Theor. Biol. **262**(2), 267–278 (2010)
Chuong, C.J., Fung, Y.C.: On residual stresses in arteries. J. Biomech. Eng. **108**, 189 (1986)
Cinar, A., Parulekar, S.J., Undey, C., Birol, G.: Batch Fermentation: Modeling Monitoring and Control. Marcel Dekker Inc., New York (2003)
Cioffi, M., Boschetti, F., Raimondi, M.T., Dubini, G.: Modeling evaulation of the fluid-dynamic microenvironment in tissue-engineered constructs: a micro-ct based model. Biotech. Bioeng. **93**(3), 500–510 (2006)
Cioffi, M., Küffer, J., Ströbel, S., Dubini, G., Martin, I., Wendt, D.: Computational evaluation of oxygen and shear stress distributions in 3d perfusion culture systems: macro-scale and micro-structured models. J. Biomech. **41**, 2918–2925 (2008)
Coletti, F., Macchietto, S., Elvassore, N.: Mathematical modeling of three-dimensional cell cultures in perfusion bioreactors. Ind. Eng. Chem. Res. **45**, 8158–8169 (2006)
Consolo, F., Bariani, C., Mantalaris, A., Montevecchi, F., Redaelli, A. Morbiducci, U.: Computational modeling for the optimization of a cardiogenic 3d bioprocess of encapsulated embryonic stem cells. Biomech. Model. Mechanobiol. **11** 1–17 (2011)
Cummings, L.J., Waters, S.L.: Tissue growth in a rotating bioreactor. part ii: fluid flow and nutrient transport problems. Math. Med. Biol. **24**, 169–208 (2006)
Cummings, L.J., Sawyer, N.B.E., Morgan, S.P., Rose, F.R.A.J., Waters, S.L.: Tracking large solid constructs suspended in a rotating bioreactor: a combined experimental and theoretical study. Biotech. Bioeng. **104**(6), 1224–1234 (2009)
Devarapalli, M., Lawrence, B.J., Madihally, S.V.: Modeling nutrient consumption in large flow-through bioreactors in tissue engineering. Biotech. Bioeng. **103**(5), 1003–1015 (2009)
Drasdo, D., Hohme, S.: A single-cell-based model of tumour growth in vitro: monolayers and spheroids. Phys. Biol. **2**(3), 133–147 (2005)
Drew, D.A.: Mathematical modelling of two-phase flow. Ann. Rev. Fluid Mech. **15**, 261–291 (1983)
Dunn, J.C.Y., Chan, W.-Y., Cristini, V., Kim, J.S., Lowengrub, J., Singh, S., Wu, B.M.: Analysis of cell growth in three-dimensional scaffolds. Tiss. Eng. **12**(4), 705–715 (2006)
Eibl, D., Eibl, R.: Bioreactors for mammalian cells: general overview. In: Eibl, R., Eibl, D., Pörtner, R., Carapano, G., Czermak, P. (eds.) Cell and Tissue Reaction Engineering: Principles and Practice. Springer, Berlin (2009)
El-Haj, A.J., Minter, S.L., Rawlinson, S.C., Suswillo, R., Lanyon, L.E.: Cellular responses to mechanical loading in vitro. J. Bone Min. Res. **5**(9), 923–932 (1990)
Fozard, J.A., Byrne, H.M, Jensen, O.E., King, J.R.: Continuum approximations of individual-based models for epithelial monolayers. Math. Med. Biol. **27**(1), 39 (2010). ISSN:1477-8599
Franks, S.J., King, J.R.: Interactions between a uniformly proliferating tumour and its surroundings: uniform material properties. Math. Med. Biol. **20**, 47–89 (2003)
Freed, L.E., Vunjak-Novakovic, G.: Culture of organized cell communities. Adv. Drug Del. Rev. **33**, 15–30 (1998)
Freed, L.E., Vunjak-Novakovic, G., Langer, R.: Cultivation of cell-polymer cartilage implants in bioreactors. J. Cell Biochem. **41**, 257–264 (1993)

Freed, L.E., Marquis, J.C., Langer, R., Vunjak-Novakovic, G.V.: Kinetics of chondrocyte growth in cell-polymer implants. Biotech. Bioeng. **43**, 605–614 (1994)

Fung, Y.: What are the residual stresses doing in our blood vessels? Ann. Biomed. Eng. **19**, 237–249 (1991). ISSN:0090-6964

Galban, C.J., Locke, B.R.: Analysis of cell growth in a polymer scaffold using a moving boundary approach. Biotech. Bioeng. **56**(4), 422–432 (1997)

Galban, C.J., Locke, B.R.: Analysis of cell growth kinetics and substrate diffusion in a polymer scaffold. Biotech. Bioeng. **65**(2), 121–132 (1999a)

Galban, C.J., Locke, B.R.: Effects of spatial variation of cells and nutrient and product concentrations coupled with product inhibition on cell growth in a polymer scaffold. Biotech. Bioeng. **64**(6), 633–643 (1999b)

Hadeler, K.P., Hillen, T., Lutscher, F.: The langevin or kramers approach to biological modeling. Math. Models Meth. Appl. Sci **14**, 1561–1583 (2004)

Hammond, T.G., Hammond, J.M.: Optimized suspension culture: the rotating-wall vessel. Physiol. Renal Physiol. **281**, F12–F25 (2001)

Han, Y., Cowin, S.C., Schaffler, M.B., Weinbaaum, S.: Mechanotransduction and strain amplification in osteocyte cell processes. Proc. Nat. Acad. Sci. **101**(47), 16689–16694 (2004)

Hollister, S.J.: Porous scaffold design for tissue engineering. Nat. Mater. **4**(7), 518–524 (2005)

Hutmacher, D.W.: Scaffolds in tissue engineering bone and cartilage. Biomaterials **21**(24), 2529–2543 (2000)

Jeong, D., Yun, A., Kim, J.: Mathematical model and numerical simulation of the cell growth in scaffolds. Biotech. Model. Mechanobiol. (2011) doi:10.1007/s10237-011-0342-y.

Johnson, E.S.: Transplant activity in the uk. activity report 2009/2010. http://www.nhsbt.nhs.uk (2010)

Jones, A.F., Byrne, H.M., Gibson, J.S., Dold, J.W.: A mathematical model of the stress induced during avascular tumour growth. J. Math. Biol. **40**(6), 473–499 (2000)

Julien, C., Whitford, W.: Bioreactor monitoring modeling and simulation. BioProcess Int. Suppl. **5**(1), 10–17 (2007)

Kim, Y., Stolarska, M.A., Othmer, H.G.: A hybrid model for tumour spheroid growth in vitro i: theoretical development and early results. Math. Models Meth. App. Sci. **17**, 1773–1798 (2007)

Klein-Nulend, J., Roelofsen, J., Sterck, J.G., Semeins, C.M., Burger, E.H.: Mechanical loading stimulates the release of transforming growth factor-beta activity by cultured mouse calvariae and periosteal cells. J. Cell Physiol. **163**(1), 115–119 (1995a)

Klein-Nulend, J., Vander Plas, A., Semeins, C.M., Ajubi, N.E., Frangos, J.A., Nijweide, P.J., Burger, E.H.: Sensitivity of osteocytes to biomechanical stress in vitro. FASEB **9**(5), 441–445 (1995b)

Kolev, N.I.: Multiphase Flow Dynamics, vol. 1—Fundamentals. Springer, Berlin (2002)

Kwon, O., Devarakonda, S.B., Sankovic, J.M., Banerjee, R.K.: Oxygen transport and consumption by suspended cells in microgravity: a multiphase analysis. Biotech. Bioeng. **99**(1), 99–107 (2008)

Lai, W.M., Hou, J.S., Mow, V.C.: A triphasic theory for the swelling and deformation behaviors of articular cartilage. J. Biomech. Eng. **113**, 245–258 (1991)

Landman, K.A., Cai, A.Q.: Cell proliferation and oxygen diffusion in a vascularising scaffold. Bull. Math. Biol. **69**(7), 2405–2428 (2007)

Landman, K.A., Please, C.P.: Tumour dynamics and necrosis: surface tension and stability. Math. Med. Biol. **18**(2), 131–158 (2001)

Lappa, M.: Organic tissues in rotating bioreactors: fluid-mechanical aspects dynamic growth models and morphological evolution. Biotech. Bioeng. **84**(5), 518–532 (2003)

Lawrence, B.J., Deverapalli, M., Madihally, S.V.: Flow dynamics in bioreactors containing tissue engineering scaffolds. Biotech. Bioeng. **102**(3), 935–947 (2008)

Lemon, G., King, J.R.: Multiphase modelling of cell behaviour on artificial scaffolds: effects of nutrient depletion and spatially nonuniform porosity. Math. Med. Biol. **24**(1), 57–83 (2007a)

Lemon, G., King, J.R.: Travelling-wave behaviour in a multiphase model of a population of cells in an artificial scaffold. J. Math. Biol. **55**(4), 449–480 (2007b)

Lemon, G., King, J.R., Byrne, H.M., Jensen, O.E., Shakesheff, K.: Multiphase modelling of tissue growth using the theory of mixtures. J. Math. Biol. **52**(2), 571–594 (2006)
Lemon, G., Howard, D., Tomlinson, M.J., Buttery, L.D., Rose, F.R.A.J., Waters, S.L., King, J.R.: Mathematical modelling of tissue-engineered angiogenesis. Math. Biosci. **221**, 101–120 (2009)
Lewis, M.C., MacArthur, B.D., Malda, J., Pettet, G., Please, C.P.: Heterogeneous proliferation with engineered cartilaginous tissue: the role of oxygen tension. Biotech. Bioeng. **91**(5), 607–615 (2005)
Lin, R.Z., Chang, H.Y.: Recent advances in three-dimensional multicellular spheroid culture for biomedical research. Biotechnology **3**(9-10), 1172–1184 (2008)
Loret, B., Simões, F.M.F.: A framework for deformation generalized diffusion mass transfer and growth in multi-species multi-phase biological tissues. Eur. J. Mech.-A/Solids **24**(5), 757–781 (2005)
Lutianov, M., Naire, S., Roberts, S., Kuiper, J.-H.: A mathematical model of cartilage regeneration after cell therapy. J. Theor. Biol. **289**, 136–150 (2011)
Marle, C.M.: On macroscopic equations governing multiphase flow with diffusion and chemical reactions in porous media. Int. J. Eng. Sci. **20**(5), 643–662 (1982)
Martin, I., Wendt, D., Heberer, M.: The role of bioreactors in tissue engineering. Trends Biotechnol **22**(2), 80–86 (2004)
McCoy, R.J., O'Brien, F.J.: Influence of shear stress in perfusion bioreactor cultures for the development of three-dimensional bone tissue constructs: a review. Tissue Eng. B **16**(6), 587–601 (2010)
Meineke, F.A., Potten, C.S., Loeffler, M.: Cell migration and organization in the intestinal crypt using a lattice-free model. Cell Prolif. **34**(4), 253–266 (2001). ISSN:0960-7722
Mow, V.C., Kuei, S.C., Lai, W.M., Armstrong, C.G.: Biphasic creep and stress relaxation of articular cartilage in compression: theory and experiments. J. Biomech. Eng. **102**, 73 (1980)
Mullender, M., El-Haj, A.J., Yang, Y., van Duin, M.A., Burger, E.H, Klein-Nulend, J.: Mechanotransduction of bone cells in vitro: mechanobiology of bone tissue. Med. Biol. Eng. Comput. **42**, 14–21 (2004)
Murray, P.J., Edwards, C.M., Tindall, M.J., Maini, P.K.: From a discrete to a continuum model of cell dynamics in one dimension. Phys. Rev. E **80**(3), 031912 (2009)
Noble, B.S., Reeve, J.: Osteocyte function osteocyte death and bone fracture resistance. Mol. Cell. Endocrinol. **159**(1–2), 7–13 (2000)
Novosel, E.C., Kleinhans, C., Kluger, P.J.: Vascularization in the key challenge in tissue engineering. Adv. Drug Del. Rev. **63**, 300–311 (2011)
Obradovic, B., Meldon, J.H., Freed, L.E., Vunjak-Novakovic, G.: Glycosaminoglycan deposition in engineered cartilage: experiments and mathematical model. AICHE J. **46**(9), 1860–1871 (2000)
O'Dea, R.D., King, J.R.: Multiscale analysis of pattern formation via intercellular signalling. Math. Biosci. **231**, 172–185 (2011a)
O'Dea, R.D., King, J.R.: Continuum limits of pattern formation in hexagonal-cell monolayers. J. Math. Biol. (2011b). doi:10.1007/s00285-011-0427-3
O'Dea, R.D., Waters, S.L., Byrne, H.M.: A two-fluid model for tissue growth within a dynamic flow environment. Eur. J. Appl. Math. **20**, 47–89 (2008)
O'Dea, R.D., Waters, S.L., Byrne, H.M.: A multiphase model for tissue construct growth in a perfusion bioreactor. Math. Med. Biol. **27**(2), 95–127 (2010)
O'Dea, R.D., Osborne, J.M., El-Haj, A.J., Byrne H.M., Waters, S.L.: The interplay between scaffold degradation tissue growth and cell behaviour in engineered tissue constructs. Submitted to J. Math. Biol. (2012)
Osborne, J.M., Whiteley, J.P.: A numerical method for the multiphase viscous flow equations. Comp. Meth. Appl. Mech. Eng. **199**(49–52), 3402–3417 (2010)
Osborne, J.M., O'Dea, R.D., Whiteley, J.P., Byrne, H.M., Waters, S.L.: The influence of bioreactor geometry and the mechanical environment on engineered tissues. J. Biomech. Eng. **132**, 051006 (2010)
Othmer, H.G., Dunbar, S.R., Alt, W.: Models of dispersal in biological systems. J. Math. Biol. **26**, 263–298 (1988)

Ouchi, N.B., Glazier, J.A., Rieu, J.P., Upadhyaya, A., Sawada, Y.: Improving the realism of the cellular potts model in simulations of biological cells. Phys. A **329**(3–4), 451–458 (2003)

Owen, M.R., Alarcon, T., Maini, P.K., Byrne, H.M.: Angiogenesis and vascular remodelling in normal and cancerous tissues. J. Math. Biol. **58**, 689–721 (2009)

Palferman, T.G.:Bone and joint diseases around the world. The UK perspective. J. Rheumatol. **67**, 33 (2003). ISSN:0315-162X

Passman, S.L., Nunziato, J.W.: A theory of multiphase mixtures. In: Truesdell, C. (ed.) Rational Thermodynamics. Springer, New York (1984)

Pitt-Francis, J., Pathmanathan, P., Bernabeu, M.O., Bordas, R., Cooper, J., Fletcher, A.G., Osborne, J.M., Walter, A., Chapman, S.J., Garny, A., Leeuwen, I.M.M., Van Maini, P.K., Rodriguez, B., Waters, S.L., Whiteley, J.P., Byrne, H.M., Gavaghan, D.: Chaste: a test-driven approach to software development for biological modelling. Comp. Phys. Comm. **180**(12), 2452–2471 (2000)

Please, G., McElwain, D.L.S.: A new approach to modelling the formation of necrotic regions in tumours. Appl. Math. Lett. **11**(3), 89–94 (1998)

Please, C.P., Pettet, G.J., McElwain, D.L.S.: Avascular tumour dynamics and necrosis. Math. Models Meth. Appl. Sci. **9**(4), 569–580 (1999)

Preziosi, L., Tosin, A.: Multiphase modelling of tumour growth and extracellular matrix interaction: mathematical tools and applications. J. Math. Biol. **58**(4), 625–656 (2009)

Raimondi, M.T., Boschetti, F., Falcone, L., Migliavacca, F., Remuzzi, A., Dubini, G.: The effect of media perfusion on three-dimensional cultures of human chondrocytes: integration of experimental and computational approaches. Biorheology **41**, 401–410 (2004)

Riccalton-Banks, L., Liew, C., Bhandari, R., Fry, J., Shakesheff, K.: Long-term culture of functional liver tissue: three-dimensional coculture of primary hepatocytes and stellate cells. Tissue Eng. **9**(3), 401–410 (2003)

Ricken, T., Bluhm, J.: Remodeling and growth of living tissue: a multiphase theory. Arch. Appl. Mech. **80**(5), 453–465 (2010)

Risbud, M.V., Sittinger, M.: Tissue engineering: advances in in vitro cartilage generation. Trends Biotech. **20**(8), 351–356 (2002)

Roelofsen, J., Klein-Nulend, J., Burger, E.H.: Mechanical stimulation by intermittent hydrostatic compression promotes bone-specific gene expression in vitro. J. Biomech. **28**(12), 1493–1503 (1995)

Roose, T., Netti, P.A., Munn, L.L., Boucher, Y., Jain, R.K.: Solid stress generated by spheroid growth estimated using a linear poroelasticity model. Microvasc. Res. **66**(3), 204–212 (2003)

Salgado, A.J., Coutinho, O.P., Reis, R.L.: Bone tissue engineering: state of the art and future trends. Macromol. Biosci. **4**, 743–765 (2004)

Sawyer, N.B.E., Worrall, L.K., Crowe, J.A., Waters, S.L., Shakesheff, K.M., Rose, F.R.A.J., Morgan, S.P.: In situ monitoring of 3d in vitro cell aggregation using an optical imaging system. Biotech. Bioeng. **100**(1), 159–167 (2007)

Schwartz, L.W.: Instability and fingering in a rotating hele-shaw cell. Phys. Fluids A **1**, 167–169 (1989)

Shakeel, M., Matthews, P.C., Waters, S.L. Graham, R.S.: A continuum model of cell proliferation and nutrient transport in a perfusion bioreactor. Math. Med. Biol. doi:10.1093/imammb/dqr022 (2011)

Shipley R.J. and Waters S.L., (2011) Fluid and mass transport modelling to drive design of cell-packed hollow fibre bioreactors for tissue engineering applications. Math. Med. Biol. Accepted.

Shipley, R.J., Jones, G.W., Dyson, R.J., Sengers, B.G., Bailey, C.L., Catt, C.J., Please, C.P., Malda, J.: Design criteria for a printed tissue engineering construct: a mathematical homogenization approach. J. Theor. Biol. **259**(3), 489–502 (2009)

Shipley, R.J., Waters, S.L., Ellis, M.J.: Definition and validation of operating equations for poly(vinyl alcohol)-poly(lactide-co-glycolide) microfiltration membrane-scaffold bioreactors. Biotech. Bioeng. **107**, 382–392 (2010)

Shipley, R.J., Davidson, A.J., Chan, K., Chaudhuri, J.B., Waters, S.L., Ellis, M.J.: A strategy to determine operating parameters in tissue engineering hollow fiber bioreactors. Biotech. Bioeng. **108**, 1450–1461 (2011)

Sipe, J.D.: Tissue engineering and reparative med. Ann. N. Y. Acad. Sci. **961**, 1–9 (2002)

Trelstad, R.L., Silver, F.H.H.: Matrix assembly. In: Hay, E.D. (ed.) Cell biology of the extracellular matrix (1981)

Treusdell, C., Noll, W.: The nonlinear field theory of mechanics. In: Flugge S. (ed.) Handbuch der physik. (1960)

Turner, S., Sherratt, J.A., Painter, K.J., Savill, N.J.: From a discrete to a continuous model of biological cell movement. Phys. Rev. E **69**(2), 021910 (2004)

Urban, J.P.G.: The chondrocyte: a cell under pressure. Rheumatology **33**(10), 901–908 (1994)

Van Leeuwen, I.M.M., Mirams, G.R., Walter, A., Fletcher, A., Murray, P., Osbourne, J., Varma, S., Young, S.J., Cooper, J., Doyle, B. et al.: An integrative computational model for intestinal tissue renewal. Cell Prolif. **42**(5), 617–636 (2009)

Waters, S.L., Cummings, L.J.: Coriolis effects in a rotating hele-shaw cell. Phys. Fluids **18**, 048101 (2005)

Waters, S.L., Cummings, L.J., Shakesheff, K.M., Rose, F.R.A.J.: Tissue growth in a rotating bioreactor. part i: mechanical stability. Math. Med. Biol. **23**, 311–337 (2006)

Weiss P., (1945) Experiments on cell and axon orientation in vitro: the role of colloidal exudates in tissue organization. J. Exp. Zool. **100**(3), 353–386. ISSN:1097-010X

Whitaker, S.: The transport equations for multi-phase systems. Chem. Eng. Sci. **28**, 139–147 (2000)

Whittaker, R.J., Booth, R., Dyson, R., Bailey, C., Parsons Chini, L., Naire, S., Payvandi, S., Rong, Z., Woollard, H., Cummings, L.J., Waters, S.L., Mawasse, L., Chaudhuri, J.B., Ellis, M.J., Michael, V., Kuiper, N.J., Cartmell, S.: Mathematical modelling of fibre-enhanced perfusion inside a tissue engineering bioreactor. J. Theor. Biol. **256**, 533–546 (2009)

Wilkinson, D.J.: Stochastic modelling for quantitative description of heterogeneous biological systems. Na. Genet. **10**(2), 122–133 (2009)

Wilson, D.J., King, J.R., Byrne, H.M.: Modelling scaffold occupation by a growing nutrient-rich tissue. Math. Models Meth. App. Sci. **17**, 1721–1750 (2007)

Yano, S., Komine, M., Fujimoto, M., Okochi, H., Tamaki, K.: Mechanical stretching in vitro regulates signal transduction pathways and cellular proliferation in human epidermal keratinocytes. J. Invest. Dermatol. **122**(3), 783–790 (2004)

You, J., Yellowley, C.E., Donahue, H.J., Zhang, Y., Chen, Q., Jacobs, C.R.: Substrate deformation levels associated with routine physical activity are less stimulatory to bone cells relative to loading-induced oscillatory fluid flow. J. Biomech. Eng. **122**, 377–393 (2000)

Yourek, G., Al-Hadlaq, A., Patel, R., McCormick, S., Reilly, G.C., Mao, J.J.: Nanophysical properties of living cells. In: Stroscio Michael, A., Mitra, D., Bin, H. (eds.) Biological Nanostructures and Applications of Nanostructures in Biology Bioelectric Engineering. Springer, New York (2004)

Yu, X., Botchwey, E.A., Levine, E.M., Pollack, S.R., Laurencin, C.T.: Bioreactor-based bone tissue engineering: the influence of dynamic flow on osteoblast phenotypic expression and matrix mineralization. PNAS **101**(31), 11203 (2004)

Zdrahala, R.J., Zdrahala, I.J.: In vivo tissue engineering: part I. Concept genesis and guidelines for its realization. J. Biomat. Appl. **14**(2), 192 (1999). ISSN:0885-3282

Multiphysics Computational Modeling in Cartilage Tissue Engineering

Manuela Teresa Raimondi, Paola Causin, Matteo Laganà, Paolo Zunino and Riccardo Sacco

Abstract A common technique for in vitro cartilage regeneration is to seed a porous matrix with cartilage cells and to culture the construct in static conditions or under medium perfusion in a bioreactor. An essential step toward the development of functional cartilage is to understand and control the tissue growth phenomenon in such systems. The growth process depends on various space- and time-varying biophysical variables of the environment surrounding the cartilage cells, primarily mass transport and mechanical variables, all involved in the cell biological response. Moreover, the growth process is inherently multiscale, since cell size (10 μm), scaffold pore size (100 μm), and cellular construct size (10 mm) pertain to three separate spatial scales. To obtain a quantitative understanding of cartilage growth in this complex multiphysics and multiscale system, advanced mathematical models and efficient scientific computing techniques have been developed. In this chapter, we discuss the existing knowledge in this field and we present the most recent advancements for the numerical simulation of cartilage tissue engineering.

M. T. Raimondi (✉) · M. Laganà
LaBS, Department of Structural Engineering, Politecnico di Milano, Milan, Italy
e-mail: manuela.raimondi@polimi.it

P. Causin
Department of Mathematics, Università degli Studi di Milano, Milan, Italy

P. Zunino
MOX, Department of Mathematics, Politecnico di Milano, Milan, Italy

R. Sacco
Department of Mathematics, Politecnico di Milano, Milan, Italy

Stud Mechanobiol Tissue Eng Biomater (2013) 10: 267–285

DOI: 10.1007/8415_2011_112

Published Online: 8 January 2012

1 Introduction

A basic concept in the design of ex vivo tissue reconstruction is to provide a proper biophysical microenvironment to cells [38]. In cartilage regeneration (Fig. 1), mechanical stimulation is being extensively evaluated as a tool to modulate extracellular matrix (ECM) synthesis, coherently with the evidence that mechanical forces play an important role in cartilage homeostasis in vivo [14]. Mechanobiology models of engineered cartilage are currently addressed at optimizing the applied mechanical stimuli, by combining different stimuli, for example flow perfusion with cyclic pressurization (Fig. 2) [25, 26, 42, 46], as a means to better mimic the complex biophysical environment of chondrocytes within native cartilage.

To gain a better insight into the quantitative relationship between the applied culture conditions and cartilage growth, advanced computational models are currently applied to interpret the results from bioreactor studies. These attempts to calculate and control the balance of mass transport and mechanical stresses exerted on cells, have proven useful in capturing a rough understanding of the conditions favoring the development of engineered cartilage [36, 37, 39, 40, 42, 26, 44].

In mechanobiology models of engineered cartilage, comparison between the experimental findings and the computational results enable the local field variables to be correlated with specific cell responses. Only from the mastering of the complex biological phenomena (cell metabolism and proliferation, substrate degradation and remodeling of the ECM) that take place during the in vitro culturing process, one can control key aspects of tissue maturation.

2 Tissue Engineering: A Multiphysics/Multiscale Problem

Bioreactors for tissue culture are complex multiphase systems composed of a scaffold portion, a culture medium and a growing biomass. For their rational design, it is thus strongly required to have a quantitative understanding of the interplay between geometry, interstitial flow field, nutrient mass transfer and cellular behavior (proliferation, migration, biosynthetic activity). These variegated phenomena encompass a wide range of embedded scales.

Figure 3 shows five distinct scales at which (at least) the considered problem can be modeled, namely:

- Macroscale: it is the scale at which the perfused scaffold is treated as a continuum and at which the Bioengineer sets the control parameters (inlet velocity, pressure drop). Its characteristic length is of the order of a few millimeters.
- Mesoscale: it is the scale corresponding to a collection of a few functional sub-units (shortly, unit cells) of the scaffold. Its characteristic length spans from tens of microns to millimeters.
- Microscale: it is the scale of the single unit cell of the polymeric scaffold, of the order of 100 microns.

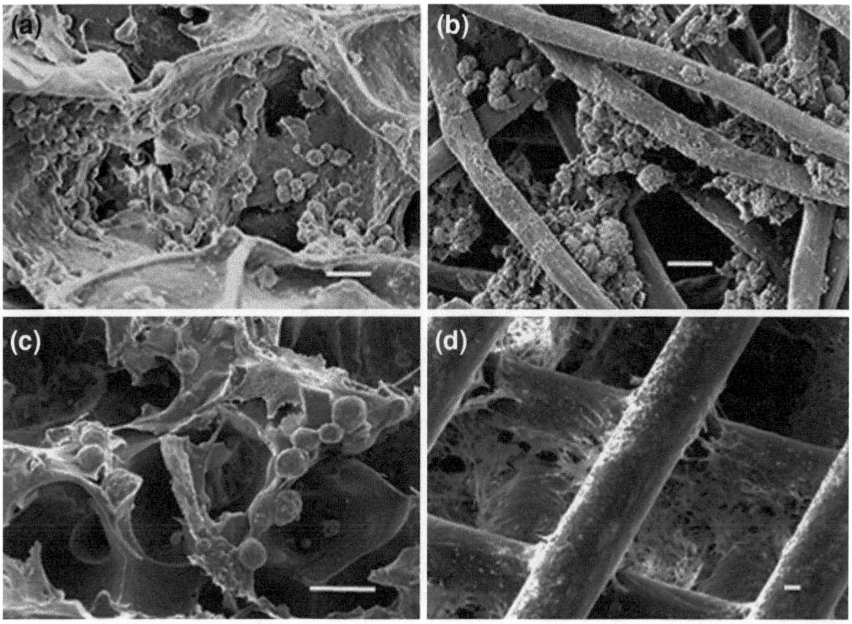

Fig. 1 Scanning electron micrographs of porous scaffolds seeded with cartilage cells. The scale bar equals 20 μm in all figures. **a** Polyurethane foam, **b** non-woven fiber mesh derived from hyaluronic acid, **c** polyurethane structure with regular spherical pores and **d** polystyrene regular grid. A coloured version of this figure is available on the online version of the book

- Cellular scale: it is the scale at which cells cannot be treated as a continuum, but must be treated as single discrete entities. Its characteristic length is of the order of microns.
- Sub-cellular scale: it is the scale accounting for all the mass transport and reaction processes that occur at the single cell membrane level. Its characteristic length is of the order of some nanometers.

We present here below a possible mathematical framework for the description of the bio-physical phenomena occurring in a bioreactor for tissue engineering. Let Ω be the bioreactor domain, composed by the time-invariant subdomains Ω_{sc}, representing a non-biodegradable scaffold, and by its complement Ω_e. This latter subdomain is, in turn, composed of a fluid portion $\Omega_{fl}(t)$ and a biomass portion $\Omega_b(t)$, both depending on time t. Notice that both Ω_{fl} and Ω_b may be, in general, composed by the union of complex, unconnected domains. The full scale approach consists of the coupled solution of the following multiphysics system:

1. model for nutrient concentration: find $c = c(\boldsymbol{x},t)$ in Ω_e such that

$$\frac{\partial c}{\partial t} + \nabla \cdot (-D\nabla c + \boldsymbol{v}c) = Q(c) \tag{1}$$

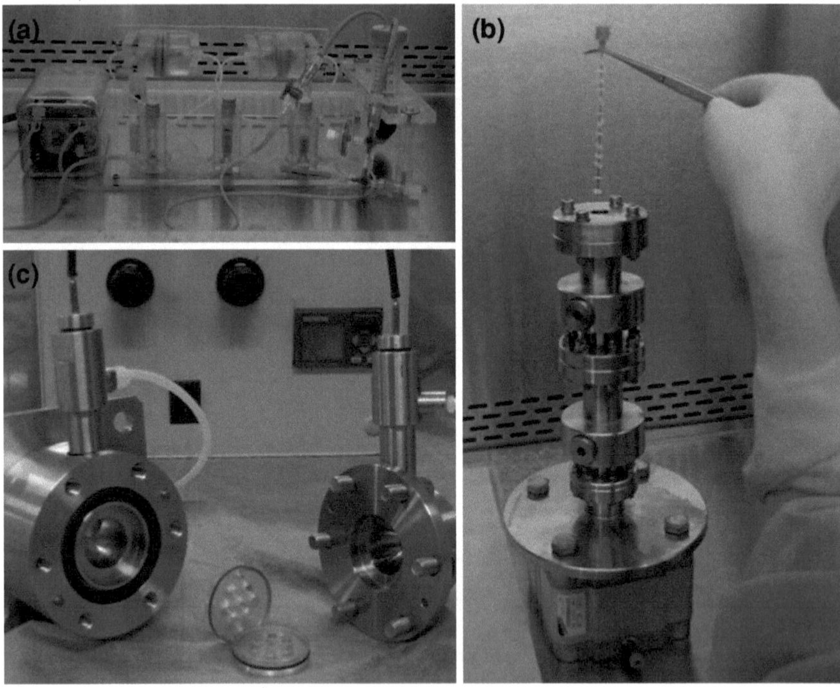

Fig. 2 Mechanical bioreactors for cartilage tissue engineering. During bioreactor culture, the cellular constructs may be subjected to **a** interstitial perfusion, **b** cyclic hydrostatic pressurization and **c** combined regimens of interstitial perfusion and cyclic hydrostatic pressurization. A coloured version of this figure is available on the online version of the book

where D is the nutrient diffusivity, equal to D_{fl} in Ω_{fl} and to D_b in Ω_b, respectively, $\boldsymbol{v}$ is the velocity field and where the mass consumption term Q, due to cellular metabolism, is given by:

$$Q(c) = \begin{cases} 0 & \text{in } \Omega_{\text{fl}}, \\ -\dfrac{V_m c}{K_m + c} & \text{in } \Omega_b \text{ (Michaelis – Menten kinetics)} \end{cases} \tag{2}$$

V_m being the maximal nutrient consumption and K_m the half saturation constant,

2. model for the fluid velocity:

- $\boldsymbol{v} = \boldsymbol{0}$ in Ω_b
- Navier-Stokes equations in Ω_{fl}: find $\boldsymbol{v} = \boldsymbol{v}(\boldsymbol{x}, t)$ such that

$$\begin{cases} \dfrac{\partial \boldsymbol{v}}{\partial t} + \boldsymbol{v} \cdot \nabla \boldsymbol{v} - \eta \Delta \boldsymbol{v} + \nabla p = 0 \\ \nabla \cdot \boldsymbol{v} = 0. \end{cases} \tag{3}$$

3. model for biomass growth (cell population and ECM accumulation):

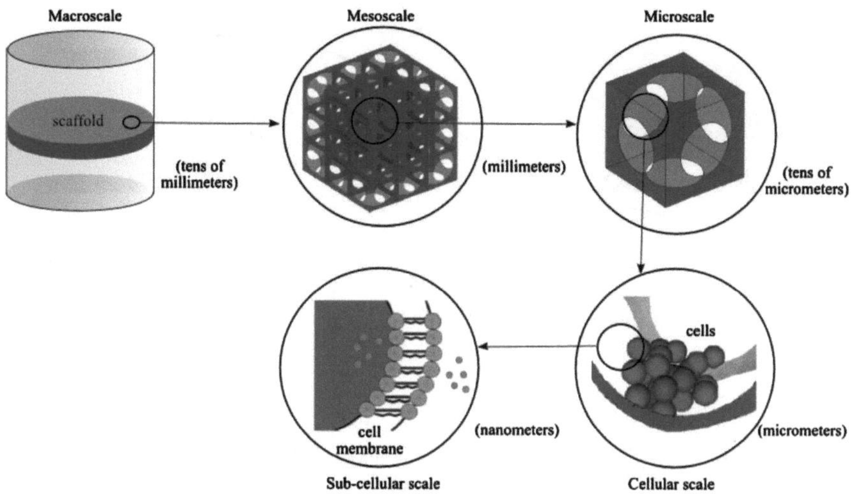

Fig. 3 Biophysical scales in a bioreactor system with their characteristic length. A coloured version of this figure is available on the online version of the book

$$\Omega_b = \Omega_b(t, c, \mathbf{v}, \text{other biophysical parameters}) \tag{4}$$

The interfaces between different materials and models have to be handled by means of suitable transmission conditions connecting the value of the variables and/or their fluxes from one domain to the other. At present, the computational challenge for solving the above three-dimensional (3D) time-dependent problem on the entire bioreactor domain with internal moving boundaries and spatial resolution at the cell (Microscale) level, is still a too demanding task, even for the more advanced numerical techniques and powerful machine resources. For this reason, computational models of cartilage tissue engineering have been proposed, addressing the problem at only a few of the above described scales, and often including only a restricted set of physical phenomena.

The biophysical admissibility of such models strongly depends on the culture time. As a matter of fact, the cell environment in the scaffold changes with culture time, at the beginning of culture there are only seeded cells, which later proliferate and synthesize ECM. Accounting for the interaction of the fluid flow with the growing biomass is a condition required to obtain a realistic simulation at high cell volume fractions, typical of long term culture. For this reason, single-physics/single-scale models might yield acceptable results for short culture times, while for longer culture times coupled multiphysics/multiscale simulations are required (Fig. 4).

In the following, we present a survey of such models, focusing on the coupling between nutrient mass transport with medium flow (Sect. 3) and biomass growth (Sect. 4). An overall description of these phenomena is provided in Sect. 5 under the perspective of homogenization, while in Sect. 6 we briefly address recent contributions to the multiscale modeling of the tissue engineering problem.

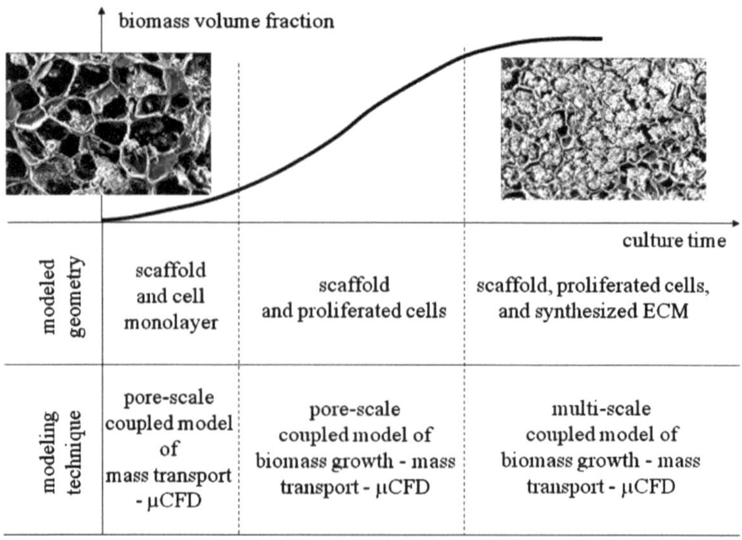

Fig. 4 Hierarchy of computational modelling techniques for cartilage tissue growth, following the evolution of the geometry of the cell environment upon culture time. A coloured version of this figure is available on the online version of the book

3 Coupled Models of Medium Flow and Mass Transport

A rough understanding of transport limitations in porous constructs has been obtained, in absence of flow, by Botchwey et al. [2, 3] and Sengers et al. [45] using homogeneous models of tissue-engineered constructs. This amounts to solving Eq. (1) with a constant nutrient diffusivity and $\boldsymbol{v} = 0$ in the entire scaffold domain Ω, instead of considering the actual microscopic structure of the domain Ω_e with its associated complex empty and filled substructures. In up-to-date recent tissue engineering bioreactors, the cell-seeded scaffold is immersed in a medium which is induced to flow through or around the scaffold surface in a perfusion flow apparatus. A simple strategy for increasing mass transport to cells would appear to be to increase the medium flow rate, but high flow rates induce high shear stresses on cells, which may be harmful instead of beneficial, at least at early stages of tissue growth.

This complexity presents serious hurdles in determining the appropriate values of flow rate and medium solute concentration. Flow around scaffolds in a concentric cylinder bioreactor has been studied by Williams et al. [49] by solving Eq. (2) coupled with Eq. (1) in the whole bioreactor domain. In this model, the biomass volume is not accounted for in the geometry, but an equivalent volumetric consumption rate is used at the right-hand side of Eq. (1). Such a model allowed to compute flow fields, shear stresses and oxygen profiles around the constructs.

Incorporating flow through the scaffold in a direct-perfusion configuration complicates the situation by establishing a velocity scale which is related to the actual fluid velocities in the scaffold, which are impossible to measure. This problem has been dealt with by using detailed pore-scale computational fluid-dynamic (CFD) simulations of fluid and chemical transport in tissue-engineering scaffolds populated with living cells [1, 10, 11, 31, 47]. In these models, the same set of equations is solved as in the above cited model, but considering only a few unit cells. This allows to obtain a much higher scale resolution and to represent the real local scaffold empty/filled structure. These simulations are able to capture flow, pressure and concentration fields resolved at the microscopic level. In particular, it is shown how the scaffold micro architecture influences the hydrodynamic shears imposed on cells within constructs. Calculations of nutrient flow indicate that inappropriately designed dynamic culture environments lead to regions of nutrient concentration insufficient to maintain cell viability. These studies provide a foundation for exploring the effects of dynamic flow on cell function and provide an important insight into the design and optimization of 3D scaffolds suitable in bioreactors for in vitro tissue engineering.

4 Coupled Models of Biomass Growth, Medium Flow and Mass Transport

A consistent mathematical description of tissue regeneration requires to provide a model of biomass growth, as indicated in Eq. (4). This is a very complex problem, involving several biophysical variables. Experimental observations and measurements [32] suggest that after seeding, cells undergo (i) a first period (5–7 days) of rapid proliferation, (ii) a second period (2–4 weeks) in which they start to significantly secrete the typical highly hydrated extracellular matrix (ECM), comprising proteoglycan monomers assembled with glycosaminoglycans (GAGs) anchored to hyaluronic acid chains, type II collagen and a small amount of other types of collagen. To our knowledge, there are no comprehensive models of (i), and (ii) in this application field, rather, only partial descriptions of (i) or (ii) have been developed.

4.1 Cell Population Dynamics

Computational tools known under the name of multicellular simulations can be used to model cell population dynamics. In these approaches, simple rules based on cell automata are adopted to describe cell behavior and the emergent trend of the cell populations is observed and analyzed. In particular, biased random walk techniques have found widespread use in biological applications like the simulation of angiogenesis [30] and, recently, have been applied to simulate cell populations that migrate, collide, and proliferate to build a tissue inside a 3D scaffold [7].

Simulation results show that the speed of cell locomotion modulates the rates of tissue regeneration by controlling the effect of contact inhibition and that the magnitude of this modulation strongly depends on the spatial distribution of the seeded cells. In our group, a strong effort has been devoted to integrate pore-scale CFD modeling and mass transport with multicellular simulation, developing computational models of cell proliferation under interstitial perfusion in a bioreactor [16, 17]. These models accounted for three physical phenomena: (1) cell proliferation and migration, simulated using established models of cell population dynamics [7, 28], (2) the hydrodynamic flow of culture medium, simulated using CFD modeling, and (3) oxygen transport from the flowing culture medium to the cells.

In these models, the increasing oxygen transfer from the culture medium to the growing cell biomass was included in the mass transport calculation, but the alteration of the fluid flow due to the growing cell volume was neglected, so that the simulation results were valid only for very low cell volume fractions, corresponding to initial culture conditions. An alternative perspective is based on homogeneous continuum approaches [29] or on porous medium representations, comprising motile cells and water inside a rigid scaffold material [8, 23]. In addition to cell growth kinetics, cell diffusion may be incorporated, to describe the effects of cell random walks. The values of some parameters used, such as the rate of cell ingrowth into the porous scaffold, are derived from the literature or by fit to published experimental data. Results suggest that the rate of tissue growth in porous scaffolds depends not only on the intrinsic rate of cell proliferation, but also on the balance of mechanical forces developed inside the tissue. This type of modeling also shows how the scaffold porosity or surface may be varied to reduce the tendency of cells to aggregate.

4.2 ECM Accumulation Dynamics

Mathematical models of ECM accumulation have been set up by treating the cell-scaffold construct either as a homogeneous continuum coupled with the nutrient field [27] or at the Microscale level [13]. In both these approaches, GAG concentration is assumed as the principal indicator of ECM secretion, distinguishing between soluble and bound GAG fractions. In Klein and Sah [21] an extension of the previous models is proposed, including in a phenomenological manner the effect of perfusion velocity on GAG release rate.

5 Homogenized Models of Biomass Growth, Medium Flow and Mass Transport

The difficulty of handling a domain with an internal time-dependent interface due to biomass growth with a robust numerical technique has brought many authors to propose homogenized-averaged approaches formulated at the Macroscale level.

These approaches are aimed at accounting for the fact that the environment in a tissue-engineered construct is strongly heterogeneous. Multiphase models allow an explicit consideration of these interactions. Each constituent is considered as a distinct phase within the multiphase system with corresponding constitutive laws and interactions with neighbouring phases, the inherent complexity of this approach, compared to that of solving the complete problem of Eqs. (1–3) can be conveniently reduced by an averaging process, yielding a single equation which holds uniformly in the material, upon characterization of effective parameters such as nutrient diffusivity, biochemical reaction rates and construct permeability.

Derivation of multiphase models applied to a wide range of problems in computational biology has been given extensive treatment by many authors, including, for example, Marle [24], Whitaker [48] and Byrne et al. [6]. In the series of papers by Galban and Locke [15], a two phase (fluid and biomass) model for cell growth and nutrient diffusion in a polymer scaffold with no perfusion is presented. A single, averaged reaction–diffusion equation for the nutrient concentration in the two phase system is derived using the volume-averaging method of Whitaker [48], Wood et al. [50] and the effective diffusion coefficient and reaction rate are calculated as a function of the local cell volume fraction which evolves according to a cell population balance equation. In the paper by Chung et al. [9] a two-phase (fluid and biomass) model analogous to the one of Galban and Locke [15] is proposed, with the inclusion of a self-consistent computation of the fluid-dynamic field via an averaged Stokes-Brinkman model stemming upon volume averaging of Stokes equations [20]. To reflect the fact that cell growth into the scaffold reduces the effective pore size, Chung et al. [9] propose to include the dependency on the cell volume fraction via a Carman-Kozeny type relation for the permeability.

In our group, an homogenized approach in the line of the above cited literature has been recently investigated in Sacco et al. [44]. The main novel contribution of this work is the systematic inclusion of experimental data, in particular the dependence of the biomass growth rate on the local fluid-dynamical shear stress, and the inclusion of the effect of the solid scaffold fraction in the characterization of the effective parameters arising from the homogenization procedure (hydraulic permeability and nutrient diffusivity). The set of results presented in Fig. 5 are obtained adopting the same model as in Sacco et al. [44] but investigating the use of a modified biomass growth model based on two contributions: a promotion and an inhibition term. This latter term is calibrated with respect to the inlet velocity modulus, that is used as an indicator of the typical fluid-dynamical shear stress in the device. In this analysis, we track the spatial and temporal evolution of the nutrient concentration and fluid-dynamical variables as well as the volume fraction occupied by the growing biomass which modifies the porosity of the scaffold matrix, thus altering the fluid flow. Computations, corresponding to different time levels, refer to the same geometry discussed in Sacco et al. [56] enforcing a Poiseuille flow profile at the inlet. Simulations show that biomass growth is enhanced close to the inlet section of the scaffold, due to a larger local availability of nutrient. In turn, this results into a higher occlusion of the porous matrix in this

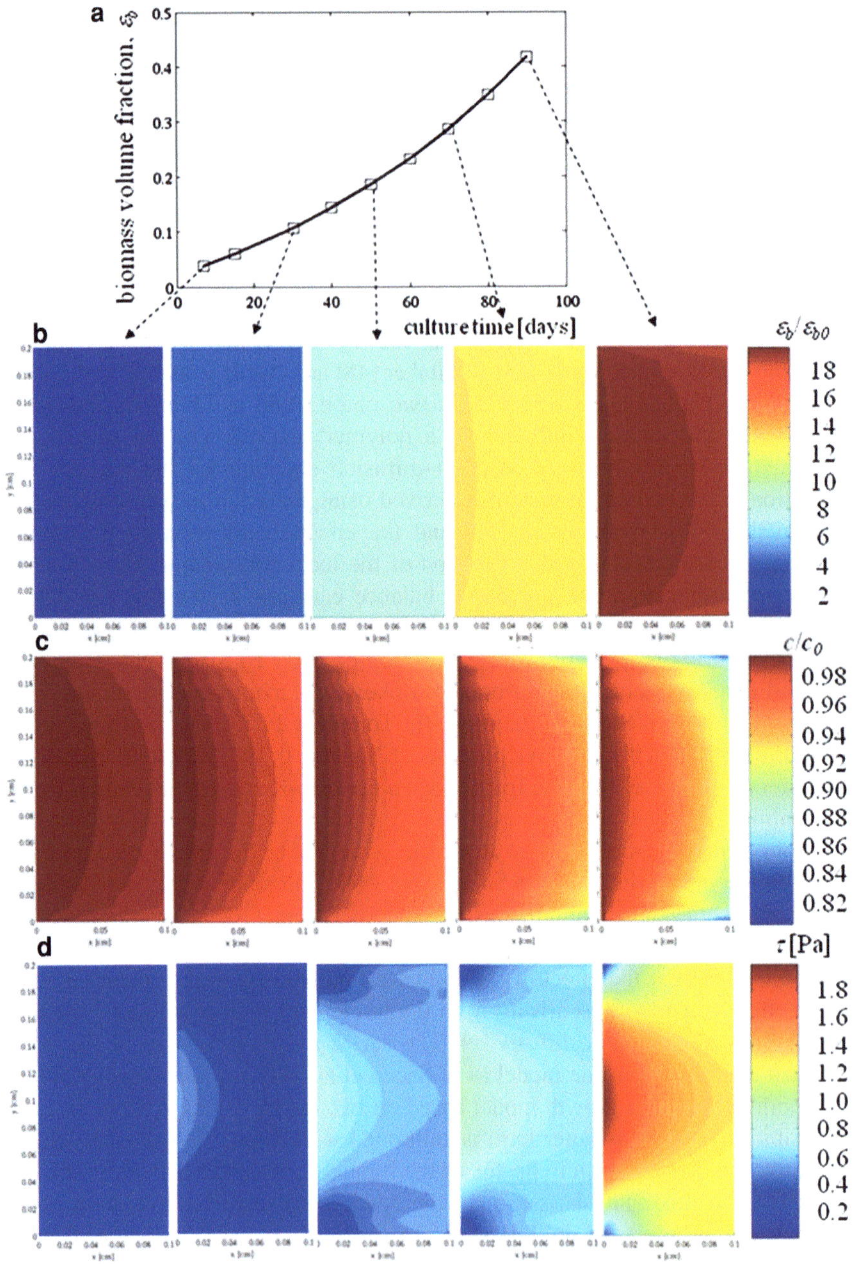

Fig. 5 Simulation of biomass growth due to mass transport in an interstitially perfused disk-shaped scaffold computed with a modification of the homogenized model presented in [44]. **a** Time evolution of the total biomass fraction. Time snapshots of the distribution of: **b** biomass, **c** nutrient concentration and **d** Darcy stress. The medium flows from left to right

area, giving rise to an increase in the local state of shear stress (represented in the model by the Darcy stress).

6 Multiscale Modeling

In the attempt of including in the mathematical description of the tissue-engineered construct a wider spectrum of phenomena, while keeping the computational complexity at an acceptable level, multiscale homogenization techniques have been applied in a certain number of recent studies. A first example in this direction is represented by the work of Cioffi et al. [11], where a Macroscale simplified model provides appropriate boundary conditions for a Microscale model. The former model is set in an axi-symmetric geometry and steady Navier-Stokes equations are solved coupled with a reaction-diffusion equation for the nutrient with a constant volumetric consumption rate. This model provides the velocity and nutrient concentration profiles at the inlet of the Microscopic model formulated on a small (sample) portion of the bioreactor which consists of the union of a few unit cells. In this Microscale computational domain the same equations as above are solved, but on the real geometry extracted from micro-CT images.

In Raimondi et al. [43], two complementary multiscale models are presented. The first approach follows a concept similar to the one described above. Namely, a 2D model of the whole scaffold seeded with a cell monolayer is considered, coupled with a 3D model of a functional sub-unit of such construct. The main novelty of this approach consists in the use of a moving boundary formulation originally proposed in Galban and Locke [15] and based on a phenomenological relation for the time evolution of the biomass-fluid interface, which consistently updates the geometry. To handle this time-evolving domains, an Arbitrary Lagrangian Eulerian (ALE) formulation is adopted and periodic remeshing is applied to adapt the computational mesh to large deformations of the computational domain. In the second approach, a lumped discrete approximation to the problem of mass transfer, nutrient uptake and biomass growth at the Macroscopic scale of the scaffold is devised. In this description, the scaffold is represented by a simplified geometry characterized by piecewise constant biophysical parameters, whose values are extracted from the Microscale model. In turn, the Macroscale model supplies the nutrient concentration boundary condition to a discrete set of Microscale problems, whose solution is the nutrient distribution and biomass evolution at the pore-size level, to be compared with real-time microscopic data.

7 Scientific Computing Techniques in Multiphysics Modeling

The aforementioned models of the engineered tissue growth feature specific difficulties with respect to their numerical approximation, which determine significant criteria to select suitable numerical schemes to come up with computational solutions of flow and mass transport problems. Since the mathematical models representing such phenomena are based on partial differential equations, approximation methods such as the finite element method or the finite difference (or finite volume) methods may be equivalently applied. However, the forthcoming description will be biased in the direction of the finite element method, which is a highly flexible approach. The main challenges in modeling cellular constructs are represented by the need to solve problems with spatially heterogeneous physical parameters or to couple different mathematical models (viscous and inviscid flows, free flows and porous media) within a genuine multiphysics framework, including: Darcy's, Brinkman's or Stokes' models for perfusion flow in the scaffold porous matrix with growing biomass, nutrient mass transport and delivery to growing cells, cell growth and metabolism. The interfaces between different materials and models have to be handled at the discrete level by a suitable treatment of the transmission conditions.

As an example, for the approximation of mass transport through heterogeneous media, which arises from the study of nutrient supply to cells, a correct quantification of the amount of nutrient reaching the target is achieved by properly capturing the balance of mass fluxes at the interface between the fluid and the growing tissue, as well as by accurately approximating the steep concentration gradients that arise in the neighborhood of the fluid/solid interface. For the former issue, an effective mathematical methodology is Domain Decomposition, for which we refer to Quarteroni and Valli [33].

Such family of methods splits a coupled multiphysics system in single subproblems, which can be then solved by standard approximation techniques and software packages. The global solution accounting for their interaction is then recovered by means of iterative strategies. Domain decomposition techniques have already been successfully applied to analyze fluid dynamics and mass transfer through biological tissues [34, 35], D'Angelo and Zunino [12]. Another important issue in the simulation of heterogeneous problems is the development of robust and conservative schemes. For the case of nutrient transport, a conservative scheme would exactly preserve the local mass balance related to nutrient concentration. Robustness, in turn, refers to the stability of the numerical scheme, which is the fundamental property that quantifies at what extent the approximate solution is sensitive with respect to perturbations on the discrete problem data. A numerical method is called robust when its stability properties are affected neither by the magnitude of the problem coefficients nor by their variation. This is achieved by means of Discontinuous Galerkin finite elements (DG). However, a major drawback of DG schemes is their increased computational cost compared to standard (conforming) finite element formulations, that in some cases makes them unsuitable for large scale realistic problems.

The previous computational techniques allow for an accurate and efficient approximation of the partial differential equations that describe the phenomena governing artificial tissue growth. However, computational methods provide quantitative results only when they are applied on the basis of realistic data for the problem at hand, through a strong interaction between simulations and

Fig. 6 (*left*) Fluorescence image of the engineered tissue growing on the fiber scaffold represented in Fig. 1b (scaffold fibers are visualized with *blue*, biomass with *green*). (*middle*) The outcome of image segmentation approach targeting the biomass. The contours of the biomass film are highlighted in *red*. (*right*) Numerical simulation of a diffusion problem where the biomass region features lower diffusivity than the surrounding fluid. Owing to the extended finite element technique, the computational mesh does not fit the complex contours of the fluid/biomass interface, although the computational accuracy is not significantly compromised

experiments. On the one hand, problem specific parameters such as nutrient diffusivity and consumption rates as well as fluid viscosities and hydraulic resistances of biomass should be provided. On the other hand, equations must be solved on realistic geometrical configurations, to allow for model validation based on direct comparison between simulations and observations.

Micro-CT and fluorescence microscopy images at the cell scale should be the starting point to reconstruct the geometrical models for numerical simulations. More precisely, the geometry of the scaffold matrix and the biomass available from images, has to be translated into quantitative terms in order to be exploited for the numerical approximation of fluid perfusion and transport processes. To address this task, several approaches are available. In principle, one can attempt to set up a parametric mathematical description of the domain boundaries or interfaces. However, the highly irregular shape of tissue-engineered constructs makes this task hardly achievable in the present context (Fig. 6). Alternatively, recent advances on numerical approximation schemes for PDEs have shown a promising way to override this difficulty. In particular, non-standard finite element schemes (such as the Extended Finite Element method, XFEM) have the ability to use Cartesian, non-boundary-fitted meshes to solve problems in complicated domains. Such approach, generally called the fictitious domain method [18], consists in embedding the computational domain into a fictitious volume with simple geometry, in such a way that the material interfaces do no longer need to be geometrically approximated. Recent developments in this direction Hansbo and Hansbo [19], Burman and Hansbo [4], Burman and Zunino [5] give rise to robust and efficient finite element schemes that are capable to approximate partial differential equations with interfaces separating between highly heterogeneous material properties even when the coefficient discontinuities are not exactly captured by the computational mesh.

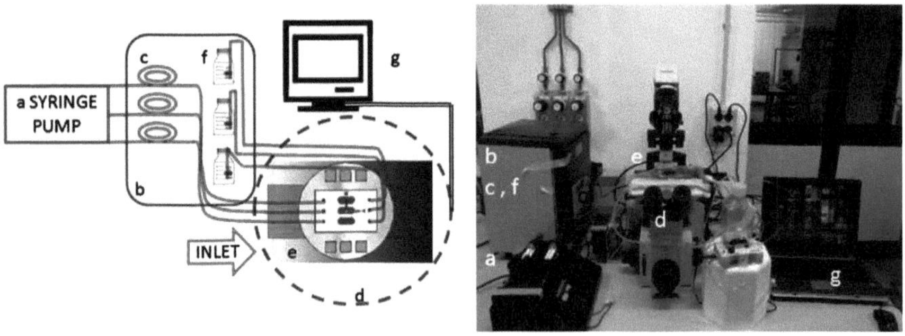

Fig. 7 Micro-bioreactor set-up developed to validate multiphysics computational models of cartilage growth. Scheme (*left*) and photograph (*right*). **a** Syringes filled with complete cell culture medium and mounted on an infusion/withdrawal programmable syringe pump, **b** cell culture incubator (37°C, 5% CO_2), **c** silicone rubber oxygenator tubes, **d** inverted microscope, **e** water-jacket heater used to maintain the micro-bioreactor chamber at 37°C, **f** cell culture medium reservoirs, and **g** laptop monitor showing a fluorescence image acquired directly on the live micro-construct with a high resolution camera. A coloured version of this figure is available on the online version of the book

8 New Tools for Experimental Validation

In mechanobiology models of engineered cartilage, comparison between the experimental findings and the computational results enables the local field variables to be correlated with specific cell responses, a crucial step towards model validation. Generally this is a very complex procedure, as most variables to be measured are not experimentally accessible in 3D and in real time. To attack this problem, a first experimental set-up was developed and validated by our group in collaboration with the University of Basel. A geometrically-defined custom made scaffold was seeded with human chondrocytes and cultured inside a miniaturised perfusion chamber, so-called mini-bioreactor, that permitted time lapse imaging of the cellular construct in real time [41].

A step forward in this regard is a new optimized set up presented in Laganà and Raimondi [22]. The new perfusion mini-bioreactor was designed and rapid-prototyped taking advantage from microfluidic know-how, so it is easy to use, cost effective, potentially disposable, and wholly mounted on a standard microscope glass slide. The experimental set-up (Fig. 7) is composed of a syringe pump, a pipeline for cell feeding, a cell culture incubator and a fluorescence microscope equipped with an imaging system. To house the custom-made polystyrene scaffolds used (3D Biotek), the perfusion chamber thickness was set at 300 μm, i.e. over the traditional microfluidic scale (Fig. 8).

The mini-bioreactor was successfully tested against leakage and used in cell culture experiments. We have used live fluorescence viable staining, DAPI (Sigma), in which the cell nuclei stain blue, and the Qtracker® Cell Labeling Kit (Invitrogen), in which the cell cytoplasm stains orange, to follow tissue growth

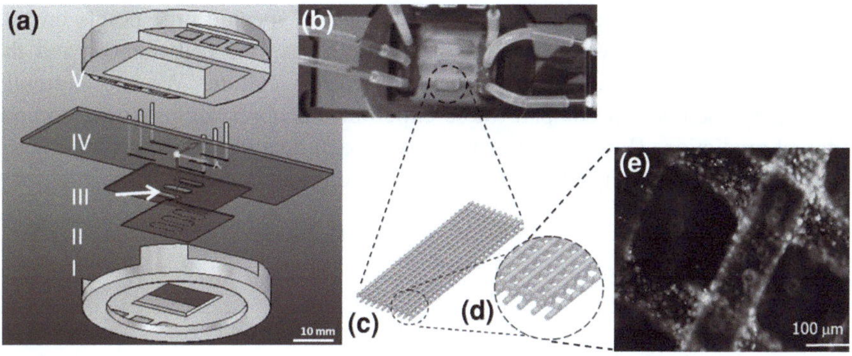

Fig. 8 Mini-bioreactor approach developed to validate multiphysics computational models of cartilage growth. **a** Exploded view showing (I) *bottom* side magnetic holder, (II) cover glass, (III) PDMS perfusion chamber (*arrow*), (IV) glass slide with CNC-milled perfusion channels, (V) *top* side magnetic holder. **b** Photograph of the mini-bioreactor placed on the microscope, **c** rendering of the polystyrene scaffold, **d** detail of the scaffold regular geometry and **e** fluorescence live image of the scaffold seeded with cells, in which cell nuclei *stain blue* and the cell cytoplasm *stains orange*. A coloured version of this figure is available on the online version of the book

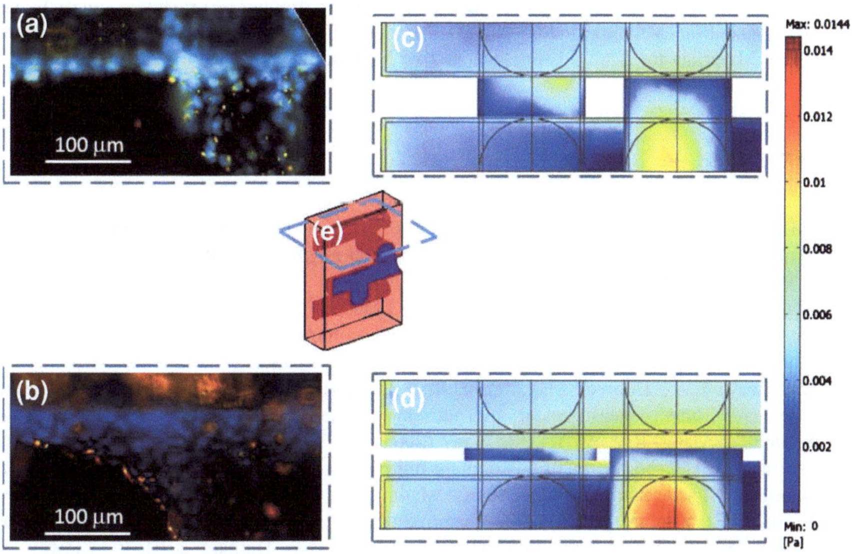

Fig. 9 Integration between experiments and simulation in cartilage tissue engineering. *Fluorescence* images of cellular constructs marked with live staining, in which the cell nuclei *stain blue* and the cell cytoplasm *stains orange*, taken **a** at the beginning and **b** the end of culture, compared to **c**, **d** the numerical output of the multiphysics model [43]. In (**c**) and (**d**) the wall fluid shear stress is mapped on the biomass surface (**c**) at the beginning and (**d**) the end of culture. **e** The 3D domain modeled

during culture. We acquired fluorescence images at specific time points on live specimens, and we developed procedures to estimate the cell number and the biomass volume fraction non-destructively during culture. We are currently developing methods to compare these experimental observations to the computational predictions (Fig. 9), in the aim to approach a validation procedure for our more advanced multiphysics models of cartilage tissue engineering.

9 Conclusion

In conclusion, the field of multiphysics computational modeling in cartilage tissue engineering is progressing rapidly. Here, we have reviewed several computational tools already available. Possible directions to improve these tools require to properly address the fundamental processes that induce growth of engineered tissue on scaffolds, i.e. cell adhesion, migration, proliferation and biosynthesis. This would open new avenues for engineering design using validated predictive tools in the field of tissue engineering.

Acknowledgments This research is funded by Politecnico di Milano, under grant 5 per Mille Junior 2009 CUPD41J10000490001 "Computational Models for Heterogeneous Media. Application to Micro Scale Analysis of Tissue-engineered Constructs", by the Italian Institute of Technology (IIT-Genoa), under grant "Biosensors and Artificial Bio-systems", and by the Cariplo Foundation (Milano), under grant 2010 "3D Micro structuring and Functionalisation of Polymeric Materials for Scaffolds in Regenerative Medicine".

References

1. Boschetti, F., Raimondi, M.T., Migliavacca, F., Dubini, G.: Prediction of the micro-fluid dynamic environment imposed to three-dimensional engineered cell systems in bioreactors. J. Biomech. **39**(3), 418–425 (2006)
2. Botchwey, E.A., Pollack, S.R., El-Amin, S., Levine, E.M., Tuan, R.S., Laurencin, C.T.: Human osteoblast-like cells in three-dimensional culture with fluid flow. Biorheology **40**, 299–306 (2003)
3. Botchwey, E.A., Dupree, M.A., Pollack, S.R., Levine, E.M., Laurencin, C.T.: Tissue-engineered bone: measurement of nutrient transport in three-dimensional matrices. J. Biomed. Mater. Res. A **67**(1), 357–367 (2003)
4. Burman, E., Hansbo. P.: Fictitious domain finite element methods using cut elements: II. A stabilized Nitsche method. Appl. Numer. Math. (2011). doi:10.1016/j.apnum.2011.01.008
5. Burman, E., Zunino, P.: Numerical approximation of large contrast problems with the unfitted Nitsche method, frontiers in numerical analysis—Durham 2010. In: Blowey, J., Jensen, M. (eds.) Lecture Notes in Computational Science and Engineering, vol. 85, Springer, Heidelberg, Germany, (2012). ISBN 978-3-642-23913-7
6. Byrne, H.M., King, J.R., McElwain, D.L.S., Preziosi, L.: A two-phase model of solid tumor growth. Appl. Math. Lett. **16**(4), 567–573 (2003)
7. Cheng, G., Youssef, B.B., Markenscoff, P., Zygourakis, K.: Cell population dynamics modulate the rates of tissue growth processes. Biophys. J. **90**(3), 713–724 (2006)

8. Chung, C.A., Yang, C.W., Chen, C.W.: Analysis of cell growth and diffusion in a scaffold for cartilage tissue engineering. Biotechnol. Bioeng. **94**(6), 1138–1146 (2006)
9. Chung, C.A., Chen, C.W., Chen, C.P., Tseng, C.S.: Enhancement of cell growth in tissue-engineering constructs under direct perfusion: modeling and simulation. Biotechnol. Bioeng. **97**(6), 1603–1616 (2007)
10. Cioffi, M., Boschetti, F., Raimondi, M.T., Dubini, G.: Modeling evaluation of the fluid-dynamic microenvironment in tissue-engineered constructs: a micro-CT based model. Biotechnol. Bioeng. **93**(3), 500–510 (2006)
11. Cioffi, M., Küffer, J., Ströbel, S., Dubini, G., Martin, I., Wendt, D.: Computational evaluation of oxygen and shear stress distributions in 3D perfusion culture systems: macro-scale and micro-structured models. J. Biomech. **41**(14), 2918–2925 (2008)
12. D'Angelo, C., Zunino, P.: Robust numerical approximation of coupled Stokes and Darcy flows applied to vascular hemodynamics and biochemical transport. ESAIM: Math. Model. Numer. Anal. (M2AN) **45**(3), 447–476 (2011)
13. DiMicco, M.A., Sah, R.L.: Dependence of cartilage matrix composition on biosynthesis, diffusion, and reaction. Transp. Porous Media **50**(1–2), 57–73 (2003)
14. El Haj, A.J., Wood, M.A., Thomas, P., Yang, Y.: Controlling cell biomechanics in orthopaedic tissue engineering and repair. Pathol. Biol. **53**(10), 581–589 (2005)
15. Galban, C.J., Locke, B.R.: Analysis of cell growth kinetics and substrate diffusion in a polymer scaffold. Biotech. Bioeng. **65**(2), 121–132 (1999)
16. Galbusera, F., Cioffi, M., Raimondi, M.T., Pietrabissa, R.: Computational modelling of combined cell population dynamics and oxygen transport in engineered tissue subject to interstitial perfusion. Comput. Methods. Biomech. Biomed. Eng. **10**(4), 279–287 (2007)
17. Galbusera, F., Cioffi, M., Raimondi, M.T.: An in silico bioreactor for simulating laboratory experiments in tissue engineering. Biomed. Microdevice **10**(4), 547–554 (2008)
18. Glowinski, R., Pan, T.W., Périaux, J.: A fictitious domain method for Dirichlet problem and applications. Comput. Methods Appl. Mech. Eng. **111**(3-4), 283–303 (1994)
19. Hansbo, A., Hansbo, P.: An unfitted finite element method, based on Nitsche's method for elliptic interface problems. Comput. Methods Appl. Mech. Eng. **191**(47-48), 5537–5552 (2002)
20. Hsu, C.T., Cheng, P.: Thermal dispersion in a porous medium. Int. J. Heat Mass Transfer **33**(8), 1587–1597 (1990)
21. Klein, T.J., Sah, R.L.: Modulation of depth-dependent properties in tissue-engineered cartilage with a semi-permeable membrane and perfusion: a continuum model of matrix metabolism and transport. Biomech. Model Mechanobiol. **6**, 21–32 (2007)
22. Laganà, M., Raimondi, M.T.: (2011) A miniaturized, optically accessible bioreactor for systematic 3D tissue engineering research. Biomedical Microdevices. doi:10.1007/s10544-011-9600-0
23. Lemon, G., King, J.R., Byrne, H.M., Jensen, O.E., Shakesheff, K.M.: Mathematical modelling of engineered tissue growth using a multiphase porous flow mixture theory. J. Math. Biol. **52**(5), 571–594 (2006)
24. Marle, C.: On macroscopic equations governing multiphase flow with diffusion and chemical reactions in porous media. Int. J. Eng. Sci. **20**(5), 643–662 (1982)
25. Mizuno, S., Tateishi, T., Ushida, T., Glowacki, J.: Hydrostatic fluid pressure enhances matrix synthesis and accumulation by bovine chondrocytes in three-dimensional culture. J. cell physiol. **193**, 319–327 (2002)
26. Moretti, M., Freed, L.E., Padera, R.F., Laganà, K., Boschetti, F., Raimondi, M.T.: An integrated experimental–computational approach for the study of engineered cartilage constructs subjected to combined regimens of hydrostatic pressure and interstitial perfusion. Bio-Med. Mater. Eng. **18**(4–5), 273–278 (2008)
27. Obradovic, B., Meldon, J.H., Lisa, E.F., Vunjak-Novakovic, G.: Glycosaminoglycan deposition in engineered cartilage: experiments and mathematical model. AIChE J. **46**, 1860–1871 (2000)

28. Palsson, E.: A three-dimensional model of cell movement in multicellular systems. Future Gener. Comput. Sys. **17**, 835–852 (2001)
29. Pisu, M., Lai, N., Cincotti, A., Concas, A., Cao, G.: Modeling of engineered cartilage growth in rotating bioreactors. Chem. Eng. Sci. **59**, 5035–5040 (2004)
30. Plank, M.J., Sleeman, B.D., Jones, P.F.: A mathematical model of tumour angiogenesis, regulated by vascular endothelial growth factor and the angiopoietins. J. Theor. Biol. **229**(4), 435–454 (2004)
31. Porter, B., Zauel, R., Stockman, H., Guldberg, R., Fyhrie, D.: 3-D computational modeling of media flow through scaffolds in a perfusion bioreactor. J. Biomech. **38**(3), 543–549 (2005)
32. Potter, H.G., Linklater, J.M., Allen, A.A., Hannafin, J., Haas, S.: Magnetic resonance imaging of articular cartilage in the knee. An evaluation with use of fast-spin-echo imaging. J. Bone Joint Surg. (Am.) **80**, 1276–1284 (1998)
33. Quarteroni, A., Valli, A.: Domain Decomposition Methods for Partial Differential Equations. Numerical Mathematics and Scientific Computation. Oxford Science Publications, The Clarendon Press, Oxford University Press, New York (1999), ISBN: 0-19-850178-1
34. Quarteroni, A., Veneziani, A., Zunino, P.: Mathematical and numerical modelling of solute dynamics in blood flow and arterial walls. SIAM J. Numer. Anal **39**(2), 1488–1511 (2002)
35. Quarteroni, A., Veneziani, A., Zunino, P.: Domain decomposition methods for blood solute dynamics. SIAM J. Sci. Comput. **23**(6), 1959–1980 (2002)
36. Raimondi, M.T., Boschetti, F., Falcone, L., Fiore, G.B., Remuzzi, A., Marinoni, E., Marazzi, M., Pietrabissa, R.: Mechanobiology of engineered cartilage cultured under a quantified fluid-dynamic environment. Biomech. Model. Mechanobiol. **1**, 69–82 (2002)
37. Raimondi, M.T., Boschetti, F., Falcone, L., Migliavacca, F., Remuzzi, A., Dubini, G.: The effect of media perfusion on three-dimensional cultures of human chondrocytes: integration of experimental and computational approaches. Biorheology **41**(3–4), 401–410 (2004)
38. Raimondi, M.T.: Engineered tissue as a model to study cell and tissue function from a biophysical perspective. Curr. Drug Discov. Technol. **3**(4), 245–268 (2006a)
39. Raimondi, M.T., Moretti, M., Cioffi, M., Giordano, C., Boschetti, F., Laganà, K., Pietrabissa, R.: The effect of hydrodynamic shear on 3D engineered chondrocyte systems subject to direct perfusion. Biorheology **43**(3–4), 215–222 (2006b)
40. Raimondi, M.T., Candiani, G., Cabras, M., Cioffi, M., Laganà, K., Moretti, M., Pietrabissa, R.: Engineered cartilage constructs subject to very low regimens of interstitial perfusion. Biorheology **45**(3–4), 471–478 (2008)
41. Raimondi, M.T., Bridgen, D.T., Laganà, M., Tonnarelli, B., Cioffi, M., Boschetti, F., Wendt, D.: In-tegration of experimental and computational microfluidics in 3D tissue engineering. In: Berthiaume, F., Morgan, J. (eds.) Methods in Bioengineering—3D Tissue Engineering. Artech House, Boston, London (2010). ISBN: 978-1-59693-458
42. Raimondi, M.T., Bonacina, E., Candiani, G., Laganà, M., Rolando, E., Talò, G., Pezzoli, D., D'Anchise, R., Pietrabissa, R., Moretti, M.: Comparative chondrogenesis of human cells in a 3D integrated experimental-computational mechanobiology model. Biomech. Model. Mechanobiol. **10**(2), 259–268 (2011a)
43. Raimondi, M.T., Causin, P., Mara, A., Nava, M., Laganà, M., Sacco, R.: Breakthroughs in computational modeling of cartilage regeneration in perfused bioreactors. IEEE Trans. Biomed. Eng. **58**(12) (2011b). doi:10.1109/TBME.2011.2163405
44. Sacco, R., Causin, P., Zunino, P., Raimondi, M.T.: A multiphysics/multiscale numerical simulation of scaffold-based cartilage regeneration under interstitial perfusion in a bioreactor. Biomech. Model. Mechanobiol. **10**(4), 577–589 (2011)
45. Sengers, B.G., van Donkelaar, C.C., Oomens, C.W., Baaijens, F.P.: Computational study of culture conditions and nutrient supply in cartilage tissue engineering. Biotechnol. Prog. **21**(4), 1252–1261 (2005)
46. Schulz, R.M., Bader, A.: Cartilage tissue engineering and bioreactor systems for the cultivation and stimulation of chondrocytes. Eur. Biophys. J. **36**, 539–568 (2007)
47. Singh, H., Teoh, S.H., Low, H.T., Hutmacher, D.W.: Flow modeling within a scaffold under the influence of uni-axial and bi-axial bioreactor rotation. J. Biotechnol. **119**, 181–196 (2005)

48. Whitaker, S.: The Method of Volume Averaging Theory and Application of Transport in Porous Media. Kluwer Academic Publishers, Dordrecht (1999)
49. Williams, K.A., Saini, S., Wick, T.M.: Computational fluid dynamics modeling of steady-state momentum and mass transport in a bioreactor for cartilage tissue engineering. Biotechnol. Prog. **18**(5), 951–963 (2002)
50. Wood, B.D., Quintard, M., Whitaker, S.: Calculation of effective diffusivities for bio films and tissues. Biotech. Bioeng. **77**(5), 495–514 (2002)

Oxygen Transport in Bioreactors for Engineered Vascular Tissues

Jason W. Bjork, Anton M. Safonov and Robert T. Tranquillo

Abstract Tissue engineered vascular grafts cultured in vitro are often done so under static conditions, which forces a diffusion-only mass transport regime for nutrient delivery and metabolite removal. Some bioreactor culture methods employ mechanical stimulation to improve material strength and stiffness; however, even with mechanical stimulation, engineered tissues are likely to operate in a diffusional transport regime for nutrient delivery and metabolite removal. In this study, we present an analysis of dissolved oxygen (DO) transport limitations that can arise in statically cultured vascular grafts and highlight bioreactor designs that improve transport, particularly by perfusion of medium through the interstitial space by transmural flow. A computational analysis is provided in conjunction with empirical data to support the models. Our goal was to investigate designs that would eliminate nutrient gradients that are evident using static culture methods in order to develop more uniform engineered vascular tissues, which could potentially improve mechanical strength and stiffness.

1 Introduction

Effective nutrient delivery and metabolite removal remain a challenge in the field of tissue engineering. Within the body, most cells are found no more than 100–200 μm from the nearest capillary to maintain viable tissue [1]; however,

J. W. Bjork · A. M. Safonov · R. T. Tranquillo (✉)
Department of Biomedical Engineering,
University of Minnesota, Minneapolis, MN, USA
e-mail: tranquillo@umn.edu

R. T. Tranquillo
Department of Chemical Engineering and Materials Science,
University of Minnesota, Minneapolis, MN, USA

Stud Mechanobiol Tissue Eng Biomater (2013) 10: 287–306
DOI: 10.1007/8415_2012_133

Published Online: 3 March 2012

engineered tissues lack this level of vasculature. In vitro mass transport limitations can impact engineered tissue constructs due to the formation of nutrient or metabolite gradients. These gradients can impact size or uniformity of engineered tissues due to changes in metabolic activity or cell viability. Transport limitations can also impact subsequent integration of these constructs in vivo.

Adequate supply of oxygen is critical due to its role in synthesis of ATP for cell energy production and is required for extracellular matrix (ECM) production. High cellular demand for oxygen coupled to low solubility and slow diffusion yield gradients in this essential nutrient and can also lead to hypoxic (low oxygen) or anoxic (zero oxygen) regions. Oxygen gradients have been measured in cartilage [2–4] and cardiac tissue [5, 6], and regions of low dissolved oxygen (DO) were associated with dead or dying cells. Stem cells, which are typically cultured in low oxygen environments, were shown to die when exposed to anoxia for 5 days [7]. Oxygen limitations have also been evaluated for their impact on cell viability and proliferation. Demol et al. recently conducted a theoretical analysis of oxygen gradients in fibrin gels and coupled DO to a model of cell growth, predicting the highly populated cell layer on the outer periphery observed in many fibrin-based engineered tissues [8].

One method aimed at improving uniformity of oxygen delivery to cells is based on bioreactor culture. Bioreactors serve to improve mass transport by flowing or mixing culture medium around cell monolayers. Mass transport is also improved during bioreactor culture of engineered tissues by either eliminating external gradients to maximize diffusional driving force or providing convective mass transfer through tissues. Bioreactors are also frequently used to provide mechanical stimulation, and while there is some degree of convective transport to the cells within the tissue construct to augment diffusional transport [9–14], improved transport is often neither intentional nor controlled. Axial perfusion bioreactors for engineered vascular grafts [15–17] cause transmural flow through the construct wall as culture medium is pulsed through the lumen, but this forced convective transport associated with transmural flow is again incidental and not controlled.

Some recent designs, however, have focused on improved nutrient delivery via controlled convective transport through the tissue. For example, Khong et al. applied interstitial flow to the culture of liver tissue [18]. Their work demonstrated enhanced enzyme activity, urea synthesis and albumin secretion in liver slices up to 1 mm thick using a needle-perfusion chamber. Forced convection has also been investigated in cardiac tissue engineering [19] to examine cardiomyocytes entrapped in Matrigel. The interstitial flow of medium improved the density of viable cells throughout the thickness, whereas static culture resulted in constructs with only a 100- to 200-μm thick surface layer containing viable cells around an acellular core. In a subsequent study using collagen constructs [20], oxygen concentration and cell viability decreased linearly from the construct surface and a physiological density of live cells was present only within the first 128 μm. However, interstitial flow of medium significantly increased oxygen concentration within the construct. Interstitial flow has also been applied to the culture of liver tissue. Chung et al. developed a detailed model to predict how changes in interstitial perfusion improved cell proliferation, nutrient delivery, and shear stress [21]. The perfusion model demonstrated improvements in cell proliferation through the scaffold, uniformity, and enhanced nutrient delivery.

Interstitial (transmural) flow for vascular tissue engineering has been investigated recently [22–24] as a means of improving uniformity and mechanical properties for these types of tissues. Transmural flow bioreactors were implemented to investigate controlled flow of culture medium to cells within the engineered tissues, whereas axial perfusion bioreactors typically impart transmural flow in an uncontrolled fashion [15–17]. Peclet numbers pertaining to these studies ranged from just over 1 to approximately 170. In all cases, significant improvements in uniformity were observed compared to diffusion alone. Models have also been developed to predict how changes in interstitial flow affected cell proliferation [25], nutrient delivery, and shear stress [17, 26]. Spatial and temporal control of the DO concentration can be used to maintain cell function and uniformity, which can ultimately improve tissue quality and mechanical properties [26].

In this work, an analysis of oxygen transport in engineered vascular tissues is performed. First, a static culture model is examined which demonstrates DO limitations that can arise in systems that depend only on diffusion. This is followed by computational and experimental analysis of several different bioreactor designs aimed at improving DO uniformity through implementation of convective transport through the tissue interstitium.

2 Static Culture Model

The static culture model was developed to understand the severity of DO gradients without any convection during incubation in a culture dish [23]. The general species conservation equation for DO in tissue is shown in Eq. 2.1:

$$\frac{\partial C_{O_2}}{\partial t} + \underline{v} \cdot \nabla C_{O_2} = D_{O_2,t} \nabla^2 C_{O_2} - R_{O_2} \tag{2.1}$$

where $D_{O_2,t}$ is the diffusivity of oxygen in tissue, C_{O_2} is the DO concentration, $\underline{v}$ is the velocity vector and R_{O_2} is the oxygen consumption term. Simplifying for steady-state, diffusion-only DO transport, and substituting Michaelis–Menten kinetics for oxygen consumption within the tissue yields Eq. 2.2:

$$D_{O_2,t} \nabla^2 C_{O_2} = \rho_{cell} \frac{V_{O_2\,\max} C_{O_2}}{K_m + C_{O_2}} \tag{2.2}$$

where ρ_{cell} is the cell density, and $V_{O_2\,\max}$ and K_m are Michaelis–Menten parameters. The system consisted of two subdomains—the tissue and the culture medium surrounding the tissue as illustrated in Fig. 1. Equation 2.2 was solved with the boundary conditions shown in Eq. 2.3–2.5:

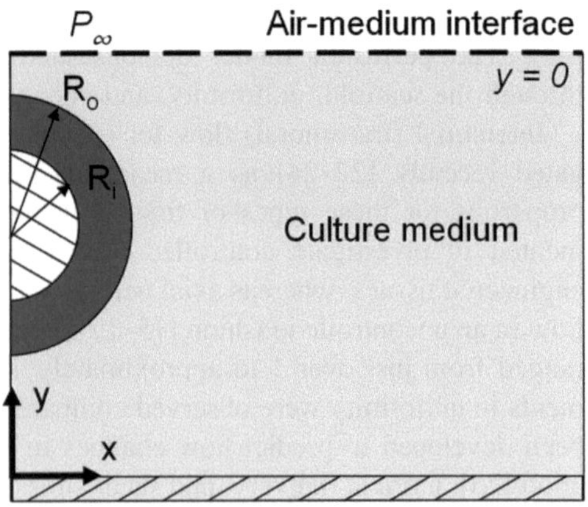

Fig. 1 Illustration of domain for DO transport problem of tubular tissue constructs cultured statically around a glass mandrel in a culture dish (not to scale). Schematic represents the symmetric domain as it was entered into COMSOL Multiphysics with two subdomains, which consisted of culture medium surrounding the DO-consuming engineered tissue

$$C_{O_2}\big|_{y=0} = \alpha P_\infty \tag{2.3}$$

$$\left.\frac{\partial C_{O_2}}{\partial r}\right|_{r=R_i} = 0 \tag{2.4}$$

$$J_{O_2,tissue}\big|_{r=R_o} = J_{O_2,fluid}\big|_{r=R_o} \tag{2.5}$$

The problem was further simplified by implementation of a symmetry plane at x = 0.

DO profiles were modeled using the finite element method (FEM) software package COMSOL Multiphysics (version 3.5a, COMSOL, Inc.). The model was solved using the UMFPACK stationary solver with a convergence tolerance of 10^{-6} and a minimum damping factor of 10^{-4}. Numerical values used in this and subsequent models are shown in Table 1.

In order to provide values for V_{max} and K_m, the oxygen consumption rate (OCR) was measured for neonatal human dermal fibroblasts (nhDFs) using a stirred microchamber and an oxygen monitoring system as described in Bjork and Tranquillo [23]. Measurements provided by this system provided changes in DO due to metabolic oxygen consumption inside an impermeable microchamber. V_{max} was determined on a per-cell basis by Eq. 2.6 [27]:

$$V_{\max} = \frac{V_{ch}\,\alpha}{N_{cell}}\left(\frac{\Delta p_{O_2}}{\Delta t}\right) \tag{2.6}$$

where V_{ch} is the volume of the microchamber, α is the Bunsen solubility coefficient (1.27 nmol/cm^3/mmHg at 37°C [28]), and N_{cell} is the number of cells in the sample as determined by DNA quantification and a cell viability assessment. $\Delta p_{O_2}/\Delta t$ is the slope of the linear portion of the curve obtained from the oxygen monitoring data. The OCR data obtained with the microchamber was used to determine K_m by plotting OCR versus the oxygen concentration. The concentration at which the OCR was one-half V_{max} was taken as K_m.

Table 1 Parameters used in DO and glucose transport models

Parameter	Description	Value	Source
Tissue properties			
L	Axial length	5 cm	Measured
H	Tissue thickness	300 μm	Measured
R_i	Inner diameter	2 mm	Measured
R_o	Outer diameter	2.3 mm	Measured
L_p	Hydraulic conductivity	1×10^{-6} cm s^{-1} Pa^{-1}	Measured
ρ_{cell}	Cell density	250×10^{6} cells mL^{-1}	Measured
V_{max,O_2}	Michaelis–Menten parameter for O_2	1.66 fmol/min/cell	Measured
K_{m,O_2}	Michaelis–Menten parameter for O_2	15.8 μM	Measured
$V_{max,\ gluc}$	Michaelis–Menten parameter for glucose	9.25 fmol/min/cell	[30]
$K_{m,\ gluc}$	Michaelis–Menten parameter for glucose	1.5 mM	[31]
$D_{O_2,t}$	Effective O_2 diffusivity in tissue	1.06×10^{-5} cm^2 s^{-1}	[33]
$D_{gluc,\ t}$	Effective glucose diffusivity in tissue	1.6×10^{-7} cm^2 s^{-1}	[34]
Fluid properties			
ρ	Density (water), 37°C	0.992 g cm^{-3}	[35]
μ	Dynamic viscosity (water)	0.0069 g cm^{-1} s^{-1}	[35]
D_f	Effective O_2 diffusivity in culture medium	3.83×10^{-5} cm^2 s^{-1}	[36]
C_∞	Equilibrium DO concentration (oxygen tension)	177 (140) nmol mL^{-1} (mmHg)	Measured
$C_{0,\ gluc}$	Culture medium glucose concentration	25 mM	[37]
ω	Pulse frequency	0.5–1 Hz	Measured
V_s	Stroke volume	0.5–1 mL	Measured

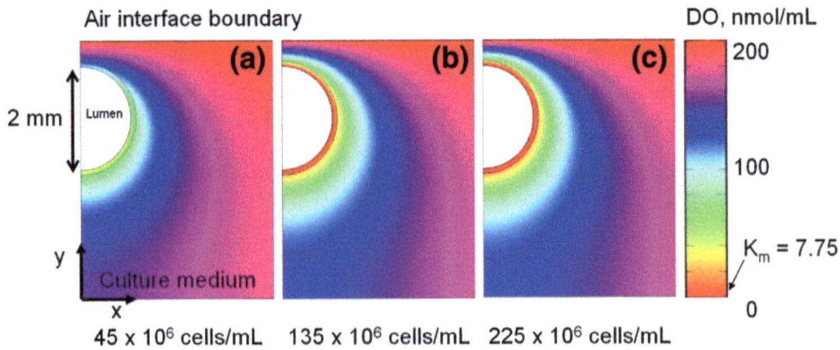

Fig. 2 DO profiles in statically-incubated tubular tissue constructs for **a** 45 × 10^6 cells/mL, **b** 135 × 10^6 cells/mL, and **c** 225 × 10^6 cells/mL. With diffusive transport alone the DO concentration falls below K_m at higher cell densities, which can result in the development of anoxic zones and cell death within the tissue. From [23]

Michaelis–Menten parameters from this method with additional numerical values, as shown in Table 1, were applied to the COMSOL and yielded DO concentration profiles for the entire domain (both culture medium and tissue subdomains) as shown in Fig. 2. The model was evaluated using cell densities ranging from 45 × 10^6 cells/mL up to 225 × 10^6 cells/mL. The top edge of each plot represents the air-medium interface where gaseous oxygen is in equilibrium with dissolved oxygen. Cells within the tissue consume oxygen which reduces the DO at increasing depths below the ablumenal surface. The minimum concentration in the tissue was at the lumenal surface (along the inner radius R_i) on the bottom side of the tissue. At 45 × 10^6 cells/mL, the minimum was 36.7 nmol/mL (3.8% O_2). As the cell density increases, the minimum DO concentration decreases to less than 1 nmol/mL. The values at this location for 135 and 225 × 10^6 cells/mL are both below K_m, which was measured to be 7.75 nmol/mL, and thus the cells in this region cannot consume oxygen at the maximum rate. The maximum concentrations occurred at the top ablumenal surface, corresponding to the location closest to P_∞. Based on these results, significant DO heterogeneity exists within the tissue for these cell densities, and there is thus an opportunity for improvement with bioreactor culture.

3 Transural Flow Bioreactor

Kitagawa et al. developed a "radial perfusion bioreactor" using steady flow of medium, where all flow into the lumen at one end was forced out through the graft wall as transmural flow [22]. They determined that the optimal flow rate for maximum cellularity and cell uniformity corresponded to a radial Peclet number, Pe_r, of approximately 170 (estimated from reported dimensions).

In our work, a similar bioreactor system as Kitagawa et al. was developed, but generalized to allow simultaneous transmural and axial flow, as well as each mode of flow alone. DO measurements were obtained in this bioreactor system, using

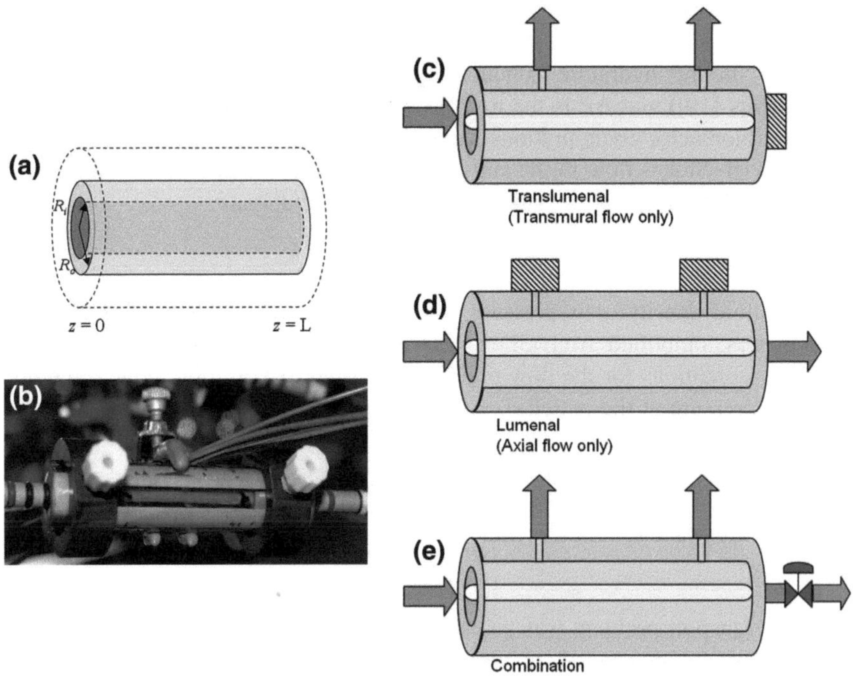

Fig. 3 Bioreactor configurations analyzed for steady flow. **a** Illustration of tissue dimensions within the bioreactor. **b** Photo of the bioreactor in operation. Schematic of **c** translumenal flow (axial outlet blocked), **b** lumenal (translumenal outlets blocked) and **c** combination of translumenal and lumenal flow. Flow proceeds from the inlet on the left side, then either through the tissue to outlet ports on the side (**c**, **e**) or axially out from the tubular tissue construct (**d**, **e**)

fibrin-based tubular tissue constructs, in order to monitor oxygen consumption of the tissue and provide data for DO model development. The model in this case was expanded to include axial and transmural flow and was coupled to the DO transport model using COMSOL. The model predicted DO profiles across the tissue thickness as a function of the transmural flow rate as well as shear rates on the cells within the tissue and the potential for construct bursting associated with pressures required to impart a prescribed transmural flow.

Multiple bioreactor configurations were implemented by blocking chamber exit ports as shown in Fig. 3. Pure transmural ("translumenal") flow was obtained by blocking the axial outflow port, pure axial ("lumenal") flow by blocking both transmural exit ports, and simultaneous transmural and axial ("combination") flow were implemented by incorporating both transmural exit ports and the axial exit port. In addition, the transmural flow rate was adjusted in this configuration by utilizing a pressure transducer and needle valve downstream of the axial exit port. The needle valve was adjusted to achieve a desired lumenal pressure which corresponded to a transmural flow velocity, v_r, given by Darcy's Law:

$$v_r = L_p \, \Delta P, \tag{3.1}$$

where L_p is the hydraulic conductivity of the tissue (measured in separate experiments [19]) and ΔP is the measured lumenal pressure.

These bioreactor configurations were implemented in COMSOL using steady-state, Navier–Stokes flow in the culture medium subdomain, as shown in Eq. 3.2, and Darcy flow from Eq. 3.1 above in the tissue subdomain.

$$\rho \, \underline{v} \cdot \nabla \underline{v} = -\nabla p + \mu \nabla^2 \underline{v} \tag{3.2}$$

ρ is the fluid density, p is pressure, and μ is the fluid viscosity. The complete domain was simplified from 3D to 2D by assuming axisymmetry. Additional boundary conditions for the flow problem consisted of Poiseuille flow at the inlet and zero pressure at the outlet(s), as shown in Eqs. 3.3 and 3.4:

$$\left.\frac{dp}{dz}\right|_{z=0} = \frac{8\,\mu \langle v_z \rangle}{R_i^2} \tag{3.3}$$

$$P\big|_{\substack{z=L \\ r=R_o}} = 0 \tag{3.4}$$

The DO transport problem was solved similarly to Eq. 2.2. The boundary conditions consisted of constant concentration, C_O, in the fluid inlet and the tissue outlet had a zero-flux boundary condition. These are shown in Eqs. 3.5 and 3.6:

$$C_{O_2}\big|_{z=0} = C_0 \tag{3.5}$$

$$\underline{n} \cdot \left(-D_{O_2} \frac{dC_{O_2}}{dr}\right)\bigg|_{r=R_o} = 0 \tag{3.6}$$

where $\underline{n}$ is the unit vector normal to the ablumenal tissue surface. The model was validated by integrating the flux over the outlet of the tissue surface for the transmural flow cases and comparing the calculated bulk-averaged DO value to measurements obtained in the outlet stream. A summary of COMSOL input values is provided in Table 1.

Outputs included the lumenal pressure field, velocity fields, and DO concentration and flux profiles along the ablumenal edge. v_r was used to assess the dominant mode of radial oxygen transport via the Peclet number and to determine the average shear stress realized by cells within the tissue according to Eq. 3.7 [26]:

$$\tau_{avg} = B \, \mu \, v_r \, \kappa^{-\frac{1}{2}} \tag{3.7}$$

where B is a shape factor (approximately equal to 1) and κ is the Darcy specific hydraulic permeability ($\kappa = L_p \, \mu \, (R_o - R_i)$). An average interstitial shear stress of 1 dyne/cm^2 was taken as a physiological reference value [26].

Model predictions for different bioreactor configurations are summarized in Fig. 4 with plots of the DO profiles along the ablumenal surface of the tubular tissue construct (the region of lowest DO). As shown in the figure, higher flow

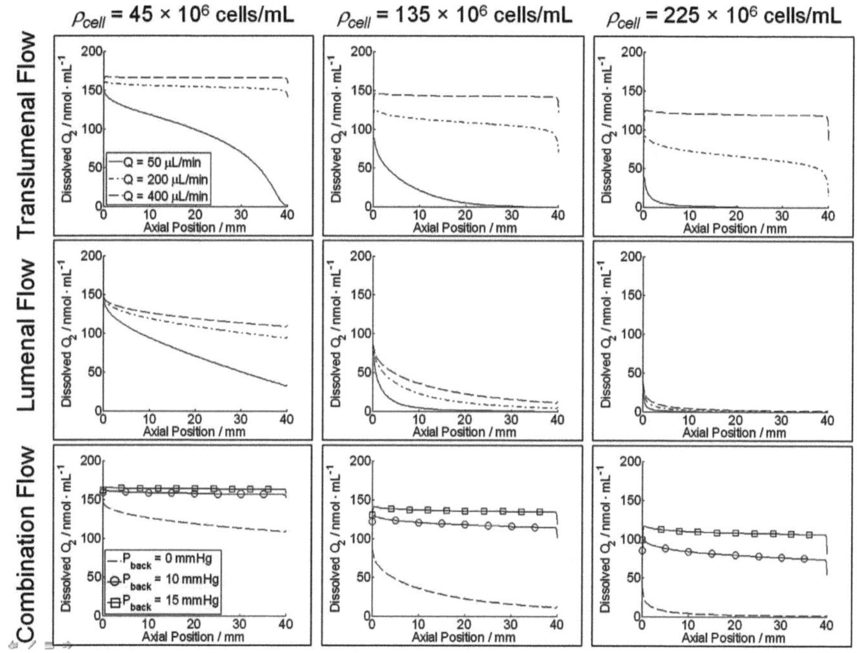

Fig. 4 Predicted DO concentration profiles along the ablumenal surface for the three different bioreactor configurations and three different cell densities. The legend in the *upper left graph* applies to each of the *top six graphs* (flow rates of 50, 200, and 400 μL/min, $Pe_r = 0.53$, 2.13, and 4.26, respectively, for translumenal flow). The legend in the *bottom left graph* applies to the *entire bottom row* (back pressures of 0, 10, and 15 mmHg), for which the inlet axial flow rate is constant at 400 μL/min. From [23]

rates result in a smaller differential from the inlet oxygen concentration (at $z = 0$) for the translumenal flow and lumenal flow configurations. Cell density also impacts the DO profile. As expected, as the cell density increases, the oxygen concentration along the ablumenal edge decreases. Furthermore, the model predicts that the v_r profile is essentially constant for a given flow rate (results not shown). Thus, increased volumetric flow will yield increased transmural flow; however, there is no difference in v_r between the lumenal inlet and outlet.

Figure 4 also reports the combination case of transmural and axial flow. Each curve for this configuration represents varied applied back pressure (0, 10 and 15 mmHg) at a fixed flow rate of 400 μL/min (no applied back pressure provides the same result as the lumenal flow configuration). Increasing the back pressure, in turn, increases the amount of flow diverted from the lumenal outlet into transmural flow. The combination case provides the benefit of having axial flow to increase the oxygen concentration—and thus the diffusional driving force for DO into the tissue—along the lumenal surface, while forcing additional oxygen through the tissue convectively.

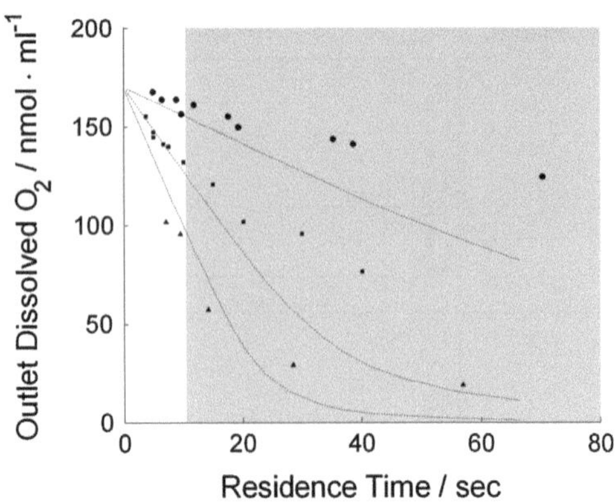

Fig. 5 Predicted and measured DO concentration versus residence time for the translumenal bioreactor configuration. The residence time here is tissue thickness divided by transmural velocity, corresponding to flow rates ranging from 50 to 400 μL/min. The *shaded region* indicates transmural flow velocities yielding $Pe_r < 1$. (- - -) Predicted DO concentrations for (*top* to *bottom*) 45×10^6, 135×10^6, and 225×10^6 cells/mL. Measured outlet DO for (*filled circle*) low, (*filled square*) medium and (*filled triangle*) high cell densities (targeted cell densities to compare to predictions). Note that the measured DO inlet in these measurements was 165 nmol mL^{-1} and is also reflected in the predictions for consistency. From [23]

The result of the combination flow case is noteworthy when also understanding the pressures required to provide DO uniformity along the ablumenal surface in comparison to the transmural flow cases, which appear very similar. In the transmural flow case, the lumenal pressures reach nearly 200 mmHg. For a developing tissue, this can be a detrimental pressure resulting in damage to the integrity of the construct. In contrast, the combination case can achieve comparable DO uniformity with much lower lumenal pressures (10–15 mmHg) since there is greater DO uniformity along the lumenal surface imparted by much higher axial flow.

Furthermore, increased transmural flow velocity increases shear stress on the cells within the tissue. Hydraulic conductivity plays a central role in these effects. As the tissue compacts due to cell traction forces and the cells deposit collagen and other extracellular matrix, the tissue becomes denser and less permeable to fluid flow, so the hydraulic conductivity, and thus κ, decreases. This, in turn, yields an increase in the shear stress according to Eq. 3.7. Given the higher lumenal pressures in the translumenal flow case, the detrimental consequences of higher interstitial shear are more likely compared to the combination flow case.

The model was verified by comparing the computed total radial molar flux of oxygen to measurements taken on the inlet and outlet streams. The total molar flux was found by integration along the ablumenal boundary in COMSOL for the case of translumenal flow. Dividing by the volumetric flow rate, which ranged from 50–400 μL/min, yielded the average DO concentration predicted in the transmural outlet lines, reflecting oxygen consumption by cells within the engineered tissue.

In practice, the length and thickness of tubular tissue constructs are variable. Residence time (tissue thickness divided by the average transmural flow velocity) was therefore used as the independent variable. Figure 5 shows the comparison of model results (lines) to measured DO concentrations. As shown in the figure, the average DO concentration decreases linearly with residence time until the consumption kinetics change (from zeroth to first order) due to low DO concentration, with the transition occurring sooner with increasing cell density. Moreover, increasing cell density has a large impact on the total DO consumed by the cells within the tissue, which is also effectively captured by the model.

4 Pulsed Stretch-Flow Bioreactor

As highlighted in the previous section, bioreactor culture incorporating controlled transmural flow can improve DO uniformity along the axial length of the cultured tubular constructs. This work was conducted in an effort to evaluate improved mass transport; however, it is known in the literature that mechanical stimulation can increase collagen deposition and thus the mechanical properties of engineered tissues so that they more closely mimic native tissue. Cyclic distension has been shown to improve these properties of fibrin-based engineered grafts seeded with nhDFs [14]. Along with high collagen content, the organization of the cell-produced collagen fibers is also of critical importance. It should recapitulate the predominant circumferential alignment of native arteries to provide burst strength and confer the natural anisotropic mechanical behavior [24].

A bioreactor system was designed to allow for culture of multiple grafts such that cyclic distention and cyclic transmural flow could be applied for stimulating tissue growth. This bioreactor system has the unique advantage of applying all four parameters that have previously been found to be beneficial for development of mechanically robust engineered tissues: cyclic stretching [9, 14, 24], transmural flow [22, 23], lumenal flow [27, 29], and ablumenal flow [15, 30]. This bioreactor system was shown to produce constructs having burst strengths with the highest values reported for engineered grafts made from biopolymer scaffolds and would be sufficient for implantation [24].

Reported here is an analysis of oxygen transport using this bioreactor system. The analysis incorporated pulsatile flow through the lumen of fibrin-based tubular tissue constructs using a combined experimental and computational approach. The aim of this study was to investigate DO in the tissue, flow profiles in the system, and to evaluate the theoretical effect of the frequency and volume of pulsatile flow on the DO concentration profiles within the tissue. Direct measurements of pressure and DO concentration both for the inlet flow and within the bioreactor were used to help guide computational analysis.

A schematic of the bioreactor is provided in Fig. 6. Tubular constructs were mounted on a custom manifold, which was placed inside a jar of cell culture medium during operation. A reciprocating syringe pump was connected to

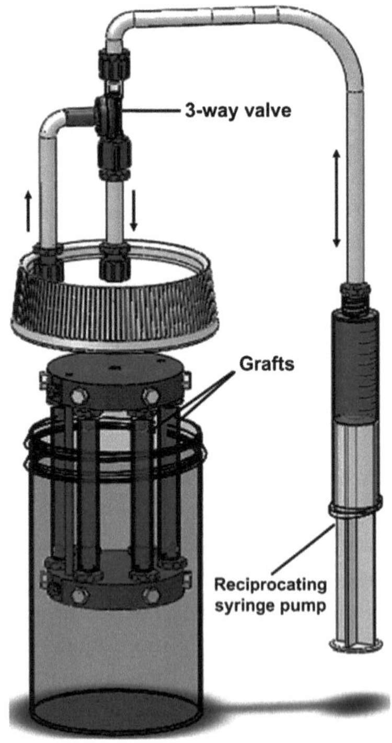

Fig. 6 Schematic of pulsed stretch-flow bioreactor. From [24]

the manifold through a three way valve. During forward motion, a fixed-volume pulse of culture medium flowed through the manifold and was distributed to each graft; consequently, each mounted graft experienced a pressure wave with resultant graft distension and transmural flow. Due to the manifold symmetry, each mounted graft was subjected to a similar pressure wave; some variations occurred since the graft properties at the time of mounting are not identical. During reverse motion, the syringe withdrew medium from the jar. An in-line pressure transducer and DO sensor were placed between the top manifold and the three way valve to provide data for use in the computational model.

Fluid flow was modeled in 3D using the transient Navier–Stokes equation in COMSOL Multiphysics. The full bioreactor geometry was simplified to a 1/6th slice based on symmetry, which reduced the number of degrees of freedom from over 250,000 to approximately 42,000. Further symmetrical simplification to 1/12th was not employed so that solutions could be more easily visualized. The true bioreactor system has a single outlet port near the upper manifold that would break symmetry; however, approximate flow velocity fields were deemed acceptable for assessing fluid mixing in the system and to maximize information gained with efficient use of computational time. The model was further simplified by neglecting the tissue deformation that occurs during pressurization of the graft, maintaining the tissue in a fixed position during pressurization.

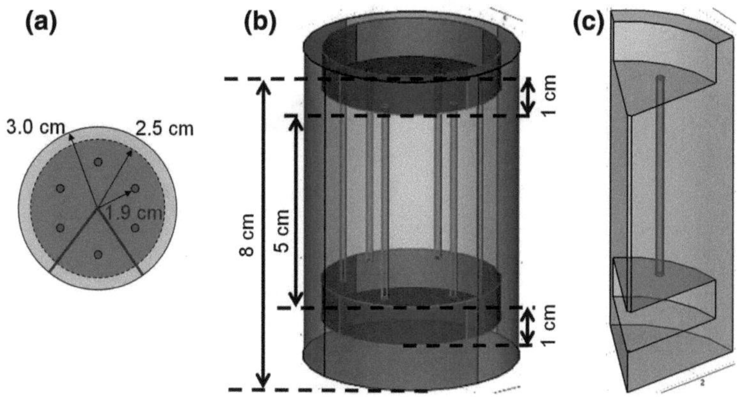

Fig. 7 Schematic of PSF bioreactor geometry used for 3D fluid flow modeling in COMSOL. **a** Top view showing dimensions of the chamber, manifold and placement of tissue constructs. **b** 3D rendering of entire bioreactor geometry in COMSOL. **c** Simplified geometry used for computation based on symmetry

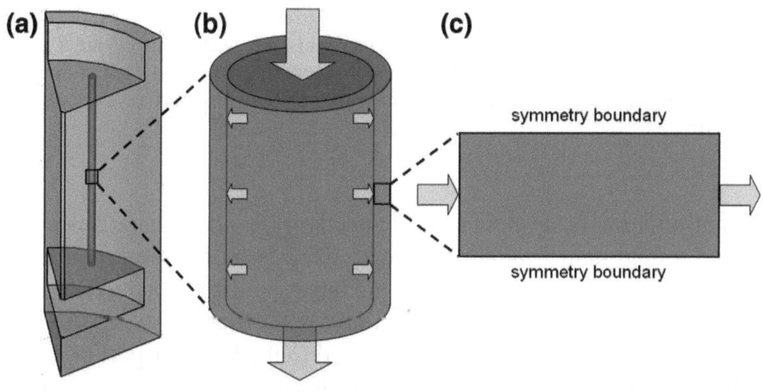

Fig. 8 Schematic of domain used for the DO model in the PSF bioreactor. **a** The domain from the flow problem was simplified to the tissue component as shown in **b** and **c**. **c** The diffusion–convection–reaction problem domain used in COMSOL consisted of a 300 μm thick slab of tissue with identical DO on both sides

A schematic of the geometrical simplification is provided in Fig. 7. The upper manifold was removed from the computational domain and its surface served as a no-slip wall boundary condition. The upper end of the construct lumen and the lower manifold central port served as fluid inlets into the system and the outlet was the annulus around the upper manifold. The vertical fluid plane on either side of the construct utilized a symmetry boundary condition. Constants used for the flow model (and subsequent DO model) are provided in Table 1. The model again accounted for flow of fluid through the tissue based on Darcy's Law (Eq. 3.1). The model equations were solved using an appropriate transient or stationary solver to

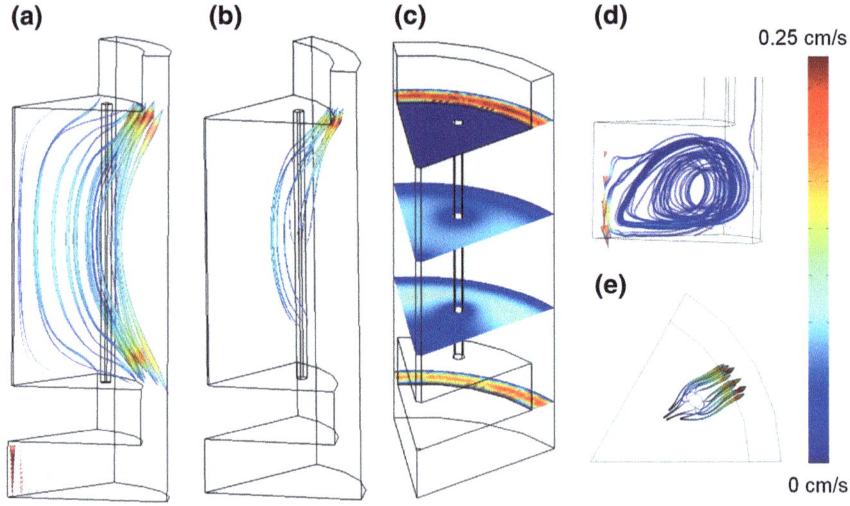

Fig. 9 Solutions to 3D flow during peak flow of PFS bioreactor. **a** Streamlines show flow around manifolds through the chamber interior. Streamline color indicates local stream velocity. **b** Solution shown in **a** isolating the tissue to highlight transmural flow. **c** Surface plot showing symmetry of flow and higher velocity around the manifolds. **d** Lower manifold outlet highlighting the swirl flow as shown by the streamlines. **e** Top view showing the tubular tissue construct surrounded by streamlines. Streamlines originating at the ablumenal surface are associated with transmural flow

examine pulsatile or steady fluid flow, respectively, with a convergence tolerance of 10^{-6} and a minimum damping factor of 10^{-4}. Solutions were typically found after 200 iterations using a biconjugate gradient stabilized solver. The mesh was refined until minimal variation in the solution was observed.

The output of the flow problem was coupled to the DO transport problem using COMSOL in order to evaluate the effect of pulsatility and stroke volume on DO profiles. Based on measurements showing no spatial variation in DO concentration in the annular region of the jar and low curvature of the construct, the 3D model was simplified to a 1D transport model as shown in Fig. 8.

The DO transport problem required solution to the species balance equation as shown in Eq. 2.1. The flow velocity was determined from Darcy's Law and was driven by a pulsatile waveform in the lumen, which was modeled by Eq. 4.1:

$$P_{pulse} = \begin{cases} P_{peak} \sin(2\pi\omega t), & \sin(2\pi\omega t) > 0 \\ 0, & \sin(2\pi\omega t) \leq 0 \end{cases} \tag{4.1}$$

where P_{peak} is the measured peak pressure and ω is the frequency of pulsatility, which results in the oscillatory pressure P_{pulse}. The model pulsatile pressure in Eq. 4.1 was compared to pressure monitoring data and found to match with suitable accuracy. The transient model was solved in COMSOL using the

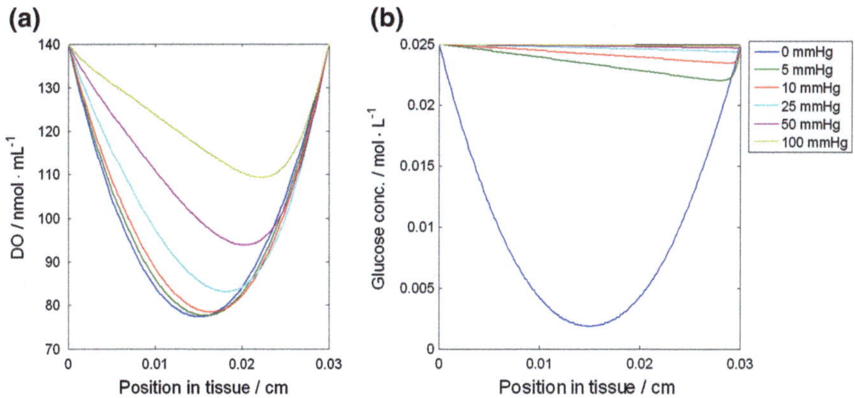

Fig. 10 Model results for **a** DO and **b** glucose during PFS bioreactor operation. Example tissue DO profiles determined for a range of peak pulse pressures induced by varying stroke volumes. For DO, the effect of flow becomes more apparent at pressures of 25 mmHg, where $\langle Pe_{r,O2} \rangle \approx 4$, and higher. Glucose profiles exhibit larger shifts at lower pulse pressures compared to DO

PARDISO solver with time steps of 0.1 s, a convergence tolerance of 10^{-6} and a minimum damping factor of 10^{-4}. Examination of step changes in the inputs, such as changing the flow rate for example, were examined over a period of at least 1 min to verify that a steady solution had been reached.

A summary of the fluid flow model is shown in Fig. 9. Model results showed velocities ranging from 15.9 cm/s at peak flow in the central port of the bottom manifold to 0.07 cm/s in the bioreactor center. Swirl flow was induced after leaving the bottom manifold, as evidenced by the streamlines in Fig. 9d. The streamlines also indicate that fluid flows around the tissue construct to the chamber interior and then toward the outlet around the top manifold. This is noteworthy since the fluid at the bottom of the chamber is well-mixed and flow around the entire construct serves to improve DO uniformity during culture. To verify the solution, calculation of mass conservation was conducted by boundary integration of the velocity field on the inlet and outlet streams. Agreement within 0.02% was obtained with the mesh geometries used to generate solutions shown in Fig. 10.

Transmural flow was evident in Fig. 9e. The vast majority of the flow was found to leave the lower manifold, with approximately 5% being pulsed through the tissue for $L_p = 1 \times 10^{-6}$ cm/s/Pa. Though stream lines are apparent in Fig. 9, ablumenal flow is approximately an order of magnitude higher than transmural flow.

For the frequency and stroke volumes within bioreactor capabilities when pressure and DO measurements were taken, the predicted DO profiles in the tissue, shown in Fig. 10a, did not vary greatly. At P_{peak} values of 5 and 10 mmHg, corresponding to a 0.5 and 1.0 mL syringe stroke volume and $\langle Pe_{r,O2} \rangle$ of 0.95 and 1.9, respectively, there was only a slight disparity from a purely diffusive transport regime. Doubling the pulse frequency from 0.5 to 1.0 Hz did not impact this result, as would be expected since transmural flow is driven by pressure. When peak pressures of 25 mmHg or greater were implemented, corresponding to $\langle Pe_{r,O2} \rangle$ of 4.7 and higher, the DO profile shifted in the direction of transmural flow as shown in Fig. 10a.

The 1D COMSOL model was extended to glucose to by altering the Michaelis–Menten constants, glucose diffusivity in tissue, and C_0 as listed in Table 1. The glucose consumption rate and K_m for glucose in dermal fibroblasts were reported to be 9.25 fmol/min/cell and 1.5 mM, respectively [30, 31]. The boundary values for glucose, C_0, were again assumed to be the same because of a high recirculating flow rate, but now computed from a macroscopic mass balance on glucose within the bioreactor in between culture medium changes. The glucose concentration profiles within the tissue are shown in Fig. 10b for $C_0 = 25$ mM (the standard glucose concentration for DMEM used in experiments). Similar to the DO profile, increasing P_{peak} yields greater uniformity in the concentration profile for glucose; however, lower pressures yield larger effects when $L_p = 10^{-6}$ cm/s/Pa. For pressures as low as 5 mmHg, a greater shift in the glucose concentration profile occurs compared to a similar pressure for DO. This difference can be attributed largely to the diffusivity of glucose being two orders of magnitude less than oxygen, as shown in Table 1, which increases the impact of transmural flow on its transport relative to oxygen.

5 Discussion

The computations and measurements presented here provide insight on DO profiles in tubular tissue constructs during operation of several bioreactor configurations, highlighting the improvement in DO uniformity over static culture. The DO transport models under steady-flow demonstrate improved uniformity when controlled transmural flow was combined with axial flow. The DO concentration profile in the tissue is greater and more uniform when transmural flow is incorporated versus a more simplistic bioreactor using only axial flow through the lumen. Model results of outlet DO values agreed with measurements collected during culture of fibrin-based tubular tissue constructs in the bioreactor system. Implementation of transmural flow must also take into consideration the lumenal pressure and interstitial shear stress, which the models predict can lead to tissue failure (bursting) and cell damage or death, respectively, at higher transmural flow rates. For these reasons, the combination of transmural and axial flow is superior to transmural flow only.

The models were further extended to assess the potential benefits of cyclic perfusion for DO delivery beyond the benefits of mechanical stimulation to the tissue. In analyzing a pulsed stretch-flow bioreactor, a symmetric 3D model revealed the extensive level of fluid mixing between the lower manifold and bottom of the chamber. Mixing helps to provide uniform nutrient delivery as the fluid flows up around the lower manifold and then around the constructs.

Assumptions used to allow for symmetry simplifications could have precluded identification of more complex flows, such as the development of secondary flows crossing symmetry planes. However, the symmetric model demonstrates substantial mixing below the lower manifold, which provides a uniform DO concentration in this region prior to reaching the tissue in the outlet flow. Since there was little DO depletion axially in the construct lumen and the chamber was a well-mixed, comparatively large volume reservoir, the lumenal and ablumenal surfaces had reasonably close DO concentrations, as was measured with DO sensors. Uniformity of the computed DO, which agreed with measurements taken at three levels within the chamber at the high flow rates being used (data not shown), allowed the mass transport model to be simplified to a 1D problem. The output of the flow problem, namely the transmural flow velocity v_r, was then used as an input to the mass transport problem to determine the effects of O_2 diffusion, convection and consumption on DO profiles within the tissue.

The model was extended to scenarios beyond the current capability of the bioreactor since it was designed for 0.5–1.0 Hz pulse frequencies and the stroke volume was limited by the allowable construct distension. Though frequency had minimal impact on DO profiles, the stroke volume was found to help improve uniformity due to greater transmural flow resulting from the higher lumenal pressure. Larger pulse pressures would induce greater transmural flow but also greater distension of the construct. Distensions above 10–15% have proven to have detrimental effects on tissue strength and stiffness [9, 14]. Pressures of approximately 25 mmHg have been shown to induce 10% strain in tubular tissue constructs [24]. The constructs also stiffen with time due to cellular deposition of collagen, thus greater pressures could be exerted without imparting detrimental strain levels assuming that L_p remains constant. Values of L_p have been measured out to 2 weeks of culture; however, the values could fluctuate depending on the rate of collagen deposition compared to the rate of fibrin degradation. This has not been fully investigated, thus the values of L_p were assumed to remain constant in this study. The pressures examined by the model are relevant since they could be achievable for bioreactor operation. The results in Fig. 10 were obtained with an L_p of 1×10^{-6} cm/s/Pa. A more porous construct could achieve higher transmural flows with lower pressures. For example, increasing L_p by an order of magnitude would cause the 100 mmHg pulse to look like a 10 mmHg pulse according to Fig. 10 since L_p, $\langle v_r \rangle$, and P are all linearly related. Likewise, decreasing L_p by an order of magnitude would have the opposite effect. In this case a 10 mmHg pulse would appear as the 100 mmHg pulse in Fig. 10. While this impacts DO uniformity, Fig. 10b shows that there is still notable improvement over a diffusion-only transport system for glucose.

Interstitial shear stress effects due to transmural flow should also be considered. At a peak pressure of 10 mmHg and L_p of 1×10^{-6} cm/s/Pa, an interstitial shear stress of 0.9 dyne/cm^2 is experienced by the cells based on Eq. 3.7 [26]. As with transmural flow velocity, higher pressures would increase the shear stresses in a linear fashion, but a decreasing L_p increases the shear stress in a non-linear manner.

The DO analysis showed that high stroke volumes, yielding high transmural flows, would be required to increase the uniformity of the DO profile. When the analysis was extended to glucose, small stroke volumes were predicted to substantially increase uniformity throughout the tissue. The models could also be extended growth factors, which could further highlight the utility of convective transport within the tissue since many growth factors are proteins with a typical diffusivity that is multiple orders of magnitude lower than oxygen [34]. The reaction kinetics are likely different compared to oxygen; regardless, the incorporation of convective transport should greatly impact the concentration profile uniformity because of the much lower diffusivity.

Acknowledgments This work has been supported by National Institutes of Health (NHLBI R01 HL083880 to RTT) and 3M Company (JWB). Furthermore, the technical assistance of Naomi Ferguson and Lee Meier is gratefully acknowledged as well as Dave Hultman for his efforts in machining and design discussions.

References

1. Jain, R.K., Au, P., Tam, J., Duda, D.G., Fukumura, D.: Engineering vascularized tissue. Nat. Biotechnol. **23**, 821–823 (2005)
2. Malda, J., Rouwkema, J., Martens, D.E., le Comte, E.P., Kooy, F.K., Tramper, J., et al.: Oxygen gradients in tissue-engineered PEGT/PBT cartilaginous constructs: measurement and modeling. Biotechnol. Bioeng. **86**, 9–18 (2004)
3. Malda, J., Woodfield, T.B.F., van der Vloodt, F., Wilson, C., Martens, D.E., Tramper, J., et al.: The effect of PEGT/PBT scaffold architecture on the composition of tissue engineered cartilage. Biomaterials **26**, 63–72 (2005)
4. Nikolaev, N.I., Obradovic, B., Versteeg, H.K., Lemon, G., Williams, D.J.: A validated model of GAG deposition, cell distribution, and growth of tissue engineered cartilage cultured in a rotating bioreactor. Biotechnol. Bioeng. **105**, 842–852 (2009)
5. Brown, D.A., MacLellan, W.R., Laks, H., Dunn, J.C.Y., Wu, B.M., Beygui, R.E.: Analysis of oxygen transport in a diffusion-limited model of engineered heart tissue. Biotechnol. Bioeng. **97**, 962–975 (2007)
6. Radisic, M., Malda, J., Epping, E., Geng, W.L., Langer, R., Vunjak-Novakovic, G.: Oxygen gradients correlate with cell density and cell viability in engineered cardiac tissue. Biotechnol. Bioeng. **93**, 332–343 (2006)
7. Potier, E., Ferreira, E., Meunier, A., Sedel, L., Logeart-Avramoglou, D., Petite, H.: Prolonged hypoxia concomitant with serum deprivation induces massive human mesenchymal stem cell death. Tissue Eng. **13**, 1325–1331 (2007)
8. Demol, J., Lambrechts, D., Geris, L., Schrooten, J., Van Oosterwyck, H.: Towards a quantitative understanding of oxygen tension and cell density evolution in fibrin hydrogels. Biomaterials **32**, 107–118 (2011)
9. Isenberg, B.C., Williams, C., Tranquillo, R.T.: Endothelialization and flow conditioning of fibrin-based media-equivalents. Ann. Biomed. Eng. **34**, 971–985 (2006)
10. Webb, A.R., Macrie, B.D., Ray, A.S., Russo, J.E., Siegel, A.M., Glucksberg, M.R., et al.: In vitro characterization of a compliant biodegradable scaffold with a novel bioreactor system. Ann. Biomed. Eng. **35**, 1357–1367 (2007)
11. Hahn, M.S., McHale, M.K., Wang, E., Schmedlen, R.H., West, J.L.: Physiologic pulsatile flow bioreactor conditioning of poly(ethylene glycol)-based tissue engineered vascular grafts. Ann. Biomed. Eng. **35**, 190–200 (2007)
12. Butcher, J.T., Barrett, B.C., Nerem, R.M.: Equibiaxial strain stimulates fibroblastic phenotype shift in smooth muscle cells in an engineered tissue model of the aortic wall. Biomaterials **27**, 5252–5258 (2006)

13. Stegemann, J.P., Hong, H., Nerem, R.M.: Mechanical, biochemical, and extracellular matrix effects on vascular smooth muscle cell phenotype. J. Appl. Physiol. **98**, 2321–2327 (2005)
14. Syedain, Z.H., Weinberg, J.S., Tranquillo, R.T.: Cyclic distension of fibrin-based tissue constructs: Evidence of adaptation during growth of engineered connective tissue. Proc. Natl. Acad. Sci. USA **105**, 6537–6542 (2008)
15. Niklason, L.E., Gao, J., Abbott, W.M., Hirschi, K.K., Houser, S., Marini, R., et al.: Functional arteries grown in vitro. Science **284**, 489–493 (1999)
16. Niklason, L.E., Abbott, W., Gao, J., Klagges, B., Hirschi, K.K., Ulubayram, K., et al.: Morphologic and mechanical characteristics of engineered bovine arteries. J. Vasc. Surg. **33**, 628–638 (2001)
17. Williams, C., Wick, T.M.: Perfusion bioreactor for small diameter tissue-engineered arteries. Tissue Eng. **10**, 930–941 (2004)
18. Khong, Y.M., Mang, J., Zhou, S.B., Cheung, C., Doberstein, K., Samper, V., et al.: Novel intra-tissue perfusion system for culturing thick liver tissue. Tissue Eng. **13**, 2345–2356 (2007)
19. Radisic, M., Yang, L., Boublik, J., Cohen, R.J., Langer, R., Freed, L.E., et al.: Medium perfusion enables engineering of compact and contractile cardiac tissue. Am. J. Physiol. Heart Circ. Physiol. **286**, H507–H516 (2004)
20. Radisic, M., Malda, J., Epping, E., Geng, W.L., Langer, R., Vunjak-Novakovic, G.: Oxygen gradients correlate with cell density and cell viability in engineered cardiac tissue. Biotechnol. Bioeng. **93**, 332–343 (2006)
21. Chung, C.A., Chen, C.W., Chen, C.P., Tseng, C.S.: Enhancement of cell growth in tissue-engineering constructs under direct perfusion: modeling and simulation. Biotechnol. Bioeng. **97**, 1603–1616 (2007)
22. Kitagawa, T., Yamaoka, T., Iwase, R., Murakami, A.: Three-dimensional cell seeding and growth in radial-flow perfusion bioreactor for in vitro tissue reconstruction. Biotechnol. Bioeng. **93**, 947–954 (2006)
23. Bjork, J.W., Tranquillo, R.T.: Transmural flow bioreactor for vascular tissue engineering. Biotechnol. Bioeng. **104**, 1197–1206 (2009)
24. Syedain, Z.H., Meier, L.A., Bjork, J.W., Lee, A., Tranquillo, R.T.: Implantable arterial grafts from human fibroblasts and fibrin using a multi-graft pulsed flow-stretch bioreactor with noninvasive strength monitoring. Biomaterials **32**, 714–722 (2011)
25. Chung, C.A., Chen, C.W., Chen, C.P., Tseng, C.S.: Enhancement of cell growth in tissue-engineering constructs under direct perfusion: modeling and simulation. Biotechnol. Bioeng. **97**, 1603–1616 (2007)
26. Wang, S., Tarbell, J.M.: Effect of fluid flow on smooth muscle cells in a 3-dimensional collagen gel model. Arterioscler. Thromb. Vasc. Biol. **20**, 2220–2225 (2000)
27. Papas, K.K., Pisania, A., Wu, H., Weir, G.C., Colton, C.K.: A stirred microchamber for oxygen consumption rate measurements with pancreatic islets. Biotechnol. Bioeng. **98**, 1071–1082 (2007)
28. Avgoustiniatos, E.: Oxygen diffusion limitation in pancreatic islet culture and immunoisolation. Massachusetts Institute of Technology, Cambridge (2001)
29. Tschoeke, B., Flanagan, T.C., Koch, S., Harwoko, M.S., Deichmann, T., Ella, V., et al.: Tissue-engineered small-caliber vascular graft based on a novel biodegradable composite fibrin-polylactide scaffold. Tissue eng. **15**, 1909–1918 (2009)
30. Hoerstrup, S.P., Zund, G., Sodian, R., Schnell, A.M., Grunenfelder, J., Turina, M.I.: Tissue engineering of small caliber vascular grafts. Eur. J. Cardiothorac. Surg. **20**, 164–169 (2001)
31. Seliktar D., Black R.A., Vito R.P., Nerem R.M.: Dynamic mechanical conditioning of collagen-gel blood vessel constructs induces remodeling in vitro. Ann. Biomed. Eng. **28**, 351–362 (2000)
32. Simpson, N.E., Han, Z.C., Berendzen, K.M., Sweeney, C.A., Oca-Cossio, J.A., Constantinidis, I., et al.: Magnetic resonance spectroscopic investigation of mitochondrial fuel metabolism and energetics in cultured human fibroblasts: Effects of pyruvate dehydrogenase complex deficiency and dichloroacetate. Mol. Genet. Metab. **89**, 97–105 (2006)

33. Decker, S., Lipmann, F.: Transport of D-glucose by membrane vesicles from normal and avian sarcoma virus transformed chicken embryo fibroblasts. Proc. Natl. Acad. Sci. USA **78**, 5358–5361 (1981)
34. Truskey, G.A., Yuan, F., Katz, D.F.: Transport Phenomena in Biological Systems. Pearson Prentice Hall, Upper Saddle River (2004)
35. Weind, K.L., Boughner, D.R., Rigutto, L., Ellis, C.G.: Oxygen diffusion and consumption of aortic valve cusps. Am. J. Physiol.-Heart Circ. Physiol. **281**, H2604–H2611 (2001)
36. Rong, Z., Cheema, U., Vadgama, P.: Needle enzyme electrode based glucose diffusive transport measurement in a collagen gel and validation of a simulation model. Analyst **131**, 816–821 (2006)
37. Dean, J.A.: Lange's Handbook of Chemistry (15th Edition), 15th edn. McGraw Hill, New York (1999)
38. Lee, J.: Biochemical Engineering. Prentice Hall, Englewood Cliffs, NJ (1991)
39. Invitrogen Life Technologies. http://products.invitrogen.com/ivgn/product/21063029#coa. Accessed 14 April 2011

Part III
Computational Tools for the Study of the In Vivo Process

Multi-Scale Modelling of Vascular Disease: Abdominal Aortic Aneurysm Evolution

Paul N. Watton, Huifeng Huang and Yiannis Ventikos

Abstract We present a fluid-solid-growth (FSG) computational framework to simulate the mechanobiology of the arterial wall. The model utilises a realistic constitutive model that accounts for the structural arrangement of collagen fibres in the medial and adventitial layers, the natural reference configurations in which the collagen fibres are recruited to load bearing and the (normalised) mass-density of the elastinous and collagenous constituents. Growth and remodelling (G&R) of constituents is explicitly linked to mechanical stimuli: computational fluid dynamic analysis produces snapshots of the frictional forces acting on the endothelial cells; a quasi-static structural analysis is employed to quantify the cyclic deformation of the vascular cells. We apply the computational framework to simulate the evolution of a specific vascular pathology: abdominal aortic aneurysm (AAA). Two illustrative models of AAA evolution are presented. Firstly, the degradation of elastin (that is observed to accompany AAA evolution) is prescribed, and secondly, it is linked to low levels of wall shear stress (WSS). In the first example, we predict the development of tortuosity that accompanies AAA enlargement, whilst in the latter, we illustrate that linking elastin degradation to low WSS leads to enlarging fusiform AAAs. We conclude that this computational framework provides the basis for further investigating and elucidating the

P. N. Watton (✉) · Y. Ventikos
Department of Engineering Science, Institute of Biomedical Engineering, University of Oxford, Oxford, UK
e-mail: Paul.Watton@eng.ox.ac.uk

Y. Ventikos
e-mail: Yiannis.Ventikos@eng.ox.ac.uk

H. Huang
Department of Mathematics, Imperial College, London, UK
e-mail: huifeng.huang@gmail.com

Stud Mechanobiol Tissue Eng Biomater (2013) 10: 309–339

DOI: 10.1007/8415_2012_143

Published Online: 18 August 2012

aetiology of AAA and other vascular diseases. Moreover, it has immediate application to tissue engineering, e.g., aiding the design and optimisation of tissue engineered vascular constructs.

1 Introduction

Tissue engineering offers the possibility of developing a biological substitute material in vitro with the inherent mechanical, chemical, biological, and morphological properties required in vivo, on an individual patient basis [1]. Computational modelling has an integral role to play in these ambitions. However, for in-silico modelling to realise its potential, i.e. to guide the design and optimisation of tissue engineered constructs, it must accurately represent the underlying mechanobiology. Whilst early modelling attempts are characterised by a substantial distance between computer and bench, the integration of biological experiments and simulation efforts are increasing [2].

The need for improved mechanobiological modelling is not restricted to the domain of tissue engineering, such research is necessary to guide our understanding of human physiological and pathophysiology, e.g. the evolution of vascular diseases. In this chapter, we illustrate a computational model for the evolution of abdominal aortic aneurysm (AAA). It combines a realistic microstructural model of the arterial wall with computational fluid dynamics and structural analyses to quantify the mechanical environment that acts on the vascular cells. Growth and remodelling algorithms simulate the cells responding to mechanical stimuli and adapting the tissue structure. The model simulates a fusiform abdominal aortic aneurysm that evolves with similar mechanical, biological and morphological properties with those observed in vivo. Whilst in need of further sophistications to more accurately reflect the mechanobiology of the arterial wall, it should be readily recognisable to the reader that this computational framework has substantial potential to be applied to aid the design and optimisation of tissue engineered vascular constructs.

Abdominal Aortic Aneurysm (AAA) is characterised by a bulge in the abdominal aorta. Development is associated with dilation of the arterial wall and the possibility of rupture [3]. Prevalence rates are estimated between 1.3 and 8.9 % in men and between 1.0 and 2.2 % in women [4]. They are more common in subjects who smoke [5] and suffer from hypertension [6]. Between 80 and 90 % of ruptured AAAs will result in death [7]; rupture is responsible for 1–2 % of all deaths in the UK each year [8]. Surgery to repair the AAA is an option; however, it is a high-risk procedure with a 5 % mortality rate [9]. Intervention is recommended when the risk of rupture exceeds the risk of surgery. Statistically, this occurs when the diameter exceeds 5.5 cm [10]. However, such a criterion fails to identify small AAAs with high risk of rupture [11] and large AAAs with low risk [12]. Thus there is a pressing need for improved diagnostic criteria to aid clinical decisions: distinguishing those AAA most at risk of rupture will yield significant improvements in patient healthcare.

To understand the aetiology of AAA requires a thorough understanding of the underlying biological mechanisms that govern arterial growth and remodelling (G&R) in both healthy and diseased states [13]. This is an extremely challenging, multidisciplinary problem: not only is the function of individual components of the system elusive, their interactions remain unclear as well. In fact, modelling the mechanical response of the healthy arterial wall alone poses significant challenges in itself: it is a highly complex integrated structure [14]. However, it is envisaged that models of AAA evolution [15–18] will lead to a greater understanding of the pathogenesis of the disease and may ultimately yield improved criteria for the prediction of rupture.

A pre-requisite to modelling AAA evolution is to model the biomechanics and mechanobiology of the healthy arterial wall. Briefly, the artery, consists of three layers: *intima, media* and *adventitia*. The innermost layer is the intima, this consists of a basement membrane and a lining of endothelial cells (ECs). ECs form a permeability barrier between the blood flow, the vessel wall and surrounding tissues, and play a significant role in regulating circulatory functions. An internal elastic laminae separates the intima from the media. The media consists of a network of elastin fibres and (approximately) circumferentially orientated vascular smooth muscle cells (VSMCs) and collagen fibres. The adventitia is an outer sheath with bundles of collagen fibres, maintained by fibroblast cells [19], arranged in helical pitches around the artery. The main load bearing constituents are elastin and collagen. Collagen is considerably stiffer than elastin; however, for a healthy large elastic artery, such as the abdominal aorta, at physiological strains elastin bears most of the load [20]. This is because collagen is tortuous in nature [21, 22] and acts as a protective sheath to prevent excessive deformation of the artery.

The structure of the artery is continuously maintained by vascular cells. The morphology and functionality of the cells are intimately linked to their extracellular mechanical environment. Haemodynamic forces due to the pulsatile blood flow give rise to cyclic stretching of the extra-cellular matrix (ECM), frictional forces acting on the inner layer of the arterial wall, normal hydrostatic pressure and interstitial fluid forces due to the movement of fluid through the ECM. Mechanosensors on the cells convert the mechanical stimuli into chemical signals. Activation of second messengers (molecules that transduce the signals from the mechanoreceptors to the nucleus) follows and leads to an increase in the activity of transcription factors. Binding of the transcriptions factors to the DNA leads to activation of genes that regulate cell proliferation, apoptosis, differentiation, morphology, alignment, migration, and synthesis and secretion of various macromolecules [23]. The living response of the artery acts to ensure it functions in an optimum manner as mechanical demands change. For instance, in response to changes in flow, ECs signal VSMCs to constrict/relax to regulate the diameter of the artery to restore wall shear stress (WSS) to homeostatic levels [24].

Fibroblasts continually maintain the collagen fabric by synthesising structural proteins and matrix degrading enzymes [25]. They work on the collagen, crawling over it and tugging on it in order to compact it into sheets and draw it out into cables [26]. In doing this, the fibroblasts attach the collagen fibres to the ECM in a state of stretch. Each fibroblast can synthesise approximately 3.5 million procollagen molecules per day. However, the amount a fibroblast secretes is regulated: between 10 and 90 % of all procollagen molecules are degraded intracellularly prior to secretion [27]. This provides a mechanism for rapid adaptation of the amount of collagen secreted and thus enables the artery to rapidly adapt in response to altered environmental conditions.

During the evolution of an AAA, it is observed that there is an accompanying loss of elastin [28] due to increased elastolysis [29]. However, the highly nonlinear mechanical response of the collagen implies that the degradation of elastin alone is insufficient to explain the large dilatations that arise as the aneurysm evolves. In fact, the collagen fabric continually remodels: collagen has a half-life of 2 months [30] whereas elastin is a relatively stable protein with a long half-life (approximately 40 years [31]). Consequently, models of AAA evolution must address both the degradation of elastin and the G&R of collagen [15].

Watton et al. [15] proposed the first mathematical model of AAA evolution. The model utilises a realistic structural model for the arterial wall [32] which is adapted to incorporate variables which relate to the *normalised mass density* (hereon referred to as *concentration*) of the elastinous and collagenous constituents and the reference configurations in which the collagenous constituents are recruited to load bearing. This enables the G&R of the tissue to be simulated as an aneurysm evolves. A key assumption of the model is that collagen fibres, which are in a continual state of deposition and degradation [30], attach to the artery in a state of stretch, denoted *the attachment stretch*, which is independent of the current configuration of the tissue. A degradation of elastin is prescribed and differential equations are employed to evolve the reference configurations and concentrations of collagen fibres to maintain the stretch of the collagen to homeostatic levels, i.e. the attachment stretch. The model predicts evolution of AAA mechanical parameters and growth-rates consistent with clinical observations [16]. The G&R framework has subsequently been applied to consider conceptual 1D models of intracranial aneurysm (IA) evolution, i.e. enlarging (and stabilising) cylindrical and spherical membranes [33] and the evolution of saccular IAs of the internal carotid artery [34]. It has been implemented into a novel Fluid-Solid-Growth (FSG) framework to couple the G&R of aneurysmal tissue to local haemodynamic stimuli [35, 36] and applied to model patient-specific IA aneurysm evolution [37].

Watton et al. [15] and Watton and Hill [16] simulated AAA evolution by prescribing the loss of elastinous constituents. However, local aortic hemodynamic conditions may influence the risk for, and the progression of, aneurysm disease [38]. Experimental models of AAA suggest an inverse relationship between WSS and aneurysm expansion [39, 40]. In fact, expansion of AAA is believed to be related to increasing macrophage infiltration secondary to low WSS in the aneurysms recirculation region [41]. Hence, spatial and temporal changes in the WSS distribution could influence the G&R of the tissue and thus the rate at which an AAA enlarges. In this study, we apply a novel FSG framework [35] to model AAA evolution. Furthermore, we sophisticate the G&R formulation (as utilised by all previous published studies with this modelling approach [15, 16, 33–37]) to explicitly link both the growth and remodelling of the collagen fabric to cyclic deformation of the arterial wall. Two novel examples of AAA development are illustrated: firstly, we prescribe the degradation of elastin secondly, we assume that the degradation of elastin is driven by low magnitudes of WSS. In both cases, collagen remodelling and collagen growth are linked to the magnitude of the local cyclic deformation of the arterial wall.

2 Fluid-Solid-Growth Model for Aneurysm Evolution

In this section, we describe our FSG computational framework for modelling AAA evolution. Figure 1 depicts the methodology. The computational modelling cycle begins with a structural analysis of the aneurysm to solve the systolic and diastolic equilibrium deformation fields for given pressure and boundary conditions (Fig. 1i). The structural analysis quantifies the stress and stretch, and the cyclic deformation, of the ECM components and the cells (each of which may have different natural reference configurations). The geometry of the aneurysm is subsequently exported to be prepared for computational fluid dynamics (CFD) analysis (Fig. 1ii): the aneurysm geometry is integrated into a physiological geometrical domain; the domain is automatically meshed; physiological flow rate and pressure boundary conditions are applied. The flow is solved assuming rigid boundaries for the haemodynamic domain. The haemodynamic quantities of interest, for example, WSS, are then exported and interpolated onto the nodes of the structural mesh: each node of the structural mesh contains information regarding the mechanical stimuli obtained from the haemodynamic and structural analyses. G&R algorithms simulate cells responding to the mechanical stimuli and adapting the tissue (Fig. 1iii). Following G&R, the constitutive model of the aneurysmal tissue is updated and the structural analysis is re-executed to calculate the new equilibrium deformation fields. The updated geometry is exported for haemodynamic analysis. The cycle continues and as the tissue adapts an aneurysm evolves. The stages of the FSG framework, i.e. the structural modelling (Fig. 1i), CFD (Fig. 1ii) and G&R methodology (Fig. 1iii) are detailed in the subsequent subsections, i.e. Sects. 2.1, 2.2 and 2.3, respectively. Note that whilst we are considering the development of an aneurysm, we could equivalently apply such a computational framework to simulate the evolving mechanical, biological and morphological properties of a tissue engineered construct. Moreover, if the mechanobiology was clearly understood and accurately modelled, such a computational framework could guide and optimise the design of a tissue-engineered construct.

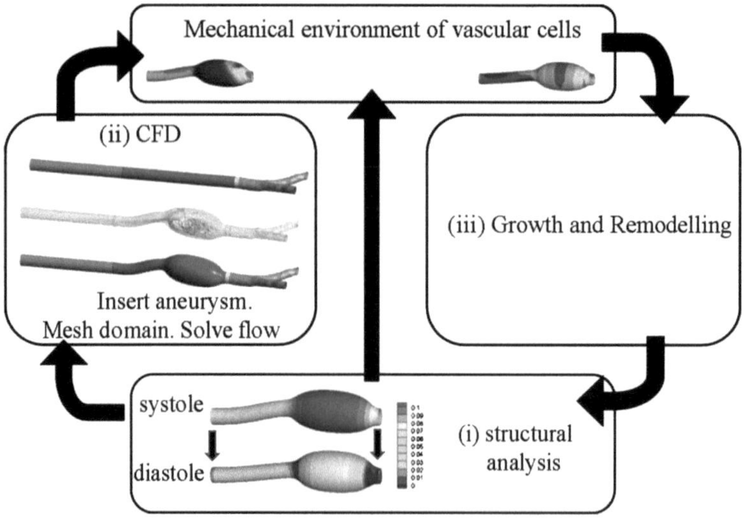

Fig. 1 Fluid-Solid-Growth computational framework for modelling aneurysm evolution consisting of (*i*) structural analysis, (*ii*) computational fluid dynamics (CFD) analysis and (*iii*) G&R algorithms. Further details are provided in Sects. 2.1, 2.2 and 2.3, respectively

2.1 Structural Model of Aneurysm Evolution

A geometric nonlinear membrane theory (see, for example, [42]) is adopted to model the steady deformation of the arterial wall. The unloaded internal abdominal aorta is treated as a thin cylinder of undeformed radius R, length L_1, and thickness H. The thickness of the media H_M is assumed to be equal to 2/3 the thickness of the arterial wall, i.e. $H_M = H/3$, and thus the thickness of the adventitia $H_A = H/3$. The artery is subject to a physiological axial pre-stretch λ_z and pressure p which causes a circumferential stretch λ. A body fitted coordinate system is used to describe the cylindrical membrane with axial and azimuthal Lagrangian coordinates $\theta_1 \in [0, L_1]$ and $\theta_2 \in [0, 2\pi R)$, respectively. Formation and development of the aneurysm is assumed to be a consequence of G&R of the material constituents of the artery. The principle of stationary potential energy is the governing equation for the steady deformation of the arterial wall. It requires that the first variation of the total potential energy vanishes,

$$\delta\Pi_{\text{int}} - \delta\Pi_{\text{ext}} = 0, \tag{1}$$

where $\delta\Pi_{\text{int}}$ represents the variation of the internal potential energy Π_{int} stored in the arterial wall, whilst $\delta\Pi_{\text{ext}}$ is the variation of the external potential energy Π_{ext} caused by the normal pressure that acts on the artery. Appropriate functional forms for the spatially and temporally heterogeneous strain-energy functions (SEFs) for the media Π_M, and the adventitia Π_A must be specified so that $\delta\Pi_{\text{int}}$ can be computed. Details of the theoretical formulation to describe the deformation of the arterial wall and the numerical formulation to solve Eq. (1) can be found in [15]: the equilibrium displacement field is solved by a finite element method coded in FORTRAN 77 [43].

2.1.1 Strain-Energy Functions for Heterogeneous Aneurysmal Tissue

The arterial wall is modelled as two layers. The inner layer models the mechanical response of the media (and intima), with contributions from the elastinous constituents (ground substance, elastin fibres and passive smooth muscle cells) and a double helical pitch of collagen fibres with orientations γ_{M_p} to the azimuthal axis: $p = +$, $p = -$ denote positively ($\gamma_{M_+} > 0$) and negatively ($\gamma_{M_-} < 0$) wound fibres, respectively. The outer layer models the mechanical response of the adventitia, which is considered to have a small elastinous contribution and a double helical pitch of collagen fibres with orientations γ_{A_p} ($p = \pm$) to the azimuthal axis. The mechanical response of each layer is modelled as the sum of a neo-Hookean strain energy function (SEF) [44] and a highly nonlinear SEF which represents the mechanical response of the collagen [32]. Spatially and temporally dependent functions are introduced for the concentration of the elastinous and collagenous constituents and the configuration in which the collagen fibres begin to be recruited to load bearing.

Recruitment stretch variables define the factor the tissue must be stretched, relative to the unloaded configuration, in the direction of a collagen fibre for it to begin to bear load. Remodelling the recruitment stretches enables the remodelling of the collagenous fabric during aneurysm evolution to be simulated. Constituent concentration variables define the ratio of the mass density of a constituent at time t to the mass density at time $t = 0$ and enable the growth/atrophy of the constituent to be simulated. For further details of the theoretical formulation to describe the G&R the interested reader is referred to [33]. The SEFs for the elastinous contributions in the medial and adventitia are multiplied by a normalised spatially and temporally dependent concentration function, denoted $m^E(\theta_1, \theta_2, t)$. This is employed to prescribe the degradation of the elastinous constituents, where $m^E(\theta_1, \theta_2, t = 0) = 1$. Fields of spatially and temporally dependent fibre recruitment stretch $\lambda^R_{J_p}(\theta_1, \theta_2, t)$ and concentration $m^C_{J_p}(\theta_1, \theta_2, t)$ variables are defined throughout the midplane of the arterial wall, where the subindex J denotes the media M or the adventitia A. The fibres within each layer are orientated at an angle of γ_{J_p} to the azimuthal axis, where p denotes the pitch $\pm\gamma_J$ relative to the azimuthal axis in the unloaded reference configuration. Hence, the SEFs are

$$\begin{aligned} \Psi_J =& m^E K^E_J (E_{11} + E_{22} + E_{33}) \\ &+ \sum_{p=\pm, E^C_{J_p} > 0} m^C_{J_p} K^C_J \left\{ \exp\left[A^C (E^C_{J_p})^2 \right] - 1 \right\}, \quad J = M, A, \end{aligned} \tag{2}$$

where the material parameters for the elastinous constituent are denoted by K^E_J, whilst K^C_J and A^C are parameters that relate to the collagen fabric. The Green-

Lagrange (GL) strains of the elastin, i.e. E_{11}, E_{22} and E_{33}, are defined relative to the unloaded configuration; in the initial cylindrical configuration, these represent strains in the axial, azimuthal and radial directions, respectively. The GL strains in the collagen fibres are denoted by $E^C_{J_p}(\theta_1, \theta_2, t)$ and are defined relative to the configuration in which the collagen fibres are recruited to load bearing. More specifically, the GL strains $E^C_{J_p}$ of the collagen fibres are a function of the GL strains of the elastin resolved in the directions of the collagen fibres, denoted E_{J_p}. Thus,

$$E^C_{J_p} = \frac{E_{J_p} - E^R_{J_p}}{1 + 2E^R_{J_p}}, \tag{3}$$

where $E^R_{J_p} = [(\lambda^R_{J_p})^2 - 1]/2$, $\lambda^R_{J_p}(\theta_1, \theta_2, t)$ are the recruitment stretches and $E_{J_p} = E_{11}\sin^2\gamma_{J_p} + E_{22}\cos^2\gamma_{J_p} + 2E_{12}\sin\gamma_{J_p}\cos\gamma_{J_p}$.

2.1.2 Geometry, Physiological Data and Material Parameters

The material parameters for the media and the adventitia as well as all other values which serve the basis for our subsequent computation are taken from [16] and are summarized in Table 1. In the systolic configuration, the artery has a radius r_s, an axial pre-stretch of 1.3, a circumferential stretch of 1.25 and we assume that the elastinous constituents bear 80 % of the load. We assume that the elastinous response of the adventitia is an order of magnitude weaker than that of the media, i.e. $K^E_A = K^E_M/10$ [32] and specify the ratio of the medial and adventitial collagen material parameters to be $K^C_A = K^C_M/4$. The remaining three independent material parameters, namely K^E_M, K^C_M, A^C are determined so that the SEFs adequately model the mechanical behaviour of the artery (see [16]).

The Attachment Stretch λ^C_{AT}

Collagen fibres are in a continual state of deposition and degradation [30, 45] and they are attached to the artery in a state of stretch. Watton et al. [15] hypothesised that *fibroblasts configure the collagen fibres to achieve a maximum stretch* λ^C_{AT} *during the cardiac cycle* and introduced the terminology *attachment stretch* to denote λ^C_{AT}.

Magnitude of Recruitment Stretches at $t = 0$

For a cylindrical artery at fixed axial stretch and subject to radial inflation, maximum collagen fibre stretches will occur in the systolic configuration. Hence, the initial values of the recruitment stretches are determined so that the stretch of the medial and adventitial collagen fibres at systole at $t = 0$ equal the attachment stretch, thus

$$\lambda^R_{J_+}|_{t=0} = \lambda^R_{J_-} = \sqrt{\lambda_s^2\cos^2\gamma_{J_+} + \lambda_z^2\sin^2\gamma_{J_+}} / \lambda^C_{AT} \tag{4}$$

Table 1 Geometry, physiological data used for modelling the human abdominal aorta artery

Length		
In vivo	L	120 mm
Reference configuration	L_1	L/λ_z
Radii		
At systole	r_s	5.7 mm
Reference configuration	R	r_s/λ_s
Wall thickness		
Total	H	$R/5$
Media	H_M	$2H/3$
Adventitia	H_A	$H/3$
Fibre orientation		
Media	$\gamma_{M_+}, \gamma_{M_-}$	$+30°, -30°$
Adventitia	$\gamma_{A_+}, \gamma_{A_-}$	$+60°, -60°$
Applied pressure, kinematics		
Systolic pressure	p_s	16 kPa
Diastolic pressure	p_d	10.67 kPa
Axial pre-stretch	λ_z	1.3
Systolic circumferential stretch at $t = 0$	λ_s	1.25
Diastolic circumferential stretch at $t = 0$	λ_d	1.13
Attachment stretch	λ^C_{AT}	1.07
Recruitment stretches $(t = 0)$		
Media	$\lambda^R_{M_+}, \lambda^R_{M_-}$	1.18
Adventitia	$\lambda^R_{A_+}, \lambda^R_{A_-}$	1.20

2.1.3 Cyclic Deformations of Aneurysmal Tissue

To quantify the cyclic deformation we determine the geometry of the aneurysm at systolic ($p = p_s$= 16 kPa) and diastolic pressures ($p = p_d$= 10.67 kPa) as it evolves. Initially, the artery has a cylindrical configuration. It can be shown (see [15]) that the governing force-balance equation for a cylindrical membrane subject to fixed axial stretch λ_z and radial inflation is

$$p = \frac{1}{R\lambda_z}\left\{ \left(h_M k_M^E + h_A k_A^E\right)\left(1 - \frac{1}{\lambda_z^2 \lambda^4}\right) + \sum_{J=M,A;p=\pm:E^C_{J_p}>0} \frac{2h_J k_J^C A^C E^C_{J_p} \exp\left\{A^C \left(E^C_{J_p}\right)^2\right\} \cos^2 \gamma_{J_p}}{\left(\lambda^R_{J_p}\right)^2} \right\}, \tag{5}$$

where the first and second terms on the right of (5) correspond to the contribution of load bearing from the elastinous and collagenous constituents, respectively. Equation (5) is derived from the governing variational equation, i.e. (1), the functional form the of the SEFs (2) and considerations of the displacement field for a cylindrical membrane. It can be solved numerically to obtain the radius as a function of the pressure. At $t = 0$, the diastolic and systolic displacement fields to satsify (1) are known, i.e.

the artery is a cylinder of length $\lambda_z L$ that has radius r_s in the systolic configuration and radius r_d in the diastolic configuration. As the material constituents evolve, the systolic/diastolic deformation fields to satisfy (1) are updated with a Newton-Raphson method using the systolic/diastolic deformation fields from the previous time-step as initial guesses; the positions of the boundaries of the domain $(\theta_1 = 0, L_1)$ in the systolic/diastolic configuration are held fixed as the aneurysm evolves.

Cyclic Areal Stretch

Let $E_{11}^D, E_{22}^D, E_{12}^D$ and $E_{11}^S, E_{22}^S, E_{12}^S$ denote the GL strains, in the diastolic (*D*) and systolic (*S*) configurations, respectively. The cyclic areal stretch, here denoted A^{CS}, is the ratio of the areal stretch at systole to the areal stretch at diastole, which expressed in terms of the Green-Lagrange strains is:

$$A^{CS} = \sqrt{\frac{(1+2E_{11}^S)(1+2E_{22}^S) - 4E_{12}^S}{(1+2E_{11}^D)(1+2E_{22}^D) - 4E_{12}^D}}. \tag{6}$$

Biaxial Stretch Index χ^{BSI}

In the initial geometrical configuration, the artery is subject to cyclic stretch that is one dimensional, i.e. the cyclic stretch is only in the circumferential direction. As the aneurysm evolves regions of the artery experience biaxial stretching. To characterise the evolution of the cyclic stretch environment, we consider a novel Biaxial Stretch Index (BSI), denoted χ^{BSI} where $0 \leq \chi^{BSI} \leq 1$. If $\chi^{BSI} = 0$, the tissue is subject to 1D cyclic stretching whereas if $\chi^{BSI} = 1$, the tissue is subject to equi-biaxial cyclic stretching. The index χ^{BSI} is calculated as follows. The principal stretches λ_1, λ_2 can be expressed in terms of the GL strains (see [46]), i.e.

$$\lambda_1 = \sqrt{\cos^2\varphi(2E_{11}+1) + \sin^2\varphi(2E_{11}+1) + 2\sin\varphi\cos\varphi E_{12}}, \tag{7}$$

$$\lambda_2 = \sqrt{\sin^2\varphi(2E_{11}+1) + \cos^2\varphi(2E_{11}+1) - 2\sin\varphi\cos\varphi E_{12}}, \tag{8}$$

where

$$\varphi = \frac{1}{2}\arctan\frac{2E_{12}}{E_{11} - E_{22}}. \tag{9}$$

We then denote the principal stretches in the diastolic configuration as $\lambda_\alpha^D = \lambda_\alpha^D(E_{11}^D, E_{22}^D, E_{12}^D, \varphi^D)$ $(\alpha = 1, 2)$ where $\varphi^D = \varphi^D(E_{11}^D, E_{22}^D, E_{12}^D)$ is calculated using (9). We then determine the magnitude of these principal stretches (with orientations defined relative to the diastolic configuration) in the systolic configuration, i.e. $\lambda_\alpha^{DS} = \lambda_\alpha^{DS}(E_{11}^S, E_{22}^S, E_{12}^S, \varphi^D)$. The cyclic variation of the (diastolic) principal stretches, denoted λ_α^{CS}, are thus:

$$\lambda_\alpha^{CS} = \frac{\lambda_\alpha^{DS}}{\lambda_\alpha^D}. \tag{10}$$

Now, let

$$\xi_1 = \max\left(\lambda_1^{CS}, \frac{1}{\lambda_1^{CS}}\right) - 1, \quad \xi_2 = \max\left(\lambda_2^{CS}, \frac{1}{\lambda_2^{CS}}\right) - 1, \tag{11}$$

so that $\xi_1 + 1, \xi_2 + 1 \geq 1$ define the factor by which the principal stretches (with orientations defined relative to the diastolic configuration) *increase during the cardiac cycle*. Note the emphasis on *cardiac cycle*, i.e. we are not distinguishing between increases that occur between diastole and systole and increases that occur between systole and diastole. This is necessary because in certain locations, principal stretches may decrease as the pressure acting on the arterial wall increases, i.e. $\lambda_\alpha^{CS} < 1$ ($\alpha = 1$ and/or $\alpha = 2$). In this scenario, the factor by which the principal stretches increase during the cardiac cycle is $(1/\lambda_\alpha^{CS}) > 1$. Hence to determine the factor by which the principal stretches increase during the cardiac cycle we consider $\max(\lambda_\alpha^{CS}, 1/\lambda_\alpha^{CS})$.

We now define the biaxial stretch index χ^{BSI} to be:

$$\chi^{BSI} = \frac{\min(\xi_1, \xi_2)}{\max(\xi_1, \xi_2)}. \tag{12}$$

Suppose $(\lambda_1^{CS} - 1) = \kappa(\lambda_2^{CS} - 1)$, i.e. the cyclic (linearised) strain in the direction of the first principal stretch is a factor κ greater than the cyclic (linearised) strain in the direction of the second principal stretch. Then it is straightforward to deduce that $\chi^{BSI} = 1/\kappa$. This measure quantifies the degree of the biaxial distortion of the tissue (during the cardiac cycle) and is independent of the magnitude of the strains.

2.2 Haemodynamic Modelling

To achieve fully developed flow in the region the aneurysm develops, extensions are attached to the structural model of the artery/aneurysm. The upstream extension is taken to be a cylinder (radius r_S and length 100 mm) whilst a patient-specific illiac artery bifurcation (obtained from Magnetic Resonance Imaging) is utilised for the downstream extension [47]. Given that the real artery is not perfectly cylindrical, a connecting patch is incorporated to join the geometries together (see Fig. 2). The methodological approach to solve the haemodynamics as the aneurysm evolves proceeds as follows. The geometry of the aneurysmal section is exported (see Fig. 1) from the structural solver to the meshing suite ANSYS ICEM (ANSYS Inc, Canonsburg, PA). ANSYS ICEM automatically integrates the aneurysmal section into the larger geometrical domain, i.e. attaches the upstream and downstream extensions, and generates an unstructured tetrahedral mesh with prism layers lining the boundary in a scripted-automated manner for the fluid domain. After meshing, appropriate boundary conditions are applied (see below) and the flow is solved by ANSYS CFX (ANSYS Inc, Canonsburg, PA) which solves the

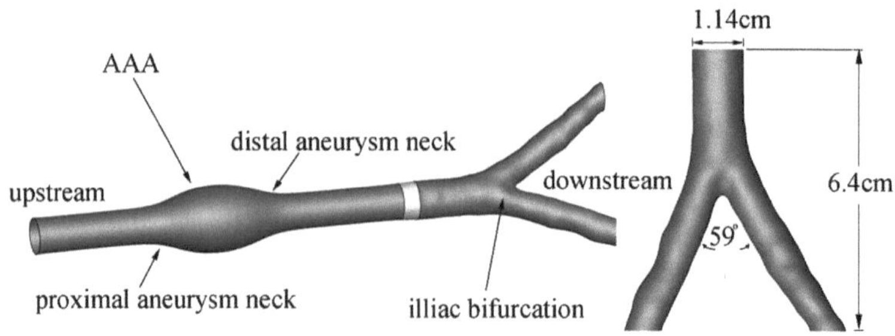

Fig. 2 A patient-specific iliac bifurcation artery is attached downstream of the AAA structural model; a connecting patch is created to join two geometries together. A cylindrical extension is attached upstream (not depicted)

incompressible Navier-Stokes equations using a finite volume formulation [48, 49]. The solver is based on a coupled approach (i.e. velocities and pressure are cast and solved as a single system) and a fully implicit time discretisation, where needed. An algebraic multigrid variant is used for convergence acceleration [50]. Blood is modelled as a Newtonian fluid [51, 52] with constant density $\rho = 1,069\,\mathrm{kgm}^{-3}$ [53, 54] and constant viscosity $\eta = 0.0035$ Pa s. At the arterial wall, no slip, no-flux conditions are applied. We adopted a steady flow analysis to reduce the cost of the computational simulations. The mean flow and pressure boundary conditions are taken from a 1D model of the arterial tree [55] which has been integrated into the software suite @neufuse which was developed for the European project @neuIST [56] (www.aneurist.org: 'Integrated Biomedical Informatics for the Management of Cerebral Aneurysms'). It solves the 1D form of the Navier-Stokes equation in a distributed model of the human systemic arteries, accounting for the ventricular-vascular interaction and wall viscoelasticity; it was recently validated through a comparison with in vivo flow measurements [57]. A flow rate of 24 cm^3/s is applied at the inlet and the pressure at the illiac arteries is set to 12760 Pa.

2.3 Growth and Remodelling

To model AAA evolution, we simulate the G&R of the load bearing constituents of the arterial wall, i.e. elastin and collagen. *Remodelling* is associated with changes in the natural reference configurations that constituents are recruited to load bearing whereas *growth* is associated with changes in the mass of the constituents.

2.3.1 Elastin Degradation

To simulate the degradation of elastin that occurs during AAA evolution, two approaches are considered. Firstly, the degradation of elastin is prescribed ([15]) and secondly it is linked to the haemodynamic environment (see [35]).

- **Case (i) Prescribed elastin degradation**
The degradation of elastin is prescribed over a representative timescale of 10 years (see [16]). To prescribe the loss of elastin, the following function is used:

$$m^E(\theta_1, \theta_2, t) = 1 - \left(1 - m^E_{\min}(t)\right) \exp\left[-\omega_1 \left(\frac{\theta_1}{L_{min}} - 1\right)^2\right], \tag{13}$$

where θ_1, θ_2 are the Lagrangian coordinates of the computational domain, i.e. $\theta_1 \in [0, L_1] \theta_2 \in [0, L_2)$, $m^E_{\min}$ represents the minimum concentration of elastin, at time t, which, by construction, is located at $(\theta_1, \theta_2) = (L_{\min}, \theta_2)$. The parameter ω_1 controls the degree of localisation of the degradation function (we adopt $\omega_1 = 20$, see [16] for influence of this parameter). We assume that the minimum concentration decays exponentially. Thus,

$$m^E_{\min}(t) = \exp\{\ln(m^E_T)(t/T)\} = (m^E_T)^{t/T}, \tag{14}$$

where $T = 10$, $0 \leq t \leq T$, $m^E_{10} = 0.05$ and $L_{\min} = 0.8L_1$. Note that at $t = 0$, $m^E_{\min}(t = 0) = 1$.
- **Case (ii) Linking Elastin degradation to Haemodynamic Environment**
Initially, the distribution of the WSS is spatially uniform. To perturb the spatial distribution of haemodynamic stimuli, a localised axisymmetric degradation of elastin is prescribed to create a small axisymmetric AAA. We assume this small aneurysm develops over several years and for numerical illustration we specifically consider a time-scale of 4 years. The functional form utilised for case (i) is adopted, i.e, (13) with $L_{\min} = 0.5L_1$ and $m^E_4 = 0.5$. Whilst the elastin is degrading, the aneurysm continues to enlarges in size and the strains in the collagen fabric are elevated above homeostatic values, i.e. $\lambda^C_J > \lambda^C_{AT}$). We stabilise the aneurysm by switching off the degradation of elastin and allowing sufficient time for the collagen fabric to adapt to a new homeostasis, i.e. $\lambda^C_{J_p} \rightarrow \lambda^C_{AT}$; numerically a period of 1 year or greater seems sufficient and so for $4 < t \leq 5$, we specify that no further elastin degrades ($\forall t : 4 < t \leq 5, m^E_{\min}(t) = m^E_4$). Subsequent degradation of elastin (for $t > 5$) is explicitly linked to deviations of haemodynamic stimuli from homeostatic levels (see [35] for further details on methodology). The concentration of elastin m^E evolves according to

$$\frac{\partial m^E}{\partial t} = -\mathscr{F}_D D_{\max} m^E \tag{15}$$

where t is in years, $D_{\max}$ specifies the maximum rate of degradation, and $\mathscr{F}_D(\theta_1, \theta_2, t) : 0 \leq \mathscr{F}_D \leq 1$ is a spatially-dependent function of the haemodynamic quantities to be linked to elastin degradation. Clearly, if $\mathscr{F}_D = 1$, elastin degrades at a maximum rate whilst if $\mathscr{F}_D = 0$, no degradation occurs.
As an illustrative example, we suppose elastin degradation is linked to low levels of WSS. More specifically, we follow [35] and assume that if the maximum value of the WSS is greater than a critical value, say τ_{crit}, no

degradation of elastin occurs whilst lower values of WSS give rise to degradation of elastin. Furthermore we suppose that there exists a value of WSS, say $\tau_X : 0 \leq \tau_X < \tau_{crit}$, at which maximum degradation occurs. We assume a simple quadratic functional form for the degradation function $\mathscr{F}_D$ that describes the relation between the local WSS and the degree of the degradation of elastin, i.e.

$$\mathscr{F}_D(\tau(\theta_1, \theta_2, t)) = \begin{cases} 0, & \tau \geq \tau_{crit}, \\ \left(\frac{\tau_{crit}-\tau}{\tau_{crit}-\tau_X}\right)^2, & \tau_X < \tau < \tau_{crit}, \\ 1, & \tau \leq \tau_X. \end{cases} \tag{16}$$

In this study, $D_{max} = 0.5$, $\tau_{crit} = 0.5$ Pa and $\tau_X = 0$ Pa.

2.3.2 Collagen Adaption

The adaption of the collagen fabric during aneurysm evolution consists of two distinct mechanisms: *growth* and *remodelling*.

2.3.3 Remodelling the Recruitment Configuration of Collagen

Evolving the reference configurations that fibres are recruited to load bearing simulates the mechanical consequences of: (i) fibre deposition and degradation in altered configurations as the aneurysm enlarges; (ii) fibroblasts configuring the collagen to achieve a maximum strain during the cardiac cycle, i.e. the *attachment strain* $E^C_{AT} = ((\lambda^C_{AT})^2 - 1)/2$. The recruitment stretches adapt so that the maximum strain in the collagen fibres remodels to E^C_{AT}. We utilise linear differential equations for the remodelling of the recruitment stretches, i.e.

$$\frac{d\lambda^R_{J_p}}{dt} = \left(\alpha \frac{E^C_{J_p}|_{max} - E^C_{AT}}{E^C_{AT}}\right), \tag{17}$$

where $\alpha = 0.3\,\text{years}^{-1}$,

$$E^C_{J_p}|_{max} = \max\left(E^C_{J_p}|_{sys}, E^C_{J_p}|_{dias}\right) \tag{18}$$

and $E^C_{J_p}|_{sys}$ and $E^C_{J_p}|_{dias}$ denote the magnitude of the collagen strains evaluated in the systolic and diastolic configurations, respectively. Note, [15] assumed that peak strains occur at systolic pressure and thus collagen remodelling is simulated by considering the steady deformations that occur in the systolic configuration. However, it was observed that for a more complex geometry, the fibres may not achieve maximum strains at systolic pressure: towards the distal and proximal ends of the AAA the axial strains increase as the pressure is reduced [15]. Hence, the subtle sophistication for the functional form for the remodelling of the recruitment stretches, i.e. (17).

2.3.4 Growth/Atrophy of the Collagen Fabric

Fibroblasts deposit collagen fibres and secrete proteases to degrade the collagenous material. They adhere to the ECM via specialised cell surface receptors, in particular integrins [58]. The integrins act as stretch sensors [59], transducing mechanical signals to the fibroblast interior, enabling fibroblasts to sense their mechanical environment. We assume that the rate of change of the concentration of the collagenous constituents is dependent on the concentration of fibroblasts $m^F_{J_p}$ (ratio of the density of the fibroblast cells at time t to the density at time $t = 0$) in the arterial wall and rate of synthesis and degradation of collagen. Let $\mathscr{F}_S$ and $\mathscr{F}_D$ be functions depicting how the rate of collagen synthesis and the secretion of matrix metalloproteases are related to the deformation of the fibroblast cell during a cardiac cycle, respectively. Rates of activity of vascular cells increase with the magnitude of the cyclic deformation [60, 61]. We assume $\mathscr{F}_S$ and $\mathscr{F}_D$ to be functions of the maximum strains $E^F_{J_p}|_{\max}$ and cyclic areal stretches A^{CS} of the cells during the cardiac cycle. Under these assumptions, the rate of change of the collagen concentration is

$$\frac{\mathrm{d}m^C_{J_p}}{\mathrm{d}t} = m^F_{J_p}\mathscr{F}_S(E^F_{J_p}|_{\max}, A^{CS}) - m^F_{J_p}\mathscr{F}_D(E^F_{J_p}|_{\max}, A^{CS}). \tag{19}$$

In response to increased stretch, fibroblasts attempt to reduce their stretch and reach a new equilibrium by restructuring their cytoskeleton and ECM contacts, i.e. fibroblasts reconfigure their natural reference configuration. Hence, to simplify the mathematical analysis, we assume that the local (natural) reference configuration of a fibroblast cell is identical to that of the collagen fabric it is maintaining. Hence the GL strain of the fibroblast cell, say $E^F_{J_p}$, is assumed to be equal to the GL strain of the collagen, i.e. $E^F_{J_p} \equiv E^C_{J_p}$. Furthermore, we assume that the concentration of fibroblasts in the arterial tissue is proportional to the concentration of collagenous constituents, i.e. $m^F_{J_p} = \xi_0 m^C_{J_p} : \xi_0 > 0$. These assumptions imply that the rate of evolution of the collagen fibre concentration $m^C_{J_p}$ is

$$\frac{\mathrm{d}m^C_{J_p}}{\mathrm{d}t} = m^C_{J_p}\mathscr{F}_G(E^C_{J_p}|_{\max}, A^{CS}) \tag{20}$$

where $\mathscr{F}_G = \xi_0(\mathscr{F}_S - \mathscr{F}_D)$. If $E^C_{J_p}|_{\max} = E^C_{AT}$ the collagen fabric is in homeostasis, i.e. the secretion of ECM is balanced by the degradation and there is no change in concentration. Hence, for $E^C_{J_p}|_{\max} = E^C_{AT}$ it is required that $\mathscr{F}_G(E^C_{J_p}) = 0$. The exact functional form of $\mathscr{F}_G$ is unknown. However, if the substrate is stretched, a net positive force acts on the cell and signalling to the nucleus results in an up-regulation of ECM protein expression and a down-regulation of collagenase expression. Conversely, relaxation of the substrate can trigger different signals resulting in a reversed pattern of protein expression [59], i.e. down regulation of ECM protein expression and up-regulation of collagenase expression. Although we do not explicitly model protein synthesis and enzymatic degradation, the net result is to stimulate increases/decreases in the ECM. The simplest functional form for $\mathscr{F}_G$ that satisfies these requirements is linear,and hence we propose the following evolution equation for the collagen fibre concentration

$$\frac{\mathrm{d}m_{J_p}^C}{\mathrm{d}t} = \beta(A^{CS})m_{J_p}^C\left(\frac{E_{J_p}^C|_{\max} - E_{AT}^C}{E_{AT}^C}\right), \tag{21}$$

where $\beta(A^{CS})$ is a phenomenological growth parameter. For a healthy abdominal aorta, vascular cells experience cyclic stretching with magnitudes of approximately 10 % whilst for older, more collagenous vessels, the magnitudes may decrease to 2 % [62]. We select a functional form for β which increases rates of production if the cyclic stretch environment increases from its initial values for the healthy artery at $t = 0$:

$$\beta(A^{CS}) = \beta_0 \exp\left[\beta^{CS}\max\left(\frac{A^{CS}}{A^{CS}|_{t=0}} - 1, 0\right)\right]. \tag{22}$$

This choice is somewhat arbitrary, however it acts to prevent unrealistically large cyclic deformations that can occur in regions of the tissue as the aneurysm evolves. For the analysis in this paper, we follow [37] and take $\beta_0 = 0.7$ years^{-1}, $\beta^{CS} = 3$.

Average fibre concentration

To simplify the presentation of the results, we define the average fibre concentration m^C of the medial and adventitial layers,where

$$m^C = \frac{1}{H}\left(H_M\frac{m_{M_+}^C + m_{M_-}^C}{2} + H_A\frac{m_{A_+}^C + m_{A_-}^C}{2}\right). \tag{23}$$

3 Physiological Growth Model Abdominal Aortic Aneurysm

We briefly review the framework of the model prior to presenting the results. Equation (2) defines the (spatially and temporally heterogeneous) SEFs for the medial and adventitial layers of the aneurysmal tissue. An AAA develops as the material constituents of the artery evolve. The variation of the total potential energy (1) governs the equilibrium displacement field, and is solved by the finite element method [15]; volume meshes of the luminal computational domain are generated and the haemodynamics are solved using finite volumes [35]. In many aneurysms, diameter enlargement is asymmetric, with primarily anterior protrusion; the posterior region is often constrained from radial expansion by the adjacent spinal column [38]. To simulate this, we follow Watton et al. [15] and model the spine as a stiff spring-backed plate [15]. Where the aneurysm wall penetrates the plate, a penalty pressure acts normal to the aneurysm. The effective pressure acting on the membrane, is given by the difference of internal physiological pressure and the penalty pressure. For further details, see [43]. Two models of AAA evolution are illustrated: case (i) prescribed elastin degradation; case (ii) the degradation of elastin is driven by low WSS. In both examples, as the geometry evolves, the collagen fabric adapts (throughout the arterial domain) to restore its strain to the homeostatic value [see (17) and (21)].

3.1 Case (i) Prescribed Elastin Degradation

First we illustrate a model of AAA evolution using a prescribed degradation of elastin. As the aneurysm evolves (see Fig. 3) it can be seen that it develops a preferential anterior bulging and becomes tortuous. Figure 3a illustrates the (prescribed) evolution of elastin concentration. In the central region of the AAA, m^E reduces to 0.05 whereas in the neck regions it is approximately 1. The average collagen concentration (see Fig. 3b) increases to compensate for loss of load borne by the elastin and the increased load acting on the wall (due to the enlargement of the geometry).

Figure 3c and d depict the evolution of the elastin Green-Lagrange (GL) strains E_{11} and E_{22}, respectively. The strains are defined with respect to the unloaded configuration, consequently they continue to increase with the enlargement of the geometry: at $t = 0$, $E_{11} = 0.345$ and $E_{22} = .281$ whereas at $t = 10$, maximum values of $E_{11} = 5.1$ and $E_{22} = 4.7$ occur in the central region of the aneurysm. Notice that E_{11} decreases in the proximal section of parent artery as the aneurysm enlarges; the axial expansion of the AAA is accentuated due to the axial retraction of the ends of the artery.

The evolution of the collagen fibre GL strains differs substantially from that of the elastin strains due to the evolution of the reference configurations that the fibres are recruited to load bearing; recall this is achieved by remodelling the fibre recruitment stretches [see (17)] which define the factor the tissue must be stretched in the direction of a fibre (relative to the unloaded reference configuration) for it to begin to bear load. Figure 3e depicts the evolution of the medial collagen GL strain E^C_{M+}. At $t = 0$, E^C_{M+} is constant throughout the domain with a magnitude of the attachment strain, i.e. $E^C_{M+} = E^C_{AT} = 0.073$. As the AAA enlarges, the collagen strains initially increase (at $t = 4$, maximum values of 0.14 occur) and then reduce as the aneurysm stabilises in size and the collagen fabric remodels to a material equilibrium, i.e. $E^C_{M+} \rightarrow E^C_{AT}$ throughout the domain. Similar results are observed for evolution of the adventitial collagen fibre GL strain E^C_{A+} (Fig. 3f).

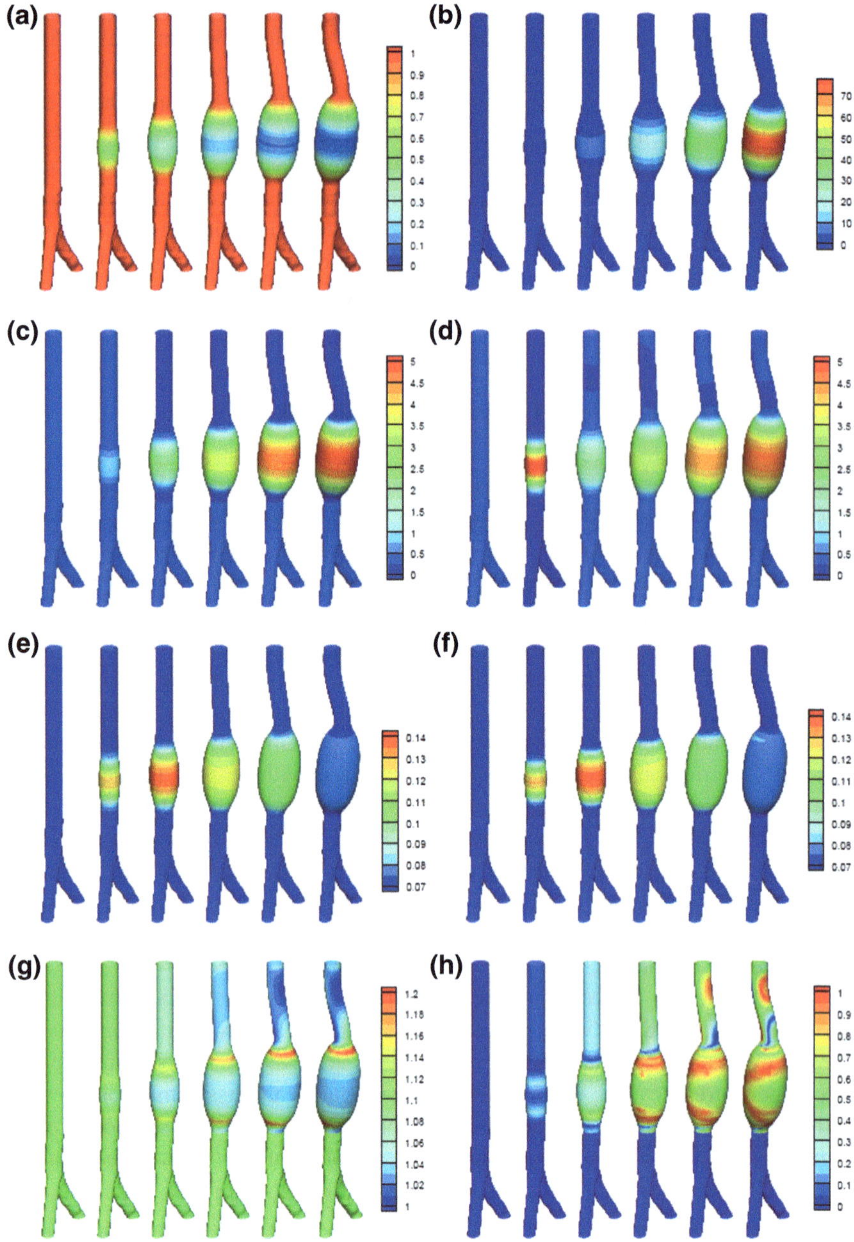

Fig. 3 Evolution, at $t = 0;2;4;6;8;10$ years, of spatial distributions of: prescribed elastin concentration mE (**a**); average collagen concentration m^C (**b**); elastin Green-Lagrange strains E11 (**c**) and E22 (**d**); medial and adventitial collagen Green-Lagrange strain E^C_{M+} (**e**) and E^C_{A+} (**f**); cyclic areal stretch ACS (**g**) and Biaxial Stretch Index cBSI (**h**)

Figure 3g illustrates the evolution of the cyclic areal stretch A^{CS}. At $t = 0, A^{CS} = 1.1$ throughout the domain, i.e. the ratio of systolic to diastolic diameters is 1.1. As the elastin degrades and the collagen, which has greater mechanical nonlinearity, takes over the load bearing, the magnitude of A^{CS} decreases, i.e. to approximately 1.04 in the AAA region. However, note that A^{CS} is slightly elevated in the proximal and distal necks of the aneurysm where $A^{CS} \approx 1.2$. Of note, the cyclic areal stretch also decreases in the proximal section of healthy, albeit tortuous artery. This is a consequence of maximum axial strains occurring in the diastolic configuration in this section of artery, i.e. the axial strains decrease between diastole and systole. Figure 3h illustrates the evolution of the biaxial stretch index χ^{BSI}. Initially, $\chi^{BSI} = 0$ throughout the domain: this indicates that the tissue is subject to 1D cyclic deformation (cyclic circumferential stretching). As the AAA enlarges, the distribution becomes more complex. At $t = 10$, the anterior region of the AAA has $\chi^{BSI} \approx 0.6$ and two regions around the aneurysm are subject to equi-biaxial stretching ($\chi^{BSI} \approx 1$). A more complex distribution of χ^{BSI} evolves in the proximal section of artery due its tortuosity.

3.2 Case (ii) Elastin Degradation Linked to Low WSS

In this example, a first-stage prescribed axisymmetric degradation of elastin (with minimum elastin concentration $m^E = 0.5$ at $t = 5$) gives rise to a small axisymmetric aneurysm. At $t = 5$, the artery is in mechanical and material equilibrium, i.e. the collagen fabric has equilibrium strains E^C_{AT} throughout the computational domain. Consequently, if there is no further loss of elastin, there is no further adaption of the collagen fabric [via the evolution equations (17) and (21)] and the aneurysm is stable in size. However, the small prescribed axisymmetric aneurysm creates a perturbation to the WSS field (see Fig. 4a): at $t = 5$ the WSS ranges from 0.02 to 0.1 Pa within the aneurysm with lower values occurring towards the proximal end; a slight elevation of the WSS is observed distal to the aneurysm. Thus, at a second stage, subsequent degradation of elastin can be linked to the haemodynamic environment: for $t > 5$, elastin degradation is linked to deviations of the WSS from normotensive values. Given the functional relationship between the degradation factor $\mathscr{F}_G$ and the magnitude of WSS [see (16)], rates of degradation are elevated towards the proximal end of the aneurysm (see Fig. 4b). Hence the elastin concentration m^E, which is initially axisymmetric, develops an axial asymmetry (Fig. 4c); minimum value of elastin concentration $m^E(t = 9) = 0.08$. Notably, the aneurysm enlarges more quickly in the upstream direction. Figure 4d illustrates the evolution of the WSSG. As the aneurysm enlarges, elevated values of WSSG are observed in the proximal (200 Pa/m) and distal necks (250 Pa/m).

Figure 5a, b illustrate the evolution of the GL strains E_{11} and E_{22} of the elastin, respectively. The strains increase from $E_{11} = 0.345$ and $E_{22} = .281$ at $t = 0$ to $E_{11} = 2.5$ and $E_{22} = 2.5$ at $t = 9$. Notice that there is a slight asymmetry in the

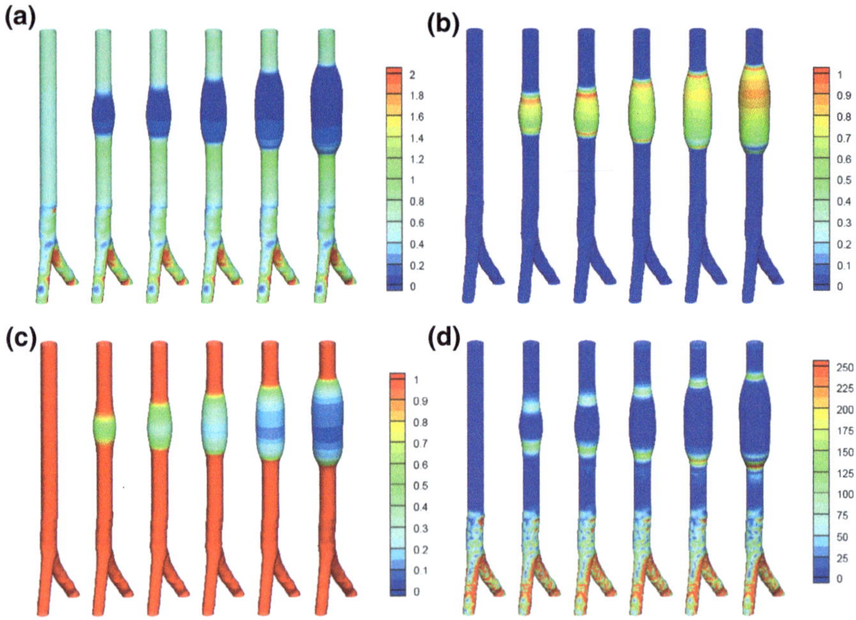

Fig. 4 Evolution of wall shear stress τ (**a**), degradation factor F^D (**b**) elastin concentration m^E (**c**) and WSSG (**d**) at $t = 0, 5, 6, 7, 8, 9$ years

spatial distribution within the aneurysm, slightly larger magnitudes are observed towards the proximal end. Also note that as with case (i), the axial strains in the parent artery decrease as the aneurysm enlarges, i.e. the axial retraction of the ends of the artery assists the axial expansion of the AAA. Figure 5c illustrates the evolution of the GL strains of positively wound collagen fibres in the media $E^C_{M_+}$. At $t = 0$, the collagen fabric is in material equilibrium, i.e. $E^C_{J_p} = E^C_{AT}$ $(J = M, A; p = \pm)$ throughout the domain. Note that at $t = 5$, even though the geometry has changed (and the elastin strains have increased) the collagen fabric is in homeostasis, i.e. $E^C_{J_p} = E^C_{AT}$; the recruitment stretches and fibre concentration have evolved to restore material equilibrium in the collagen fabric. As the elastin degrades and aneurysm enlarges ($t > 5$), the magnitude of the collagen strains increase; notice though that the increases are small relative to the large deformation that is occurring—a consequence of the remodelling of the reference configurations that the fibres are recruited to load bearing. The average collagen concentration m^C increases to compensate for the loss of load borne by the elastin (Fig. 5d).

Lastly, we consider the evolution of the cyclic stretch environment. Initially the cyclic areal stretch is equal to 1.1 throughout the domain. At $t = 5$ slightly elevated cyclic areal stretches ($A^{CS} = 1.12$) are present in the proximal and distal necks of the aneurysm (see Fig. 5e). As the elastin degrades and the collagen takes over the load bearing, the cyclic areal stretch reduces to 1.04 within the aneurysm region and elevated values are observed in the aneurysm neck (1.08). The cyclic stretch environment evolves from uniaxial ($t = 0$) to almost equi-biaxial cyclic stretching, i.e. in the aneurysm region $\chi^{BSI} = 0.7$ (see Fig. 5f). Transition regions occur in the upstream and downstream necks where the cyclic stretch environment changes from almost uniaxial ($\chi^{BSI} = 0.03$) to equi-biaxial ($\chi^{BSI} = 0.83$). As with case (i), as the aneurysm evolves, the proximal and distal regions of the artery experience biaxial stretching. This is a consequence of proximal/distal regions of the artery developing cyclic axial strain due to the axial expansion of the aneurysm during the cardiac cycle.

4 Discussion

We have presented a fluid-solid-growth (FSG) model for evolution of AAA. The model of the arterial wall accounts for the structural arrangement of collagen fibres in the medial and adventitial layers, the natural reference configuration that collagen fibres are recruited to load bearing and the concentrations (normalised mass-densities) of the load bearing constituents. To simulate the development of AAA we adopted two approaches: (i) we prescribed an axisymmetric degradation of elastin; (ii) we linked degradation of elastin to local haemodynamic stimuli, i.e. low WSS. In both examples, as the elastin degrades, the collagen fabric adapts (via G&R) to restore its strain to the attachment strain E^C_{AT}. The reference configurations of the collagen fibres evolve to simulate the effect of fibre deposition (with fibres attaching in a state of strain E^C_{AT}) and fibre degradation in altered configurations as the aneurysm enlarges; this simulates *remodelling*. The concentration of collagen fibres evolves to compensate for the loss of load borne by the elastin; this simulates *growth*. In the first example, we illustrated a AAA that stabilises in size and develops tortuosity; to our knowledge, this is the first model of AAA evolution to predict the formation of tortuosity. In the second example, we illustrated that linking elastin degradation to low WSS predicts the evolution of enlarging fusiform aneurysms.

Computational fluid dynamic (CFD) studies of aneurysms often emphasise the role of WSS (or WSSG) on pathogenesis of the disease, sometimes extrapolating conclusions from other conditions, namely atherosclerosis However, in vivo, vascular cells are also subject to cyclic stretching due to the pulsatility of the blood pressure. Cell functionality [63] and vascular homeostasis [64, 65] are influenced by cyclic stretching. Hence, to address the G&R of the tissue that occurs during aneurysm evolution, in addition to quantifying the haemodynamic stimuli that act on the ECs, it is important to quantify the cyclic stretch environment of the vascular cells. In this paper we propose a novel FSG computational framework: rigid walls for the purpose of the CFD analysis combined with a quasi-static analysis to determine the cyclic deformation of the arterial wall. Moreover, we extended the existing G&R framework utilised to model AAA evolution [15, 16]

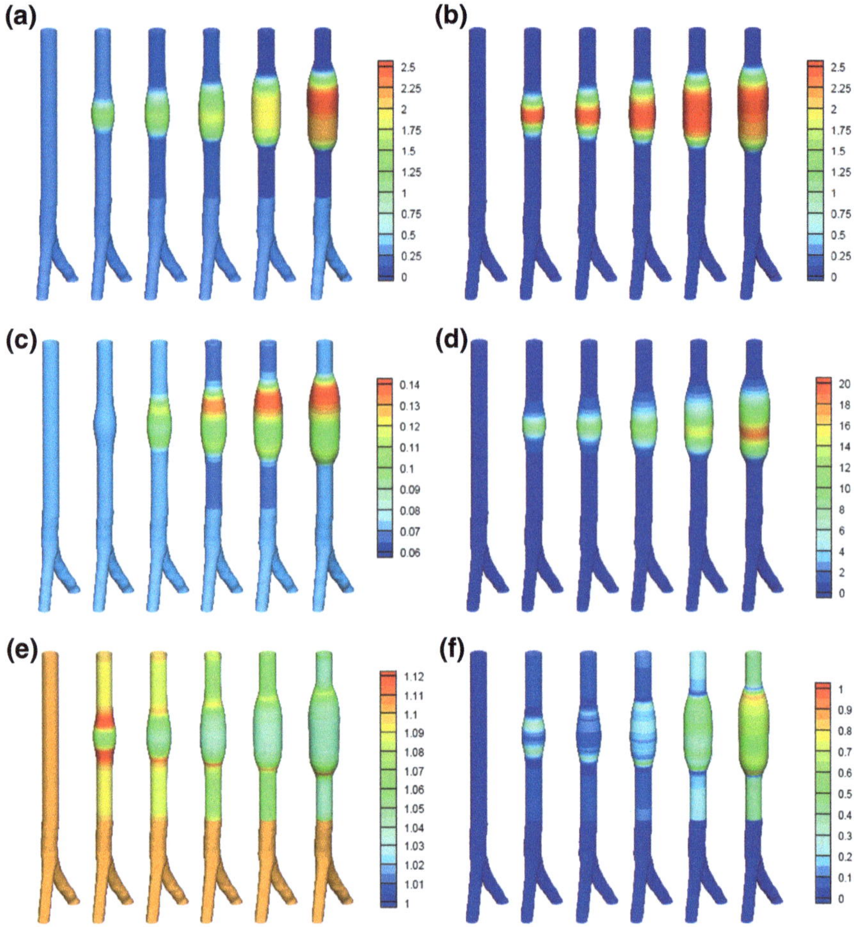

Fig. 5 Evolution of Elastin Green-Lagrange strains E_{11} (**a**) and E_{22} (**b**), medial collagen Green-Lagrange strain E^{C}_{M+} (**c**), average collagen concentration m^{C} (**d**), cyclic areal stretch A^{CS} (**e**) and the Biaxial Stretch Index χ^{BSI} (**f**) at $t = 0, 5, 6, 7, 8, 9$ years

and IA evolution [33–37, 66, 67] to link both *growth* and *remodelling* to cyclic deformation of vascular cells.

Increased tortuosity is associated with disease or vessel ageing [68]. Expanding AAAs develop specific, quantifiable shapes that can be expressed as a quantitative tortuosity index that may be relevant to their natural history [69]. Moreover, it is suggested that centreline tortuosity may become a useful addition to maximum diameter in the decision-making process of AAA treatment [70]. In fact, Shum et al. [71] analysed the tortuosity of 10 ruptured and 66 unruptured AAAs and observed that the mean tortuosity of the ruptured aneurysms (1.33 ± 0.12) was greater than the mean tortuosity of the unruptured aneurysms (1.23 ± 0.11). The mechanism for generating tortuosity is incompletely understood [72]. However, it appears to be partly attributed to failure of elastin: in an experimental study on canine carotid arteries, Dobrin et al. [73] observed that degradation of elastin caused aneurysmal dilatation and a marked decrease in longitudinal retractive force which permitted the development of tortuosity; failure of collagen causes vessels to rupture but it was observed not to facilitate the development of tortuosity. These findings are consistent with our computational model: we observed that elastin degradation gave rise to aneurysmal expansion and a reduced axial stretch of the proximal section of artery; coupling with a model of spinal contact and collagen G&R, tortuosity naturally develops. Hence, we suggest tortuosity is influenced by several factors: elastin degradation, remodelling of collagen and interaction with perivascular structures.

The structural arrangement of the cells is influenced by their local mechanical environment. For instance, ECs align with the mean direction of flow and if they are grown on a deformable substrate and subjected to cyclic uniaxial stretching, they reorient away from the stretching direction [74, 75]; in fact, it is observed that the alignment of ECs subjected to WSS appears to be significantly enhanced by the addition of cyclic stretching [76]. VSMCs align in the direction of cyclic stretching and it is observed that in 2D monolayers they align perpendicular to the applied haemodynamic WSS [77]. Fibroblasts align perpendicular to the direction of interstitial flow [78] and their orientation may be influenced by the direction of maximum principal stress or strain [79] or cyclic stretch [80]. Recently, a novel parameter was proposed to characterise the biaxial cyclic stretch environment [36], i.e. the biaxial stretch index χ^{BSI}: $\chi^{BSI} = 0$ denotes 1D cyclic stretch and $\chi^{BSI} = 1$ denotes equi-biaxial stretching. In this study, it was observed that as the AAA evolved, the BSI distribution changed significantly: within the aneurysm, the tissue experiences almost equi-biaxial cyclic stretching; however, in certain regions of the tissue, e.g. the necks, rapid transition regions were observed. If it is assumed that the AAA should evolve to achieve a particular distribution of χ^{BSI} then analysing the evolving distributions of χ^{BSI} could be used to test competing hypotheses for remodelling of cell/fibre alignment and dispersion. Elucidating the relationship between the mechanical environment of vascular cells and their orientation and functionality will pave the way for more sophisticated computational models.

Following Watton et al. [15] we do not explicitly remodel the collagen fibre orientations. However, it is important to appreciate that the collagen fibre orientations are defined relative to the undeformed reference configuration. Consequently, the fibres reorientate in the loaded configuration towards directions of increasing principal stretch as the aneurysm evolves; Zeinali-Davarani et al [17] use an identical approach, i.e. in the vivo configuration, newly deposited collagen fibres align along the direction of the exisiting collagen fibres. There is some justification for this approach. Fibroblasts crawl along the existing ECM matrix thus the orientation in which they deposit and degrade collagen is partly dependent on the existing ECM structure [81]. However, given that as an AAA evolves, the collagen fabric plays an increasingly important load bearing role, it is important to simulate how the collagen structure adapts to predict stress distributions. Implementing more sophisticated models to represent the remodelling of fibre alignment [82–84], collagen fibril distribution [85], fibre dispersion [86] or proteoglycan cross-bridges [87] may prove useful in this respect. Furthermore, given that VSMCs secrete connective tissue and matrix degrading enzymes [88] and are subject to apoptosis during AAA evolution [89], explicitly modelling the VSM cells with a suitable constitutive model [90–92] is needed to better understand the aetiology of AAA evolution.

We modelled the abdominal aorta as a cylindrical nonlinearly elastic membrane subject to an axial pre-stretch and uniform internal pressure. The distal and proximal ends of the abdominal aorta are fixed to simulate vascular tethering by the renal and iliac arteries. Formation and development of AAA is assumed to be a consequence of the material constituents of the artery remodelling. At physiological pressures the radius of a typical abdominal aorta is approximately 10 mm and the thickness is 1 mm. Neglecting thrombus formation, the ratio of the thickness of the wall to the diameter of the AAA will decrease as the aneurysm enlarges, therefore the deformation of the three dimensional arterial wall of the developing aneurysm is closely related to the deformation of its midplane. The residual strain that is present in the unloaded configuration gives rise to an approximately uniform strain field through the thickness of the arterial wall at physiological pressures. If it is assumed that the physiological mechanism by which collagen fibres attach to the artery is independent of both the current configuration of the artery and the radial position in the arterial wall, then the remodelling process may naturally maintain a uniform strain field (in the collagen fibres) through the thickness of the arterial wall as the AAA develops. These considerations thus support the suitability of a membrane model to model the development of a AAA at physiological pressures. Nevertheless, the G&R framework of Watton et al. [15] has recently been extended to consider transmural variations of G&R for a thick-walled model of the artery [66, 67]: the influence of transmural variations in biochemomechanical stimuli on G&R and AAA evolution will be explored in future studies.

We modelled the healthy abdominal aorta as cylindrical. However, in reality the abdominal aorta is slightly tortuous and tapers. Although, most aneurysm evolution models to date have used conceptual geometrical models for the healthy artery, recently, Zeinali-Davarani et al. [17] applied a stress-mediated constrained mixture FEM approach [93] to model AAA evolution [18] using patient-specific geometries with a nonlinear membrane formulation [65, 94]. Modelling the exact geometry of the healthy abdominal aorta would yield physiologically realistic spatial distributions of haemodynamic stimuli and thus may enable more accurate prediction of the evolution of AAA geometries. However, given that the abdominal aorta is almost cylindrical, hypotheses can be explored and insight obtained with (simpler) non-patient specific geometries. For instance, we observed that the linking elastin degradation to low WSS gives rise to enlarging fusiform aneurysms.

We adopted a steady flow analysis to reduce the cost of the computational simulations. The similarity of the spatial WSS/WSSG distributions for steady and pulsatile flow [95–97] implies that this is a reasonable approach for the purposes of investigating phenomenological hypotheses that explore the link between G&R and deviations of the WSS/WSSG from homeostatic levels [36]. However, temporal changes in WSS distribution during the cardiac cycle affect the functionality of ECs: oscillatory shear is elevated in regions of disturbed flow and is associated with proatherogenic patterns of gene expression [98, 99]. A pulsatile flow analysis yields additional quantification of the haemodynamic stimuli that act on ECs and thus is necessary to explore more sophisticated G&R hypotheses related to the haemodynamics.

Approximately 75 % of AAAs have an associated intraluminal thrombus (ILT) [100]. ILT alters the stress distribution and reduces peak wall stress in AAA. However, presence of ILT leads to regional wall weakening [101]: platelet activation, fibrin formation, binding of plasminogen and its activators, and trapping of erythrocytes and neutrophils, leads to oxidative and proteolytic injury of the arterial wall [102]. Consequently, ILT must play an important role in the aetiology of AAA and thus the biochemomechanical roles of ILT must be understood and modelled better [103]. However, incorporating a model for ILT evolution is challenging; it would require integration of a model for thrombus evolution [104–106] combined with a constitutive model describing its mechanical response [107–109]; this may merit investigation with conceptual mathematical models with simpler geometries first. We note also that our model does not include calcification of the arterial wall which are often present in AAAs [110]. Calcifications influence stress distributions [111] and thus may influence G&R and AAA evolution [112].

5 Conclusion

In order to sophisticate computational models to more accurately represent mechanobiology, guidance is needed from experiments. In return, computational models assist in the interpretation of experimental data and in the identification of questions that need to be addressed by experiments. Moreover, they can play a vital role in guiding our understanding of mechanobiology: they serve as an in silico testbed for exploration of hypotheses; enable underlying mechanisms to be evaluated with incomplete data sets and yield insight which would be impossible from in vitro/vivo experimental set-up alone. We have extended the FSG model proposed by [35] to link both *growth* and *remodelling* to cyclic deformation of vascular cells and applied it to simulate the evolution of abdominal aortic aneurysm. The model predicts an aneurysm that evolves with similar mechanical, biological and morphological properties with those observed in vivo. Whilst in need of further sophistications to more accurately reflect the underlying mechanobiology, this computational framework has clear potential to be applied to aid the design and optimisation of tissue engineered vascular constructs.

Acknowledgments Paul Watton is funded by The Centre of Excellence in Personalized Healthcare (funded by the Wellcome Trust and EPSRC, grant number WT 088877/Z/09/Z). This support is gratefully acknowledged.

References

1. Barron, V., Lyons, E., Stenson-Cox, C., McHugh, P.E., Pandit, A.: Bioreactors for cardiovascular cell and tissue growth: a review. Ann. Biomed. Eng. **31**, 1017–1030 (2003)
2. Geris, L.: In vivo, in vitro, in silico: computational tools for product and process design in tissue engineering. Springer, Heidelberg (2012)
3. Vorp, D.A.: Review: Biomechanics of abdominal aortic aneurysm. J. Biomech. **40**, 1887–1902 (2007)
4. Sakalihasan, N., Kuivaniemi, H., Nusgens, B., Durieux, R., Defraigne, J.G.: Aneurysm: epidemiology aetiology and pathophysiology, biomechanics and mechanobiology of aneurysms. Springer, Heidelberg (2011)
5. Baxter, B.T., Terrin, M.C., Dalman, R.L.: Medical management of small abdominal aortic aneurysms. Circulation **117**, 1883 (2008)
6. Davies, M.J.: Aortic aneurysm formation: lessons from human studies and experimental models. Circulation **98**, 193–195 (1998)
7. Wilmink, W.B.M., Quick, C.R.G., Hubbard, C.S., Day, N.E.: The influence of screening on the incidence of ruptured abdominal aortic aneurysms. J. Vasc. Surg. **30**, 203–208 (1999)
8. Carrell, T.W.G., Smith, A., Burnand, K.G.: Experimental techniques and models in the study of the development and treatment of abdominal aortic aneurysms. J. Surg. **86**, 305–312 (1999)
9. Raghavan, M.L, Vorp, D.A.: Toward a biomechanical tool to evaluate rupture potential of abdominal aortic aneurysm: identification of a finite strain constitutive model and evaluation of its applicability. J. Biomech. **33**, 475–482 (2000)
10. Powell, J.T, Brady, A.R.: Detection, management and prospects for the medical treatment of small abdominal aortic aneurysms. Arteriosclerosis Thromb. Vasc. Biol. **24**, 241–245 (2004)
11. Powell, J.T., Gotensparre, S.M., Sweeting, M.J., Brown, L.C., Fowkes, F.G., Thompson, S.G.: Rupture rates of small abdominal aortic aneurysms: a systematic review of the literature. Eur. J. Vasc. Endovasc. Surg. **41**, 2–10 (2011)
12. Darling, R.C., Messina, C.R., Brewster, D.C., Ottinger, L.W.: Autopsy study of unoperated abdominal aortic aneurysms: the case for early resection. Circulation **56**, 161–164 (1977)
13. Humphrey, J.D, Taylor, C.A.: Intracranial and abdominal aortic aneurysms: Similarities, differences, and need for a new class of computational models. Ann. Rev. Biomed. Eng. **10**, 221–246 (2008)
14. O'Connell, M.K., Murthy, S., Phan, S., Xu, C., Buchanan, J., Spilker, R., Dalman, R.L., Zarins, C.K., Denk, W., Taylor, C.A.: The three-dimensional micro- and nanostructure of the aortic medial lamellar unit measured using 3D confocal and electron microscopy imaging. Matrix Biology **27**, 171–181 (2008)
15. Watton, P.N., Hill, N.A., Heil, M.: A mathematical model for the growth of the abdominal aortic aneurysm. Biomech. Model. Mechanobiol. **3**, 98–113 (2004)
16. Watton, P.N., Hill, N.A.: Evolving mechanical properties of a model of abdominal aortic aneurysm. Biomech. Model. Mechanobiol. **8**, 25–42 (2009)
17. Zeinali-Davarani, S., Sheidaei, A., Baek, S.: A finite element model of stress-mediated vascular adaptation: application to abdominal aortic aneurysms. Comput. Methods Biomech. Biomed. Eng. **14**, 803–817 (2011)
18. Sheidaei, A., Hunley, S.C., Zeinali-Davarani, S., Raguin, L.G., Baek, S.: Simulation of abdominal aortic aneurysm growth with updating hemodynamic loads using a realistic geometry. Med. Eng. Phys. **33**, 80–88 (2011)

19. Chiquet, M.: Regulation of extracellular matrix gene expression by mechanical stress. Matrix Biology **18**, 417–426 (1999)
20. Armentano, R., Barra, J., Levenson, J., Simon, A., Pichel, R.: Arterial wall mechanics in conscious dogs: assessment of viscous, inertial and elastic moduli to characterize aortic wall behaviour. Circ. Res. **76**, 468–478 (1995)
21. Shadwick, R.: Mechanical design in arteries. J. Exp. Biology **202**, 3305–3313 (1999)
22. Raghavan, M.L., Webster, M., Vorp, D.A.: Ex-vivo bio-mechanical behavior of AAA: assessment using a new mathematical model. Ann. Biomed. Eng. **24**, 573–582 (1999)
23. Kakisis, J.D., Liapis, C.D., Sumpio, B.E.: Effects of cyclic strain on vascular cells. Endothelium **11**, 17–28 (2004)
24. Gleason, R.L, Humphrey, J.D.: A 2d constrained mixture model for arterial adaptations to large changes in flow, pressure and axial stretch. Math. Med. Biology **22**, 347–369 (2005)
25. Gupta, V., Grande-Allen, K.J.: Effects of static and cyclic loading in regulating extracellular matrix synthesis by cardiovascular cells. Cardiovasc. Res. **72**, 375–383 (2006)
26. Alberts B., Bray D., Lewis J., Raff M., Roberts K., Watson J.D., (1994) Molecular biology of the cell, 4th edn. Garland Publishing, New York
27. McAnulty, R.J.: Fibroblasts and myofibroblasts: their source, function and role in disease. Int. J. Biochem. Cell Biology **39**, 666–671 (2007)
28. He, C.M, Roach, M.: The composition and mechanical properties of abdominal aortic aneurysms. J. Vasc. Surg. **20**, 6–13 (1993)
29. Shimizu, K., Mitchell, R.N., Libby, P.: Inflammation and cellular immune responses in abdominal aortic aneurysms. Arteriosclerosis Thromb. Vasc. Biology **26**, 987–994 (2006)
30. Nissen, R., Cardinale, G.J., Udenfriend, S.: Increased turnover of arterial collagen in hypertensive rats. Proc. Natl. Acad. Sci. U.S.A. Med. Sci. **75**, 451–453 (1978)
31. Arribas, S.M., Hinek, A., Gonzalez, M.C.: Elastic fibres and vascular structure in hypertension. Pharm. Ther. **111**, 771–791 (2006)
32. Holzapfel, G.A., Gasser, T.C., Ogden, R.W.: A new constitutive framework for arterial wall mechanics and a comparative study of material models. J. Elast. **61**, 1–48 (2000)
33. Watton, P.N., Ventikos, Y., Holzapfel, G.A.: Modelling the growth and stabilisation of cerebral aneurysms. Math. Med. Biology **26**, 133–164 (2009)
34. Watton, P.N, Ventikos, Y.: Modelling evolution of saccular cerebral aneurysms. J. Strain Anal. **44**, 375–389 (2009)
35. Watton, P.N., Raberger, N.B., Holzapfel, G.A., Ventikos, Y.: Coupling the hemodynamic environment to the evolution of cerebral aneurysms: computational framework and numerical examples. ASME J. Biomech. Eng. **131**, 101003 (2009)
36. Watton, P.N., Selimovic, A., Raberger, N.B., Huang, P., Holzapfel, G.A., Ventikos, Y.: Modelling evolution and the evolving mechanical environment of saccular cerebral aneurysms. Biomech. Model. Mechanobiol. **11**, 109–132 (2011)
37. Watton, P.N., Ventikos, Y., Holzapfel, G.A.: Modelling cerebral aneurysm evolution, biomechanics and mechanobiology of aneurysms. Springer, Heidelberg (2011)
38. Dua, M.M, Dalman, R.L.: Hemodynamic influences on abdominal aortic aneurysm disease: application of biomechanics to aneurysm pathophysiology. Vasc. Pharm. **53**, 11–21 (2010)
39. Nakahashi, T.K., Tsao, K.H.P.S., Sho, E., Sho, M., Karwowski, J.K., Yeh, C., Yang, R.B., Topper, J.N., Dalman, R.L.: Flow loading inducesmacrophage antioxidative gene expression in experimental aneurysms. Arteriosclerosis Thromb. Vasc. Biology **22**, 2017–2022 (2002)
40. Hoshina, K., Sho, E., Sho, M., Nakahashi, T.K., Dalman, R.L.: Wall shear stress and strain modulate experimental aneurysm cellularity. J. Vasc. Surg. **37**, 1067–1074 (2003)
41. Sho, E., Sho, M., Hoshina, K., et al.: Hemodynamic forces regulate mural macrophage infiltration in experimental aortic aneurysms. Exp. Mol. Pathol. **76**, 108–116 (2004)
42. Heil, M.: The stability of cylindrical shells conveying viscous flow. J. Fluids Struct. **10**, 173–196 (1996)
43. Watton, P.N.: Mathematical modelling of the abdominal aortic aneurysm. Ph.D. Thesis, Department of Mathematics, University of Leeds, Leeds, UK (2002)

44. Watton, P.N., Ventikos, Y., Holzapfel, G.A.: Modelling the mechanical response of elastin for arterial tissue. J. Biomech. **42**, 1320–1325 (2009)
45. Humphrey, J.D.: Remodelling of a collagenous tissue at fixed lengths. J. Biomech. Eng. **121**, 591–597 (1999)
46. Gruttmann, F., Taylor, R.L.: Theory and finite element formulation of rubberlike membrane shells using principal stretches. Int. J. Numer. Methods Eng. **35**, 1111–1126 (1992)
47. Huang, H.: Haemodynamics in diseased arteries: effects on plaque and aneurysm progression by advanced imaging and modelling techniques. Ph.D. Thesis, Department of Engineering Science, University of Oxford, Oxford, UK (2010)
48. Patakar, S.V.: Numerical heat transfer and fluid flow. Hemisphere Publishing Corporation, Washington – New York – London. McGraw Hill Book Company, New York (1980)
49. Ferziger J.H, Peric M., (2002) Computational methods for fluid dynamics, 3rd edn. Springer, Heidelberg
50. Hutchinson, B.R, Raithby, G.D.: A multigrid method based on the additive correction strategy. Numer. Heat Transf. **9**, 511–537 (1986)
51. Cebral, J.R., Castro, M.A., Appanaboyina, S., Putman, C.M., Millan, D., Frangi, A.F.: Efficient pipeline for image-based patient-specific analysis of cerebral aneurysm hemodynamics: technique and sensitivity. IEEE Trans. Med. Imaging **24**, 457–467 (2005)
52. Fisher, C., Rossmann, J.S.: Effect of non-newtonian behavior on hemodynamics of cerebral aneurysms. ASME J. Biomech. Eng. **131**, 091004 (2009)
53. Chatziprodromou, I., Tricoli, A., Poulikakos, D., Ventikos, Y.: Haemodynamics and wall remodelling of a growing cerebral aneurysm: a computational model. J. Biomech. **40**, 412–426 (2007)
54. Oshima, M., Torii, R., Kobayashia, T., Taniguchic, N., Takagid, K.: Finite element simulation of blood flow in the cerebral artery. Comput. Methods Appl. Mech. Eng. **191**, 661–671 (2001)
55. Reymond, P., Merenda, F., Perren, F., Rufenacht, D., Stergiopulos, N.: Validation of a one-dimensional model of the systemic arterial tree. Am. J. Physiol. Heart Circ. Physiol. **297**, H208–H222 (2009)
56. Villa-Uriol M.C., Berti G., Hose D.R., Marzo A., Chiarini A., Penrose J., Pozo J., Schmidt J.G., Singh P., Lycett R., Larrabide I., Frangi A.F.: @neurist complex information processing toolchain for the integrated management of cerebral aneurysms. Interface Focus **1**, 308–319 (2011)
57. Reymond P., Bohraus Y., Perren F., Lazeyras F., Stergiopulos N., (2011) Validation of a patient-specific one-dimensional model of the systemic arterial tree. Am. J. Physiol. Heart Circ. Physiol. **301**, H1173–H1182
58. Wang, J.H.C, Thampaty, B.P.: An introductory review of cell mechanobiology. Biomech. Model. Mechanobiol. **5**, 1–16 (2006)
59. Chiquet, M., Renedo, A.S., Huber, F., Flück, M.: How do fibroblasts translate mechanical signals into changes in extracellular matrix production?. Matrix Biology **22**, 73–80 (2003)
60. Sotoudeh, M., Jalali, S., Usami, S., Shyy, J.Y., Chien, S.: A strain device imposing dynamic and uniform equi-biaxial strain to cultured cells. Ann. Biomed. Eng. **26**, 181–189 (1998)
61. Shin, H.Y., Gerritsen, M.E., Bizios, R.: Regulation of endothelial cell proliferation and apoptosis by cyclic pressure. Ann. Biomed. Eng. **30**, 297–304 (2002)
62. Länne, T., Sonesson, B., Bergqvist, D., Bengtsson, H., Gustafsson, D.: Diameter and compliance in the male human abdominal aorta: influence of age and aortic aneurysm. Eur. J. Vasc. Surg. **6**, 178–184 (1992)
63. Cummins, P.M., von Offenberg Sweeney, N., Killeen, M.T., Birney, Y.A., Redmond, E.M., Cahill, P.A.: Cyclic strain-mediated matrix metalloproteinase regulation within the vascular endothelium: a force to be reckoned with. Am. J. Physiol. Heart Circ. Physiol. **292**, H28–H42 (2007)
64. Hsiai, T.K.: Mechanical transduction coupling between endothelial and smooth muscle cells: role of hemodynamic forces. Am. J. Physiol. Cell Physiol. **294**, C695–C661 (2008)
65. Zeinali-Davarani, S., Raguin, L.G., Baek, S.: An inverse optimization approach toward testing different hypotheses of vascular homeostasis using image-based models. Int. J. Struct. Changes Solids Mech. Appl. **3**, 33–34 (2011)

66. Schmid, H., Watton, P.N., Maurer, M.M., Wimmer, J., Winkler, P., Wang, Y.K., Roehrle, O., Itskov, M.: Impact of transmural heterogeneities on arterial adaptation: application to aneurysm formation. Biomech. Model. Mechanobiol. **9**, 295–315 (2010)
67. Schmid, H., Grytsan, A., Postan, E., Watton, P.N., Itskov, M.: Influence of differing material properties in media and adventitia on arterial adaption: application to aneurysm formation and rupture. Comput. Methods Biomech. Biomed. Eng. (2011). doi:10.1080/10255842.2011.603309
68. Rodriguez, Z.M., Kenny, P., Gaynor, L.: Improved characterisation of aortic tortuosity. Med. Eng. Phys. **33**, 1712–1719 (2011)
69. Pappu, S., Dardik, A., Tagare, H., Gusberg, R.J.: Beyond fusiform and saccular: a novel quantitative tortuosity index may help classify aneurysm shape and predict aneurysm rupture potential. Ann. Vasc. Surg. **22**, 88–97 (2008)
70. Georgakarakos E., Ioannou C.V., Kamarianakis Y., Papaharilaou Y., Kostas T., Manousaki E., Katsamouris A.N.: The role of geometric parameters in the prediction of abdominal aortic aneurysm wall stress. Eur. J. Vasc. Endovasc. Surg. **39**, 42–48 (2010)
71. Shum, J., Martufi, G., Di Martino, E., Washington, C.B., Grisafi, J., Muluk, S.C., Finol, E.A.: Quantitiatve assessment of abdominal aortic aneurysm geometry. Ann. Biomed. Eng. **39**, 277–286 (2011)
72. Dougherty, G., Johnson, M.J.: Clinical validation of three-dimensional tortuosity metrics based on the minimum curvature of approximating polynomial splines. Med. Eng. Phys. **30**, 190–198 (2008)
73. Dobrin, P.B., Schwarz, T.H., Baker, W.H.: Mechanisms of arterial and aneurysmal tortuosity. Surgery **104**, 568–571 (1998)
74. Wang, J.H.C., Goldschmidt-Clermont, P., Yin, F.C.P.: Contractility affects stress fiber remodeling and reorientation of endothelial cells subjected to cyclic mechanical stretching. Ann. Biomed. Eng. **28**, 1165–1171 (2000)
75. Owatverot, T.B., Oswald, S.J., Chen, Y., WIllie, J.J.: Effect of combined cyclic stretch and fluid shear stress on endothelial cell morphological responses. ASME J. Biomech. Eng. **127**, 374–382 (2005)
76. Moore, J.E. Jr, Blirki, E., Sucui, A., Zhao, S., Burnier, M., Brunner, H.R., Meister, J.J.: A device for subjecting vascular endothelial cells to both fluid shear stress and circumferential cyclic stretch. Ann. Biomed. Eng. **22**, 416–422 (1994)
77. Lee, A.A., Graham, D.A., Dela Cruz, S., Ratcliffe, A., Karlon, W.J.: Fluid shear stress-induced alignment of cultured vascular smooth muscle cells. J. Biomech. Eng. **124**, 37–43 (2002)
78. Ng, C.P., Schwartz, M.A.: Mechanisms of interstitial flow-induced remodeling of fibroblast collagen cultures. Ann. Biomed. Eng. **34**, 446–454 (2006)
79. Wagenseil, J.E.: Cell orientation influences the biaxial mechanical properties of fibroblast populated collagen vessels. Ann. Biomed. Eng. **32**, 720–731 (2004)
80. Neidlinger-Wilke, C., Grood, E., Claes, L., Brand, R.: Fibroblast orientation to stretch begins within three hours. J. Orthop. Res. **20**, 953–956 (2002)
81. Dallon, J., Sherratt, J.A.: A mathematical model for fibroblast and collagen orientation. Bulletin Math. Biology **60**, 101–130 (1998)
82. Driessen, N.J.B., Wilson, W., Bouten, C.V.C., Baaijens, F.P.T.: A computational model for collagen fibre remodelling in the arterial wall. J. Theor. Biology **226**, 53–64 (2004)
83. Baaijens, F., Bouten, C., Driessen, N.: Modeling collagen remodeling. J. Biomech. **43**, 166–175 (2010)
84. Creane, A., Maher, E., Sultan, S., Hynes, N., Kelly, D.J., Lally, C.: A remodelling metric for angular fibre distributions and its application to diseased carotid bifurcations. Biomech. Model. Mechanobiol. **11**(6), 869–882 (2012). doi:10.1007/s10237-011-0358-3
85. Zulliger, M.A., Fridez, P., Hayashi, K., Stergiopulos, N.: A strain energy function for arteries accounting for wall composition and structure. J. Biomech. **37**, 989–1000 (2004)
86. Gasser, T.C., Ogden, R.W., Holzapfel, G.A.: Hyperelastic modelling of arterial layers with distributed collagen fibre orientations. J. Royal Soc. Interface **3**, 15–35 (2006)

87. Gasser, T.C.: An irreversible constitutive model for fibrous soft biological tissue: a 3-d microfiber approach with demonstrative application to abdominal aortic aneurysms. Acta Biomaterialia **7**, 2457–2466 (2011)
88. Asanuma, K., Magid, R., Johnson, C., Nerem, R.M., Galis, Z.S.: Uniaxial strain regulates matrix-degrading enzymes produced by human vascular smooth muscle cells. Am. J. Physiol. Heart Circ. Physiol. **284**, H1778–H1784 (2003)
89. Zhang, J., Schmidt, J., Ryschich, E., Schumacher, H., Allenberg, J.R.: Increased apoptosis and decreased density of medial smooth muscle cells in human abdominal aortic aneurysms. Chin. Med. J. **116**, 1549–1552 (2003)
90. Zulliger, M.A., Rachev, A., Stergiopulos, N.: A constitutive formulation of arterial mechanics including vascular smooth muscle tone. Am. J. Physiol. Heart Circ. Physiol. **287**, H1335–1343 (2004)
91. Baek, S., Valentín, A., Humphrey, J.D.: Biochemomechanics of cerebral vasospasm and its resolution: II constitutive relastions and model simulations. Ann. Biomed. Eng. **35**, 1498–1509 (2007)
92. Murtada, S., Kroon, M., Holzapfel, G.A.: A calcium-driven mechanochemical model for prediction of force generation in smooth muscle. Biomech. Model. Mechanobiol. **9**, 749–762 (2010)
93. Baek, S., Rajagopal, K.R., Humphrey, J.D.: A theoretical model of enlarging intracranial fusiform aneurysms. J. Biomech. Eng. **128**, 142–149 (2006)
94. Zeinali-Davarani, S., Raguin, L.G., Vorp, D.A., Baek, S.: Identification of in vivo material and geometric parameters of a human aorta: toward patient-specific modeling of abdominal aortic aneurysm. Biomech. Model. Mechanobiol. **10**, 689–699 (2010)
95. Myers, J.G., Moore, J.A., Ojha, M., Johnston, K.W., Ethier, C.R.: Factors influencing blood flow patterns in the human right coronary artery. Ann. Biomed. Eng. **29**, 109–120 (2001)
96. Mantha, A., Karmonik, C., Benndorf, G., Strother, C., Metcalfe, R.: Hemodynamics in a cerebral artery before and after the formation of an aneurysm. Am. J. Neuroradiol. **27**, 1113–1118 (2006)
97. Doenitz, C.M.S.K., Zoephel, R., Brawanski, A.: A mechanism for rapid development of intracranial aneurysms: a case study. Neurosurgery **67**, 1213–1221 (2010)
98. Chien, S.: Mechanotransduction and endothelial cell homeostasis: the wisdom of the cell. Am. J. Physiol. Heart Circ. Physiol. **292**, H1209–H1224 (2007)
99. Chatzizisis, Y.S., Coskun, A.U., Jonas, M., Edelman, E.R., Feldman, C.L., Stone, P.H.: Role of endothelial shear stress in the natural history of coronary atherosclerosis and vascular remodeling: molecular, cellular, and vascular behavior. J. Am. Coll. Cardiol. **49**, 2379–2393 (2007)
100. Wang, D.H.J., Makaroun, M.S., Webster, M.W., Vorp, D.A.: Effect of intraluminal thrombus on wall stress in patient-specific models of abdominal aortic aneurysms. J. Vasc. Surg. **36**, 598–604 (2002)
101. Vorp, D.A., Lee, P.C., Wang, D.H., Makaroun, M.S., Nemoto, E.M., Ogawa, S., Webster, M.W.: Association of intraluminal thrombus in abdominal aortic aneurysm with local hypoxia and wall weakening. J. Vasc. Surg. **34**, 291–299 (2001)
102. Michel, J.B., Martin-Ventura, J.L., Egido, J., Sakalihasan, N., Treska, V., Lindholt, J., Allaire, E., Thorsteinsdottir, U., Cockerill, G., Swedenborg, J.: Novel aspects of the pathogenesis of aneurysms of the abdominal aorta in humans. Cardiovasc. Res. **90**, 18–27 (2011)
103. Humphrey, J.D., Holzapfel, G.A.: Mechanics, mechanobiology, and modeling of human abdominal aorta and aneurysms. J. Biomech. **45**(5), 805–814 http://dx.doi.org/10.1016/j.jbiomech.2011.11.021
104. Xu, Z., Chen, N., Kamocka, M.M., Rosen, E.D., Alber, M.S.: Multiscale model of thrombus development. J. Royal Soc. Interface **5**, 705–722 (2008)
105. Xu, Z., Kamocka, M., Alber, M., Rosen, E.D.: Computational approaches to studying thrombus development. Arteriosclerosis Thromb. Vasc. Biology **31**, 500–505 (2011)
106. Biasetti, J., Hussain, F., Gasser, T.C.: Blood flow and coherent vortices in the normal and aneurysmatic aortas: a fluid dynamical approach to intra-luminal thrombus formation. J. Royal Soc. Interface **8**, 1449–1461 (2011)

107. Vande Geest, J.P., Sacks, M.S., Vorp, D.A.: A planar biaxial constitutive relation for the luminal layer of intra-luminal thrombus in abdominal aortic aneurysms. J. Biomech. **39**, 2347–2354 (2006)
108. van Dam, E.A., Dams, S.D., Peters, G.W.M., Rutten M., C.M., Schurink, G.W.H., Buth, J., van de Vosse, F.N.: Non-linear viscoelastic behavior of abdominal aortic aneurysm thrombus. Biomech. Model. Mechanobiol. **7**, 127–137 (2008)
109. Tong, J., Sommer, G., Regitnig, P., Holzapfel, G.A.: Dissection properties and mechanical strength of tissue components in human carotid bifurcations. Ann. Biomed. Eng. **39**, 1703–1719 (2011)
110. Sakalihasan, N., Michel, J.B.: Functional imaging of atherosclerosis to advance vascular biology. Eur. J. Vasc. Endovasc. Surg. **37**, 728–734 (2009)
111. Speelman, L., Bohra, A., Bosboom, E.M., Schurink, G.W., van de Vosse, F.N., Makaroun, M.S., Vorp, D.A.: Effects of wall calcifications in patient- specific wall stress analyses of abdominal aortic aneurysms. J. Biomech. Eng. **129**, 105–109 (2007)
112. Maier, A., Gee, M.W., Reeps, C., Eckstein, H.H., Wall, W.A.: Impact of calcifications on patient-specific wall stress analysis of abdominal aortic aneurysms. Biomech. Model. Mechanobiol. **9**, 511–521 (2010)

Computational Mechanobiology in Cartilage and Bone Tissue Engineering: From Cell Phenotype to Tissue Structure

Thomas Nagel and Daniel J. Kelly

Abstract This chapter gives a short overview of computational models in cartilage and bone tissue engineering with a focus on how mechanical cues can regulate tissue regeneration on multiple levels, from cell phenotype to tissue architecture. The chapter begins with a brief review of single cell models with a focus on cell-substrate interactions and cytoskeletal remodelling. After summarising a number of current theories for mechanoregulated tissue differentiation, we explain how such hypotheses can either be corroborated or rejected by attempting to simulate in vivo regenerative events. We then outline a recently introduced model for MSC differentiation based on substrate stiffness and oxygen tension as well as how tissue phenotype and organisation can be explored simultaneously within a computational model. The application of computational models to aid in the design of scaffolds for bone and cartilage repair is demonstrated. We also outline how such models can be used in the design and analysis of bioreactors, demonstrating how changes in tissue structure in response to mechanical loading during bioreactor culture can potentially impact the mechanical properties of the final engineered constructs. The chapter closes with a short overview of multiscale models with relevance to tissue engineering.

T. Nagel (✉) · D. J. Kelly
Centre for Bioengineering, Trinity Biomedical Sciences Institute,
Trinity College Dublin, 152–160 Pearse Street, Dublin 2, Ireland
e-mail: nagelt@tcd.ie

D. J. Kelly
e-mail: kellyd9@tcd.ie

T. Nagel · D. J. Kelly
Department of Mechanical and Manufacturing Engineering,
School of Engineering, Trinity College Dublin, Dublin 2, Ireland

Stud Mechanobiol Tissue Eng Biomater (2013) 10: 341–377

DOI: 10.1007/8415_2012_131

Published Online: 15 February 2012

1 Introduction

> "Classical science is a conversation between theory and experiment. A scientist can start at either end – with theory or experiment – but progress usually demands the union of both a theory to make sense of the experiments and data to verify the theory. Technological novelties such as computer models are neither here nor there. A really good dynamic computer model – of the global atmosphere, for example – is like a theory that throws off data, or data with a built-in theory. It's easy to see why such technological worlds are regarded with such wariness by science – they seem corrupted coming and going. But in fact, these models yield a third kind of truth, an experiential synthesis – a parallel existence, so to speak" [57].

The quantitative evaluation of any experiment is benefited by a model that defines a consistent set of parameters allowing comparisons to be made among results from different experiments. In practice, the experiments or systems in question quickly attain a high degree of complexity that is reflected in the models used to describe them. Intuitive understanding or even analytic solutions of such models are limited to a few and practically usually irrelevant cases. Computer technology in combination with numerical mathematics provides the means of solving complex models under practically relevant boundary conditions. Since the necessary tools have become widely available and affordable, engineers and decision makers in the industrial and public sectors have come to rely on quantitative computer analyses in order to develop products or assess risks more cost effectively and quickly by partially eliminating expensive experiments [86]. Cost-effectiveness and possible liabilities require thoroughly validated and verified models with a sufficient quantitative accuracy and their application domain usually has a high degree of overlap with their validation domain, the only exception being high-consequence systems where full-scale physical testing is never an option [86]. Models in contemporary bioengineering are at a different developmental stage with different modelling challenges taking precedence. The particular difficulties in the interpretation of experimental data related to cells and biological tissues are rooted in the large sample variabilities and the nonlinear, multiphasic, heterogeneous, anisotropic, viscoelastic and often active nature of these tissues as well as the usually large deformations they undergo. However, computational models are being used to analyse the mechanical behaviour of biological tissues leading to the development of a wide range of constitutive models and furthering our understanding of the structure-function relationships of the main tissues prevalent in the musculoskeletal and cardiovascular system.

Besides this tight synergistic coupling of experiment and model *analyses*, computer *simulations* are becoming more wide-spread. In contrast to pure quantitative analyses of a known, i.e. experimentally well defined, system, simulations aim to go a step further. Simulations create a virtual reality in which the modeller has the opportunity to alter certain aspects of the system and its environment to test different hypotheses by observing the effect they have on the system. Therefore, simulations often focus on the time course, i.e. the transient development, of a system rather than analysing one specific state and involve theories from a spectrum of scientific areas—the focus shifts from parameter identification to explanation. In essence, the modeller tries to ask the system of interest a specific question and seeks to approach a possible answer by conducting in silico experiments. In mechanobiology, where the central question is how cells and tissues respond and adapt to mechanical forces, these simulations can provide a valuable extension of classical experimental methods. Even under very defined and simplified experimental conditions the inherent heterogeneity and adaptivity of biological tissues presents a major hurdle in evaluating and quantifying the stimuli and mechanisms responsible for an observed response. The role of computer simulations in this quest has been defined by van der Meulen and Huiskes [111] as follows:

> "Computational mechanobiologists hypothesize a potential rule and determine if the outcome of this hypothesis produces realistic tissue structures and morphologies, hence *trial-and-error*. If the results correspond well, they might be an explanation for the mechanism being modeled. This method of research is common practice and productive in physics, less common in biology (Huiskes, 1995); although *theoretical biology* is based on this type of approach."

In other words, due to the complexity of mechanobiological systems simple answers are often unlikely from experiments alone and alterations to the experimental system tend to introduce other unknowns. In such cases, simulation environments can help unravel part of the mysteries by approaching a problem in a more systematic and quantitative way. For that purpose, computational engineering methods have been introduced into the field and aim to tackle the multiphysics problem by linking mechanics and various branches of biology.

As the opening quote stated and our short introduction outlined, the approaches, intentions and philosophies behind theoretical models differ between the various fields of science and engineering. In this review, we focus on hypothesis driven simulations that try to explain certain experimental phenomena, the assumptions made in them and the conceptual ideas relevant for the understanding of both their potentials and limitations. For this reason we have excluded purely quantitative analyses of biomechanical behaviour. For computer aided tissue engineering with a focus on CAD, image processing and various computer aided scaffold manufacturing techniques we further refer the reader to the review articles by Sun et al. [109, 110]. Since the field of computational mechanobiology is growing rapidly and the wealth of publications is significant, we narrowed our focus on studies that have relevance to cartilage and bone tissue engineering applications. In particular, our interest lies with the mechanoregulation of mesenchymal stem cell differentiation and the incorporation of tissue architecture into simulations of tissue engineering and regeneration. This chapter is organised into several sections that are intended to highlight different conceptual modelling aspects relevant to tissue engineering applications and as such are often very selective in the literature cited:

- Section 2 introduces some examples of mechanobiological single cell models and exemplary questions currently under investigation, namely the interaction of cells with their substrate. This is done for the purpose of showing how

theoretical models can help overcome particular experimental hurdles and in order to illustrate some aspects of structural reorganisation in cell biology.
- Section 3 introduces the concept of testing theories of mechanoregulated tissue differentiation by performing simulations, where the hypothesis under investigation can be corroborated or rejected by comparing in silico predictions with in vivo results. The main focus will be on simulations of fracture healing and osteochondral defect healing. The section is kept brief with a focus on providing essential historical and conceptual background information. It closes with some recent work from our lab on differentiation guided by substrate stiffness and oxygen tension as well as the incorporation of tissue structure into simulations of tissue regeneration.
- Section 4 is concerned with the application of computational models to tissue engineering and regeneration approaches involving bioreactors and scaffolds. The focus here is on understanding how scaffold heterogeneity and architecture affect mechanical stimuli imposed on cells and the role of modelling in complex design challenges. We further provide an example of tissue adaptation during bioreactor culture demonstrating how modelling approaches somewhat akin to those introduced in Sects. 2 and 3 can be used to determine the role of collagen remodelling on the mechanical properties of engineered cartilage subjected to dynamic compression.
- Section 5 finally bridges the gap between the scales by giving a short overview over multiscale simulations in tissue engineering and regeneration.

2 Single Cell Models and Cell-Substrate Interactions

2.1 Mechanosensing, Cell Rheology and Substrate Effects

When descending to the cellular level the main mechanobiological question relates to the fundamental aspect of mechanosensing—how does the cell sense, integrate and translate mechanical signals of different kinds to evoke a biological response. But cells are not merely passive sensors. Cells actively interact with their environment, probe it, sense it and adapt to it. Cytoskeletal components such as actin filaments and microtubuli undergo continuous polymerisation and depolymerisation depending on cell activity, attachment and external cues. Hence, an inherent problem of elucidating cellular properties and behaviours is the living nature of the cell itself—it interacts with the commonly applied measurement tools such that the measurement itself alters the cellular properties that are to be measured. For that reason, intrinsic properties are difficult to obtain by experimental methods alone and constitutive models that feature the salient aspects of cellular behaviour are required for a thorough evaluation [27]. Since the understanding of mechanotransduction pathways requires knowledge on the mechanics of the cell and its stress or strain generating mechanisms, the lack of such constitutive models is particularly problematic for the elucidation of relevant signalling cascades [53]. Cellular rheology is therefore a very active field of research where phenomena unknown from traditional engineering materials play a role, e.g. the ability of the "material" to perform work via ATP hydrolysis. While classical equilibrium materials exhibit thermal fluctuations, the cell presents a nonequilibrium material with additional actively driven fluctuations [69]. For a recent review on mechanosensing, candidate signalling mechanisms and their relation to cell rheology see the article by Janmey and McCulloch [53].

Most tissue cells are anchorage dependent, probe their environment by actively pulling on it and respond with cytoskeletal (re)organisation and rearrangement of focal adhesion complexes [29]. Apart from this anchorage dependent behaviour of differentiated cells, mesenchymal stem cells (MSCs) have recently been shown to differentiate into neurons, myoblasts and osteoblasts depending on the stiffness of the substrate onto which they adhere [31]. This differentiation mechanism was shown to be dependent on nonmuscle myosin II and is hence linked to the active contractility of the cells [31]. In contrast to the response to external forces relying on outside-in signalling pathways, sensitivity to passive properties of the extracellular space requires a two-step inside-out outside-in mechanism [29]. Cell contraction and stress-fibre formation are thus fundamental aspects of both the behaviour of differentiated cells and stem cell differentiation itself and are of direct relevance for tissue engineering applications. In this review we chose to highlight a class of models that incorporates some of the active mechanisms described above and appears useful in the investigation of cell-ECM or cell-substrate interactions. Understanding these interactions is important not only for explaining fundamental cell functions, cell remodelling and cell differentiation, but also for the design of biomaterials that support cells and are used in pharmacology, cell culture and tissue engineering.

2.2 *Modelling the Environmentally Regulated Dynamics of the Cytoskeleton*

Early models of cell contraction relied on static discrete sets of stress-fibres [77] or classical continuum modelling strategies such as thermoelasticity [83], neglecting the dynamic and anisotropic nature of the cell. Deshpande et al. [27] developed a continuum model that incorporates "three key biochemical processes: (i) an activation signal that triggers actin polymerization and myosin phosphorylation, (ii) the tension-dependent assembly of the actin and myosin into stress fibres, and (iii) the cross-bridge cycling between the actin and the myosin filaments that generates the tension." The finite strain constitutive model for the cell was developed based on the homogenisation of the activation and deformation behaviour of a single cellular stress fibre. It has been shown to capture the scaling of the cellular forces with the substrate stiffness, the development of structural anisotropy dependent on cell shape and boundary conditions and the co-localisation of high concentrations of stress-fibres with focal adhesion sites or externally induced local stress concentrations. From a modelling perspective it is noteworthy that this type of model allows a representation of the cytoskeleton in some ways similar to that in tensegrity models [48] and not usually captured in continuum models. Additionally, the continuum approach offers the advantage that the cytoskeletal arrangement does not have to be pre-defined and is dynamic in nature, i.e. the cytoskeleton can remodel, which is a non-trivial task for tensegrity models.

Cells are attached to the ECM through discrete multi-protein complexes called focal adhesions. These focal adhesions assemble and disassemble depending on the forces exerted upon them and are linked to the formation and dissociation of stress-fibres in the cytoskeleton. The adhesion sites undergo maturation from dynamic to fully reinforced static adhesion sites, passing the stages of initial adhesions, adhesion complexes and focal adhesions. Based on continuum mechanics arguments it has been suggested that with maturation into the static state the adhesion sites loose their sensitivity to substrate stiffness; for more details on adhesion sites in relation to mechanosensitivity see the articles by Fereol et al. [32], Nicolas et al. [85], and Nicolas and Safran [84]. The cell contractility model [27] described previously has been extended and applied to the investigation of focal adhesion dynamics using a thermodynamical approach and uniaxial example problems [28]. The authors of this study stated that: "The fact that many proteins perform both structural and signaling functions hinders the ability of traditional genetic approaches to parse the mechanisms that regulate [focal adhesion] dynamics, since knocking out a protein by genetic manipulation may also eliminate an essential structural component. Consequently, to understand how specific molecular features give rise to the observed behavior it is essential to combine experimental studies of adherent cells—including the use of microscopy to observe the structure and dynamics of the cytokskeleton and [focal adhesions]—with computational models that include the salient mechanics" [28]. Among the successful predictions made by the model was the focal adhesion concentration around the cell periphery as well as the dependence of focal adhesion intensity on cell size and contractility.

The coupled mechanosensitive focal adhesion/stress-fibre model was extended to two dimensions and applied to simulate the effect that a micro-patterned shape of ligands would have on cytoskeletal remodelling and focal adhesion distributions [88]. Both convex ligand patterns where the entire cell periphery adheres to the substrate and various concave patterns with free non-adhered edges were investigated. The model's predictions in terms of stress-fibre alignment and concentration were regarded as satisfactory while some discrepancies in the vinculin[1] distribution were observed.

[1] A protein in the membrane that correlates with the concentration of high-affinity integrins and is involved in their binding to cytoskeletal components.

Uniaxial and biaxial cyclic stretching of cells by 3 and 10% at 0.1, 1.0 and 10.0 Hz were simulated using the Deshpande model in Wei et al. [114]. Corresponding to experimental observations the model predicted stress-fibre alignment perpendicular to the stretch direction and its dependence on the stretch magnitude, the amount of lateral contraction and the frequency. Isotropic stress-fibre architectures were predicted in the biaxially stretched cells. An earlier model [113] had investigated cytoskeletal reorganisation in response to cyclic loading by assuming acting filament remodelling in response to normal strain and filament disassembly when their strain energies reach certain levels below or above their basal attachment values. While that model was capable of predicting experimentally observed alignment patterns, it is incapable of reproducing frequency dependent effects. This is rooted in the purely elastic (i.e. scleronomous) approach taken that neglected transient biochemical signals. While the model by Deshpande and coworkers [27, 28] represents a three dimensional finite strain constitutive framework suitable for finite element implementation, this early model was based on two in-plane normal strains and assumed linear elastic actin filaments [113].

Arrays of flexible micro posts have been used to measure forces exerted by the cells onto a substrate and their distribution within the cell. Due to the dynamic nature of the cells the interpretation of the results and the extraction of meaningful parameters has proven difficult. McGarry et al. [75] simulated smooth muscle cells, mesenchymal stem cells and fibroblasts on different beds of micro posts using the model by Deshpande and coworkers [27, 28]. Certain model parameters such as the maximum tensile stress of a fibre bundle were cell type dependent which has been attributed to the expression of different isoforms of actin and myosin among those cell types and potentially presents meaningful parameters for the interpretation of cell type dependent experimental results. For a single cell type, namely smooth muscle cells, the model could predict a number of experimental observations with the same set of parameters: "(i) the scaling of the force exerted by the cells with the number of posts; (ii) actin distributions within the cells, including the rings of actin around the micro-posts; (iii) the curvature of the cell boundaries between the posts; and (iv) the higher post forces towards the cell periphery" [75]. The experimental and computed actin stress-fibre architecture of a cell on a micropost array is depicted in Fig. 1.

Motivated by observations of fibroblast alignment with directions of tensile strain, cell locomotion along rigidity gradients towards areas of higher stiffness or tensile strain and enhanced cell spreading and cytoskeletal organisation with increasing substrate stiffness, Bischofs and Schwarz [6] developed a model based on the idea that cells have a preference for large effective stiffnesses in their local environment which drives their positioning and orientation. Linear elasticity continuum theory was adopted to model the ECM in order to keep the presented calculations feasible. Cells pull on the ECM with their contractile machinery to extract mechanical information that encompasses the effects of both rigidity and prestrain in the substrate. In order to capture this behaviour, mechanical work was chosen as the fundamental stimulus of the mechanoregulation algorithm. By pulling on the ECM cells build up a force at a focal adhesion site that in

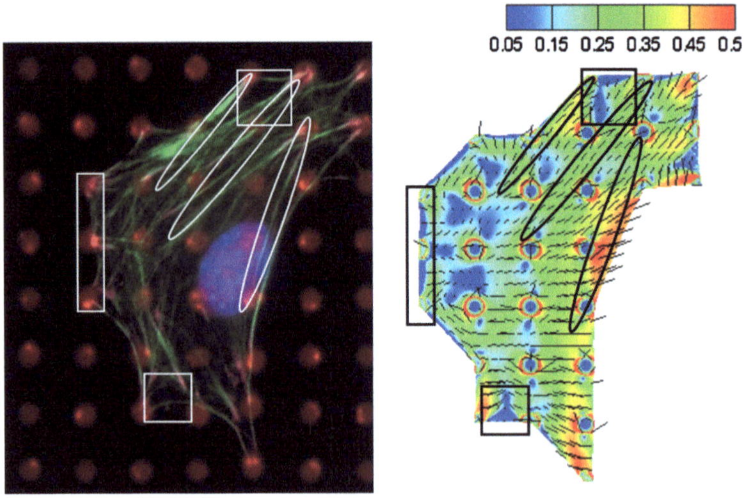

Fig. 1 Steady-state actin stress-fibre distribution in a fibroblast on a micropost array (*left*) and corresponding simulation results based on McGarry et al. [75] (*right*). Contours illustrate the degree of stress fibre formation, vectors correspond to the dominant stress fibre direction. Images courtesy of Dr. Patrick McGarry, Mechanical and Biomedical Engineering, NUI Galway, Ireland

conjunction with a displacement produces that mechanical work. This work can now be minimised by the cell in different ways: If the substrate stiffness increases, the cell has to strain the ECM less to create a given level of force which corresponds to less work being invested. A prestrain of the substrate is shown to lower the work as well and hence corresponds to an effective stiffening. In terms of the wider concepts of modelling the authors make a statement that holds for phenomenological models in general and thus fits the general context of this chapter: "It is important to note that conceptually the principle suggested here does not imply that the cell actually minimizes the work W invested into its soft environment. Instead we suggest here that calculating the quantity W for different situations of interest is an appropriate measure for the kind of information a cell can extract from its elastic environment through active mechanosensing. The real justification of our model will be its success in explaining a large body of experimental data" [6]. Nevertheless, the considerations imply a mechanism by which higher matrix stiffness leads to an increased efficiency in the up-regulation of cell-matrix contact points. Based on the work optimisation principle, the model predicted cell alignment parallel and perpendicular to free and clamped edges, respectively, cell alignment with applied tensile strain and coalignment of cells into strings. Cells were further predicted to have a tendency of moving away from free surfaces and the model can potentially explain certain aspects of cell behaviour that cannot be established based on contact guidance models [3]. Cell alignment in response to ECM alignment, i.e. contact guidance, is relevant for tissue engineering applications and soft scaffold contraction. To allow a theoretical investigation of contact guidence effects an extensive modelling framework has been introduced by Barocas and Tranquillo [3]. It relies on mixture theory to capture the biphasic nature of the soft tissue equivalents under consideration and accounts for collagen alignment in response to anisotropic deformation, subsequent cell alignment, i.e. contact guidance, and the resulting anisotropic traction exerted by the cells onto the ECM as well as their anisotropic migration.

The described models are taking a major step towards the coupling of a cell's response to its external as well as internal mechanical environment. They have the potential of facilitating a better understanding of the cell's interaction with its environment, including intrusive measurement tools. This is crucial for the adequate interpretation of experimental results and for examining cell rheology. Several aspects might be put onto the research agenda: The incorporation of actual signalling pathways into the models will not only increase their predictive capability, but also allow the investigation of pathological or experimental (e.g. drug related) inhibition of those very pathways. Further, while the models described in this section link mechanics to cellular remodelling and are capable of rationalising a number of experimental observations, the link between biology and mechanics needs to be extended.

3 Stem Cell Differentiation During In Vivo Regeneration

3.1 Testing Mechanoregulation Hypotheses In Silico

The idea that mechanics affects tissue differentiation has been discussed in the scientific community for a long time [89, 95]. By incorporating hypotheses regarding mechanoregulated tissue differentiation into computer simulations of regenerative events (such as fracture healing) that exhibit well defined and repeatable temporal and spatial patterns of tissue differentiation with alternative healing paths depending on mechanical stimulation, these hypotheses can be corroborated or rejected by comparing the in silico predictions to the experimental observations. These approaches therefore constitute a valuable tool in the assessment of potential biophysical regulators of tissue differentiation and may have specific relevance for the incorporation of biophysical stimulation into bioreactor designs for tissue engineering applications. Similarly, such corroborated hypotheses might serve to form the basis for the evaluation of the mechanical environment in scaffolds for tissue engineering (see Sect. 4) The ultimate goal of tissue engineering is to replace or repair damaged tissues in vivo. Quantitatively evaluating the in vivo environment and its effects on the tissues during regeneration therefore has direct relevance to tissue engineering itself.

In accordance with the scope of this chapter we restrict this overview on relatively recent work where theories regarding mechanoregulated tissue differentiation have been tested using modern simulation techniques. Specifically we focus on some recent developments, namely (i) a tissue differentiation algorithm based on substrate stiffness and oxygen tension; (ii) the simultaneous prediction of tissue structure and phenotype during regenerative events; (iii) performing non-deterministic simulations to capture variability.

Among the mechanical stimuli proposed as potential mechanoregulators are

- Octahaedral shear stress/strain or maximum principal strain and hydrostatic stress (Carter and co-workers [15–17]). Quantitative boundaries were only proposed later by others [50].
- Principal strain and hydrostatic stress (Claes and Heigele [22, 23])
- Octahaedral shear strain and fluid velocity (Prendergast and Huiskes [46, 91])
- Others include certain strain invariants [34, 94] and strain energy density [1].

The model by Carter and colleagues [15–17] was used among others in a study where pseudoarthrosis formation was studied in oblique fractures [71]. Furthermore, this model has been applied frequently in tendon mechanobiology, such as fibrocartilaginous metaplasia formation in tendons wrapping around bony prominences [39, 116].

The theory by Claes and Heigele [22, 23], which is based on Pauwel's ideas [89], was mainly applied to problems of fracture healing. A recent model for tissue differentiation and revascularisation in fracture healing by Simon and co-workers [107] implemented a set of rules into a fuzzy-logic controller to simulate regenerative events in the fracture callus. The limiting factors for healing were found to be revascularisation in stable fractures and an inadequate mechanical environment in unstable fractures. A similar fuzzy logic model was able to predict all stages of trabecular fracture healing including the final remodelling stages to re-establish a trabecular structure depending on the loading direction [106].

The model by Prendergast et al. [91] accounts for the biphasic nature of most biological tissues and as such incorporates fluid flow as a stimulus. Its first implementation into an automated feedback algorithm to simulate the time course of fracture healing was performed by Lacroix et al. [68]. The origin of MSCs, whose migration into the fracture callus was modelled as a diffusive process, was predicted to have a significant impact on the rate at which healing progressed. The model was further able to predict the spatio-temporal sequence of phenotypes occurring during fracture healing. The results provided support for the hypothesis that fluid flow and shear strain are regulators of MSC differentiation during fracture healing. Following this initial corroboration, the model was applied to analyse the influence of fracture gap size and loading on the emerging phenotypes and the healing outcome [67]. Further corroboration was achieved by applying the tissue differentiation model in simulations of other regenerative events. Besides fracture healing [49, 67, 68], it has been successfully used to predict key events during distraction osteogenesis [8, 9, 51] osteochondral defect healing [55, 56], implant integration [36, 46] and pseudoarthrosis formation [43, 81]. Many studies on scaffold aided tissue repair or scaffold design rely on this algorithm since fluid perfusion is thought to play a major role in porous scaffolds (see Sect. 4)

Several comparison studies between the various algorithms named above have been performed with no full corroboration achieved for any of them [35, 49, 50] although the influence of torsional loading on fracture healing could only be captured by considering fluid flow and shear strain as regulators of MSC differentiation [49]. Further reviews of in silico approaches to regenerative medicine, namely for bone and wound healing, can be found elsewhere in the literature [10, 38].

3.2 MSC Differentiation Regulated by Substrate Stiffness

As outlined in Sect. 2, the profound effect of substrate stiffness on cells including the stiffness-dependent differentiation of mesenchymal stem cells into various lineages [31] has attracted considerable attention and is an important factor in the choice of an appropriate biomaterial carrier for tissue engineering applications. Cells furthermore depend on the supply of oxygen and nutrients and different phenotypes thrive under different ambient oxygen tensions. Perhaps most importantly, oxygen tension itself has been shown to be a regulator of stem cell fate [105]. In regenerative events or during progenitor cell based tissue engineering, osteogenesis, for example, relies heavily on sufficient vascular supply, while chondrogenesis has been shown to be favoured under low oxygen conditions [76]. Various tissue differentiation models therefore include an angiogenic component [19, 37, 107].

A recent mechanoregulation model for tissue differentiation [12] developed in our lab is based on two fundamental components: The influence of substrate stiffness and oxygen tension. The model relies on a four-step analysis approach:

1. MSC infiltration into the regenerating tissue is modelled as a diffusive process.
2. A biphasic analysis is performed to evaluate the mechanical environment throughout the regenerating domain.
3. Angiogenesis is modelled as a mechanoregulated diffusive process where the formation of new capillaries is inhibited in regions of high shear strain as determined in analysis step 2.
4. Based on the vascular supply another diffusion analysis additionally accounts for cellular consumption and determines the levels of oxygen tension in the regenerating tissue.

Based on the current distribution of tissues, information on both the local oxygen tension and substrate stiffness is now available in each point. This information is used to predict the cell phenotypes for the next iteration based on the theory described in Fig. 2. The mechanical properties, state of vascular and cellular infiltration as well as oxygen perfusion can now be updated for the next time increment of the analysis. In contrast to most mechanoregulation models, mechanical stimuli such as strain, stress or fluid flow do not exert their influence on MSC differentiation directly in this model. Instead, neovascularisation is inhibited in regions of high strain and the resulting low

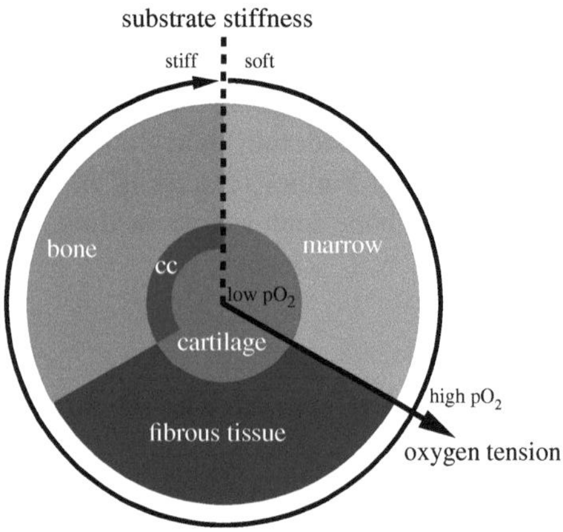

Fig. 2 Regulation of tissue differentiation by substrate stiffness and oxygen tension. Cartilage forms in a low oxygen environment and turns into calcified cartilage (CC) in close proximity to bone. MSCs will differentiate to form marrow (stroma and adipogenesis), fibrous tissue or bone when attached to soft, medium or stiff extracellular materials, respectively. Colour version available online

oxygen concentration results in cartilage formation. Thus, the effect of deformation on stem cell fate is more indirect. Rather than responding to external strain, the cells are assumed to actively probe the ECM in order to determine its stiffness and differentiate accordingly.

Like several other theories, this regulatory model has successfully captured many of the key stages of fracture healing [12]. This illustrates that the corroboration of one hypothesis in a simulation does not allow one to reject others. Extensive assessment of the efficacy of a proposed theory via the simulation of a broad range of experiments is needed to further corroborate the proposed hypothesis.

3.3 Accounting for Tissue Architecture During Skeletal Regeneration

A tissue's structure is crucial for its mechanical fitness. Understanding how the structure evolves during development and regeneration can have important implications for how tissue engineering protocols, scaffolds and bioreactors are designed in order to mimic a desired tissue architecture. The extension of the predictive power of tissue differentiation models to include anisotropy will benefit the investigation of specifically those healing processes and tissue engineering strategies where recapitulating normal tissue architecture is important. For example chondral and osteochondral defect repair critically depends on achieving a native-like zonal structure within the cartilage tissue so that the tissue can endure in-vivo loads.

Cullinane et al. [25, 26] have demonstrated that the mechanical environment during bone defect healing can influence both tissue differentiation and the molecular organisation (collagen fibre architecture) of the repair tissue. They showed that cyclic bending applied daily to an experimental mid-femoral defect

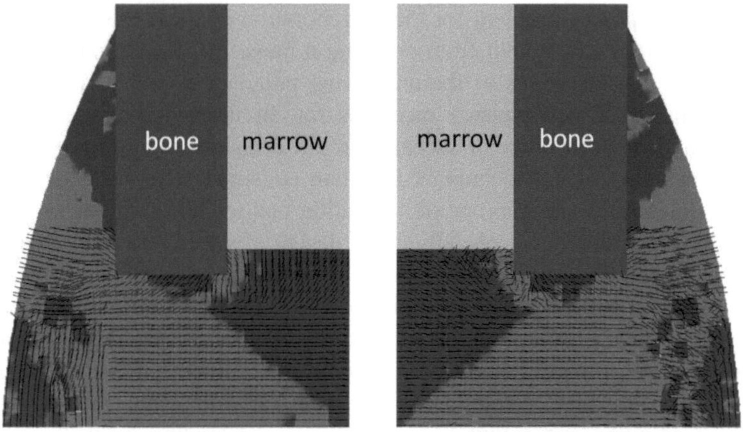

Fig. 3 Distribution of phenotypes (*yellow* fibrous tissue; *red* cartilage) in an idealised model of a fracture callus subjected to cyclic bending. Vector plots show predicted collagen fibre orientation in the region of interest when collagen alignment was regulated by stress (*left*) or deformation (*right*). The predicted orientations of collagen fibres [81] were comparable to those observed experimentally [25]. Colour version available online

results in the formation of cartilage as opposed to bone tissue. These neoarthroses exhibited preferred fibre angles consistent with those seen in articular cartilage. Hayward and Morgan [43] further demonstrated that the patterns of tissue differentiation observed experimentally could be predicted using the mechanoregulation theory of Prendergast et al. [91], providing further evidence to suggest that both strain and fluid flow are key regulators of tissue differentiation. Nagel and Kelly [81] extended the mechanoregulation model of Prendergast et al. [91], as implemented by Lacroix et al. [68], to include a biphasic fibre reinforced constitutive model for soft tissues, where the organisation of the fibre network is regulated by the mechanical environment. Collagen fibres where assumed to align between the principal directions of various stress and deformation measures and depending on the local maximum principal values occurring during a bending cycle. By simulating the effect of bending on bone defect repair, the predicted patterns of differentiation and collagen fibre orientations in the repair tissue could be compared to those observed experimentally [25]. It was demonstrated that mechanoregulation models can be used to successfully predict both tissue differentiation and organisation during skeletal tissue regeneration (Fig. 3).

3.4 Variability in Tissue Differentiation

A high degree of variability is observed in the outcome of tissue engineering as well as in vivo experiments. In the context of tissue differentiation models, this variability can pose a problem for the corroboration of these theories. Similarly, model predictions can benefit from covering a range of possible outcomes based on statistical considerations of the underlying populations and variabilities. Predicting a range of performance measures for an implanted tissue engineered construct and its probability of success can aid decisions regarding its design or even just whether or not to implant a certain construct. The need for nondeterministic simulations and appropriate validation metrics has long been recognised in traditional engineering, especially in the design of safety relevant systems [86]. These validation metrics should be based on a statistical evaluation of both the experiments and the computations. In bioengineering, the variations that occur are usually larger than those observed in controlled technical systems. Focussing on scenarios for which mechanoregulatory aspects are of importance, for example for load bearing implants, variability arises from environmental factors, such as implant positioning and activity levels, i.e. external loading, as well as genetic factors such as the level of mechanosensitivity.

The variability of bone geometry and mechanosensitivity observed in a population was modelled to simulate the clinical trial of low and high-stiffness intramedullary prostheses in 100 patients [92]. Variable mechanosensitivity was included by modulating the parameter values that determine the onset of bone resorption/deposition in the bone remodelling algorithm used. When only one averaged ideal patient was simulated, i.e. no variability included, the deterministic simulation predicted that the low stiffness implant would migrate less. The simulation of the clinical trial with 100 different patients, however, predicted no statistically significant difference between the two implants. The study concluded, that stochastic simulations might be more validatable against an actual clinical trial and may eventually overcome the lack of falsifiability of this type of model [92].

Modelling certain cellular processes in a stochastic fashion [90] introduces only a minor degree of nondeterminism in the simulation outcomes. Khayyeri et al. [59] simulated tissue differentiation inside an in vivo bone chamber in which the tissue could be loaded in compression and investigated the influence of the load magnitude, implant positioning, blood pressure and bone marrow MSC density on the distribution of phenotypes. The study concluded that these environmental factors were not sufficient to fully explain the high degree of variability observed in the experiments, namely the emergence of two distinct experimental groups. In a later study [61], the authors thus hypothesised that a variable mechanosensitivity between individuals could be responsible for the variability observed in the tissue differentiation experiments. Simulations of the bone chamber experiment were performed and the effect of stochastically up- or down-regulating the cellular process rates for proliferation, differentiation and apoptosis in an individual-specific manner was investigated. Simulation results were highly sensitive to MSC activity. The simulations not only predicted the general patterns of tissue differentiation but also produced a variability akin to that observed in the experiment. Specifically, the simulations predicted the dichotomy, i.e. the emergence of two distinct differentiation patterns. Results such as this can have important implications for the corroboration of mechanoregulation theories.

Instead of focussing on variability, other studies take a subject specific approach. A bone remodelling algorithm based on open system thermodynamics was applied to simulate the pattern of bone mineral density in the proximal tibia in response to gait loading [87]. A subject specific geometry and load, derived from gait analysis, were used. The predicted bone density distribution showed excellent qualitative and quantitative agreement with the subject specific pattern measured by X-ray absorptiometry. In another study, Wolff's law was "inverted" to estimate the loading history of a murine vertebral body via an optimisation approach based on a set of unit loads applied to a subject specific model of a murine vertebral body, the internal structure of which was modelled based on μCT scans [21]. Based on the assumption that bone tissue remodels such that the strain energy density approaches a target value and that the tissue is loaded uniformly, the dominant load case could be extracted both in direction and magnitude. The strain energy density in the trabecular architecture remained highly non-uniform nevertheless. The review article by Geris et al. [38] includes a discussion of patient specificity and variability in in silico approaches.

4 Scaffolds in Bone and Cartilage Repair

4.1 Computer Aided Design of Scaffolds and Bioreactors

Scaffold design is constrained by a large number of design criteria that often compete. This competition implies that there is some optimum design that represents the best compromise between the various requirements. Relevant aspects to be considered in scaffold design include:

- The *material* has to be non-toxic, biocompatible, sterilisable, manufacturable into arbitrary shapes accommodating patient specific defect geometries, and possess biologically favourable surface properties.
- Considerations regarding the intrinsic *mechanical properties* of the material include compliance-matching with the host tissue, sufficient strength during implantation, cellular response to substrate stiffness (see Sect. 2), and expected deformations that potentially act as external stimuli.
- The scaffold *architecture* can serve as a template to guide tissue architecture, including anisotropy [30]. The *porosity* is linked to the apparent properties of the scaffold and is therefore mechanically constrained such that the scaffold maintains its structural integrity. However, higher porosities are favourable for cell seeding and proliferation, tissue ingrowth and mass transport. Porosity also affects the permeability and hence fluid velocities during seeding as well as mechanical loading. Depending on the target tissue, the pore architecture and its interconnectivity has to be suitable for *vascularisation* of scaffold and tissue.

- The *dissolution rate* of biodegradable scaffolds should be optimised to match the speed of differentiation and matrix synthesis by cells within the scaffold. The dissolution rate can be modulated via the relative amounts of the components that make up the biomaterial, which will in turn affect intrinsic properties such as stiffness.

Alterations to the properties above likely affect the mechanical environment in the scaffold and hence the mechanoregulation of biological processes. Due to the multitude of design criteria and their strong coupling, mechanobiological modelling can contribute significantly to finding optimal scaffold properties or limiting the number of physical prototypes to be tested. Provided the knowledge, efficiency and resources in the future, this computational optimisation of scaffolds could be envisaged routinely on a patient-specific basis to maximise clinical success and stream-line the automated production process [56, 72].

Bioreactors are a fundamental part of tissue engineering for two main reasons [72]. First, one approach to tissue engineering involves cell biopsies, ex vivo expansion and culture in a bioreactor to achieve desired tissue properties and subsequent reimplantation. More indirectly, bioreactors bear significant relevance for mechanobiological research. They present model systems in which environmental aspects such as cytokine concentrations, medium composition, pH, oxygen environment and mechanical loading can be much more tightly controlled than in vivo.

Computational models can not only help in the design of scaffolds, but also in the design of bioreactors and the choice of their operating parameters. Several bioreactor designs exist that induce fluid flow in or around the scaffolds, such as direct perfusion, spinner flask or rotating wall bioreactors. Their prime purpose is to enhance culture medium transport to provide the optimal biochemical environment for the cells. However, fluid flow induced shear is a potent mechanical stimulus that can influence the metabolic activity of the cells or even their phenotype. Hence it has been suggested [72] that computational fluid dynamics methods rather than trial and error should be used to establish the flow conditions for the various bioreactor systems, scaffold architectures and cell types. The same concept can of course be applied to related aspects of bioreactor performance such as direct mechanical stimulation by deformation. Other applications include the derivation of culture and rehabilitation protocols derived from the analysis of in vivo conditions [13], the derivation of the cellular environment from the macroscopic bioreactor input profiles [72], the online integration of computational models into tightly monitored commercial tissue culture systems to improve automated control and surgical planning [72] and hence enable a more cost-effective and scalable approach to tissue engineering. In the context of cartilage tissue engineering and bioreactor design we wish to draw attention to the contributions by Raimondi et al. [93] and Bjork et al. [7] in this book for further information.

4.2 *Tissue Differentiation in Scaffolds*

4.2.1 Optimisation of Scaffold Properties

Significant efforts have been directed at engineering bone and cartilage or enhancing the natural reparative processes by a variety of interventional methods. Scaffolds implanted into osteochondral defects have been used to enhance repair tissue quality but a major unknown are the ideal scaffold properties. The earliest attempt to determine optimal mechanical properties of a scaffold using a computational mechanobiological model was for the repair of an osteochondral defect [56]. The main assumptions of the model were that MSC differentiation as well as cell proliferation and apoptosis were determined by a combination of fluid flow and shear strain [91] and that cell migration could be modelled as a diffusive process. By systematically varying the Young's modulus and the permeability of homogeneous scaffolds it could be shown that optimal values for these parameters exist, with both higher and lower values leading to decreased amounts of cartilage tissue. This result was due to the differential influence of fluid flow and shear strain on cell differentiation. To further enhance repair, i.e. reduce the amount of fibrous tissue and stop the subchondral bone from progressing into the cartilage zone, a bilayered scaffold with inhomogeneous properties was then optimised. The osseous part of the scaffold was modelled as homogeneous while the properties of the chondral part were depth dependent. The study found that the optimal scaffold should have a stiffness that decreases with depth from the superficial to the deep zone and a permeability that increases with depth, i.e. has its minimum at the articular surface. Given the inhomogeneous properties of articular cartilage this scaffold has to a certain extend biomimetic qualities. The study closes with the remark that scaffolds optimised for the in vitro setting may not be well suited for the in vivo environment and vice versa.

A similar approach has been followed to investigate design parameters in bone scaffolds by Byrne et al. [14]. Tissue differentiation was again regulated by fluid flow and shear strain [91], while cellular activities were modelled using the lattice approach [90]. Tissue differentiation inside a bioresorbable scaffold with regular geometry typical for 3D printing techniques was simulated under low and high loading conditions varying three key design variables: Young's modulus, porosity and dissolution rate. The authors were able to determine unique combinations of these parameters that maximised bone formation depending on the loading condition. Especially under high loading conditions the initial porosity and dissolution rate had to be chosen conservatively as to not compromise the structural integrity of the scaffold. High porosities, a medium dissolution rate and a high stiffness gave the optimal results under low loading conditions. This suggests that scaffold manufacturing should be tailored towards the loading conditions at the implantation site.

Scaffold based tissue repair of critical sized defects is often hindered by inadequate vascular supply resulting in a degraded scaffold core and peripheral tissue formation. Due to its importance in tissue regeneration, the lattice model approach has been extended to include angiogenesis [19] in an attempt to better model the process of bone regeneration. Checa and Prendergast [20] closely followed the approach of Byrne et al. [14] to investigate the effect of cell seeding and mechanical loading on vascularisation and tissue formation inside a scaffold. High and low loading conditions and different seeding densities were simulated and homogeneous seeding compared to only peripheral seeding. It was predicted that reducing the MSC seeding density from 1 to 0.5% greatly improved vascular infiltration and bone tissue formation, whereas under high seeding density conditions only peripheral vascularisation and ossification occurred. Likewise, peripheral seeding was predicted to allow scaffold core vascularisation and subsequent bone formation. While low loading conditions were generally beneficial for bone formation, increasing the applied stress above a certain threshold value was predicted to result in decreased vascularisation and a mainly cartilaginous tissue.

4.2.2 Incorporating Realistic Scaffold Architectures

While the previous two studies modelled a regularly shaped scaffold, an irregularly shaped scaffold inside a loaded in vivo bone chamber was implemented on the lattice level by Khayyeri et al. [60]. Its geometrical representation was based on μCT scans of a highly porous collagen-glycosaminoglycan (collagen-GAG) scaffold. On the level of the regularly structured finite element mesh material properties were homogenised by a rule of mixtures approach using the volume fractions of tissues and scaffold material present in the element's lattice. This approach allows for an easy implementation of the scaffold geometry into the lattice. It is computationally efficient and hence suitable for larger scaffold structures at the expense of micromechanical accuracy of the solution due to the application of the rule of mixtures. The effect of a varying scaffold stiffness on tissue differentiation patterns was investigated. Soft scaffolds were predicted to result in mainly fibrous tissue formation while increasing stiffness values of up to 1 GPa increased the amount of cartilage and bone present in the bone chamber significantly.

A micromechanically more accurate assessment of biophysical stimuli in a μCT based model of an irregular scaffold is necessarily computationally very expensive since a large number of elements is required to mesh the scaffold structure (and the pore space). This approach was followed in Sandino et al. [99] for calcium phosphate (CaP) and glass scaffolds by performing a solid analysis of applied compression and a steady-state Newtonian fluid flow analysis to assess stress, strain, fluid pressure, fluid velocity and flow induced surface shear stress in the scaffold during early bioreactor culture. The heterogeneity was not only reflected in the strain pattern, but also in the fluid velocities: Some pores were never perfused despite sufficient interconnectivity while in others the fluid velocity was

1000 times higher than the inlet velocity. Due to the considerable computational cost only a part of the full scaffold was modelled.

Sandino et al. [100] included angiogenesis and tissue differentiation into a μCT based model of a CaP scaffold section by filling the irregular pore space with a lattice to simulate cell activities. In vitro seeding was simulated by attaching the initial MSC population to the struts throughout the scaffold and compared to In vivo colonisation, where the scaffold was initially cell free and the original MSC population was modelled to reside at the outside faces of the pore network where endothelial cells had their origin in both cases as well. In vivo colonisation led to faster migration and proliferation but affected final tissue distribution and vascularisation of the scaffold only marginally. Despite 70% of the pore volume having a stimulus favourable for ossification, only 40% of that space filled with bone due to vascularisation being restricted to the periphery of the scaffold. Deeper vascular penetration was inhibited by the scaffold walls in the model. When the applied compression was doubled from 0.5 to 1% strain, cartilage was also predicted in external pores. Apoptosis, induced by a mechanical stimulus twice as high as that for fibrous tissue formation, increased from 17 to 22% as the applied strain increased from 0.5 to 1%.

Stops et al. [108] simulated MSC differentiation in a collagen-GAG scaffold based on μCT scans using a combination of FE and CFD analysis. Strain dependent cell proliferation was included and octahaedral shear strain in combination with fluid velocity used to determine MSC fate. Applied strains of 1% and above led to predictions of lower cell densities. By comparing different scaffold strains and fluid inlet velocities the authors were able to determine specific combinations of these two parameters that favoured certain phenotypes. While certain conditions favourable for osteoblasts (final cell fraction 84.9%) and fibroblasts (final cell fraction 73.9%) could be established, none of the combinations proved particularly suitable for robust chondrocyte differentiation (maximum cell fraction achieved 56.7%).

Besides its influence on the mechanical conditions inside the scaffold, the pore size influences cell attachment morphologies. In a combined fluid-elastostatic analysis, Jungreuthmayer et al. [54] simulated the effect of cell attachment modes on the experienced stimuli. Three random sub-volumes of the μCT reconstruction of a collagen-GAG scaffold were numerically seeded with cells that either attached flatly to one strut or bridged two struts by means of cellular processes. A steady-state incompressible Newtonian-fluid flow analysis revealed that pressures and wall shear stresses experienced by the cells were largely independent of attachment mode. A subsequent linear elastic analysis of cells subjected to the pressure and shear loads derived from the CFD simulation revealed that bridging cells underwent approximately 500 times higher displacements than flatly attached cells. Van Mises stresses in bridging cells were about 26 times higher than in flatly attached cells. These results can potentially explain why cells seeded onto 3D scaffolds with different morphologies, which promote various modes of cellular attachment, elicit a dramatically different biological response when subjected to similar levels of inlet flow. They could also explain the higher flow rates necessary to stimulate cells in 2D compared to a 3D environment [4, 52].

This brief selection of numerical studies on tissue differentiation in scaffolds seeded with mesenchymal stem cells provides a small insight into how quantitative engineering analyses can help understand the heterogeneity of microscopically induced deformations with peak values significantly exceeding macroscopically applied loads. They present a valuable tool to balance the various competing design criteria for scaffolds in a more systematic way compared to experimental trial and error to achieve optimal tissue growth and regeneration. The application will dictate whether a micromechanically accurate representation of the scaffold is necessary or whether the use of a smeared continuum treatment is sufficient.

4.3 Modelling the Bioreactor Environment: From Structure to Function in Engineered Cartilage

4.3.1 Importance of Structure and Composition for Mechanical Fitness

While these studies demonstrate a role of models in predicting phenotypical changes, the mechanical environment also influences the organisation of the tissue. A biomimetic recapitulation of tissue structure has relevance for those tissues in which the architecture is optimised for their loadbearing duties, such as articular cartilage. Cartilage tissue engineering is based either on chondrocytes or progenitor cells induced to undergo chondrogenesis and differentiate towards a chondrocyte-like cell. The chondrocytes then synthesise cartilage specific extracellular matrix components to establish a functional and viable tissue. Common assessment of this functionality is usually based on measuring glycosaminoglycan and collagen II content as well as performing compression tests to determine tissue mechanical properties. Both in native and in engineered cartilage various tissue stiffness measures have been found to correlate with the amount of these two main components constituting the cartilage extracellular matrix. Yet, this correlation of composition and functionality is an oversimplification and neglects the structure-function relationships present. This becomes apparent when considering experiments [5, 44, 45, 70, 73] that report a stiffer cartilage matrix when the constructs were mechanically loaded during bioreactor culture than when it was kept unloaded, but found no differences in the collagen and sulphated GAG content. Yan et al. [117] investigated structure-function relations in tissue engineered cartilage, including minor ECM components and cross-linking proteins into their analysis. They found strong correlations between these usually omitted constituents and the mechanical properties of the constructs as well as an increase in their concentration due to loading. They concluded that low levels of collagen IX and mature collagen cross-linking are a major contributing factor to poor mechanical properties of in vitro engineered cartilage. Thus it could be shown experimentally that basing composition-function relationships on major ECM components alone does not provide a complete picture. The importance of cross-linking collagen to achieve mechanically mature network has also been shown in cardiovascular tissue engineering [2, 96].

There is another key factor that contributes to tissue mechanical properties. Several experimental studies have suggested that enhanced structural organisation of the collagen network could contribute to the elevated stiffness values observed in bioreactor studies [44, 58, 70, 74]. The effect of structural alterations is however difficult to investigate experimentally, since tissue structure is not easily altered in a tightly controlled manner, biochemical and biomechanical alterations are commonly not easily uncoupled and certain structural features aside from fibre orientation are difficult to quantify. However, from extensive studies on a wide range of load bearing soft tissues, it has become apparent that the collagen architecture in these tissues is not only optimised for the particular load bearing duties but is also able to adapt to changes in the loading environment [33, 63, 97]. Understanding the mechanisms of tissue remodelling will potentially allow us to design bioreactors to control the structure and organisation of engineered tissues.

4.3.2 Modelling the Influence of the Collagen Architecture in Tissue Engineered Cartilage

Several studies exist that model cartilage structure. Collagen remodelling algorithms developed in the cardiovascular biomechanics field have been successfully applied to predict the collagen organisation in tibial plateau cartilage [115]. Khoshgoftar et al. [62] evaluated deformation fields in cartilaginous constructs undergoing various loading protocols: unconfined compression, sliding indentation and combined compression-sliding indentation. Based on the predicted strain fields the latter was hypothesised to provide the most suitable stimulation for the development of a Benninghoff-like zonal structure in the engineered constructs. This native cartilage architecture, in which collagen fibres arcade from an alignment perpendicular to the articular surface in the deep zone to an orientation parallel to the surface in the superficial zone, is crucial for cartilage structure-function relationships and has proven difficult to recapitulate in engineered tissues [64]. The influence of postnatal collagen reorientation on the confined compression behaviour of articular cartilage was investigated with a composition based constitutive model in van Turnhout et al. [112]. Klisch et al. [65] developed a cartilage growth mixture model in which proteoglycans and collagen can grow independently of each other via volumetric mass deposition but are constrained to move together during deformation. More advanced constitutive relations were incorporated later on that included the balance between stresses generated by the proteoglycans and those in the collagen network [66]. The model was successfully validated against biochemical content, tissue volume and tensile moduli from in vitro growth experiments [66]. The effect of collagen orientation on the equilibrium properties of charged and neutral biphasic tissues was investigated in Nagel and Kelly [78]. While in uncharged tissues, fibres perpendicular to the loading direction cause the highest stiffness in unconfined compression, fibres aligned with the loading direction produce the highest initial stiffness values in charged tissues. This is routed in the pre-stress of the collagen network due to the swelling pressures induced by the fixed negative charges of the GAG molecules. None of these studies, however, investigated potential collagen remodelling mechanisms during bioreactor culture and their effect on the mechanical properties of the engineered cartilage.

In a recent study from our lab we applied a newly developed collagen remodelling algorithm [80] to test the hypothesis that structural changes to the collagen network in response to loading could contribute to enhanced biomechanical properties of dynamically compressed cartilaginous constructs even in the absence of biochemical differences between the free-swelling (FS) and dynamically loaded (DL) groups [79]. Fifty-six days of bioreactor culture were simulated during which proteoglycans and collagen were synthesised leading to a steady increase in material parameters that was equal in both the FS and DL groups. The collagen remodelling algorithm used [79, 80] allowed for changes in the local collagen orientation and its stress-free configuration (i.e. the transition from a buckled fibre with insignificant stress contribution to a tensed cable like configuration with a high stress contribution where the transition point is called recruitment stretch, Fig. 4a) via a multiplicative decomposition of the deformation gradient (Fig. 4b). Collagen reorientation alone was predicted to lead to decreased construct stiffnesses. Remodelling the stress-free configuration of the collagen network increased swelling pressures and altered its state of pre-stress leading to increased mechanical properties. Only when combining both mechanisms could the increased Young's moduli, decreased Poisson's ratios and altered construct geometries be predicted in accordance with experimental observations (Fig. 5). These results provide support for the hypothesis that in addition to various mechanisms such as cross-linking, a structural reorganisation of the collagen network potentially contributes to enhanced mechanical properties, and provides mechanistic insight into the effects of different structural phenomena [79]. The study further demonstrates how constitutive models can add insight to experimental results via a theoretical decoupling of physical mechanisms.

5 Bridging the Gap: Multiscale Models

5.1 The Multiscale Approach

Extensive knowledge on biological processes and their dependence on mechanics is being acquired at all relevant length scales from the biomolecular level up to the tissue and organ level. One challenge is to data mine, integrate, present and distil this huge amount of information and computational information technology is an

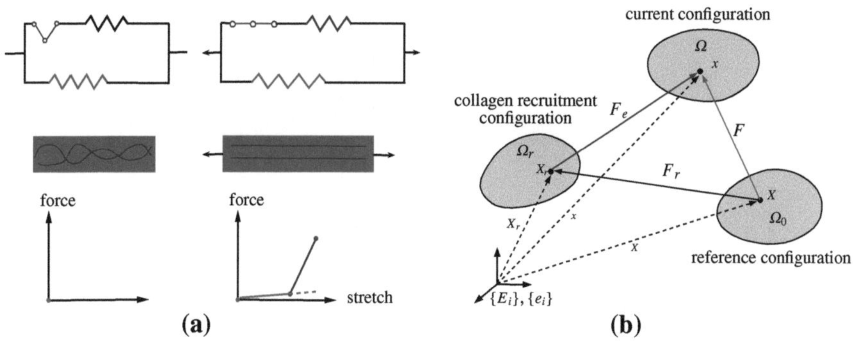

Fig. 4 Illustration of the collagen network's state of tension-compression transition in 1D (**a**) and 3D (**b**). **a** The stress-free configuration of a collagen fibre can be visualised by a one dimensional rheological analogy (*top row*) and the concept of crimp (*middle row*). Once the fibres are uncrimped and the angular joint in the assembly is straightened out, the top spring representing the collagen fibres begins to bear load. After this (recruitment) point the force curve of the idealised material has a higher slope (*bottom row*). **b** A multiplicative decomposition of the deformation gradient F into a recruitment F_r and elastic part F_e allows the definition of appropriate free energy potentials for various constituents and a remodelling of the collagen network's stress-free configuration in 3D. Colour version available online

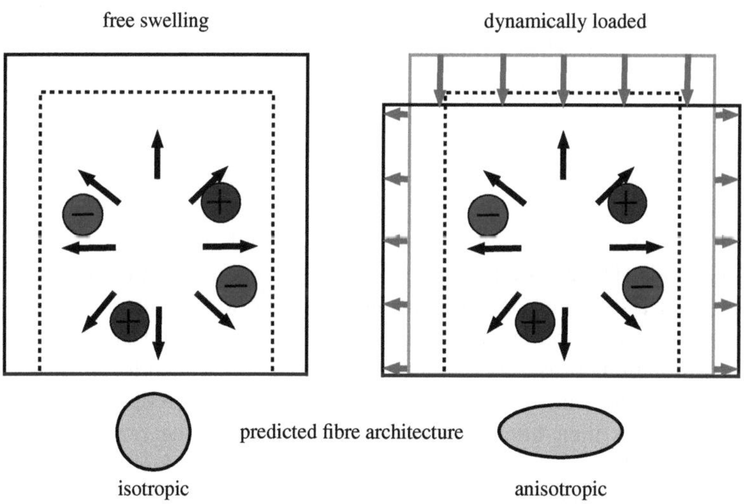

Fig. 5 The fixed negative charges of the glycosaminoglycans lead to an isotropic swelling of the constructs. The FS group therefore has an isotropic tissue architecture. Dynamic compression induces axial compressive strains and lateral expansion. Remodelling leads to horizontal fibre alignment and a similarly anisotropic distribution of the recruitment stretch. In combination, this leads to higher Young's moduli, lower Poisson's ratios as well as lower, wider and more compact samples. Colour version available online

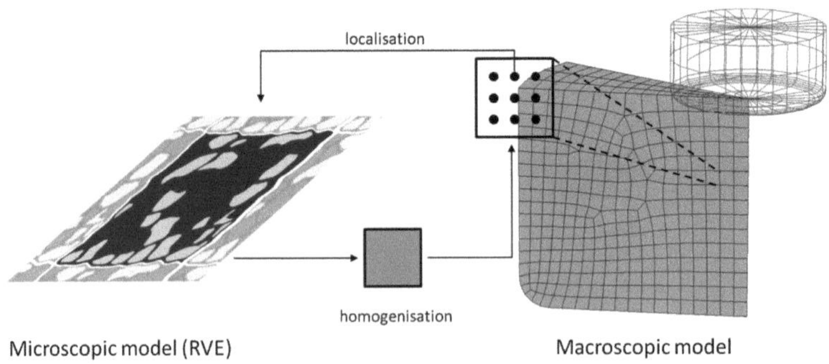

Fig. 6 Schematic illustration of the multiscale approach. Instead of a constitutive model at the integration point level of the macroscopic domain, a localisation rule is used to determine the boundary conditions for a representative volume element (RVE, here example with periodic boundary conditions [82]). The solution of the microscale problem is then homogenised (smeared) and passed up to the macroscopic level again. Colour version available online

indispensable tool in this process [104]. Another challenge is to bridge the scales and integrate the information into multiscale models that are predictive across a range of scales. This presents challenges on all levels: experimental, theoretical and computational. The computational methods for scale bridging and coarse graining have not come to a close and neither have mechanoregulation theories for tissue differentiation. Thus, while a significant body of literature exists at both ends, little has been done at the interface, i.e. multiscale mechanobiological tissue differentiation models.

Bone properties are determined by the intrinsic material properties and morphology of the underlying trabecular structure. Since a single computational model of a full bone that is detailed enough to capture the underlying trabecular architecture does not seem feasible, the problem lends itself to hierarchical multiscale methods. The classic approach (Fig. 6) is computationally very expensive: Starting from a deformation estimate on the macro-model, local boundary conditions are derived for the micro-models (following a localisation rule). Finite element simulations of often complex representative volume elements have to be then solved for each integration point location of the macroscopic domain and the stress response is then homogenised and passed up to the macroscopic model replacing pre-defined constitutive models at this level. This loop has to be iterated until equilibrium is achieved and only then can the next increment be processed. For models of a practically relevant size this approach is only feasible on large parallel computing facilities where the RVE simulations are distributed and solved simultaneously. Hambli et al. [42] introduced neural network computation into multiscale simulations of bone remodelling. Five μCT based models of the trabecular structure served as RVEs and a remodelling algorithm [41] was implemented. These RVEs were loaded with boundary conditions derived from the macroscopic model under known inputs (the amplitude, orientation and frequency of the applied stress) to derive five outputs (the averaged bone density, damage, elastic modulus and stimulus). These inputs and outputs from 100 factorial designs were used to train a neural network, i.e. determine the weights of its connections. Once trained, the network, which does not contain any a priori rules, is then able to very quickly predict outputs when given a set of input values. This neural network thus eliminates the need for the RVE FE simulations and can be incorporated as a material formulation at the integration point level which speeds up simulations tremendously. The model was applied to the remodelling of a femoral head loaded with 7000 cycles per day for 100 days. Simulations were about 1000 times faster than with the classical approach.

Theoretical models of cells that account for their interaction with the pericellular (PCM) and extracellular matrix (ECM) and are coupled to tissue level models "can provide information on biophysical parameters that cannot be measured experimentally in situ at the cellular level, e.g., the stress–strain, fluid flow, physicochemical, and electrical states in the immediate vicinity of the cell" [40]. By investigating cell-matrix interactions using a biphasic multiscale model, the transient heterogeneity of the microscopic mechanical environment due to the differences in mechanical properties between cells and their ECM could be shown. Adding a PCM to the model had a significant effect on the stress and strain fields within the chondrocytes suggesting a biomechanically functional role for the PCM [40].

A method to couple macroscopic FE and microscopic Voxel-FE simulations to investigate bone regeneration was presented in Sanz-Herrera et al. [101]. Bone remodelling and cell migration were solved at the macroscopic level. The information was passed to the microscopic level and used to simulate bone growth and scaffold resorption in the representative volume elements. Using homogenisation techniques, the mechanical properties derived from the representative volume elements were then passed up again to the macroscopic model. The model was applied to the formation of immature and mature bone in a non-resorbing ceramic scaffold in a rabbit femoral defect. The principal model behaviour and the effect of different macroscopic locations on the evolution of local RVEs was investigated. Later, Sanz-Herrera et al. [102] studied scaffold-aided bone tissue regeneration in a rabbit femur in more detail. Variations of scaffold stiffness, porosity, pore size, degradation mode and seeding were simulated. The femur was modelled macroscopically only and its mechanical properties were dependent on the apparent density of the bone that was allowed to remodel depending on the local strain energy density. Macroscopic properties in the repair zone were derived from microscale analyses of both the solid and the fluid domains of the scaffold. In turn, stimuli and cell densities derived at the macroscopic level were used to determine the rate of microscopic bone formation in the scaffold. The mechanical stimulus used was based on local strain energy density and biomaterial stiffness. Degradation via scaffold hydrolysis was also included. The highest rate of bone formation was predicted in the stiffest scaffold and no bone formation in the softest. Compared to a non-seeded scaffold, pre-seeding led to higher rates of earlier bone formation. Higher rates were also observed for increased mean pore sizes while fast resorption kinetics were predicted to lead to scaffold collapse.

A multiscale approach to scaffold design spanning a number of length scales and hierarchies has been presented in Chan et al. [18] for hydroxyapatite-collagen composite materials. The elastic properties of the HA nanoparticles were determined from first (i.e. quantum mechanical) principles. The constitutive behaviour of the HA-collagen composite was then determined for various volume ratios using microstructural 2D unit cell FE modelling of a HA particle embedded in a collagen matrix. The collected data was used to investigate the behaviour of a 3D RVE of the scaffold architecture in order to optimise pore size, pore density and HA volume fraction for cortical and cancellous bone. While the properties of cancellous bone could be matched, those of cortical bone could not be reproduced. Layer-by-layer scaffold fabrication based on printing techniques was envisioned to manufacture the computationally optimised scaffold.

A middle-out multiscale approach to assess fracture risk in a proximal femur and an outline for appropriate verification and validation methods was introduced by Cristofolini et al. [24] spanning several scales from the body to the cell level. The central part of the approach is the organ model of the bone where the fracture occurs. This model and its boundary conditions are informed by the body level model establishing the musculoskeletal loads during daily activity. Downstream, the organ level model also relies on the tissue level model where constitutive relations are used to derive stresses and strains as well as the risk for material failure (fracture). The tissue level model is coupled to the cell level model, that provides the material properties for the tissue level model and relies on the input of biophysical stimuli from the tissue level model to simulate bone cell activity and remodelling. This type of hierarchical approach potentially allows the investigation of a multitude of factors as diverse as pharmacological treatment and subject specific gait patterns and activity levels on the fracture risk of a bone. Currently however, many questions regarding the individual component models remain unresolved and under investigation. For any clinical use the various levels of the model also need to be coupled in a fully automated and robust fashion to improve usability.

5.2 Tissue Structure in Multiscale Simulations of Tissue Regeneration

Bone and cartilage are anisotropic tissues and as such scaffolds can provide environmental cues that guide subsequent anisotropic tissue formation [30, 47]. Microscopic tissue structure has been incorporated into models investigating the biomechanics of tissue behaviour. Sander et al. [98] report a recent example in the context of tissue equivalent mechanics. Based on orientation and alignment measured using polarised light imaging unique 3D fibre networks were created in each point of the cell-compacted collagen gels. Based on constitutive parameters defined at the fibre level and fit to macroscopic off-axis hold tests, the model was then able to predict both the macroscopic response to equibiaxial stretch tests as well as the non-affine and heterogeneous fibre network rearrangement as measured during deformation using polarimetric fibre alignment imaging. The model thus allowed for the simultaneous prediction of experimental data acquired at multiple length scales and provides insight into how macroscopically applied loads translate into highly heterogeneous fibre deformations, both tensile and compressive, which have relevance to cellular mechanotransduction.

Relatively few studies have combined architectural considerations with models of tissue differentiation and regeneration of osseous and chondral tissues. The multiscale bone growth model by Sanz-Herrera et al. [101, 102] was extended to study the effect of pore structure on directional bone growth [103]. Scaffolds with idealised isotropic (spherical) and anisotropic (ellipsoid) pore structures as well as realistic pore structures were simulated and anisotropic elasticity and permeability tensors derived from the homogenisation process and passed to the macroscopic scaffold domain. While the net rate of bone formation was found to be similar for the various architectures once they were sufficiently interconnected, the initial anisotropy determined the directions of bone growth. This was caused by the influence of the pore architecture both on the microscopic deformation and on cell migration/fluid perfusion.

Trabecular healing in a vertebral body was investigated using a multiscale approach in Boccaccio et al. [11]. The spinal segment L3-L4-L5 with a mild wedge fracture in the L4 vertebra was used to derive the poroelastic boundary conditions for representative volume elements in the repair zone. An idealised trabecular structure with a 0.5 mm diastasis served as the microdomain and was used to predict the patterns of tissue differentiation based on fluid flow and shear strain [91] during the first 100 days after fracture. Equivalent material properties of the RVEs were then derived an passed up to their locations in the fracture gap. The predicted tissue differentiation patterns were consistent with those observed in vivo. The primary bone formation mode was endochondral ossication with new woven bone occupying most of the space within the fracture site about 7–8 weeks post fracture. The final stage predicted by the model was bone remodelling leading to the formation of a new trabecular architecture.

6 Conclusion

Computational engineering methods penetrate biology at many levels. Experimental and computational data is continuously being acquired at all levels from the molecular to the organism (or even population) level. Ultimately a coupling of the relevant time and length scales is envisioned but significant challenges remain at each individual level. Experimentation and simulation have to inform one another in an iterative and closely coupled fashion. Once confidence in a model for a certain phenomenon is established, the number of experiments needed can be reduced drastically to the level necessary for parameter identification. Furthermore, neural network approaches that to a certain extent mimic learning processes can be used to reduce computational costs significantly by substituting complex simulations at one or more levels of a multiscale model. Since they can easily incorporate additional data as it becomes available from a variety of sources and adapt to them (via re-training) they present a powerful tool for the integration of information from mechanical, biological and other sources. Neural networks do not rely on any a-priori defined rules but "discover" them and as such may aid in the development of potent mechanobiological theories based on large amounts of diverse data, both experimental and theoretical, that may be difficult to condense analytically.

The advances made have provided and continue to provide insight into many aspects relevant for tissue engineering: Cellular models accounting for the dynamic nature of the cytoskeleton and focal adhesions improve our understanding of the cell's interaction with its immediate surrounding. This includes measurement tools so that these active models can help interpret experimental data. Models of regenerative processes in vivo allow the investigation of environmental factors on tissue differentiation processes. These frameworks help to uncouple individual mechanisms, such as the contribution of tissue structure to bulk mechanical properties, and test hypotheses related to these mechanisms that could not be tested directly experimentally. Computational techniques and engineering approaches to problem solving in general can benefit in particular processes such as scaffold design, where a large number of coupled criteria have to be balanced in order to achieve a desired product. Here, simulation techniques serve both purposes: Investigation and understanding of the biophysical environment and its biological consequences on the one hand, and, more traditionally yet not simpler, a robust quantification to facilitate decisions on design parameters on the other. The diversity of current modelling techniques reflects that of the length scales, questions, applications and challenges in the field of tissue engineering.

6.1 Outlook

Challenges in computational mechanobiology that offer opportunities for further development are abundant. From a basic research perspective, models need to be developed to advance our understanding of how cells and tissues perform in a mechanically challenging environment as well as why certain regenerative approaches work and others don't. In order to be able to provide such insights several challenges need to be addressed:

- Models should reflect individual aspects of a problem more accurately and mechanistically to provide quantitative understanding, predictive ability and allow the testing of hypotheses via e.g. numerical *knock-out* simulations.
- Models addressing problems in tissue engineering need to consider multiple concepts, such as transport, electromechanical stimulation, effects of the chemical environment and reactions occurring therein.

- These mechanisms need to be coupled, i.e. thermodynamical considerations (energy and mass exchange) affect growth and remodelling, which in turn alters the mechanical properties of the tissue and might affect the environment in which chemical reactions take place, etc.
- Time and length scales need to be bridged and knowledge from the individual models effectively linked.

When viewed from a clinical perspective other considerations enter the agenda. While in basic research the immediate goal is to gain knowledge, models are only justified and necessary in a clinical environment where there is a measurable patient benefit. Therefore, questions on whether a particular computational tool can aid in making treatment decisions, plan surgical interventions or choose an appropriate implant need to be answered under the pressure of health care budget constraints. Another major challenge lies before the realisation of computer-assisted patient specific regenerative medicine: Verification and validation. When the welfare of human beings is affected by decisions based on computational predictions, the underlying models will need to be rigorously tested and their creation, validation and application officially regulated.

Finally, the commercial sector can contribute to both the clinical and academic worlds. For example, one can envisage the development of adaptive bioreactors that feature the online integration of predictive models for the assessment of otherwise unknown state variables of the cultured tissue and the cells within it. Robustness, ease of use and validation are some of the principal requirements. The challenges and opportunities are huge and provide space for a large variety of modelling approaches and their development towards routine applications. It seems equally important to discover new approaches as it is to learn from established fields to avoid re-inventing the wheel. The enthusiasm in exploring these new developments needs to be paired with a critical assessment of their potential benefits.

Acknowledgments We thank Dr. Patrick McGarry for image material. Funding was provided by IRCSET (G30345) and a SFI PIYRA award (08/YI5/B1336).

References

1. Ament, C., Hofer, E.P.: A fuzzy logic model of fracture healing. J. Biomechanics **33**(8), 961–968 (2000). doi:10.1016/S0021-9290(00)00049-X. http://www.sciencedirect.com/science/article/pii/S002192900000049X
2. Balguid, A., Rubbens, M.P., Mol, A., Bank, R.A., Bogers, A.J.J.C, van Kats, J.P., de Mol, B.A.J.M., Baaijens, F.P.T., Bouten, C.V.C.: The role of collagen cross-links in biomechanical behavior of human aortic heart valve leaflets—relevance for tissue engineering. Tissue Eng. **13**(7), 1501–1511 (2007). doi:10.1089/ten.2006.0279. http://dx.doi.org/10.1089/ten.2006.0279
3. Barocas, V.H., Tranquillo, R.T.: An anisotropic biphasic theory of tissue-equivalent mechanics: the interplay among cell traction, fibrillar network deformation, fibril alignment, and cell contact guidance. J. Biomech. Eng. **119**(2), 137–145 (1997)

4. Batra, N.N., Li, Y.J., Yellowley, C.E., You, L., Malone, A.M., Kim, C.H., Jacobs, C.R.: Effects of short-term recovery periods on fluid-induced signaling in osteoblastic cells. J. Biomech. **38**(9), 1909–1917 (2005). doi:10.1016/j.jbiomech.2004.08.009. http://dx.doi.org/10.1016/j.jbiomech.2004.08.009
5. Bian, L., Fong, J.V., Lima, E.G., Stoker, A.M., Ateshian, G.A., Cook, J.L., Hung, C.T.: Dynamic mechanical loading enhances functional properties of tissue-engineered cartilage using mature canine chondrocytes. Tissue Eng. Part A **16**(5), 1781–1790 (2010). doi:10.1089/ten.TEA.2009.0482. http://dx.doi.org/10.1089/ten.TEA.2009.0482
6. Bischofs, I.B., Schwarz, U.S.: Cell organization in soft media due to active mechanosensing. Proc. Natl Acad. Sci. U S A **100**(16), 9274–9279 (2003). http://www.pnas.org/content/100/16/9274.abstract
7. Bjork, J., Safonov, A., Tranquillo, R.: Computational Modeling in Tissue Engineering. Oxygen Transport in Bioreactors for Engineered Vascular Tissues. Springer, Berlin-Heidelberg (2012)
8. Boccaccio, A., Pappalettere, C., Kelly, D.J.: The influence of expansion rates on mandibular distraction osteogenesis: a computational analysis. Ann. Biomed. Eng. **35**(11), 1940–1960 (2007). doi:10.1007/s10439-007-9367-x. http://dx.doi.org/10.1007/s10439-007-9367-x
9. Boccaccio, A., Lamberti, L., Pappalettere, C., Cozzani, M., Siciliani, G.: Comparison of different orthodontic devices for mandibular symphyseal distraction osteogenesis: a finite element study. Am. J. Orthod. Dentofac. Orthop. **134**(2), 260–269 (2008). doi:10.1016/j.ajodo.2006.09.066. http://dx.doi.org/10.1016/j.ajodo.2006.09.066
10. Boccaccio, A., Ballini, A., Pappalettere, C., Tullo, D., Cantore, S., Desiate, A.: Finite element method (fem), mechanobiology and biomimetic scaffolds in bone tissue engineering. Int. J. Biol. Sci. **7**(1), 112–132 (2011)
11. Boccaccio, A., Kelly, D.J., Pappalettere, C.: A mechano-regulation model of fracture repair in vertebral bodies. J. Orthop. Res. **29**(3), 433–443 (2011). doi:10.1002/jor.21231. http://dx.doi.org/10.1002/jor.21231
12. Burke, D., Kelly, D.: Could substrate stiffness and oxygen tension regulate stem cell differentiation during fracture healing? In: Proceedings of the ASME 2011 Summer Bioengineering Conference, Farmington, Pennsylvania, USA, 22–25 June 2011
13. Butler, D.L., Goldstein, S.A., Guilak, F.: Functional tissue engineering: the role of biomechanics. J. Biomech. Eng. **122**(6), 570–575 (2000)
14. Byrne, D.P., Lacroix, D., Planell, J.A., Kelly, D.J., Prendergast, P.J.: Simulation of tissue differentiation in a scaffold as a function of porosity, young's modulus and dissolution rate: application of mechanobiological models in tissue engineering. Biomaterials **28**(36), 5544–5554 (2007). doi:10.1016/j.biomaterials.2007.09.003. http://dx.doi.org/10.1016/j.biomaterials.2007.09.003
15. Carter, D.R., Blenman, P.R., Beauprcé, G.S.: Correlations between mechanical stress history and tissue differentiation in initial fracture healing. J. Orthop. Res. **6**(5), 736–748 (1988). doi:10.1002/jor.1100060517. http://dx.doi.org/10.1002/jor.1100060517
16. Carter, D.R., Wong, M., Orr, T.E.: Musculoskeletal ontogeny, phylogeny, and functional adaptation. J. Biomech. **24**(Suppl 1), 3–16 (1991). doi:10.1016/0021-9290(91)90373-U. http://www.sciencedirect.com/science/article/pii/002192909190373U (proceedings of the NASA Symposium on the Influence of Gravity and Activity on Muscle and Bone)
17. Carter, D.R., Beauprcé, G.S., Giori, N.J., Helms, J.A.: Mechanobiology of skeletal regeneration. Clin. Orthop. Relat. Res. **355**(355 Suppl), S41–S55 (1998)
18. Chan, K.S., Liang, W., Francis, W.L., Nicolella, D.P.: A multiscale modeling approach to scaffold design and property prediction. J. Mech. Behav. Biomed. Mater. **3**(8), 584–593 (2010). doi:10.1016/j.jmbbm.2010.07.006. http://dx.doi.org/10.1016/j.jmbbm.2010.07.006
19. Checa, S., Prendergast, P.J.: A mechanobiological model for tissue differentiation that includes angiogenesis: a lattice-based modeling approach. Ann. Biomed. Eng. **37**(1), 129–145 (2009). doi:10.1007/s10439-008-9594-9. http://dx.doi.org/10.1007/s10439-008-9594-9
20. Checa, S., Prendergast, P.J.: Effect of cell seeding and mechanical loading on vascularization and tissue formation inside a scaffold: a mechano-biological model using

a lattice approach to simulate cell activity. J. Biomech. **43**(5), 961–968 (2010). doi:10.1016/j.jbiomech.2009.10.044. http://dx.doi.org/10.1016/j.jbiomech.2009.10.044

21. Christen, P., van Rietbergen, B., Lambers, F.M., Müller, R., Ito, K.: Bone morphology allows estimation of loading history in a murine model of bone adaptation. Biomech. Model Mechanobiol. **10**(5), 663–670 (2011). doi:10.1007/s10237-011-0327-x. http://dx.doi.org/10.1007/s10237-011-0327-x
22. Claes, L.E., Heigele, C.A.: Magnitudes of local stress and strain along bony surfaces predict the course and type of fracture healing. J. Biomech. **32**(3), 255–266 (1999)
23. Claes, L.E., Heigele, C.A., Neidlinger-Wilke, C., Kaspar, D., Seidl, W., Margevicius, K.J., Augat, P.: Effects of mechanical factors on the fracture healing process. Clin. Orthop. Relat. Res. **355**(355 Suppl), S132–S147 (1998)
24. Cristofolini, L., Taddei, F., Baleani, M., Baruffaldi, F., Stea, S., Viceconti, M.: Multiscale investigation of the functional properties of the human femur. Philos. Trans. A Math. Phys. Eng. Sci. **366**(1879), 3319–3341 (2008). doi:10.1098/rsta.2008.0077. http://dx.doi.org/10.1098/rsta.2008.0077
25. Cullinane, D.M., Fredrick, A., Eisenberg, S.R., Pacicca, D., Elman, M.V., Lee, C., Salisbury, K., Gerstenfeld, L.C., Einhorn, T.A.: Induction of a neoarthrosis by precisely controlled motion in an experimental mid-femoral defect. J. Orthop. Res. **20**(3), 579–586 (2002). doi:10.1016/S0736-0266(01)00131-0. http://dx.doi.org/10.1016/S0736-0266(01)00131-0
26. Cullinane, D.M., Salisbury, K.T., Alkhiary, Y., Eisenberg, S., Gerstenfeld, L., Einhorn, T.A.: Effects of the local mechanical environment on vertebrate tissue differentiation during repair: does repair recapitulate development? J. Exp. Biol. **206**(Pt 14), 2459–2471 (2003)
27. Deshpande, V.S., McMeeking, R.M., Evans, A.G.: A model for the contractility of the cytoskeleton including the effects of stress-fibre formation and dissociation. Proc. Royal Soc. A Math. Phys. Eng. Sci. **463**(2079), 787–815 (2007) http://rspa.royalsocietypublishing.org/content/463/2079/787.abstract
28. Deshpande, V.S., Mrksich, M., McMeeking, R.M., Evans, A.G.: A bio-mechanical model for coupling cell contractility with focal adhesion formation. J. Mech. Phys. Solids **56**(4), 1484–1510 (2008). doi:10.1016/j.jmps.2007.08.006. http://www.sciencedirect.com/science/article/B6TXB-4PJM9VR-1/2/e406bd345af9e0173c1b96ef7698463f
29. Discher, D.E., Janmey, P., Wang, Y.l.: Tissue cells feel and respond to the stiffness of their substrate. Science **310**(5751), 1139–1143 (2005). http://www.sciencemag.org/content/310/5751/1139.abstract
30. Engelmayr, G.C., Papworth, G.D., Watkins, S.C., Mayer, J.E., Sacks, M.S.: Guidance of engineered tissue collagen orientation by large-scale scaffold microstructures. J. Biomech. **39**(10), 1819–1831 (2006). doi:10.1016/j.jbiomech.2005.05.020. http://dx.doi.org/10.1016/j.jbiomech.2005.05.020
31. Engler, A.J., Sen, S., Sweeney, H.L., Discher, D.E.: Matrix elasticity directs stem cell lineage specification. Cell **126**(4), 677–689 (2006). http://linkinghub.elsevier.com/retrieve/pii/S0092867406009615
32. Fereol, S., Fodil, R., Pelle, G., Louis, B., Laurent, V., Planus, E., Isabey, D.: Understanding adhesion sites as mechanosensitive cellular elements. In: Chapman & Hall/CRC Mathematical and Computational Biology, pp. 221–241. Chapman & Hall/CRC, London (2010). http://dx.doi.org/10.1201/9781420094558-c8
33. Foolen, J., van Donkelaar, C.C., Soekhradj-Soechit, S., Ito, K.: European society of biomechanics s.m. perren award 2010: an adaptation mechanism for fibrous tissue to sustained shortening. J. Biomech. **43**(16), 3168–3176 (2010). doi:10.1016/j.jbiomech.2010.07.040. http://www.sciencedirect.com/science/article/B6T82-50XJC0S-1/2/47b646146d92bb178404711cc88ec093
34. García-Aznar, J.M., Kuiper, J.H., Gómez-Benito, M.J., Doblaré, M., Richardson, J.B.: Computational simulation of fracture healing: influence of interfragmentary movement on the callus growth. J. Biomech. **40**(7), 1467–1476 (2007). doi:10.1016/j.jbiomech.2006.06.013 . http://dx.doi.org/10.1016/j.jbiomech.2006.06.013

35. Geris, L., Oosterwyck, H.V., Sloten, J.V., Duyck, J., Naert, I.: Assessment of mechanobiological models for the numerical simulation of tissue differentiation around immediately loaded implants. Comput. Methods Biomech. Biomed. Eng. **6**(5–6), 277–288 (2003). doi:10.1080/10255840310001634412. http://dx.doi.org/10.1080/10255840310001634412
36. Geris, L., Andreykiv, A., Oosterwyck, H.V., Sloten, J.V., van Keulen, F., Duyck, J., Naert, I.: Numerical simulation of tissue differentiation around loaded titanium implants in a bone chamber. J. Biomech. **37**(5), 763–769 (2004). doi:10.1016/j.jbiomech.2003.09.026. http://www.sciencedirect.com/science/article/B6T82-4B0PPFB-2/2/2a52f7f66103bdeb80efe3cda53aed30
37. Geris, L., Gerisch, A., Sloten, J.V., Weiner, R., Oosterwyck, H.V.: Angiogenesis in bone fracture healing: a bioregulatory model. J. Theor. Biol. **251**(1), 137–158 (2008). doi:10.1016/j.jtbi.2007.11.008. http://dx.doi.org/10.1016/j.jtbi.2007.11.008
38. Geris, L., Schugart, R., Van Oosterwyck, H.: In silico design of treatment strategies in wound healing and bone fracture healing. Philos. Trans. A Math. Phys. Eng. Sci. **368**(1920), 2683–2706 (2010). doi:10.1098/rsta.2010.0056. http://dx.doi.org/10.1098/rsta.2010.0056
39. Giori, N.J., Beaupr, G.S., Carter, D.R.: Cellular shape and pressure may mediate mechanical control of tissue composition in tendons. J. Orthop. Res. **11**(4), 581–591 (1993). doi:10.1002/jor.1100110413. http://dx.doi.org/10.1002/jor.1100110413
40. Guilak, F., Mow, V.C.: The mechanical environment of the chondrocyte: a biphasic finite element model of cell-matrix interactions in articular cartilage. J. Biomech. **33**(12), 1663–1673 (2000)
41. Hambli, R., Soulat, D., Gasser, A., Benhamou, C.L.: Strain-damage coupled algorithm for cancellous bone mechano-regulation with spatial function influence. Comput. Methods Appl. Mech. Eng. **198**(33–36), 2673–2682 (2009). doi:10.1016/j.cma.2009.03.014. http://www.sciencedirect.com/science/article/pii/S004578250900139X
42. Hambli, R., Katerchi, H., Benhamou, C.L.: Multiscale methodology for bone remodelling simulation using coupled finite element and neural network computation. Biomech. Model Mechanobiol. **10**(1), 133–145 (2011). doi:10.1007/s10237-010-0222-x. http://dx.doi.org/10.1007/s10237-010-0222-x
43. Hayward, L., Morgan, E.: Assessment of a mechano-regulation theory of skeletal tissue differentiation in an in vivo model of mechanically induced cartilage formation. Biomech. Model Mechanobiol. **2**(2), 109–126 (2009). doi:10.1007/s10237-009-0148-3. http://dx.doi.org/10.1007/s10237-009-0148-3
44. Hoenig, E., Winkler, T., Mielke, G., Paetzold, H., Schuettler, D., Goepfert, C., Machens, H.G., Morlock, M.M., Schilling, A.F.: High amplitude direct compressive strain enhances mechanical properties of scaffold-free tissue-engineered cartilage. Tissue Eng. Part A **17**(9–10), 1401–1411 (2011). doi:10.1089/ten.tea.2010.0395. http://www.liebertonline.com/doi/abs/10.1089/ten.tea.2010.0395, http://www.liebertonline.com/doi/pdf/10.1089/ten.tea.2010.0395
45. Huang, A., Farrell, M., Kim, M., Mauck, R.: Long-term dynamic loading improves the mechanical properties of chondrogenic mesenchymal stem cell-laden hydrogel. Eur. Cells Mater. **19**, 72–85 (2010)
46. Huiskes, R., Driel, W.D.V., Prendergast, P.J., Søballe, K.: A biomechanical regulatory model for periprosthetic fibrous-tissue differentiation. J. Mater. Sci. Mater. Med. **8**(12), 785–788 (1997)
47. Hutmacher, D.W.: Scaffolds in tissue engineering bone and cartilage. Biomaterials **21**(24), 2529–2543 (2000)
48. Ingber, D.E.: Tensegrity: the architectural basis of cellular mechanotransduction. Annu. Rev. Physiol. **59**(1), 575–599 (1997). doi:10.1146/annurev.physiol.59.1.575. http://www.annualreviews.org/doi/abs/10.1146/annurev.physiol.59.1.575, http://www.annualreviews.org/doi/pdf/10.1146/annurev.physiol.59.1.575
49. Isaksson, H., van Donkelaar, C.C., Huiskes, R., Ito, K.: Corroboration of mechanoregulatory algorithms for tissue differentiation during fracture healing: comparison with in vivo results. J. Orthop. Res. **24**(5), 898–907 (2006). doi:10.1002/jor.20118. http://dx.doi.org/10.1002/jor.20118

50. Isaksson, H., Wilson, W., van Donkelaar, C.C., Huiskes, R., Ito, K.: Comparison of biophysical stimuli for mechano-regulation of tissue differentiation during fracture healing. J. Biomech. **39**(8), 1507–1516 (2006). doi:10.1016/j.jbiomech.2005.01.037. http://dx.doi.org/10.1016/j.jbiomech.2005.01.037
51. Isaksson, H., Comas, O., van Donkelaar, C.C., Mediavilla, J., Wilson, W., Huiskes, R., Ito, K.: Bone regeneration during distraction osteogenesis: mechano-regulation by shear strain and fluid velocity. J. Biomech. **40**(9), 2002–2011 (2007). doi:10.1016/j.jbiomech.2006.09.028 . http://dx.doi.org/10.1016/j.jbiomech.2006.09.028
52. Jaasma, M.J., O'Brien, F.J.: Mechanical stimulation of osteoblasts using steady and dynamic fluid flow. Tissue Eng. Part A **14**(7), 1213–1223 (2008). doi:10.1089/tea.2007.0321. http://dx.doi.org/10.1089/tea.2007.0321
53. Janmey, P.A., McCulloch, C.A.: Cell mechanics: integrating cell responses to mechanical stimuli. Annu. Rev. Biomed. Eng. **9**(1), 1–34 (2007). doi:10.1146/annurev.bioeng.9.060906.151927. http://www.annualreviews.org/doi/abs/10.1146/annurev.bioeng.9.060906.151927, http://www.annualreviews.org/doi/pdf/10.1146/annurev.bioeng.9.060906.151927
54. Jungreuthmayer, C., Jaasma, M.J., Al-Munajjed, A.A., Zanghellini, J., Kelly, D.J., O'Brien, F.J.: Deformation simulation of cells seeded on a collagen-gag scaffold in a flow perfusion bioreactor using a sequential 3d cfd-elastostatics model. Med. Eng. Phys. **31**(4), 420–427 (2009). doi:10.1016/j.medengphy.2008.11.003. http://dx.doi.org/10.1016/j.medengphy.2008.11.003
55. Kelly, D.J., Prendergast, P.J.: Mechano-regulation of stem cell differentiation and tissue regeneration in osteochondral defects. J. Biomech. **38**(7), 1413–1422 (2005). doi:10.1016/j.jbiomech.2004.06.026. http://dx.doi.org/10.1016/j.jbiomech.2004.06.026
56. Kelly, D.J., Prendergast, P.J.: Prediction of the optimal mechanical properties for a scaffold used in osteochondral defect repair. Tissue Eng. **12**(9), 2509–2519 (2006). doi:10.1089/ten.2006.12.2509. http://dx.doi.org/10.1089/ten.2006.12.2509
57. Kelly, K.: The third culture. Science **279**(5353), 992–993 (1998). doi:10.1126/science.279.5353.992. http://www.sciencemag.org/content/279/5353/992.short
58. Kelly, T.A.N., Ng, K.W., Wang, C.C.B., Ateshian, G.A., Hung, C.T.: Spatial and temporal development of chondrocyte-seeded agarose constructs in free-swelling and dynamically loaded cultures. J. Biomech. **39**(8), 1489–1497 (2006). doi:10.1016/j.jbiomech.2005.03.031. http://www.sciencedirect.com/science/article/B6T82-4GH4B00-1/2/76d2371137ab41375f98c24852d1fd7d
59. Khayyeri, H., Checa, S., Tgil, M., Prendergast, P.J.: Corroboration of mechanobiological simulations of tissue differentiation in an in vivo bone chamber using a lattice-modeling approach. J. Orthop. Res. **27**(12), 1659–1666 (2009). doi:10.1002/jor.20926. http://dx.doi.org/10.1002/jor.20926
60. Khayyeri, H., Checa, S., Tägil, M., O'Brien, F., Prendergast, P.: Tissue differentiation in an in vivo bioreactor: in silico investigations of scaffold stiffness. J. Mater. Sci. Mater. Med. **21**, 2331–2336 (2010). doi:10.1007/s10856-009-3973-0. http://dx.doi.org/10.1007/s10856-009-3973-0
61. Khayyeri, H., Checa, S., Tgil, M., Aspenberg, P., Prendergast, P.J.: Variability observed in mechano-regulated in vivo tissue differentiation can be explained by variation in cell mechano-sensitivity. J. Biomech. **44**(6), 1051–1058 (2011). doi:10.1016/j.jbiomech.2011.02.003. http://www.sciencedirect.com/science/article/pii/S0021929011000844
62. Khoshgoftar, M., van Donkelaar, C.C., Ito, K.: Mechanical stimulation to stimulate formation of a physiological collagen architecture in tissue-engineered cartilage: a numerical study. Comput. Methods Biomech. Biomed. Eng. **14**(2), 135–144 (2011). doi:10.1080/10255842.2010.519335. http://www.tandfonline.com/doi/abs/10.1080/10255842.2010.519335, http://www.tandfonline.com/doi/pdf/10.1080/10255842.2010.519335
63. Kjaer, M.: Role of extracellular matrix in adaptation of tendon and skeletal muscle to mechanical loading. Physiol. Rev. **84**(2), 649–698 (2004). doi:10.1152/physrev.00031.2003. http://dx.doi.org/10.1152/physrev.00031.2003
64. Klein, T.J., Rizzi, S.C., Reichert, J.C., Georgi, N., Malda, J., Schuurman, W., Crawford, R.W., Hutmacher, D.W.: Strategies for zonal cartilage repair using hydrogels. Macromol.

Biosci. **9**(11), 1049–1058 (2009). doi:10.1002/mabi.200900176. http://dx.doi.org/10.1002/mabi.200900176
65. Klisch, S.M., Chen, S.S., Sah, R.L., Hoger, A.: A growth mixture theory for cartilage with application to growth-related experiments on cartilage explants. J. Biomech. Eng. 125(2), 169–179 (2003)
66. Klisch, S.M., Asanbaeva, A., Oungoulian, S.R., Masuda, K., Thonar, E.J.M., Davol, A., Sah, R.L.: A cartilage growth mixture model with collagen remodeling: validation protocols. J. Biomech. Eng. **130**(3), 031006 (2008). doi:10.1115/1.2907754. http://dx.doi.org/10.1115/1.2907754
67. Lacroix, D., Prendergast, P.J.: A mechano-regulation model for tissue differentiation during fracture healing: analysis of gap size and loading. J. Biomech. **35**(9), 1163–1171 (2002)
68. Lacroix, D., Prendergast, P.J., Li, G., Marsh, D.: Biomechanical model to simulate tissue differentiation and bone regeneration: application to fracture healing. Med. Biol. Eng. Comput. **40**(1), 14–21 (2002)
69. Lau, A.W.C., Hoffman, B.D., Davies, A., Crocker, J.C., Lubensky, T.C.: Microrheology, stress fluctuations, and active behavior of living cells. Phys. Rev. Lett. **91**(19), 198101 (2003). doi:10.1103/PhysRevLett.91.198101
70. Lima, E.G., Bian, L., Ng, K.W., Mauck, R.L., Byers, B.A., Tuan, R.S., Ateshian, G.A., Hung, C.T.: The beneficial effect of delayed compressive loading on tissue-engineered cartilage constructs cultured with tgf-beta3. Osteoarthr. Cartil **15**(9), 1025–1033 (2007). doi:10.1016/j.joca.2007.03.008. http://dx.doi.org/10.1016/j.joca.2007.03.008
71. Loboa, E.G., Beaupr, G.S., Carter, D.R.: Mechanobiology of initial pseudarthrosis formation with oblique fractures. J. Orthop. Res. **19**(6), 1067–1072 (2001). http://dx.doi.org/10.1016/S0736-0266(01)00028-6
72. Martin, I., Wendt, D., Heberer, M.: The role of bioreactors in tissue engineering. Trends Biotechnol. **22**(2), 80–86 (2004). doi:10.1016/j.tibtech.2003.12.001. http://www.sciencedirect.com/science/article/B6TCW-4B7231G-2/2/2db5486ab02127892b5708fc9b97c5bf
73. Mauck, R.L., Wang, C.C.B., Oswald, E.S., Ateshian, G.A., Hung, C.T.: The role of cell seeding density and nutrient supply for articular cartilage tissue engineering with deformational loading. Osteoarthr. Cartil. **11**(12), 879–890 (2003). doi:10.1016/j.joca.2003.08.006. http://www.sciencedirect.com/science/article/B6WP3-49SWBCC-1/2/f33484507367eff3b262cb7434c5fa1d
74. Mauck, R.L., Wang, C.C.B., Oswald, E.S., Ateshian, G.A., Hung, C.T.: The role of cell seeding density and nutrient supply for articular cartilage tissue engineering with deformational loading. Osteoarthr. Cartil. **11**(12), 879–890 (2003)
75. McGarry, J.P., Fu, J., Yang, M.T., Chen, C.S., McMeeking, R.M., Evans, A.G., Deshpande, V.S.: Simulation of the contractile response of cells on an array of micro-posts. Philos. Trans. A Math. Phys. Eng. Sci. **367**(1902), 3477–3497 (2009). doi:10.1098/rsta.2009.0097. http://dx.doi.org/10.1098/rsta.2009.0097
76. Meyer, E.G., Buckley, C.T., Thorpe, S.D., Kelly, D.J.: Low oxygen tension is a more potent promoter of chondrogenic differentiation than dynamic compression. J. Biomech. **43**(13), 2516–2523 (2010). doi:10.1016/j.jbiomech.2010.05.020. http://www.sciencedirect.com/science/article/pii/S0021929010002939
77. Mohrdieck, C., Wanner, A., Roos, W., Roth, A., Sackmann, E., Spatz, J.P., Arzt, E.: A theoretical description of elastic pillar substrates in biophysical experiments. ChemPhysChem **6**(8), 1492–1498 (2005). http://dx.doi.org/10.1002/cphc.200500109
78. Nagel, T., Kelly, D.: The influence of fibre orientation on the equilibrium properties of neutral and charged biphasic tissues. J. Biomech. Eng. **132**(11), 114506 (2010) (7 pages)
79. Nagel, T., Kelly, D.: Mechanically induced structural changes during dynamic compression of engineered cartilaginous constructs can potentially explain increases in bulk mechanical properties. J. Royal Soc. Interface (2011). doi:10.1098/rsif.2011.0449. http://rsif.royalsocietypublishing.org/content/early/2011/09/07/rsif.2011.0449.abstract, http://rsif.royalsocietypublishing.org/content/early/2011/09/07/rsif.2011.0449.full.pdf+html

80. Nagel, T., Kelly, D.: Remodelling of collagen fibre transition stretch and angular distribution in soft biological tissues and cell-seeded hydrogels. Biomech. Model Mechanobiol. (2011). doi:10.1007/s10237-011-0313-3. http://dx.doi.org/10.1007/s10237-011-0313-3
81. Nagel, T., Kelly, D.J.: Mechano-regulation of mesenchymal stem cell differentiation and collagen organisation during skeletal tissue repair. Biomech. Model Mechanobiol. **9**(3), 359–372 (2010). doi:10.1007/s10237-009-0182-1. http://dx.doi.org/10.1007/s10237-009-0182-1
82. Nagel, T., Loerakker, S., Oomens, C.W.J.: A theoretical model to study the effects of cellular stiffening on the damage evolution in deep tissue injury. Comput. Methods Biomech. Biomed. Eng. p. 1 (2009). doi:10.1080/10255840902788603. http://dx.doi.org/10.1080/10255840902788603
83. Nelson, C.M., Jean, R.P., Tan, J.L., Liu, W.F., Sniadecki, N.J., Spector, A.A., Chen, C.S.: Emergent patterns of growth controlled by multicellular form and mechanics. Proc. Natl Acad. Sci. U S A **102**(33), 11594–11599 (2005). doi:10.1073/pnas.0502575102. http://www.pnas.org/content/102/33/11594.abstract, http://www.pnas.org/content/102/33/11594.full.pdf+html
84. Nicolas, A., Safran, S.A.: Limitation of cell adhesion by the elasticity of the extracellular matrix. Biophys. J. **91**(1), 61–73 (2006). http://www.pubmedcentral.nih.gov/articlerender.fcgi?artid=1479082&tool=pmcentrez&rendertype=abstract
85. Nicolas, A., Geiger, B., Safran, S.A.: Cell mechanosensitivity controls the anisotropy of focal adhesions. Proc. Natl Acad. Sci. U S A **101**(34), 12520–12525 (2004). http://www.pnas.org/content/101/34/12520.abstract
86. Oberkampf, W.L., Trucano, T.G., Hirsch, C.: Verification, validation, and predictive capability in computational engineering and physics. Appl. Mech. Rev. **57**(5), 345–384 (2004). doi:10.1115/1.1767847. http://link.aip.org/link/?AMR/57/345/1
87. Pang, H., Shiwalkar, A.P., Madormo, C.M., Taylor, R.E., Andriacchi, T.P., Kuhl, E.: Computational modeling of bone density profiles in response to gait: a subject-specific approach. Biomech. Model Mechanobiol. (2011). doi:10.1007/s10237-011-0318-y. http://dx.doi.org/10.1007/s10237-011-0318-y
88. Pathak, A., Deshpande, V.S., McMeeking, R.M., Evans, A.G.: The simulation of stress fibre and focal adhesion development in cells on patterned substrates. J. Royal Soc. Interface **5**(22), 507–524 (2008). http://rsif.royalsocietypublishing.org/content/5/22/507.abstract
89. Pauwels, F.: Eine neue Theorie über den Einflu mechanischer Reize auf die Differenzierung der Stützgewebe. Anat. Embryol. **121**(6), 478–515 (1960). http://dx.doi.org/10.1007/BF00523401
90. Perez, M., Prendergast, P.: Random-walk models of cell dispersal included in mechanobiological simulations of tissue differentiation. J. Biomech. **40**, 2244–2253 (2007). http://linkinghub.elsevier.com/retrieve/pii/S0021929006003988
91. Prendergast, P., Huiskes, R., Sballe, K.: Esb research award 1996. Biophysical stimuli on cells during tissue differentiation at implant interfaces. J. Biomech. **30**(6), 539–548 (1997)
92. Prendergast, P.J., Galibarov, P.E., Lowery, C., Lennon, A.B.: Computer simulating a clinical trial of a load-bearing implant: an example of an intramedullary prosthesis. J. Mech. Behav. Biomed. Mater. **4**(8), 1880–1887 (2011). doi:10.1016/j.jmbbm.2011.06.005. http://dx.doi.org/10.1016/j.jmbbm.2011.06.005
93. Raimondi, M., Causin, P., Lagana, M., Zunino, P., Sacco, R.: Computational Modeling in Tissue Engineering. Multiphysics Computational Modeling in Cartilage Tissue Engineering. Springer, Berlin-Heidelberg (2012)
94. Reina-Romo, E., Gómez-Benito, M., García-Aznar, J., Domínguez, J., Doblaré, M.: Modeling distraction osteogenesis: analysis of the distraction rate. Biomech. Model. Mechanobiol. **8**, 323–335 (2009). doi:10.1007/s10237-008-0138-x. http://dx.doi.org/10.1007/s10237-008-0138-x
95. Roux, W.: Beiträge zur Morphologie der funktionellen Anpassung. 3. Beschreibung und Erluterung einer knöchernen Kniegelenksankylose. Arch. Anat. Entwicklungsgeschichte **10**, 120–158 (1885)

96. Rubbens, M.P., Mol, A., van Marion, M.H., Hanemaaijer, R., Bank, R.A., Baaijens, F.P.T., Bouten, C.V.C.: Straining mode-dependent collagen remodeling in engineered cardiovascular tissue. Tissue Eng. Part A **15**(4), 841–849 (2009). doi:10.1089/ten.tea.2008.0185. http://dx.doi.org/10.1089/ten.tea.2008.0185
97. Sacks, M.S., Smith, D.B., Hiester, E.D.: The aortic valve microstructure: effects of transvalvular pressure. J. Biomed. Mater. Res. **41**(1), 131–141 (1998). http://dx.doi.org/10.1002/(SICI)1097-4636(199807)41:1<131::AID-JBM16>3.0.CO;2-Q
98. Sander, E.A., Stylianopoulos, T., Tranquillo, R.T., Barocas, V.H.: Image-based multiscale modeling predicts tissue-level and network-level fiber reorganization in stretched cell-compacted collagen gels. Proc. Natl Acad. Sci. U S A **106**(42), 17675–17680 (2009). doi:10.1073/pnas.0903716106. http://dx.doi.org/10.1073/pnas.0903716106
99. Sandino, C., Planell, J., Lacroix, D.: A finite element study of mechanical stimuli in scaffolds for bone tissue engineering. J. Biomech. **41**(5), 1005–1014 (2008). doi:10.1016/j.jbiomech.2007.12.011. http://www.sciencedirect.com/science/article/pii/S0021929007005428
100. Sandino, C., Checa, S., Prendergast, P.J., Lacroix, D.: Simulation of angiogenesis and cell differentiation in a cap scaffold subjected to compressive strains using a lattice modeling approach. Biomaterials **31**(8), 2446–2452 (2010). doi:10.1016/j.biomaterials.2009.11.063. http://dx.doi.org/10.1016/j.biomaterials.2009.11.063
101. Sanz-Herrera, J., Garca-Aznar, J., Doblar, M.: Micro-macro numerical modelling of bone regeneration in tissue engineering. Comput. Methods Appl. Mech. Eng. **197**(33–40), 3092–3107 (2008). doi:10.1016/j.cma.2008.02.010. http://www.sciencedirect.com/science/article/pii/S0045782508000704
102. Sanz-Herrera, J.A., García-Aznar, J.M., Doblaré, M.: On scaffold designing for bone regeneration: a computational multiscale approach. Acta. Biomater. **5**(1), 219–229 (2009). doi:10.1016/j.actbio.2008.06.021. http://dx.doi.org/10.1016/j.actbio.2008.06.021
103. Sanz-Herrera, J.A., Doblaré, M., García-Aznar, J.M.: Scaffold microarchitecture determines internal bone directional growth structure: a numerical study. J. Biomech. **43**(13), 2480–2486 (2010). doi:10.1016/j.jbiomech.2010.05.027. http://dx.doi.org/10.1016/j.jbiomech.2010.05.027
104. Semple, J.L., Woolridge, N., Lumsden, C.J.: In vitro, in vivo, in silico: computational systems in tissue engineering and regenerative medicine. Tissue Eng. **11**(3–4), 341–356 (2005). doi:10.1089/ten.2005.11.341. http://dx.doi.org/10.1089/ten.2005.11.341
105. Sheehy, E.J., Buckley, C.T., Kelly, D.J.: Oxygen tension regulates the osteogenic, chondrogenic and endochondral phenotype of bone marrow derived mesenchymal stem cells. Biochem. Biophys. Res. Commun. **63**(11), 3284–3293 (2011). doi:10.1016/j.bbrc.2011.11.105. http://www.sciencedirect.com/science/article/pii/S0006291X11021267
106. Shefelbine, S.J., Augat, P., Claes, L., Simon, U.: Trabecular bone fracture healing simulation with finite element analysis and fuzzy logic. J. Biomech. **38**(12), 2440–2450 (2005). doi:10.1016/j.jbiomech.2004.10.019. http://dx.doi.org/10.1016/j.jbiomech.2004.10.019
107. Simon, U., Augat, P., Utz, M., Claes, L.: A numerical model of the fracture healing process that describes tissue development and revascularisation. Comput. Methods Biomech. Biomed. Eng. **14**(1), 79–93 (2011). doi:10.1080/10255842.2010.499865. http://dx.doi.org/10.1080/10255842.2010.499865
108. Stops, A.J.F., Heraty, K.B., Browne, M., O'Brien, F.J., McHugh, P.E.: A prediction of cell differentiation and proliferation within a collagen-glycosaminoglycan scaffold subjected to mechanical strain and perfusive fluid flow. J. Biomech. **43**(4), 618–626 (2010). doi:10.1016/j.jbiomech.2009.10.037. http://dx.doi.org/10.1016/j.jbiomech.2009.10.037
109. Sun, W., Darling, A., Starly, B., Nam, J.: Computer-aided tissue engineering: overview, scope and challenges. Biotechnol. Appl. Biochem. **39**(1), 29–47 (2004). doi:10.1042/BA20030108. http://dx.doi.org/10.1042/BA20030108
110. Sun, W., Starly, B., Darling, A., Gomez, C.: Computer-aided tissue engineering: application to biomimetic modelling and design of tissue scaffolds. Biotechnol. Appl. Biochem. **39**(1), 49–58 (2004). doi:10.1042/BA20030109. http://dx.doi.org/10.1042/BA20030109

111. van der Meulen, M.C.H., Huiskes, R.: Why mechanobiology? a survey article. J. Biomech. **35**(4), 401–414 (2002)
112. van Turnhout, M., Kranenbarg, S., van Leeuwen, J.: Contribution of postnatal collagen reorientation to depth-dependent mechanical properties of articular cartilage. Biomech. Model. Mechanobiol. **10**(2), 269–279 (2010). http://dx.doi.org/10.1007/s10237-010-0233-7, 10.1007/s10237-010-0233-7
113. Wang, J.: Substrate deformation determines actin cytoskeleton reorganization: a mathematical modeling and experimental study. J. Theor. Biol. **202**(1), 33–41 (2000). doi:10.1006/jtbi.999.1035. http://www.sciencedirect.com/science/article/B6WMD-45CWX3K-7T/2/6c255986d529b28f0ff948b2bc2d63ac
114. Wei, Z., Deshpande, V.S., McMeeking, R.M., Evans, A.G.: Analysis and interpretation of stress fiber organization in cells subject to cyclic stretch. J. Biomech. Eng. **130**(3), 031009 (2008). doi:10.1115/1.2907745. http://dx.doi.org/10.1115/1.2907745
115. Wilson, W., Driessen, N.J.B., van Donkelaar, C.C., Ito, K.: Prediction of collagen orientation in articular cartilage by a collagen remodeling algorithm. Osteoarthr. Cartil. **14**(11), 1196–1202 (2006). doi:10.1016/j.joca.2006.05.006. http://dx.doi.org/10.1016/j.joca.2006.05.006
116. Wren, T.A., Beaupr, G.S., Carter, D.R.: Mechanobiology of tendon adaptation to compressive loading through fibrocartilaginous metaplasia. J. Rehabil. Res. Dev. **37**(2), 135–143 (2000)
117. Yan, D., Zhou, G., Zhou, X., Liu, W., Zhang, W.J., Luo, X., Zhang, L., Jiang, T., Cui, L., Cao, Y.: The impact of low levels of collagen ix and pyridinoline on the mechanical properties of in vitro engineered cartilage. Biomaterials **30**(5), 814–821 (2009). doi:10.1016/j.biomaterials.2008.10.042. http://www.sciencedirect.com/science/article/pii/S0142961208008168

Mechanobiological Modelling of Angiogenesis: Impact on Tissue Engineering and Bone Regeneration

Esther Reina-Romo, Clara Valero, Carlos Borau, Rafael Rey, Etelvina Javierre, María José Gómez-Benito, Jaime Domínguez and José Manuel García-Aznar

Abstract Angiogenesis is essential for complex biological phenomena such as tissue engineering and bone repair. The ability to heal in these processes strongly depends on the ability of new blood vessels to grow. Capillary growth and its impact on human health has been focus of intense research from an in vivo, in vitro and in silico perspective. In fact, over the last decade many mathematical models have been proposed to understand and simulate the vascular network. This review addresses the role of the vascular network in well defined and controlled processes such as wound healing or distraction osteogenesis and covers the connection between vascularization and bone, starting with the biology of vascular ingrowth,

E. Reina-Romo (✉) · J. Domínguez
Department of Mechanical Engineering, University of Seville,
41092 Seville, Spain
e-mail: erreina@us.es

J. Domínguez
e-mail: jaime@us.es

C. Valero · C. Borau · R. Rey · E. Javierre ·
M. J. Gómez-Benito · J. M. García-Aznar
Aragon Institute of Engineering Research (I3A),
University of Zaragoza, C/María de Luna 5,
Agustín de Betancourt Building, 50018 Zaragoza, Spain

M. J. Gómez-Benito
e-mail: gomezmj@unizar.es

J. M. García-Aznar
e-mail: jmgaraz@unizar.es

E. Javierre
Centro Universitario de la Defensa, Academia General Militar Ctra,
Huesca s/n, 50090 Zaragoza, Spain
e-mail: etelvina.javierre@unizar.es

Stud Mechanobiol Tissue Eng Biomater (2013) 10: 379–404
DOI: 10.1007/8415_2011_111

Published Online: 8 January 2012

moving through its impact on tissue engineering and bone regeneration, and ending with repair. Furthermore, we also describe the most recent in-silico models proposed to simulate the vascular network within bone constructs. Finally, discrete as well as continuum approaches are analyzed from a computational perspective and applied to three distinct phenomena: wound healing, distraction osteogenesis and individual cell migration in 3D.

1 Introduction

Angiogenesis, i.e. the formation of new blood vessels from the existing vasculature is essential for complex biological phenomena in healthy and diseased individuals, including tissue reconstruction [53] and tumor growth [36]. In the case of tissue reconstruction, the ability of bone defects (resulting from trauma, congenital deficiency, infection or tumor resection) to heal depends on the ability of new blood vessels to proliferate from surrounding vessels. Since bone defects represent a medical and socioeconomical challenge [65] it is very important to understand the role of the vascular system in bone regeneration. Therefore, particular attention is paid to surgical techniques to bridge bone injuries. Different types of biomaterials, internal plates and fixation devices are applied according to the reconstruction interests but in all cases vascular supply is crucial. Within the field of tissue engineering and in the case of large skeletal defects, capillary growth appears to be particularly complex, possibly because of the slow rate of infiltration of blood vessels into the scaffold. It makes the process insufficient to provide adequate nutrients to the cells in the interior of large constructs, which results in peripheral tissue formation and central scaffold necrosis [65, 83, 86]. In addition, the in vivo studies on bioengineered tissue vascularization have not been performed until recently and are focus of intense research activity nowadays. In contrast, the robust nature of bone induction in fracture healing and distraction osteogenesis has made it an important system to study the role of angiogenesis in bone regeneration [32, 56, 75]. Therefore this review addresses the role of the vascular network in these well defined and controlled processes. It aims to cover the connection between vascularization and bone, starting with the biology of vascular ingrowth, moving through its impact on tissue engineering and bone regeneration, and ending with repair. Furthermore, we will review the most recent in-silico models proposed to simulate the vascular network within bone constructs. Mathematical as well as algorithmic discrete approaches, which are implemented through continuous representations and operational descriptions respectively, will be described.

2 Vascular Ingrowth

Blood vessels form from either de novo, via vasculogenesis or by sprouting or splitting of existing blood vessels via angiogenesis. Angiogenesis may be categorized into angiogenic remodelling and sprouting [125]. In the former, an initial

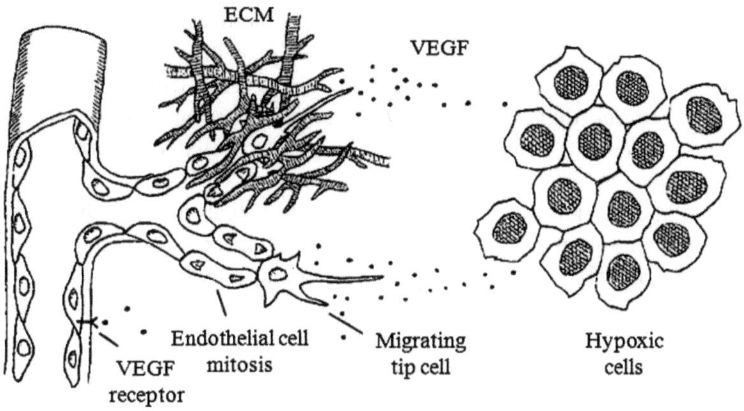

Fig. 1 Early events in sprouting angiogenesis (adapted from Bauer et al. [7])

vascular network is modified through network enlargement or pruning. In the latter, new capillaries form from existing vessels and are directed toward a vascular tissue. Here we focus on the latter. In the case of tissue engineering, wound healing and thrombosis, newly formed capillaries grow in a controlled manner and stop growing once the pathological condition has been alleviated; this is, however not the case for tumor-induced angiogenesis [39].

2.1 Biology of Vascular Ingrowth

Angiogenesis is a complex regulated process, in which endothelial cell (EC) growth, migration and tube formation are regulated by pro and anti-angiogenic factors, matrix degrading proteases and cell extracellular matrix interactions. It includes several basic steps [38] (see Fig. 1):

- *Activation of endothelial cells* and vasodilatation of the parental vessel. Endothelial cells are activated by angiogenic growth factors. This switches the endothelial cells from the quiescent endothelial phenotype to an active mesenchymal phenotype, which secretes enzymes and growth factors and can migrate and proliferate. The initial vasodilatation of parental vessels is accompanied by the increase in permeability which is frequently driven by an excess of vascular endothelial growth factor (VEGF), nitric oxide, or other mediators.
- *Local degradation of the basement membrane and surrounding extracellular matrix* of the vessel by proteases (matrix metalloproteinases and serine proteases). The basement membrane coats the exterior of blood capillaries and is believed to maintain endothelial stability, impermeability, and resistance against the hydrostatic pressure of the blood flow within vessels [105]. Thus, the local

degradation not only provides room for the migrating endothelial cells but also amplifies the angiogenic stimulus by releasing growth factors, including bFGF, VEGF and IGF-1 which otherwise remain sequestered within the matrix [26].

- *ECs proliferation and migration.* Degradation of surrounding matrix involves physical barriers to be dissolved allowing endothelial cells to migrate and proliferate.
  Endothelial cell migration responds to chemical signals through chemotaxis (the directional migration along a gradient of soluble chemoattractants) as well as to mechanical signals via haptotaxis (the directional migration along a gradient of immobilized ligands) [69]. Chemotaxis, is driven by growth factors such as VEGF and basic fibroblast growth factors (bFGF) (Fig. 1). Several additional factors are implicated such as platelet derived growth factors. Haptotaxis is associated with increased endothelial cell migration activated in response to integrins binding to ECM component [64]. In fact, the presence of fibers enhances the migration speed of ECs as the fibers provide integrin receptors promoting mesenchymal motion of the tip cells [40].
  By the proliferation process, ECs located initially behind the migrating tip cells extend the sprouting blood vessel [43]. After the initial sprouts have extended into the extracellular matrix, repeated branching of the tips can be observed. Sprout tips approaching others may fuse and form loops in a process called anastomosis. Anastomosis can be observed to occur either by sprout tips fusing with other sprout tips thus arresting their migration, or by sprout tips fusing with already established sprouts at some distance behind the tip.
- *Generation of the capillary lumen* and formation of tube-like structures. As endothelial cell migrate into the extracellular matrix, they assemble into solid cords, and subsequently acquire a lumen which will establish a network that allows blood circulation. This is accomplished by either intracellular vacuolar fusion or by arrangement of cells around a central lumen. Lumen diameter is regulated by several factors such as VEGF, Ang1 and multiple integrins [26].
- *Maturation.* It incorporates the rebuilding of a basal lamina and the recruitment of pericytes and vascular smooth muscle cells to stabilize the vessel walls [59]. Ang-1 expression may be involved in the vessel maturation of the newly formed vasculature through the reassociation of ECs with supporting periendothelial cells (PCs) [73].

Once new vessels have assembled, the endothelial cells become remarkably resistant to exogenous factors and are quiescent, with survival measured in years. Therefore ECs, which have an active role in angiogenesis, are assumed to occur in three states: quiescent lining regular vessel walls, migrating cells located at the tip of the outgrowing sprouts and proliferating ECs located at a small region behind the migrating tip cell.

The regulation of angiogenesis is complex involving a combination of factors, which include adequate compression forces, nutrients and growth factors, extracellular matrix components and physiological modulators such as pH and oxygen tension [76], which will be reviewed next.

2.1.1 The Role of Hypoxia

It has been proved that hypoxia is the major initiator of both fetal and adult angiogenesis [74] (Fig. 1). In fact, there is a close link between hypoxia and the upregulation of proangiogenic factors, such as VEGF [110]. The transcription factor responsible is named hypoxia inducible factor-1 (HIF-1) which activates genes involved in anaerobic metabolism, angiogenesis and erythopoiesis. However, severe hypoxia, as is often noted in infected open wounds, cannot sustain the growth of functional blood vessels [52, 87, 106].

2.1.2 Chemical Stimulation

Chemical stimulation of angiogenesis is performed by various angiogenic proteins, including several angiogenic growth factors.

Vascular Endothelial Growth Factors

It is the most studied growth factor and the major regulator of neovascularization under physiological and pathological conditions [8]. It was initially defined by its ability to induce vascular leak and permeability, and thus promote EC migration and proliferation [73, 125]. It is produced by numerous cell types, including endothelial cells, macrophages, fibroblasts, smooth muscle cells, osteoblasts and hypertrophic chondrocytes [9, 28, 63].

Numerous studies have shown that the expression of VEGF is tightly associated with blood vessel invasion of hyaline cartilage, growth plate morphogenesis, cartilage remodelling and bone repair [42, 95, 119, 128]. VEGF is expressed in much the same temporal and spatial pattern in the fracture callus as that occurs during long bone development [33, 46]. The human fracture hematoma, present after injury but not during development, has potent angiogenic activity that appears to be predominantly due to VEGF [117, 118].

Inhibition of VEGF activity disrupts repair of femoral fractures and cortical bone defects in mice [119] and decreases blood flow and leads to non-unions in rabbit radial fractures [24]. These findings support the idea that VEGF production is the major mechanism by which angiogenesis and osteogenesis are tightly coupled during bone repair [95, 119]. Given the importance of VEGF in normal bone repair, treatment with exogenous VEGF is expected to promote angiogenesis. In fact VEGF treatment increases bone blood flow in rabbit radial fractures [24] and in rabbit tibiae during distraction osteogenesis [30].

In addition to contributing to vasculogenesis and angiogenesis during embryonic development and bone regeneration processes, VEGF is an essential factor in angiogenesis-dependent diseases [16, 34]. For instance, VEGF is a key angiogenic factor in tumors, which express it at high levels. Overexpression of VEGF can also cause vascular disease in the retina of the eye and other parts of the body. Drugs such as bevacizumab can inhibit VEGF and control or slow those diseases [122].

Other Factors Contributing to Angiogenesis

There are many other factors that play a role in the reestablishment of vascular supply to bones during the healing process such as the *fibroblastic growth factors* (FGF). Basic FGF is a strong fibroblast mitogen. It is also a chemotactic factor for EC and is highly angiogenic. In contrast to VEGF, it is not specific for EC. FGF receptors are expressed in EC and in smooth muscle cells, fibroblasts, myoblasts and tumor cells [6].

Thrombin, found in fracture hematoma, is inherently angiogenic through its action on protease activated receptors (PAR). Thrombin has been shown in vivo and in vitro to act as a chemotactic factor for endothelial cells and to increase the production of $\alpha_v\beta_3$ integrin on the surface of these cells. This integrin facilitates migration, attachment and survival of endothelial cells [121].

The *bone morphogenetic proteins* (BMP), known for their ability to promote mesenchymal stem cell differentiation that leads to bone formation, also promote angiogenesis. Deckers et al. [28] found that BMP2, BMP4 and BMP6 were capable of simulating murine derived osteoblasts to secrete VEGF in culture and stimulated angiogenesis in fetal mice bone explants. The action of BMPs on osteoblasts establishes a positive feedback loop where the BMP-induced VEGF release causes vessel ingrowth, leading to the delivery of osteogenic precursor cells on which BMPs will act to further increase VEGF concentrations at the fracture sire.

The *extracellular matrix* is critical for angiogenesis, and should be viewed as a dynamic player in the process. In addition to providing a site for storage of growth factors and pro-enzymes, for release and activation, its components serve as binding sites and targets for endothelial and mesenchymal cell-derived integrins and growth factors [26]. It also plays an important role in cell migration. Fibers such as collagen, laminin and fibrillin are distributed through the ECM occupying roughly 30% of the ECM. These fibers form bundles that serve as guiding structures for migrating cells [27].

2.1.3 Mechanical Stimulation

The effect of mechanical stimulation in angiogenesis has been widely studied in distraction osteogenesis [17, 23, 32, 54, 55, 92, 126] and bone healing [51, 119]. In a fixated fracture, the vascular network is rapidly reestablished and healing occurs mainly by intramembranous ossification [33, 66]. Conversely in an unstable mechanical environment the spreading capillaries are disrupted and the hypoxic environment promotes the differentiation of chondrocytes that stabilize the fracture by cartilage formation. In distraction osteogenesis the effect of tension activates biologic processes of the interfragmentary connective tissue. This includes the prolongation of angiogenesis with increased tissue oxygenation and the increased fibroblast proliferation with intensification of biosynthetic activity (Fig. 2). In both distraction osteogenesis and bone healing, lack of angiogenesis has been pointed out as one of the main reasons for non-healing bone [32].

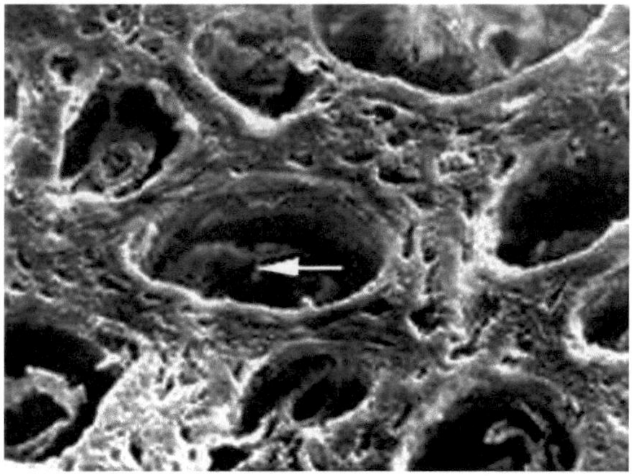

Fig. 2 Newly formed vessels in the distraction regenerate, electron micrograph x500 [57]

2.2 Impact on Tissue Engineering and Regeneration

The majority of fractures heal well under standard conservative or surgical therapy. However extended bone defects following trauma or cancer resection or non-union of fractures may require more sophisticated treatment. In these cases, bone grafting procedures, segmental bone transport, distraction osteogenesis or biomaterials are applied for reconstruction [65]. In all these processes, the impact of angiogenesis on bone regeneration is crucial. For instance, in the United States 5–10 % (250,000–500,000) of fractures go onto nonunion or delayed union [104]. Negative effects on the vascular system might be the mechanism whereby many other risk factors delay or impair bone healing [48]. In distraction osteogenesis many studies have suggested that blood supply has a significant influence on the shape and mass of the resulting bone [3, 55]. If blood supply is inadequate to support normal or increased mechanical loading, then the bone cannot respond favorably, leading to atrophic or degenerative changes. By contrast, if blood supply is adequate to support increased mechanical loading, the bone will demonstrate compensatory hypertrophic changes. This has motivated efforts aimed at better understanding the vasculature around fractures/osteotomies, as well as advancing in the therapeutic measures that may improve these processes.

Regarding wound healing, chronic wounds represent a major public health problem affecting 6.5 million people in the United States. Chronic wounds, such as pressure or venous ulcers, are frequently suffered by patients with diabetes and obesity. Treating chronic wounds costs an estimated US$25 billion each year and is growing rapidly due to increasing health care costs, an aging population and a rise in the incidence of diabetes and obesity worldwide [111]. Ischemia, which does not fall into the category of chronic wound and is primarily caused by peripheral artery diseases, also represents a major complicating factor in cutaneous wound healing [124].

In the field of tissue engineering, one of the outstanding challenges is to accurately mimic tissues and to create an environment within the scaffold and its associated cells and growth factors that promotes the regeneration of functional tissues. Bone grafts are one of the most successful products on the market. In USA, the number of bone graft procedures approximates to 500,000 per year. They represent a worldwide market estimated at around $300 million for bone replacement and repair and include autologous bone grafts, allogenic and xenogenic grafts, as well as synthetic bone materials [50]. Skin repair is also an important field of the tissue engineering, especially in the case of extended third-degree burns. Every year in the world millions of patients are disabled and require hospitalization due to the burn injuries, among which more than 300,000 persons die ultimately [13, 120]. So far three types of skin substitutes, i.e. epidermal equivalents, dermal equivalents and composite equivalents, have been used in burn treatment [96].

3 In-Silico Approaches

Angiogenesis modelling has gained increased interest over the last decades due to a greater availability of experimental data as well as further developed computational capabilities. There are three principal approaches for the mathematical modelling of vascular ingrowth: continuum models which face the problem from a macroscopic point of view, at the cell density level; discrete models in which mathematical rules are formulated at the mesoscopic or cellular scale and hybrid models which combine both continuum and discrete overviews (see [4, 73, 94] for a review).[1]

3.1 Continuum Models

Continuum models aim to describe the average behavior of cell populations with continuum variables. Accordingly, capillary networks are described in terms of endothelial cell densities. Populations of cells are usually represented in a macroscopic point of view not preserving the identity and properties of individual cells [84]. The main drawbacks of continuum models are that the individual movements of the cells and capillary tips cannot be tracked, that they require more assumptions than discrete models, and that behaviors at the cell level cannot be predicted. In contrast, these models are less computationally expensive and provide useful insights into the ways in which different physical mechanisms, such as the strength

[1] The authors apologize any related reference with the present work which remained uncited along the text.

of the chemotactic response or the rate of capillary tip formation influence angiogenesis.

Models of the continuum type are usually derived from mass conservation equations and chemical kinetics. This results in a system of nonlinear partial differential equations, modelling macroscopic quantities such as cell density and chemical concentrations. They have been extensively studied in tumor induced angiogenesis and wound healing.

Tumour-induced angiogenesis has received a lot of attention since 1985, when Balding and McElwain [5] proposed a simple one dimensional model to describe the vascular network through the introduction of a chemotactic stimulus. Therefore, they concentrated only on the chemotactic response of the ECs to angiogenic factors. Later, models were extended to two dimensions and included the haptotatic response of ECs to adhesive gradients, usually of fibronectin, as well as the role of the inhibitors in the angiogenic process [2, 89]. Following Orme and Chaplain [90], Levine et al. [67, 68] have developed highly complex biochemically-based models that account for specific angiogenic factors, interactions between ECs, and cells such as pericytes and macrophages that are also involved in angiogenesis. Whereas in these previous models, vascular remodelling has been neglected, it has been considered in recent studies [1, 20, 71, 77, 78, 91, 114, 115]. The models of angiogenesis mentioned above do not account for the effect of blood flow, which has stimulated the development of hybrid models.

Given the recognized biological similarities between tumour and wound angiogenesis [18, 37], the models for tumor induced angiogenesis may also be relevant to wound healing, wherein growth factors released at the wound site by inflammatory cells perform analogous roles to the tumour angiogenesis factor [88]. There are several continuum mathematical models of wound healing which incorporate the effect of angiogenesis [15, 35, 41, 88, 98, 100, 108, 124]. As in tumour angiogenesis these models consist on a system of nonlinear partial differential equations. Olsen et al. [88] and Gaffney et al. [41] proposed a simple 2-species model. The former aimed to investigate the role of the extracellular matrix on the proliferative and migratory response of endothelial cells whilst the latter described the passive motion of capillary sprouts following their leading tip cell. In 1996, Pettet et al. [98, 100] developed two models of wound healing angiogenesis where they incorporated the chemotactic response of endothelial cells to angiogenic factors. Later, Byrne et al. [15] with a three species-model also incorporated chemotaxis and compared their results with experimental data. More recently, Schugart et al. [108] developed a seven species model of acute wound healing angiogenesis, using a similar approach to Gaffney et al. [41] and recognizing the role of tissue oxygenation in wound care. Xue et al. [124] developed an 8-variables model which involves the concentration of oxygen, PDGF and VEGF, the densities of macrophages, fibroblasts, capillary tips and sprouts, and the density and velocity of the ECM. Some models additionally incorporated the oxygen [35].

The angiogenic process involved during the formation of the gap vasculature network during fracture healing has been less well studied. Geris et al. [44] proposed a continuous mathematical model that describes healing in terms of densities of the most important cell types, generic growth factor families and tissues. The process of angiogenesis is modelled through the spatiotemporal evolution of an endothelial cell concentration and a vascular density. Shefelbine et al. [112] proposed a fuzzy logic model to simulate fracture healing based on the mechanoregulation theory proposed by Claes and Heigele [25] and the vascularity. They developed fuzzy rules and considered a single variable representing the vascularity. Using a similar model, Simon et al. [113] included blood perfusion as a spatio-temporal state variable to simulate the revascularisation process. Chen et al. [22] modified the set of fuzzy logic rules and simulated nutrition supply instead of the vascularity.

3.2 Discrete Models

Discrete models aim to determine the system behavior at the cellular level, instead of averages responses of continuum variables. This has the advantage of enabling microscopic properties of the capillary network, such as vessel branching and looping, to be captured. They are based on phenomenological descriptions of single cells either by solving systems of differential equations at discrete locations (i.e. points on a lattice) or by "if-then" computational algorithms [94]. There are several different types of discrete models: cellular automata, agent-based, cellular potts and cell force-based models.

3.2.1 Cellular Automata Models

Cellular automata (CA) models capture complex biological phenomena by defining a series of simple rules that are easy to quickly compute and that can even be parallelized [4]. Most cellular automata models reduce the simulation process into a set of discrete states, discrete time cycles and space coordinates [4]. CA are usually divided in three main categories: deterministic automata, lattice-gas models and solidification models [31].

In the *deterministic automata*, the space is divided in a fixed lattice and each lattice point has a state associated with it. The state at the next time step is only function of the earlier states of the cell and its neighbors. *Lattice gas models* consist of a discrete spatial grid in which particles, in contrast to the deterministic automata, are driven by random events. Therefore they contain a stochastic element in the sense that they can allow the computation of different outcomes across different simulations. *Solidification models* are very similar to lattice ones except that moving particles may be irreversibly bound at grid points, or cells may undergo irreversible state transitions.

Within the field of tumour induced angiogenesis, Anderson and Chaplain [2] derived a discrete model using a cellular automaton approach. In particular they proposed a lattice based model in which the endothelial cell migration depends on chemotactic and haptotactic probabilistic rules. An extension to three dimensions is presented in Chaplain [19].

Recently, Checa and Prendergast [21] have proposed a lattice-based mechanobiological model for tissue differentiation and blood vessel growth to determine the influence of cell seeding on vascular network development and tissue growth inside a regular-structured bone scaffold under different loading conditions. They combined the mechanoregulation theory proposed by Prendergast et al. [101] with a stochastic model for cell migration and proliferation based on a random walk theory [97].

3.2.2 Cellular Potts Models

The Glazier–Graner–Hogeweg model, also known as the cellular potts model (CPM) is a discrete lattice Monte Carlo model developed by Glazier and Graner and is based on an energy minimization principle [47]. Therefore, in these models migration occurs in those directions that minimize a local energy function [82] and the aggregation/interaction between cells is defined by cell–cell contact energies [4]. CPM have been used to model vasculogenesis by Merks et al. [79–81] and angiogenesis by Bauer et al. [7] and Merks et al. [81]. Bauer et al. [7] described the first cell-based model of tumor-induced angiogenesis. At the cellular level, the model uses the cellular Potts model and describes endothelial cell migration, growth, division, cellular adhesion, and the evolving structure of the stroma. Merks et al. [81] proposed a CPM that simulated both vasculogenesis and angiogenesis with a single biophysical mechanism.

3.2.3 Agent-Based Models

In contrast to previous models, agent based models go beyond the lattice-based space representation and aim to incorporate information of the shape and size of the cells [84]. Stokes and Lauffenburger [116] have based their modelling of angiogenesis on stochastic differential equations accounting for chemotaxis and random motion with an agent based model (lattice-free). They were able to model tip cell migration by imposing rules for branching and anastomosis. However, other important features such as the interaction of cells and the ECM and the brush border effect were not taken into account.

Wood et al. [123] developed an agent field model to simulate the angiogenic vascular network formation. The model consists of an inhomogeneous population Markov chain where transition probabilities vary according to the state of the gel matrix field.

3.2.4 Cell Force-Based Models

These models normally use individual cells as basic units, each of which is characterized by its location and orientation, its state of stress, and the active forces it can exert in response to the local micro-environment. Knowing this for each cell, the movement of each single cell, and hence of one collection of cells can be estimated through the force equilibrium equation. Normally, the effect of inertia is ignored in this equation and two kind of forces are considered: interaction forces and viscous drag forces. Two main interaction forces are considered, coming from cell-matrix and cell–cell interactions. The viscous drag force is generated from the cell movement through a viscous environment and is dependent on the cell velocity of each cell. Therefore, establishing the force equilibrium the velocity of each cell can be evaluated.

Using this approach, a mathematical model for cell movement in multicellular systems has been developed by [93] to simulate in 3D cell movement during aggregation and slug stage of Dictyostelium discoideum, embryogenesis, limb formation and wound healing. The movement of individual cells in 3D considering a more realistic cell-matrix interaction under the consideration of the dependency on the receptor-ligand adhesivity has been simulated by Zaman et al. [127].

More recently, a cell force-based modelling framework for bone tissue engineering applications has been presented [45]. In this study, cell–cell and cell–environment interactions such as adhesion, repulsion and drag forces have been considered to model cell aggregate behavior, although not angiogenesis.

Although in 2D, a more realistic mechanical behavior of the cell body has been considered to model collective cell behaviour incorporating sub-cellular effects with multiple viscoelastic elements [60].

Nevertheless, these approaches are really focused on modelling collective cell migration and not angiogenesis. As far as we know, there are not previous works that have simulated angiogenesis using this approach.

4 Examples of Application

Next, discrete as well as continuum approaches are presented which are analyzed from a computational perspective and applied to three distinct phenomena: wound healing, distraction osteogenesis and individual cell migration in 3D. The results shown in these examples of application are preliminar results.

4.1 Continuum Models: Wound Healing

As it has been previously mentioned, a number of mathematical and numerical angiogenesis models have been developed in the last few years. Angiogenesis is present in the second stage of the wound healing process, the proliferative phase, so it is very important to understand how every factor present in angiogenesis affects the evolution of the wound.

Table 1 List of normalised model parameters

Parameter	Dimensionless value	Parameter	Dimensionless value	Parameter	Dimensionless value
D_1	$4.32 \cdot 10^{-3}$	$\lambda_{3,1}$	0.864	$\lambda_{1,1}$	0.864
D_2	0.0864	$\lambda_{1,2}$	2.592	$\lambda_{2,2}$	0.864
D_3	$4.32 \cdot 10^{-3}$	$\lambda_{3,2}$	4.32	$u_{3^{eq}}$	1.0
D_n	$2 \cdot 10^{-2}$	$\lambda_{n,n}$	0.7488	r_n	1.664
$u_1^{\theta_2}$	0.25	K	10.0	K_1	2

4.1.1 Mathematical Model

We propose an angiogenesis model that incorporates biochemical factors, consisting of a number of coupled reaction-diffusion equations. The primary variables of the model are the concentrations of oxygen (u_1), macrophage-derived growth factors -such as VEGF or others, henceforth MDGF- (u_2), capillary density (u_3) and fibroblasts (n). We have modified the model proposed by Javierre et al. [61], based on the model of Maggelakis [72], in order to introduce the effect of fibroblasts in the process. The set of equations that describe the behavior of angiogenesis is:

$$\frac{\partial u_1}{\partial t} = D_1 \nabla^2 u_1 + \lambda_{3,1} u_3 - \lambda_{1,1} u_1 \tag{1}$$

$$\frac{\partial u_2}{\partial t} = D_2 \nabla^2 u_2 + \lambda_{1,2} Q(u_1) - \lambda_{2,2} u_2 \tag{2}$$

$$\frac{\partial u_3}{\partial t} = D_3 \nabla^2 u_3 + \lambda_{3,2} u_2 u_3 \left(1 - \frac{u_3}{u_e^{eq}}\right) \tag{3}$$

$$\frac{\partial n}{\partial t} = D_n \nabla^2 u_n - \lambda_{n,n} n + r_n n \left(1 - \frac{n}{K}\right) u_1 \left(1 - \frac{u_1}{K_1}\right) \tag{4}$$

where Q represents the function describing the production of MDGF when the levels of oxygen are low,

$$Q(u_1) = \begin{cases} 1 - \dfrac{u_1}{u_1^{\theta_2}} & \text{if } u_1 < u_1^{\theta_2}; \\ 0 & \text{otherwise.} \end{cases} \tag{5}$$

It is known that fibroblasts are responsible for ECM synthesis, and serve as an indicator of ECM maturity. Hence, as a novelty from the previous model [61], we have incorporated the fibroblasts equation (Eq. 4) in which fibroblast proliferation is modelled with a logistic term with an oxygen dependent rate [108]. The values of the (dimensionless) model parameters are included in Table 1.

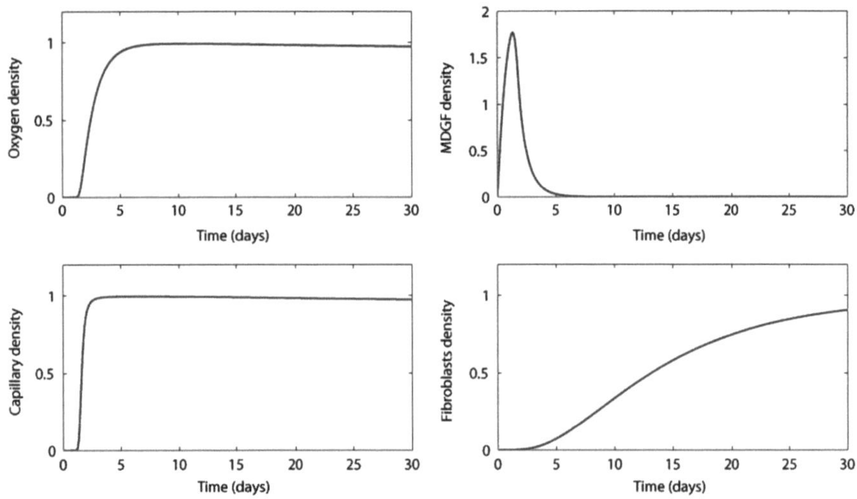

Fig. 3 Evolution of the oxygen density (*top left*), MDGF density (*top right*), capillary density (*bottom left*) and fibroblasts density (*bottom right*) in the centre of the wound

The computational domain consists of two different parts: the undamaged tissue and the wound. The initial concentration of every species in the wound site is null. The undamaged tissue is full of all the species except MDGF, which is not present.

We have solved the resulting problem using a finite element analysis. The primary unknowns are interpolated from nodal values through shape functions and the times derivatives are approximated with a trapezoidal method. The resulting nonlinear system of equations is solved using a standard explicit method. This numerical method is analogue to the one used in Javierre et al. [62] for wound contraction alone. The interested reader is referred to [62] to find the details of the implementation.

4.1.2 Numerical Results

A circular wound of radius 1 cm has been studied. We present the obtained results, showing the evolution of the species concentration in the wound centre along time (see Fig. 3).

We observe how the oxygen concentration regulates the rest of the variables. When there is no oxygen in the centre of the wound the level of MDGF increases rapidly, which causes the appearance of capillaries. When the capillary density grows, more oxygen is supplied to the wound centre and the MDGF decreases. The capillary density grows rapidly to the capillary density of undamaged tissue (u_3^{eq}) when the oxygen begins to increase. Finally, it is clear from the results that oxygen affects the fibroblasts concentration (see Fig. 4). When oxygen is not included in

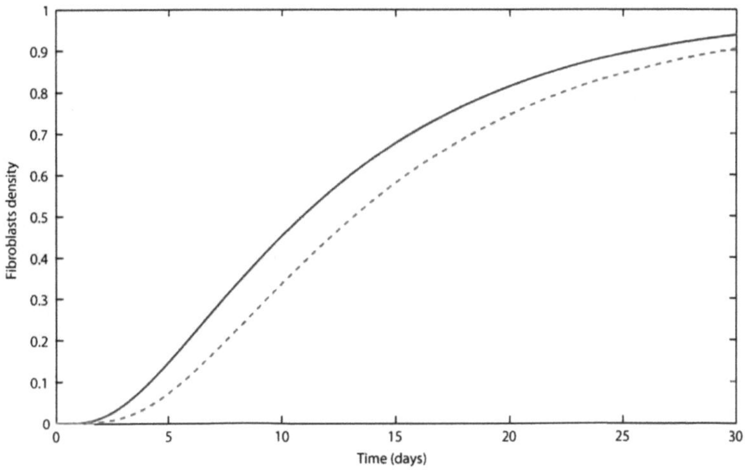

Fig. 4 Evolution of the fibroblasts density in the wound centre. Without the influence of oxygen (*continuous line*) and under the oxygen effect (*dashed line*)

the fibroblasts kinetics, the model predicts fibroblasts invasion into the wound faster than reality, and hence wound healing rate is overestimated.

After a few days the concentrations of all the species are homogeneous in the whole domain. Both, oxygen and capillary density reach a concentration of 1 (the normalized density of undamaged dermis), while MDGF appears in the wound site during the first days but then disappears. The fibroblasts concentration progress more slowly, but finally reaches the fibroblast density corresponding to the undamaged dermis.

The distribution of every species one day post-wounding is shown in Fig. 5. Results show an elevated density of MDGF in the wound area, whereas the densities of oxygen, capillaries and fibroblasts grow within the wound more slowly. In fact, the advance of the so-called healing unit, with the capillaries invading the wound in first place (i.e. earlier in time) and being followed by the fibroblasts slightly later in time, is clearly visible.

4.2 Lattice-Based Discrete Models: Distraction Osteogenesis

Distraction osteogenesis is inherently a multi-scale process that involves coupled phenomena that occur across different spatial and temporal scales. However, as far as the authors know, this process has only been computationally investigated on a macroscopic scale, from a continuum perspective [10, 11, 58, 102, 103]. Existing continuum models consider as state variables the population of cells or tissue distributions and predict their macroscopic evolution according to different

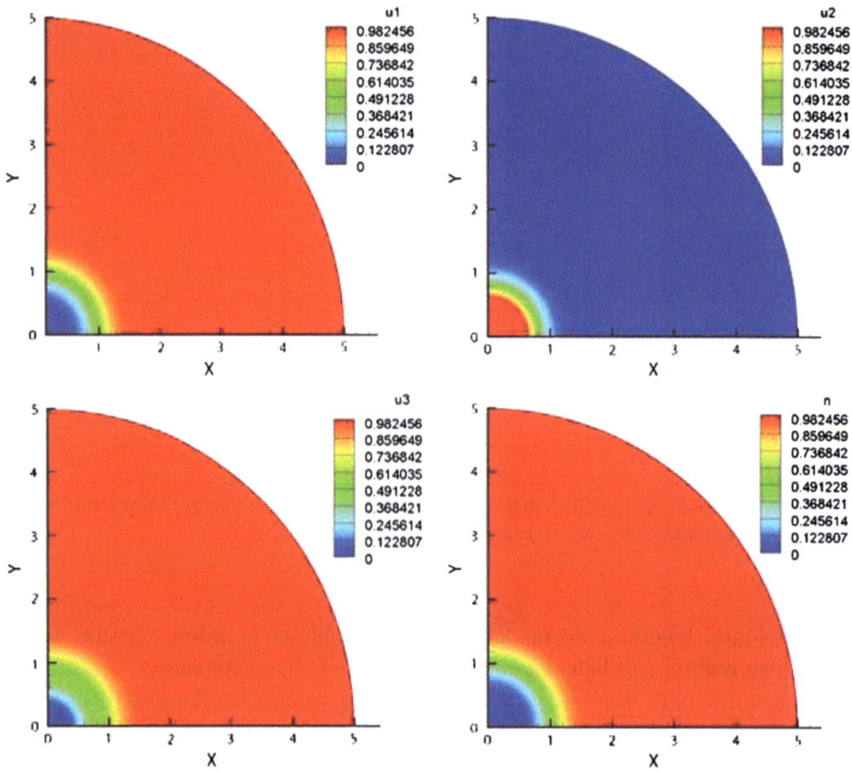

Fig. 5 Distribution of oxygen (*top left*), MDGF (*top right*), capillary (*bottom left*) and fibroblasts (*bottom right*) in a quarter of the whole domain after one day

mechanoregulation theories [49, 101]. To incorporate the cellular scale with a more realistic approach, a lattice based formulation is being developed by the authors and is explained next.

A two-dimensional lattice based model is simulated to predict the outcome of the distraction osteogenesis process. At the macroscopic scale a poroelastic analysis is performed to compute the mechanical stimulus within the distracted gap. An axisymmetric finite element model of a sheep tibia has been established. The macroscopic scale of the model includes the bone marrow, the cortical bone, the periosteum, the nail and the distracted gap [14, 102]. The distracted gap, is treated with a lattice-based approach and therefore each macrocopic element is modelled at the cellular scale with a regular cartesian grid. The different cells (chondrocytes, osteoblasts, mesenchymal stem cells, fibroblats and dead cells) are constrained to move on this grid and their evolution is assumed to depend on different migration, proliferation and differentiation rules [49, 102]. Cell migration and proliferation were modelled as random-walk processes following the random

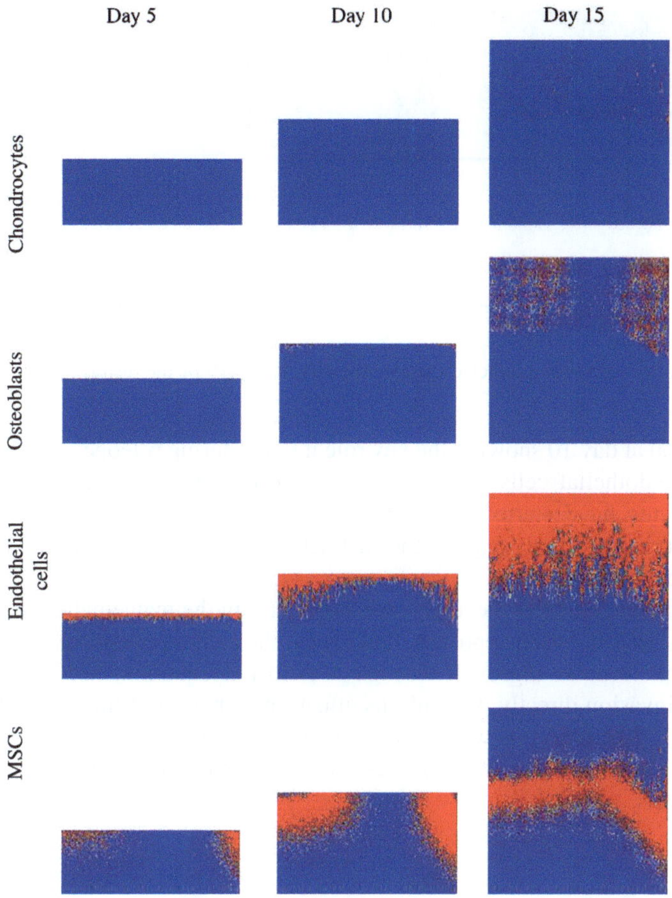

Fig. 6 Cell distribution prediction within the distracted gap during the first 15 days of the process

walk model proposed by Pérez and Prendergast [97]. In addition the model incorporates the vascular ingrowth in order to be able to capture the relationship between bone formation and angiogenesis [21]. This capillary network is described in terms of endothelial cell densities.

Figure 6 shows the evolution of four cell fates over the first 15 days of the process of distraction: chondrocytes, osteoblasts, endothelial cells and mesenchymal stem cells (MSCs). It can be observed how this cell distribution varies significantly. Initially, during the first days of the distraction procedure the gap is gradually filled by MSCs that migrate from the marrow cavity, periosteum and surrounding soft tissues to the interfragmentary gap. By day 10, osteoblasts could be distinguished close to the old bone matrix. Osteoblasts differentiation is assumed to be regulated by the vascular network. Figure 7 depicts the blood vessel

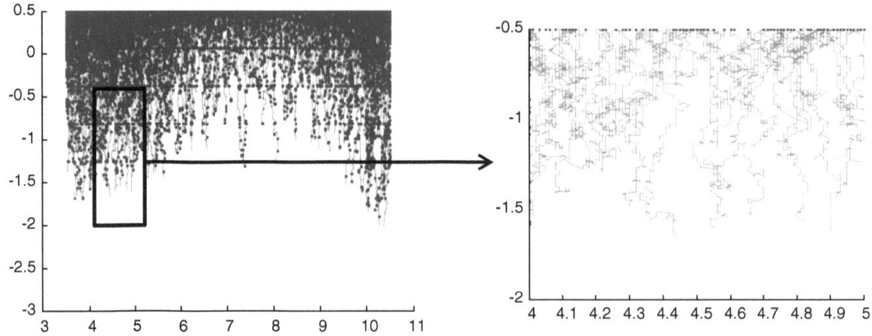

Fig. 7 Vascular network predicted computationally at day 10 of the distraction process

distribution at day 10 showing the key role it plays during osteogenesis. Finally, by day 15, endothelial cells have progressed further and considerably amounts of chondrocytes have started to differentiate (Fig. 6). Regarding the ossification pattern during the first stage of the distraction procedure only intramembranous ossification occurs.

This model considers the angiogenesis directly in the mathematical formulation through a random-based approach. Previously, the authors considered this complex process of vascular ingrowth through a diffusion equation, not modelling the vascular invasion directly, but only its effect: the advance of the ossification front [49, 102]. Therefore, this work represents an important step-forward in the modelling of angiogenesis in bone regeneration and specifically in distraction osteogenesis.

4.3 Cell Force-Based Discrete Models: Cell Migration in 3D

Cell migration is essential for tissue development in different physiological and pathological conditions. Cell movement is guided by input signals from the surroundings. The physical–chemical nature of the extracellular matrix (ECM) medium and its binding with the transmembrane, allows cells to probe the mechanical properties of their environment and react to it in an specific way. This ability of cells to sense ECM flexibility or pre-strain is responsible for cell behavior and its understanding is crucial. It has become clear in the last few years that cells extend some lamellipodia and filopodia which attach to the substrate. They then exert contractile forces in order to explore the properties of their environment in a process called mechanosensing [29]. These active forces are generated by the myosin motors and transmitted to the ECM by means of transmembrane proteins (integrins) in so-called focal adhesions.

Here the authors present a 3D migration model of a single cell, which is regulated by a mechano-sensing mechanism (more details can be found in [12]).

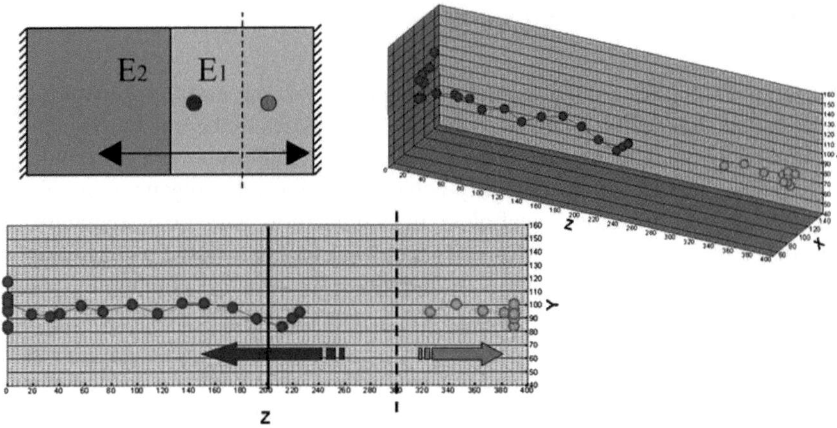

Fig. 8 Example of cell migration in a substrate with two different stiffness and boundary conditions

The main relevant components of the cell, the traction forces exerted and the major phases in cell migration are taken into account. The model includes important parameters such ECM stiffness and viscosity and the boundary conditions. With all these, it is able to predict some features such as cell movement tendencies, traction forces and cell speeds in several substrates with different stiffness and under different constraints. The influence of the stiffness and topography of the (ECM) has been recently investigated both computationally and experimentally [70, 85, 99, 107]. Some important experimental findings are that cells prefer to migrate to the stiffer part of a substrate [70, 109]. Furthermore focal adhesions are more stable and traction forces increase on stiffer substrates [70, 109]. The model presented here reproduces these findings (see Fig. 8). For example, a case of an elastic substrate with two different rigidities ($E_2 > E_1$) is simulated. Note that both the stiffer and the softer side, are constrained. Depending on the initial position of the cell, it migrates in different directions according to the mechano-sensing process described above. In general the cell tends to move toward the stiffer side (E_2) and never goes from the stiffer to the softer one. However, if the cell is initially placed close enough to the softer side constraint, the cell senses it and moves in that direction. If the cell reaches the constraint, it begins to move randomly but does not deviate far from the boundary. This model appears to agree with experimental data, and its parameters can be adjusted for several types of cells and substrates, nevertheless, additional factors such as matrix-remodeling, mechano-chemistry, cell–cell interactions or a complete focal adhesion model must be introduced to improve the results and get closer to a more realistic approach.

5 Discussion and Conclusions

Providing a functional angiogenesis is required for successful regeneration of many tissues, but has yet to be achieved for critical sizes of the trauma. The ability to build or manipulate the microenvironment is critically important for regulating angiogenesis in order to successfully achieve in-vivo tissue regeneration. Tissue engineering uses materials (scaffolds) in combination with other regulators to control the microenvironment that cells sense and facilitate tissue regeneration in-vivo. Achieving this goal requires an interdisciplinary effort combining cellular and molecular biology, matrix engineering, biomechanics/mechanobiology, and micro/nanofabrication.

The development of in-vitro and in-vivo models is the typical way of analyzing and understanding the role of each factor in the final outcome of tissue regeneration. However, the control of these factors and the limitation of the information obtained from these models clearly limit their use and this gives the opportunity to other kind of models to make a contribution: computer or mathematical models. Therefore, computer modelling is currently a potent tool to unravel and improve the understanding of biological phenomena.

In this work, we have reviewed many computer models that focus on the mechano-biological modelling of angiogenesis in tissue regeneration and specifically in tissue engineering following different approaches: continuum and discrete. In addition, the authors have shown recent contributions of their own work in this field:

- A wound healing simulation is developed under a continuum approach, clearly showing that it is highly regulated by angiogenesis.
- Bone distraction is a biological process where angiogenesis plays a key role. In this example, a lattice cell-based model is used to simulate time-evolution of angiogenesis during distraction.
- Cell migration in 3D under different environment conditions is fundamental for understanding many key fundamental processes in tissue engineering and particularly in angiogenesis. Here, a first approach using a cell force-based approach is proposed able to incorporate mechanosensing phenomenon [29] in the process of cell migration.

Acknowledgments This research was supported by the project part financed by the European Union (European Regional Development Fund) through the grant DPI 2009-14115-C03-01.

References

1. Alarcón, T., Byrne, H.M., Maini, P.K.: A cellular automaton model for tumour growth in inhomogeneous environment. J. Theor. Biol. **225**, 257–274 (2003)
2. Anderson, A., Chaplain, M.: Continuous and discrete mathematical models of tumor-induced angiogenesis. Bull. Math. Biol. **60**, 857–899 (1998)

3. Aronson, J.: Experimental and clinical experience with distraction osteogenesis. Cleft Palate Craniofac. J. **31**, 473–81 (1994)
4. Azuaje, F.: Computational discrete models of tissue growth and regeneration. Brief Bioinform. **12**, 64–77 (2011)
5. Balding, D., McElwain, D.L.S.: A mathematical model of tumour-induced capillary growth. J. Theor. Biol. **114**, 53–73 (1985)
6. Basilico, C., Moscatelli, D.: The FGF family of growth factors and oncogenes. Adv. Cancer Res. **59**, 115–165 (1992)
7. Bauer, A.L., Jackson, T.L., Jiang, Y.: A cell-based model exhibiting branching and anastomosis during tumor-induced angiogenesis. Biophys. J. **92**, 3105–21 (2007)
8. Beamer, B., Hettrich, C., Lane, J.: Vascular endothelial growth factor: an essential component of angiogenesis and fracture healing. HSS J. **6**, 85–94 (2010)
9. Bluteau, G., Julien, M., Magne, D.: VEGF and VEGF receptors are differentially expressed in chondrocytes. Bone **40**, 568–576 (2007)
10. Boccaccio, A., Pappalettere, C., Kelly, D.J.: The influence of expansion rates on mandibular distraction osteogenesis: a computational analysis. Ann. Biomed. Eng. **35**, 1940–1960 (2007)
11. Boccaccio, A., Prendergast, P.J., Pappalettere, C., Kelly, D.J.: Tissue differentiation and bone regeneration in an osteotomized mandible: a computational analysis of the latency period. Med. Biol. Eng. Comput. **46**, 283–298 (2008)
12. Borau, C., Kamm, R.D., García-Aznar, J.M.: Mechano-sensing and cell migration: a 3D model approach. Phys. Biol. **8**, Article No:066008 (2011)
13. Boucard, N., Viton, C., Agay, D., Mari, E., Roger, T., Chancerelle, Y., Domard, A.: The use of physical hydrogels of chitosan for skin regeneration following third-degree burns. Biomaterials **28**, 3478–3488 (2007)
14. Brunner, U.H., Cordey, J., Schweiberer, L., Perren, S.M.: Force required for bone segment transport in the treatment of large bone defects using medullary nail fixation. Clin. Orthop. Relat. Res. **301**, 147–155 (1994)
15. Byrne, H.M., Chaplain, M.A.J., Evans, D.L., Hopkinson, I.: Mathematical modelling of angiogenesis in wound healing: Comparison of theory and experiment. J. Theor. Med. **2**, 175–197 (2000)
16. Carmeliet, P., Jain, R.K.: Angiogenesis in cancer and other diseases. Nature **407**, 249–257 (2000)
17. Carvalho, R.S., Einhorn, T.A., Lehmann, W., Edgar, C., Al-Yamani, A., Apazidis, A., Pacicca, D., Clemens, T.L., Gerstenfeld, L.C.: The role of angiogenesis in a murine tibial model of distraction osteogenesis. Bone **34**, 849–861 (2004)
18. Chaplain, M.A.J., Byrne, H.M.: Mathematical modelling of wound healing and tumour growth-2 sides of the same coin. Wounds Compend. Clin. Res. Pract. **8**, 42–48 (1996)
19. Chaplain, M.A.: Mathematical modeling of angiogenesis. J. Neurooncol. **50**, 37–51 (2000)
20. Chaplain, M.A.J., McDougall, S.R., Anderson, A.R.A.: Mathematical modelling of tumorinduced angiogenesis. Annu. Rev. Biomed. Eng. **8**, 233–257 (2006)
21. Checa, S., Prendergast, P.J.: A mechanobiological model for tissue differentiation that includes angiogenesis: a lattice-based modeling approach. Ann. Biomed. Eng. **37**, 129–145 (2009)
22. Chen, G., Niemeyer, F., Wehner, T., Simon, U., Schuetz, M., Pearcy, M., Claes, L.: Simulation of the nutrient supply in fracture healing. J. Biomech. **42**, 2575–2583 (2009)
23. Choi, I.H., Ahn, J.H., Chung, C.Y., Cho, T.J.: Vascular proliferation and blood supply during distraction osteogenesis: a scanning electron microscopic observation. J. Orthop. Res. **18**, 698–705 (2000)
24. Chu, T.W., Wang, Z.G., Zhu, P.F., Jiao, W.C., Wen, J.L., Gong, S.G.: Effect of vascular endothelial growth factor in fracture healing. Zhongguo Xiu Fu Chong Jian Wai Ke Za Zhi **16**, 75–78 (2002)
25. Claes L.E., Heigele C.A. (1999) Magnitudes of local stress and strain along bony surfaces predict the course and type of fracture healing. J. Biomech. 32(3):255–266

26. Conway, E.M., Collen, D., Carmeliet, P.: Molecular mechanisms of blood vessel growth. Cardiovasc. Res. **49**, 507–21 (2001)
27. Davis, G.E., Senger, D.R.: Endothelial extracellular matrix: biosynthesis, remodeling, and functions during vascular morphogenesis and neovessel stabilization. Circ. Res. **97**, 1093–1097 (2005)
28. Deckers, M.M., van Bezooijen, R.L., van der Horst, G., Hoogendam, J., van Der Bent, C., Papapoulos, S.E., Löwik, C.W.: Bone morphogenetic proteins stimulate angiogenesis through osteoblast-derived vascular endothelial growth factor A. Endocrinology **143**, 1545–1553 (2002)
29. Discher, D.E., Janmey, P., Wang, Y.L.: Tissue cells feel and respond to the stiffness of their substrate. Science **310**(5751), 1139–1143 (2005)
30. Eckardt, H., Bundgaard, K.G., Christensen, K.S., Lind, M., Hansen, E.S., Hvid, I.: Effects of locally applied vascular endothelial growth factor (VEGF) and VEGF-inhibitor to the rabbit tibia during distraction osteogenesis. J. Orthop. Res. **21**, 335–340 (2003)
31. Ermentrout, G.B., Edelstein-Keshet, L.: Cellular automata approaches to biological modeling. J. Theor. Biol. **160**, 97–133 (1993)
32. Fang, T.D., Salim, A., Xia, W., Nacamuli, R.P., Guccione, S., Song, H.M., Carano, R.A., Filvaroff, E.H., Bednarski, M.D., Giaccia, A.J., Longaker, M.T.: Angiogenesis is required for successful bone induction during distraction osteogenesis. J. Bone Miner. Res. **20**, 1114–1124 (2005)
33. Ferguson, C., Alpern, E., Miclau, T., Helms, J.A.: Does adult fracture repair recapitulate embryonic skeletal formation? Mech. Dev. **87**, 57–66 (1999)
34. Ferrara, N.: Vascular endothelial growth factor: basic science and clinical progress. Endocr. Rev. **25**, 581–611 (2004)
35. Flegg, J.A., McElwain, D.L.S., Byrne, H.M., Turner, I.W.: A three species model to simulate application of hyperbaric oxygen therapy to chronic wounds. PLoS Comput. Biol. **5**, e1000451 (2009)
36. Folkman, J.: Tumour angiogenesis: therapeutic implications. New Engl. J. Med. **285**, 1182–1186 (1971)
37. Folkman, J., Shing, Y.: Angiogenesis. J. Biol. Chem. **267**, 10931–10934 (1992)
38. Folkman, J., Brem, H.: Angiogenesis and inflamation. In: Gallin, J.I., Goldstein, I.M., Snyderman, R. (eds) Inflamation: Basic Principles and Clinical Correlates, pp. 821–839. Raven Press, New York (1992)
39. Folkman, J.: Angiogenesis: an organizing principle for drug discovery? Nat. Rev. Drug Discov. **6**, 273–286 (2007)
40. Friedl, P., Bröcker, E.B.: The biology of cell locomotion within three-dimensional extracellular matrix. Cell. Mol. Life Sci. **57**, 41–64 (2000)
41. Gaffney, E.A., Pugh, K., Maini, P.K., Arnold, F.: Investigating a simple model of cutaneous wound healing angiogenesis. J. Theor. Biol. **45**, 337–374 (2002)
42. Gerber, H.P., Vu, T.H., Ryan, A.M.: VEGF couples hypertrophic cartilage remodeling, ossification and angiogenesis during endochondral bone formation. Nat. Med. **5**, 623–628 (1999)
43. Gerhardt, H., Golding, M., Fruttiger, M., Ruhrberg, C., Lundkvist, A., Abramsson, A., Jeltsch, M., Mitchell, C., Alitalo, K., Shima, D., Betsholtz, C.: VEGF guides angiogenic sprouting utilizing endothelial tip cell filopodia. J. Cell. Biol. **161**, 1163–1177 (2003)
44. Geris, L., Gerisch, A., Vander Sloten, J., Weiner, R., Van Oosterwyck, H.: Angiogenesis in bone fracture healing: a bioregulatory model. J. Theor. Biol. **25**, 137–158 (2008)
45. Geris, L., Van Liedekerke, P., Smeets, B., Tijskens, E., Ramon, H.: A cell based modelling framework for skeletal tissue engineering applications. J. Biomech. **43**(5), 887–892 (2010)
46. Gerstenfeld, L.C., Cullinane, D.M., Barnes, G.L., Graves, D.T., Einhorn, T.A.: Fracture healing as a post-natal developmental process: molecular, spatial, and temporal aspects of its regulation. J. Cell. Biochem. **88**, 873–884 (2003)
47. Glazier, J.A., Graner, F.: Simulation of the differential adhesion driven arrangement of biological cells. Phys. Rev. E **47**, 2128–2154 (1993)

48. Glowacki, J.: Angiogenesis in fracture repair. Clin. Orthop. **355**, S82–S89 (1998)
49. Gómez-Benito, M.J., García-Aznar, J.M., Kuiper, J.H., Doblaré, M.: Influence of fracture gap size on the pattern of long bone healing: a computational study. J. Theor. Biol. **235**, 105–119 (2005)
50. Grellier, M., Bordenave, L., Amedee, J.: Cell-to-cell communication between osteogenic and endothelial lineages: implications for tissue engineering. Trends Biotechnol. **27**, 562–571 (2009)
51. Hausman, M.R., Schaffler, M.B., Majeska, R.J.: Prevention of fracture healing in rats by an inhibitor of angiogenesis. Bone **29**, 560–564 (2001)
52. Hopf, H.W., Gibson, J.J., Angeles, A.P., Constant, J.S., Feng, J.J., Rollins, M.D., Zamirul Hussain, M., Hunt, T.K.: Hyperoxia and angiogenesis. Wound Repair Regen. **13**, 558–564 (2005)
53. Hunt, T.K., Knighton, D.R., Thakral, K.K., Goodson, W.H., Andrews, W.S.: Studies on inflammation and wound healing: angiogenesis and collagen synthesis stimulated in vivo by resident and activated macrophages. Surgery **96**, 48–54 (1984)
54. Ilizarov, G.A., Soibel'man, L.M.: Clinical and experimental data on bloodless lengthening of lower extremities. Eksp Khir Anesteziol **14**, 27–32 (1969)
55. Ilizarov, G.A.: The tension-stress effect on the genesis and growth of tissues. Part I: the influence of stability of fixation and soft-tissue preservation. Clin. Orthop. **238**, 249–281 (1989)
56. Ilizarov, G.A., Ledyaev, V.I.: The replacement of long tubular bone defects by lengthening distraction osteotomy of one of the fragments. Clin. Orthop. **280**, 7–10 (1992)
57. Ilizarov, G.A.: Transosseous Osteosynthesis. Springer, Heidelberg (1992)
58. Isaksson, H., Comas, O., Van Donkelaar, C.C., Mediavilla, J., Wilson, W., Huiskes, R., Ito, K.: Bone regeneration during distraction osteogenesis: mechano-regulation by shear strain and fluid velocity. J. Biomech. **40**, 2002–2011 (2007)
59. Jain, R.K.: Molecular regulation of vessel maturation. Nat. Med. **9**, 685–693 (2003)
60. Jamali, Y., Azimi, M., Mofrad, M.R.: A sub-cellular viscoelastic model for cell population mechanics. PLoS ONE **5**(8), pii:e12097 (2010)
61. Javierre, E., Vermolen, F.J., Vuik, C., Zwaag, S.: Numerical modelling of epidermal wound healing. In: Kunisch, K., Of, G., Steinbach, O. (eds) Numerical Mathematics and Advanced Applications, pp. 83–90. Springer, Berlin (2008)
62. Javierre, E., Moreo, P., Doblar, M., Garca-Aznar, J.M.: Numerical modeling of a mechano-chemical theory for wound contraction analysis. Int. J. Solids Struct. **46**, 3597 3606 (2009)
63. Josko, J., Gwozdz, B., Jedrzejowska-Szypulka, H.: Vascular endothelial growth factor (VEGF) and its effect on angiogenesis. Med. Sci. Monit. **6**, 1047–1052 (2000)
64. Klemke, R.L., Cai, S., Giannini, A.L., Gallagher, P.J., de Lanerolle, P., Cheresh, D.A.: Regulation of cell motility by mitogen-activated protein kinase. J. Cell. Biol. **137**, 481–492 (1997)
65. Kneser, U., Stangenberg, L., Ohnolz, J., Buettner, O., Stern-Strater, J., Möbest, D., Horch, R.E., Stark, G.B., Schaefer, D.J.: Evaluation of processed bovine cancellous bone matrix seeded with syngenic osteoblasts in a critical size calvarial defect rat model. J. Cell. Mol. Med. **10**, 695–707 (2006)
66. Le, A.X., Miclau, T., Hu, D., Helms, J.A.: Molecular aspects of healing in stabilized and non-stabilized fractures. J. Orthop. Res. **19**, 78–84 (2001)
67. Levine, H.A., Sleeman, B.D., Nilsen-Hamilton, M.: A mathematical model for the roles of pericytes and macrophages in angiogenesis. Part I: the role of protease inhibitors in preventing angiogenesis. Math. Biosci. **168**, 77–115 (2000)
68. Levine, H.A., Tucker, A.L., Nilsen-Hamilton, M.: A mathematical model for the role of cell signal transduction in the initiation and inhibition of angiogenesis. Growth Factors **20**, 155–175 (2002)
69. Li, S., Huang, N.F., Hsu, S.: Mechanotransduction in endothelial cell migration. J. Cell. Biochem. **96**, 1110–1126 (2005)

70. Lo, C., Wang, H., Dembo, M., Wang, Y.: Cell movement is guided by the rigidity of the substrate. Biophys. J. **79**, 144–152 (2000)
71. Macklin, P., McDougall, S.R., Anderson, A.R.A., Chaplain, M.A.J., Cristini, V., Lowengrub, J.: Multiscale modelling and nonlinear simulation of vascular tumour growth. J. Math. Biol. **58**, 765–798 (2009)
72. Maggelakis, S.: A mathematical model of tissue replacement during epidermal wound healing. Appl. Math. Model. **27**, 189–196 (2003)
73. Mantzaris, N., Webb, S., Othmer, H.: Mathematical modeling of tumor-induced angiogenesis. J. Math. Biol. **49**, 111–187 (2004)
74. Marti, H.H.: Angiogenesis–a self-adapting principle in hypoxia. In: Clauss, M., Breier, G. (eds) Mechanisms of Angiogenesis, pp. 163–180. Birkhauser, Switzerland (2005)
75. McCarthy, J.G., Schreiber, J., Karp, N., Thorne, C.H., Grayson, B.H.: Lengthening the human mandible by gradual distraction. Plast. Reconstr. Surg. **89**, 1–8 (1992)
76. McCarthy, I.: The physiology of bone blood flow: a review. J. Bone Joint Surg. Am. **88**, 4–9 (2006)
77. McDougall, S.R., Anderson, A.R.A., Chaplain, M.A.J., Sherratt, J.A.: Mathematical modelling of flow through vascular networks: implications for tumour-induced angiogenesis and chemotherapy strategies. Bull. Math. Biol. **64**, 673–702 (2002)
78. McDougall, S.R., Anderson, A.R.A., Chaplain, M.A.J.: Mathematical modelling of dynamic adaptive tumour-induced angiogenesis: clinical implications and therapeutic targeting strategies. J. Theor. Biol. **241**, 564–589 (2006)
79. Merks, R.M.H., Glazier, J.A.: Dynamic mechanisms of blood vessel growth. Nonlinear **19**, C1–C10 (2006)
80. Merks, R.M.H., Brodsky, S.V., Goligorksy, M.S., Newman, S.A., Glazier, J.A.: Cell elongation is key to in silico replication of in vitro vasculogenesis and subsequent remodeling. Dev. Biol. **289**, 44–54 (2006)
81. Merks, R.M., Perryn, E.D., Shirinifard, A., Glazier, J.A.: Contact-inhibited chemotaxis in de novo and sprouting blood-vessel growth. PLoS Comput. Biol. **4**, e1000163 (2008)
82. Merks, R.M.H., Koolwijk, P.: Modeling Morphogenesis in silico and in vitro: towards quantitative, predictive, cellbased modeling. Math. Model. Nat. Phenom. **4**, 149–171 (2009)
83. Mikos, A.G., Leite, S.M., Vacanti, J.P., Langer, R.: Prevascularization of porous biodegradable polymers. Biotechnol. Bioeng. **42**, 716–723 (1993)
84. Milde, F., Bergdorf, M., Koumoutsakos, P.: A hybrid model for three-dimensional simulations of sprouting angiogenesis. Biophys. J. **95**, 3146–3160 (2008)
85. Moreo, P., Garca-Aznar, J.M., Doblar, M.: Modeling mechanosensing and its effect on the migration and proliferation of adherent cells. Acta Biomaterialia **4**, 613–621 (2008)
86. Nomi, M., Atala, A., Coppi, P.D., Soker, S.: Principals of neovascularization for tissue engineering. Mol. Aspects Med. **23**, 463–483 (2002)
87. Oberringer, M., Jennevein, M., Matsch, S.E., Pohlemann, T., Seekamp, A.: Different cell cycle responses of wound healing protagonists to transient in vitro hypoxia. Histochem. Cell. Biol. **123**, 595–603 (2005)
88. Olsen, L., Sherratt, J.A., Maini, P.K., Arnold, F.: A mathematical model for the capillary endothelial cell-extracellular matrix interactions in wound-healing angiogenesis. IMA J. Math. Appl. Med. Biol. **14**, 261–281 (1997)
89. Orme, M.E., Chaplain, M.A.: A mathematical model of the first steps of tumour-related angiogenesis: capillary sprout formation and secondary branching. IMA J. Math. Appl. Med. Biol. **13**, 73–98 (1996)
90. Orme, M.E., Chaplain, M.A.: Two-dimensional models of tumour angiogenesis and anti-angiogenesis strategies. IMA J. Math. Appl. Med. Biol. **14**, 189–205 (1997)
91. Owen, M.R., Alarcón, T., Maini, P.K., Byrne, H.M.: Angiogenesis and vascular remodelling in normal and cancerous tissues. J. Math. Biol. **58**, 689–721 (2009)
92. Pacicca, D.M., Patel, N., Lee, C., Salisbury, K., Lehmann, W., Carvalho, R., Gerstenfeld, L.C., Einhorn, T.A.: Expression of angiogenic factors during distraction osteogenesis. Bone **33**, 889–898 (2003)

93. Palsson, E.: A three-dimensional model of cell movement in multicellular systems. Future Gener. Comput. Syst. **17**, 835–852 (2001)
94. Peirce, S.M.: Computational and mathematical modeling of angiogenesis. Microcirculation **15**, 739–751 (2008)
95. Peng, H., Wright, V., Usas, A., Gearhart, B., Shen, H.C., Cummins, J., Huard, J.: Synergistic enhancement of bone formation and healing by stem cell-expressed VEGF and bone morphogenetic protein-4. J. Clin. Invest. **110**, 751–759 (2002)
96. Pereira, C., Gold, W., Herndon, D.: Review paper: burn coverage technologies: current concepts and future directions. J. Biomater. Appl. **22**, 101–121 (2007)
97. Pérez, M., Prendergast, P.J.: Random-walk model of cell-dispersal included in mechanobiological simulation of tissue differentiation. J. Biomech. **40**, 2244–2253 (2007)
98. Pettet, G.J., Byrne, H.M., Mcelwain, D.L.S., Norbury, J.: A model of wound-healing angiogenesis in soft tissue. Math. Biosci. **136**, 35–63 (1996)
99. Pelham, R.J., Wang, Y.: Cell locomotion and focal adhesions are regulated by substrate flexibility. Proc. Natl. Acad. Sci. USA **94**, 13661–13665 (1997)
100. Pettet, G., Chaplain, M.A.J., Mcelwain, D.L.S., Byrne, H.M.: On the role of angiogenesis in wound healing. Proc. R Soc. Lond. Ser. B **263**, 1487–1493 (1996)
101. Prendergast, P.J., Huiskes, R., Soballe, K.: Biophysical stimuli on cells during tissue differentiation at implant interfaces. ESB Research Award 1996. J. Biomech. **30**, 539–548 (1997)
102. Reina-Romo, E., Gómez-Benito, M.J., García-Aznar, J.M., Domínguez, J., Doblaré, M.: Modeling distraction osteogenesis: analysis of the distraction rate. Biomech. Model. Mechanobiol. **8**, 323–335 (2009)
103. Reina-Romo, E., Gómez-Benito, M.J., García-Aznar, J.M., Domínguez, J., Doblaré, M.: Growth mixture model of distraction osteogenesis: effect of pre-traction stresses. Biomech. Model. Mechanobiol. **9**, 103–115 (2010)
104. Rockwood, C.A., Green, D.P., Bucholz, R.W., Heckman, J.D., Court-Brown, C.M., Koval, K.J., Tornetta, P.: Rockwood and green's fractures. In: Adults: Rockwood, Green, and Wilkins' Fractures. Lippincott Williams & Wilkins, ISBN 0781746361 (2006)
105. Ryan, T.J.: Biochemical consequences of mechanical forces generated by distension and distortion. J. Am. Acad. Dermatol. **21**, 115–130 (1989)
106. Safran, M., Kaelin, W.G.J.: HIF hydroxylation and the mammalian oxygen-sensing pathway. J. Clin. Invest. **111**, 779–783 (2003)
107. Sanz-Herrera, J.A., Moreo, P., Garca-Aznar, J.M., Doblar, M.: HOn the effect of substrate curvature on cell mechanics. Biomaterials **30**(34), 6674–6686 (2009)
108. Schugart, R.C., Friedman, A., Zhao, R., Sen, C.K.: Wound angiogenesis as a function of tissue oxygen tension: a mathematical model. Proc. Natl. Acad. Sci. USA **105**, 2628–2633 (2008)
109. Schwarz, U.S., Bischofs, I.B.: Physical determinants of cell organization in soft media. Med. Eng. Phys. **27**, 763–772 (2005)
110. Semenza, G.L., Wang, G.L.: A nuclear factor induced by hypoxia via de novo protein synthesis binds to the human erythropoietin gene enhancer at a site required for transcriptional activation. Mol. Cell. Biol. **12**, 5447–5454 (1992)
111. Sen, C.K., Gordillo, G.M., Roy, S., Kirsner, R., Lambert, L., Hunt, T.K., Gottrup, F., Gurtner, G.C., Longaker, M.T.: Human skin wounds: a major and snowballing threat to public health and the economy. Wound Repair Regen. **17**, 763–771 (2009)
112. Shefelbine, S.J., Augat, P., Claes, L., Simon, U.: Trabecular bone fracture healing simulation with finite element analysis and fuzzy logic. J. Biomech. **38**, 2440–2450 (2005)
113. Simon, U., Augat, P., Utz, M., Claes, L.: A numerical model of the fracture healing process that describes tissue development and revascularisation. Comput. Methods Biomech. Biomed. Eng. **14**, 79–93 (2011)
114. Stéphanou, A., McDougall, S.R., Anderson, A.R.A., Chaplain, M.A.J.: Mathematical modelling of flow in 2D and 3D vascular networks: applications to anti-angiogenic and chemotherapeutic drug strategies. Math. Comput. Model. **41**, 1137–1156 (2005)

115. Stéphanou, A., McDougall, S.R., Anderson, A.R.A., Chaplain, M.A.J.: Mathematical modelling of the influence of blood rheological properties upon adaptative tumour-induced angiogenesis. Math. Comput. Model. **44**, 96–123 (2006)
116. Stokes, C.L., Lauffenburger, D.A.: Analysis of the roles of microvessel endothelial cell random motility and chemotaxis in angiogenesis. J. Theor. Biol. **152**, 377–403 (1991)
117. Street, J., Winter, D., Wang, J.H., Wakai, A., McGuinness, A., Redmond, H.P.: Is human fracture hematoma inherently angiogenic? Clin. Orthop. **378**, 224–237 (2000)
118. Street, J.T., Wang, J.H., Wu, Q.D., Wakai, A., McGuinness, A., Redmond, H.P.: The angiogenic response to skeletal injury is preserved in the elderly. J. Orthop. Res. **19**, 1057–1066 (2001)
119. Street, J., Bao, M., deGuzman, L., Bunting, S., Peale, F.V. Jr, Ferrara, N., Steinmetz, H., Hoeffel, J., Cleland, J.L., Daugherty, A., van Bruggen, N., Redmond, H.P., Carano, R.A., Filvaroff, E.H.: Vascular endothelial growth factor stimulates bone repair by promoting angiogenesis and bone turnover. Proc. Natl. Acad. Sci. USA **99**, 9656–9661 (2002)
120. Torpy, J.M., Lynm, C., Glass, R.M.: JAMA patient page. Burn Inj JAMA 302:1828 (2009)
121. Tsopanoglou, N.E., Andriopoulou, P., Maragoudakis, M.E.: On the mechanism of thrombin-induced angiogenesis: involvement of alphavbeta3-integrin. Am. J. Physiol. Cell. Physiol. **283**, C1501–C1510 (2002)
122. Willett, C.G., Boucher, Y., di Tomaso, E., Duda, D.G., Munn, L.L., Tong, R.T., Chung, D.C., Sahani, D.V., Kalva, S.P., Kozin, S.V., Mino, M., Cohen, K.S., Scadden, D.T., Hartford, A.C., Fischman, A.J., Clark, J.W., Ryan, D.P., Zhu, A.X., Blaszkowsky, L.S., Chen, H.X., Shellito, P.C., Lauwers, G.Y., Jain, R.K.: Direct evidence that the VEGF-specific antibody bevacizumab has antivascular effects in human rectal cancer. Nat. Med. **10**, 145–147 (2004)
123. Wood, L., Kamm, R., Asada, H.: Stochastic modeling and identification of emergent behaviors of an Endothelial Cell population in angiogenic pattern formation. Int. J. Robot. Res. **30**, 659–677 (2011)
124. Xue, C., Friedman, A., Sen, C.K.: A mathematical model of ischemic cutaneous wounds. Proc. Natl. Acad. Sci. USA **106**, 16782–16787 (2009)
125. Yancopoulos, G.D., Davis, S., Gale, N.W., Rudge, J.S., Wiegand, S.J., Holash, J.: Vascular-specific growth factors and blood vessel formation. Nature **407**, 242–248 (2000)
126. Yasui, N., Sato, M., Ochi, T., Kimura, T., Kawahata, H., Kitamura, Y., Nomura, S.: Three modes of ossification during distraction osteogenesis in the rat. J. Bone Joint Surg. Br. **79**, 824–830 (1997)
127. Zaman, M.H., Kamm, R.D., Matsudaira, P., Lauffenburger, D.A.: Computational model for cell migration in three-dimensional matrices. Biophys. J. **89**, 1389–1397 (2005)
128. Zelzer, E., McLean, W., Ng, Y.S., Fukai, N., Reginato, A.M., Lovejoy, S., D'Amore, P.A., Olsen, B.R.: Skeletal defects in VEGF (120/120) mice reveal multiple roles for VEGF in skeletogenesis. Development **129**, 1893–1904 (2002)

Mathematical Modelling of Regeneration of a Tissue-Engineered Trachea

Greg Lemon, John R. King and Paolo Macchiarini

Abstract One of the most promising recent achievements in the field of regenerative medicine was the first successful transplantation of a tissue-engineered trachea (Macchiarini et al. The Lancet 372(9655), 2023–2030). This land-mark operation has paved the way for developing a host of successful stem cell-based therapies for treating disease, and the exciting possibility of the tissue engineering of whole organs. It has also provided the opportunity for new directions in mathematical and computational modelling for tissue engineering. By way of describing an approach to modelling the regeneration of a tissue-engineered trachea seeded with cells in situ this chapter will highlight some of the opportunities and challenges involved in applying mathematical models to these new therapies.

1 Introduction

A significant medical breakthrough was made in 2008 with the first successful implantation of a tissue-engineered windpipe[1] which saved the life of a young woman with end-stage tuberculosis. The operation, described in [47], utilised a decellularised donor trachea (Fig. 1a) which was seeded *ex vivo*, using a special bioreactor, with stem cell derived chondrocytes and epithelial cells (EPCs) taken from the recipient. With the transplanted trachea in situ the seeded cells assembled

[1] BBC News, "Windpipe transplant breakthrough" Nov. 2008, `http://news.bbc.co.uk/1/hi/7735696.stm`

G. Lemon (✉) · J. R. King · P. Macchiarini
Advanced Center for Translational Regenerative Medicine, Karolinska Institutet, Stockholm, Sweden
e-mail: greg.lemon@ki.se

Stud Mechanobiol Tissue Eng Biomater (2013) 10: 405–439
DOI: 10.1007/8415_2012_145

Published Online: 14 September 2012

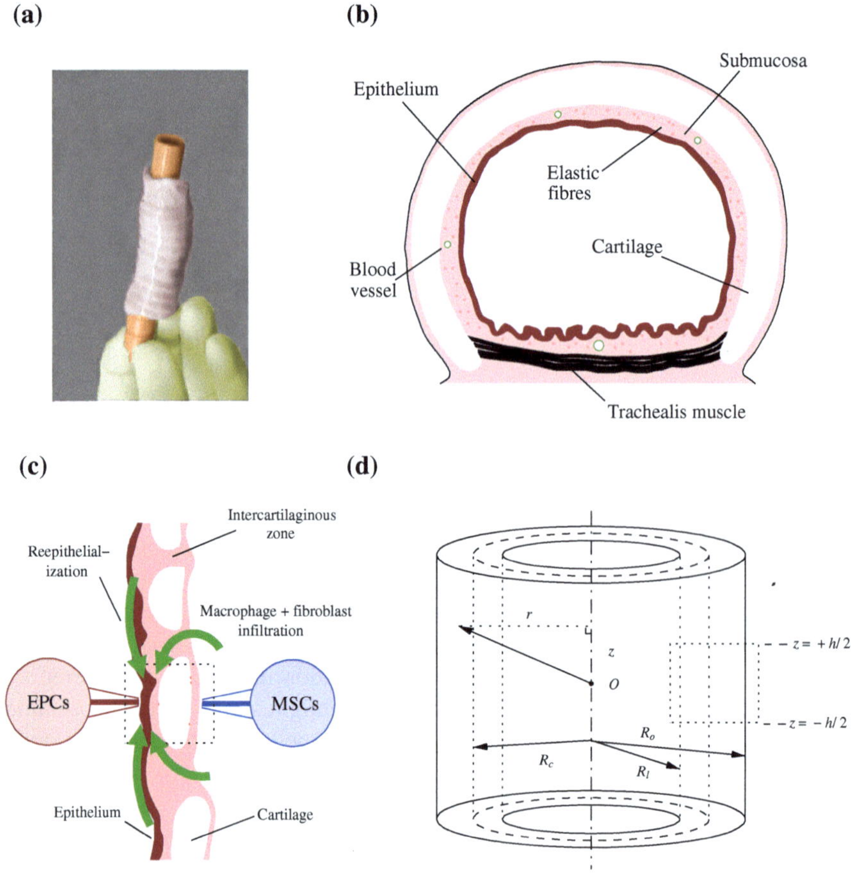

Fig. 1 Images of **a** a decellularised trachea, **b** the radial cross section of a healthy trachea, **c** an axial cross-section of the trachea showing topical application of EPCs and MSCs, reepithelization and infiltration of cells, and **d** a schematic diagram representing the cylindrical geometry of the trachea, showing the position coordinates used in the mathematical model

to form a healthy functioning trachea, whereby the chondrocytes regenerated the cartilage ring and the EPCs provided the lining of the airway (Fig. 1b). The therapy has been further developed[2] in the case of a young boy suffering from congenital stenosis (abnormally narrow airway at birth) by seeding the donor windpipe with autologously derived stem cells and EPCs in situ without the need for seeding *ex vivo*. Improvements to the therapy are evolving rapidly, with the most recent advance being the implantation of the first synthetic bio-engineered trachea [P. Macchiarini, private communication].

[2] BBC News, "Windpipe transplant success in UK child" Mar. 2010, `http://news.bbc.co.uk/1/hi/8576493.stm`

With their ability to self-renew and differentiate into a variety of cell types, the use of adult stem cells has been proposed as a treatment for a wide range of different diseases [11], and their application to organ transplantation represents an exciting development in the field of regenerative medicine [40]. Hence there are valuable opportunities now arising to develop mathematical models for the emerging field of stem cell-based therapies for tissue engineering. Indeed the main part of this chapter concerns a description and analysis of a mathematical model for tracheal regeneration using mesenchymal stem cells (MSCs). But what should be the goals in developing a mathematical model of tracheal regeneration? Clearly if a living trachea could be represented faithfully using a mathematical or computational model i.e. an *in silico* trachea, virtual experiments could be performed on the model to develop therapies without the expense and ethical concerns of using live subjects. But is this realistic and if not what are the obstacles to achieving this? How could a simplified mathematical model that omits much detail of the biology be of any use for studying regeneration of a tissue-engineered trachea? Some consideration will be given to such questions in this chapter, which of course exemplify issues that arise much more generally in systems biology.

The main thrust of this chapter is to (i) present a possible approach to modelling the regeneration of a tissue-engineered trachea in situ, in the course of which more general modelling principles will be illustrated, and (ii) to use the model to gain some insight into the possible causes of success or failure of the transplantation therapy. Consideration is given to the implantation of a trachea with intra-operative seeding using different combinations of cells i.e. with or without EPCs applied to the lumenal surface, and with or without MSCs applied to the outer surface of the trachea (Fig. 1c).

In Sect. 2 of this chapter some description of tissue-engineered tracheal transplantation therapies and the relevant biology is given. This motivates the formulation of the mathematical model which is described in Sect. 3. The numerical methods used to solve the model equations are given in Sect. 4, and in Sect. 5 the solutions of the equations are described, representing the cases of operations involving seeding with different combinations of EPCs and MSCs. Discussion of the modelling approach used and the results, as well as suggestions for possible future work, are given in Sect. 6

2 Biological Considerations

The first successful operation involving implantation of a tissue-engineered trachea into a human is reported in [47], with further reviews and description of the procedures given in [6, 7, 37]. These works form the basis of the following description of the therapy and the associated biology given below, which is then used in the design of the mathematical model in the following section. Due to the enormous complexity of the tracheal regeneration process the following description represents only a brief and simplified summary.

An important first step in the therapy is to decellularise the donor trachea. This is done to prevent rejection by the host and has the potential advantage over synthetic tracheal prostheses of presenting cells with the appropriate guidance clues for growth and differentiation. Decellularisation removes all the cells from the trachea while leaving the extra-cellular matrix (ECM) components largely intact [36]. Then the donor trachea is seeded with correct types of cells from the patient, crucially EPCs to regenerate the lining of the tracheal lumen and MSC-derived chondrocytes to regenerate the cartilage ring [26]. Decellularised tracheas implanted into laboratory animals without both cell types present have been shown to fail [66]. Seeding of the trachea is performed so as to direct each type of cell into the appropriate layer of the trachea. For the first tracheal operation this was done *ex vivo* using a special bioreactor [4] to distribute the cells through the trachea mechanically.

A different approach can be used whereby MSCs are applied to the exterior of the scaffold in situ [54] along with appropriate growth factors [6]. This technique aims to harness the reparative potential of MSCs, which have been shown to be able to invade cartilage and differentiate into chondrocytes [62]. The contrast between the two approaches is similar to that between 'dynamic seeding' and 'static seeding' of tissue engineering scaffolds. Static seeding, where cells are applied to the exterior surface of the scaffold, can be problematic because of the time it takes for cells to migrate through the scaffold. A mathematical modelling study by our group of the differences between the two approaches is described in [45]. In terms of revascularisation of the tissue-engineered trachea, where blood vessels grow from surrounding tissue into the implant, such a delay could lead to failure of the graft if the seeded cells react adversely to the hypoxic conditions within the trachea [20]; mathematical modelling of the problem of vascularisation of porous biomaterials by our group is described in [43].

The different types of cells that participate in the regeneration must be rapidly directed to the correct layer within the tracheal cross section (Fig. 1b) and this layering must be maintained indefinitely. A pertinent question is: to what extent does morphogenesis during development relate to the process of regeneration? The growth of some embryonic tissues are guided by global polarity signals so it is interesting to speculate as to whether such clues persist in the adult trachea to maintain its shape and structure. On the other hand, the layered structure of the trachea and the presence of certain growth factors within the underlying ECM, established during development, may serve as a template to guide the positioning of cells during episodes of healing in the adult tissue. The mechanical environment experienced by cells in situ will also serve as a guide for cell positioning: for example EPCs favour a fluid-air interface [90], which makes them preferentially grow to cover the lining of the tracheal lumen.

Pathological states can arise after implantation of the trachea, which are characterised by fibrotic growth (excessive proliferation of fibroblasts and production of ECM in the submucosa) and malacia (softening of the cartilage rings) [10, 26], resulting in the initially normal structure of the donor trachea or graft to be overrun and destroyed. The presence of both EPCs and MSCs appear to be necessary for successful regeneration [26], and if the epithelium is lost or damaged inflammation and fibrosis can occur [49, 70]. Fibrosis is associated with epithelial-mesenchymal transition (EMT) [15, 85] whereby EPCs differentiate into myofibroblasts, with tissue growth factor-β1 (TGF-β1) being the main switch [86]. TGF-$\beta1$ also suppresses EPC proliferation [50].

Shortly after implantation, the trachea becomes inflamed due to immune cells that invade the implant to remove pathogens and instigate healing. The migration of monocytes into damaged tissues is initiated by cytokines released during the degranulation of blood platelets. Once inside the wound, monocytes differentiate into macrophages and secrete pro-inflammatory cytokines including TGF-$\beta1$ [5]. However EPCs attached to the lining of the tracheal lumen secrete interleuken-10 (IL-10) which inhibits macrophage activity [14], and its absence is associated with fibrotic disease [13]. Indeed the loss of epithelium and excessive numbers of macrophages are associated with fibrosis of tracheal allografts [25]. If inflammation persists for prolonged periods healthy cells may be destroyed by the action of pro-inflammatory cytokines secreted by macrophages [55].

Macrophages play an important role in the repair of tissues by secreting growth factors, including vascular endothelial growth factor (VEGF) and basic fibroblast growth factor (bFGF): these stimulate the growth of blood vessels (angiogenesis) in wounds [71] and secrete factors that induce migration of fibroblasts into wounds [73].

However macrophages are known to play a central role in fibrotic disease [87, 88] by triggering excessive fibroblast proliferation and collagen production [61]. Indeed TGF-$\beta1$ secreted by macrophages stimulates secretion and remodelling of ECM by fibroblasts, and is upregulated in fibrotic tissue [83] and under conditions of airway stenosis [38]. Macrophages also produce the pro-inflammatory cytokine TNF-α which is known to play a role in the loss of GAGs in diseases of cartilage such as arthritis by causing apoptosis of chondrocytes [3] and inducing chondrocytes to degrade GAGs [84]. Resolution of inflammation requires the egress and apoptosis of macrophages from tissue after healing. This can also involve the conversion of the pro-inflammatory 'classically-activated' macrophage (CAM) into an 'alternatively-activated' macrophage subtype (AAM) that can ameliorate the effects of, and suppress, the pro-inflammatory subtype [41]. Pro-inflammatory macrophage activity is reduced by anti-inflammatory cytokines, such as IL-10 which induces apoptosis of macrophages [8] and reduces their secretion of pro-inflammatory cytokines including TNF-α [24].

In [66] it was shown that GAGs in the cartilage rings (which give the trachea mechanical strength) were left intact after decellularisation but that residual chondrocytes remained, possibly serving as progenitors of new chondrocytes. It is also possible that biochemical clues remaining behind in the decellularised cartilage, or secreted by residual chondrocytes, guide cells during the regeneration process. Interestingly, ECM degradation products can have mitotic and chemotactic effects on progenitor cells [65]. Thus pro-inflammatory cells causing degradation of proteoglycans within the cartilage [28] could be a mechanism by which MSCs invade and regenerate the cartilage rings.

Effectively harnessing the multipotent and self-renewal properties of stem cells is crucial for their use in the regeneration and healing of tissue-engineered organs. MSCs have the ability to differentiate into a variety of cells that constitute tracheal tissue including chondrocytes [81], endothelial cells (ECs) [57], fibroblasts [23], and EPCs [82]. MSCs contribute to the repair of diseased and damaged lung tissues [76] also by their ability to suppress inflammation [75].

MSCs migrate to, and localise at, sites of inflammation due to their chemotactic and proliferative response to pro-inflammatory cytokines such as TNF-α [60]. They can down-regulate the secretion of pro-inflammatory cytokines by macrophages [1], and are associated with the upregulation of anti-inflammatory cytokines including IL-10, which may be secreted by MSCs directly [17] or by macrophages that have been converted into their alternatively-activated form by MSCs [93]. MSCs are recruited from the bone marrow of the host to the lung tissue to bring about healing, and this is associated with increased levels of granulocyte macrophage colony-stimulating factor (GM-CSF) and granulocyte stimulating factor (G-CSF) [68]. The mechanisms by which MSCs are released from the bone marrow and migrate through the body to damaged tissues are similar to those of immune cells such as leukocytes [89]. GM-CSF and G-CSF are known to be produced endogenously by macrophages, ECs and fibroblasts in wounds. The exogenous application of these substances to the tissue-engineered trachea [6] can further boost the recruitment of MSCs and aid the regeneration process. However the numbers of endogenous MSCs arriving at the implant site may be small so intra-operative application of exogenous MSCs is performed to aid regeneration.

In addition to applying stem cells to regenerate the trachea, a 'cocktail' of growth factors and hormones are added to boost the healing process [6]. These include the hormone erythropoietin (EPO) which is used to increase the numbers of MSCs and stimulate angiogenesis [39]. EPO also dampens down inflammation by down-regulating the production of cytokines by pro-inflammatory cells. EPO prevents cell apoptosis [78], stimulates the secretion of angiogenic factors by MSCs, and mobilizes endothelial progenitor cells from the bone marrow [32].

To enhance proliferation and differentiation of MSCs into chondrocytes, the growth factor TGF-β (TGF-$\beta 1, 2, 3$) [12] is applied to the tissue-engineered trachea intraoperatively [6]. Chondrocytic differentiation of MSCs can be caused by factors produced by chondrocytes themselves [34], which includes TGF-$\beta 1$ produced by MSCs and MSC-derived chondrocytes [94]. Certain ECM components have been shown to induce differentiation of MSCs into chondrocytes [22] including hyaluronic acid and autologous synovial fluid [29]. A variety of other intrinsic factors, such as hypoxic stress caused by the initial absence of a vasculature, can also influence commitment.

Complete regeneration of the trachea requires the ingrowth of blood vessels to supply nutrients and remove waste products. This process, angiogenesis, involves a complex sequence of events commencing with the secretion of growth factors such as VEGF and bFGF that induces chemotaxis of ECs and their assembly into hollow tubes that join to form an interconnected network of vessels [2]. In a normal healthy trachea blood vessels cover the outer surface and connect the submucosa to the host tissue through inter-cartilaginous zones which lie between the cartilage rings (see Fig. 1c). Blood vessels cannot penetrate the rings because of anti-angiogenesis factors present in the cartilage [31]. Hence the chondrocytes must rely on the delivery of oxygen by passive diffusion through distances in the order of mm. However if the hypoxic conditions are prolonged the cartilage can degenerate due to cell necrosis and the migration of chondrocytes up oxygen gradients [20]. The decellularisation process may leave behind angiogenic factors in the trachea, such as bFGF, that stimulate angiogenesis [67]. Also ECs can be preseeded into the tracheal implant to encourage revascularisation [92]. In the tissue-engineered trachea this revascularisation occurs rapidly, with new blood vessels forming inside the trachea 24 h after implantation [P. Macchiarini, private communication].

The preceding description identifies the major roles for the different types of cells, ECM components and cytokines that participate in regeneration of the tissue-engineered trachea *in-situ*. This is used as the basis for formulating the mathematical model in the following section.

3 Mathematical Model

The previous section illustrates how the tissue engineering of a trachea typifies many of the challenges that arise in the mathematics of regenerative medicine, such as describing cell division, signalling and differentiation within mixed cell-type populations. The first decision to be made is to decide on which scale, e.g. cell, tissue or organ, the model should operate—here the intermediate (tissue-scale) approach is adopted.

This section begins with a summary of those mechanisms described in Sect. 2 that are included in the mathematical model of tracheal regeneration, and key assumptions and simplifications that are made in the model formulation are set out. This is followed in Sect. 3.2 by a description of the general methodology used to construct the model. In Sect. 3.3 the mathematical model is set out in detail, whereby equations appropriate for each of the species are given. The model is then nondimensionalized in Sect. 3.4 with appropriate initial conditions given in Sect. 3.5 and a discussion of the choice of parameter values provided in Sect. 3.6

3.1 Summary of Included Mechanisms

The complexity of tracheal regeneration poses significant challenges for modelling the process in its entirety and a crucial element of a useful mechanistic model is that it includes sufficient, but not too much, of the relevant details. For simplicity the mathematical model presented in this chapter provides a simplified description of reepithelization, inflammation, and regeneration of the cartilage. Moreover, consideration will not be given to revascularisation of the trachea.

In the model a fictitious surgical procedure is considered whereby a decellularised trachea is grafted so as to replace the native trachea of the subject (either a human or laboratory animal). This is immediately preceded by applying EPCs to the lining of the tracheal lumen and MSCs being applied to the outer surface (see Fig. 1c), but with no application of exogenous growth factors (TGF-β1 and G-CSF are usually applied locally during the operation). The wound is then sealed and the subject is allowed to recover without further medical intervention.

The mechanisms that are included in the model are depicted in Fig. 2 which shows the species that interact within a small cylindrical segment of the trachea (the dashed rectangles in Fig. 1c, d), and is divided into radial compartments representing the epithelium, submucosa and cartilage ring. The arrows represent migration of cells into the segment: EPCs axially across the surface of the lumen, macrophages and fibroblasts axially into the submucosa, and MSCs radially by being applied to the outer surface of the trachea during the operation. The contribution of endogenous MSCs to the regeneration is assumed to be negligible. For simplicity the migration of macrophages and fibroblasts into the segment is assumed to be passive i.e. their arrival at the segment boundaries is not in response to the presence of chemokines within.

Macrophages are assumed to enter the trachea in their (classically) activated state but undergo apoptosis due to the presence of IL-10 which is produced by EPCs (the AAM subtype is not included). Consistent with their anti-inflammatory effect MSCs are also assumed to secrete IL-10. TGF-β1, secreted by macrophages, suppresses the proliferation of EPCs (EMT is not considered in the present model). The mutually antagonistic effects of TGF-β1 (pro inflammatory) and IL-10 (anti inflammatory) are central to the mechanism of inflammation and its resolution, as discussed further in Sect. 5 Fibroblasts migrate into the submucosa to maintain the ECM, assumed to be composed mainly of collagen; however, the rate at which fibroblasts secrete the collagen is assumed to be increased by TGF-β1. Hence if inflammation persists, so there are high levels of TGF-β1, excess collagen will be deposited, resulting in thickening of the submucosa. The rate of proliferation of MSCs, and their rate of differentiation into chondrocytes, are assumed to be increased by TGF-β1. The pro-inflammatory cytokine TNF-α, which is produced by macrophages, is also assumed to stimulate proliferation of MSCs. The MSC-derived chondrocytes are responsible for maintaining the GAGs in the cartilage but in the presence of TNF-α they undergo apoptosis. The MSC-derived chondrocytes are assumed to secrete TGF-β1, thereby creating a cycle of self-regeneration through stimulating MSC proliferation.

The cytokines IL-10, TGF-β1 and TNF-α are allowed to freely diffuse through the submucosal and cartilage regions. Once inside the trachea the macrophages, fibroblasts and MSCs migrate randomly but no chemotactic effects on cell migration by the cytokines is included in the model (see comments at the end of

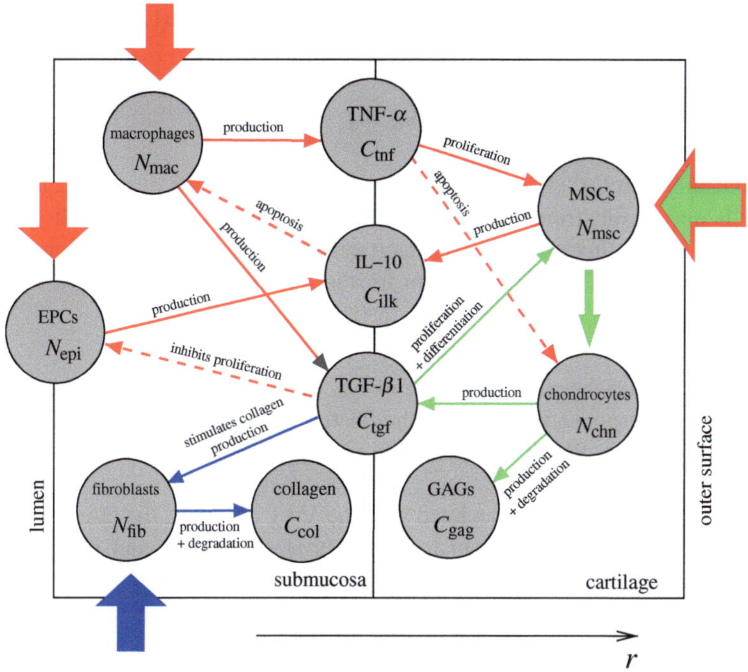

Fig. 2 Diagram showing the network of relationships between the different species in the mathematical model. Below the name of each species is the corresponding variable name in the model. Large *thick arrows* represent inward migration of cells into the trachea, and differentiation of MSCs into chondrocytes. The *thin arrows* represent activation (*solid*) or inhibition (*dashed*) of one species by another. *Red arrows* represent interactions that are relevant to inflammation, *blue* represents maintenance of ECM in the submucosa, *green* represents maintenance of ECM in the cartilage

Sect. 3.2). MSCs are assumed to be able to migrate through both layers of the trachea, whereas the fibroblasts and macrophages are confined to the submucosal layer.

3.2 Modelling Framework

The model is constructed as a set of continuum reaction diffusion equations for each of the species [51]. However, the need to describe stenosis of the implant suggests the need for a moving boundary approach, so the reaction diffusion equations will be considered to be on an evolving domain [19], whose boundary must be calculated at the same time as the other unknowns. In such an approach the derivation of the equation for the amount of a species, u, is applied in a reference frame that moves with the tissue with velocity $\mathbf{v}$ (Lagrangian frame of reference) leading to

$$\frac{\partial u}{\partial t} = -\nabla \cdot (\mathbf{v}u + \mathbf{J}_u) + f(u, \mathbf{x}, t). \tag{1}$$

The quantity f in Eq. (1) is the local rate of change of mass of the species, which for cells may result from division or differentiation from another cell type. The quantity $\mathbf{J}_u$ is the flux of the species as it pertains to a stationary reference frame; for a cell species an appropriate form for the flux is

$$\mathbf{J}_u = -D_u \nabla u + u\chi_u \nabla a. \tag{2}$$

The two terms on the right hand side of Eq. (2) respectively model diffusion and chemotaxis in response to a generic chemoattractant with concentration a (see for example [51, Chap. 9]). If u is a soluble chemical messenger or chemokine only the first term is used.

The dilation term in (1) associated with the $\nabla \cdot \mathbf{v}$ describes the rate of expansion of the bulk of the tissue to which a species is embedded, and is specified by consideration of the dynamics of the extra cellular matrix (ECM) as follows. Tissue in the submucosa comprises cells attached to a collagen (type I) mesh that is actively secreted and degraded by fibroblasts. As collagen is secreted it rapidly polymerises and expands to form a fibrous mesh. Because the collagen matrix serves as a substrate for cell attachment, as the matrix expands it takes cells with it. The cartilage ring on the other hand consists of collagen (type II) and GAGs (chondroitin sulfate). The mechanical properties of the cartilage are dominated by the GAGs which impart structural rigidity and compressive strength to the tissue. If the GAGs are lost (malacia) there will be a loosening and expansion of the cartilage which takes embedded cells. Such changes may be contributed to by the loss of collagen II which imparts cohesiveness to the cartilage however for simplicity only the effects of GAG degradation will be considered. The preceding considerations motivates making the dilation of the submucosa and cartilage tissue elements a scalar function of the net rate of secretion of ECM in such a way that that the tissue domain can expand (stenosis) or contract (involution/atrophy). Further details of how this is done are given in Sect. 3.3 A similar idea has been used to model size increase of tissue engineering scaffolds in a bioreactor [56].

In a 3D domain the velocity, $\mathbf{v}$, has three components and in general additional assumptions have to be made to close the model. This could be done, for example, by specifying the spatial distribution in the rate of expansion of the ECM. Since the velocity of the matrix in the vicinity of a particular point in the domain $\mathbf{r} - \mathbf{r}_0$ relative to that point is $d\mathbf{v} = (\nabla \mathbf{v})|_{\mathbf{r}_0} \cdot d\mathbf{r}$, the principal axes of $\nabla \mathbf{v}$ could be chosen to reflect some preferred direction in the tissue. However, care needs to be taken that such a choice is compatible with physical constraints imposed on the growth. This issue can be avoided by considering a 1D geometry thereby requiring only one component of $\mathbf{v}$ to be determined. The geometry of the trachea is highly suggestive of the use of cylindrical polar coordinates (see Fig. 1d) and neglecting any dependence on θ and z reduces the system to 1D with coordinate r. This simplification means that the cylindrical shape of the trachea is imposed in the model rather than arising from the emergent behaviour of the cells within it.

The present model is concerned with just one cylindrical segment of the trachea, lying wholly within a single cartilage ring, and is lined internally with epithelium and a sub-mucosal layer as shown in Fig. 1d. The segment does not include the inter-cartilaginous zones. For simplicity, the muscular posterior wall (see Fig. 1b) is considered to be part of the cartilage ring and not flattened, so the tracheal segment can be assumed to be cylindrical. It is assumed that the tracheal segment does not deform axially, and so the displacement of its top and bottom edges are held fixed at $z = \pm h/2$. Also the outer curved surface of the trachea, at $r = R_o$, is held fixed while the inner curved surface, at $r = R_l$, and the boundary between the cartilage and the submucosa, at $r = R_c$, are free to move.

To model the exchange of cells between the trachea and the surrounding tissue, the flux of cells into the domain is prescribed by some function, I_u, on the boundary. Continuity of the flux of cells across the boundary implies $I_u = -\hat{\mathbf{n}} \cdot \mathbf{J}_u$ where $\hat{\mathbf{n}}$ is the outward pointing unit normal on the boundary hence from Eq. (2), assuming $\hat{\mathbf{n}} \cdot \nabla a = 0$ (see below), the boundary conditions for the cells are

$$D_u \hat{\mathbf{n}} \cdot \nabla u = I_u. \tag{3}$$

This boundary condition is also valid for the diffusible chemical species but with $I_u = 0$, thereby reducing to $\hat{\mathbf{n}} \cdot \nabla u = 0$ in those cases.

Because of the cylindrical symmetry I_u need only be specified on the tracheal surface at $r = R_l$ and $r = R_o$; indeed the flux of cells and growth factors is generally assumed to be zero there. However, a non-zero flux of cells entering the tracheal segment at $z = \pm h/2$ needs to be taken into account. This is done by assuming negligible changes in the variables in the z direction i.e. $u = u(r,t), \mathbf{v} = v(r,t)\hat{\mathbf{r}}$ and $\mathbf{J}_u - J_r(r,t)\hat{\mathbf{r}} + J_z(r,t)\hat{\mathbf{z}}$, where $\hat{\mathbf{r}}$ and $\hat{\mathbf{z}}$ are unit vectors corresponding to the coordinates r and z respectively. Integrating Eq. (1) from $z = -h/2$ to $z = +h/2$ and dividing by h gives, in cylindrical polar coordinates,

$$\frac{\partial u}{\partial t} = -\frac{1}{r}\frac{\partial}{\partial r}(r(uv + J_r)) + \frac{2}{h} I_u(r, \pm h/2, t) + f(u, r, t), \tag{4}$$

where I_u is assumed to be the same on both boundaries. This way of reducing the problem to 1D is tantamount to making the assumption that species mobility in the axial direction is large compared to that in the radial direction. In cylindrical polar coordinates the radial component of Eq. (2) is

$$J_r = -D_u \frac{\partial u}{\partial r} + u\chi_u \frac{\partial a}{\partial r}. \tag{5}$$

Although I_u specifies an influx of cells that may in real tissues have a strong chemotactic causation, in the model chemotaxis of cells is neglected once they are inside the tracheal segment hence $\chi_u = 0$.

3.3 The Detailed Formulation

The modelling framework described in Sect. 3.2 is used below to set out the model equations in full, whereby Eqs. (4) and (5) are used to formulate equations for each of the cellular and chemical species together with appropriate boundary conditions. For the sake of brevity only the salient features of the equation for each species is described. A detailed guide to the techniques for modelling biochemical reactions and tissue growth using reaction-diffusion equations can be found in [51].

In what follows a variable of the form $N_\times$ is used to denote a cell density while $C_\times$ denotes a concentration. Besides the parameters introduced in Sect. 3.2, a parameter of the form $b_\times$ appearing in an equation refers to a rate of cell division or a rate of production of a chemical species whereas $d_\times$ refers to a rate of cell death or apoptosis, or a rate of chemical degradation. Parameters of the form $K_\times$ characterise the strength of the effect of one type of species on another.

Macrophages are assumed to be excluded from the cartilage ring, but enter and migrate randomly within the submucosal region in their classically-activated form while undergoing apoptosis at a rate which increases with increasing IL-10 concentration. The equation governing their density is, applying Eq. (4),

$$\frac{\partial N_{\mathrm{mac}}}{\partial t} = \frac{1}{r}\frac{\partial}{\partial r}\left(-rvN_{\mathrm{mac}} + rD_{\mathrm{mac}}\frac{\partial N_{\mathrm{mac}}}{\partial r}\right) + \frac{2}{h}I_{\mathrm{mac}} - d_{\mathrm{mac}}\left(1 + \frac{C_{\mathrm{ilk}}^2}{K_{\mathrm{mi}}^2}\right)N_{\mathrm{mac}}, \quad (6)$$

for $R_l \leq r \leq R_c$. Macrophages cannot enter or leave the tracheal segment radially hence the boundary conditions are, from Eq. (3),

$$\frac{\partial N_{\mathrm{mac}}}{\partial r}(R_l, t) = \frac{\partial N_{\mathrm{mac}}}{\partial r}(R_c, t) = 0. \quad (7)$$

Epithelial cells are assumed to remain attached to the basement membrane ECM. The radial symmetry assumption implies that the surface density of EPCs on the lumenal surface, N_{epi}, and the speed $v = v(R_l, t)$ depend only on t and, because cells cannot migrate radially, $J_r = 0$. Hence Eq. (4) for the EPCs reduces to

$$\frac{dN_{\mathrm{epi}}}{dt} = -\frac{v(R_l, t)}{R_l}N_{\mathrm{epi}} + \frac{2}{h}I_{\mathrm{epi}} + b_{\mathrm{epi}}N_{\mathrm{epi}}\left(N_{\mathrm{epi}}^*\left(1 - \frac{C_{\mathrm{tgf}}(R_l, t)}{K_{\mathrm{et}}}\right) - N_{\mathrm{epi}}\right). \quad (8)$$

The second term in the right-hand side of Eq. (8) models infiltration of EPCs that migrate into the implant from where it is grafted to the patients own trachea. The third term in the right-hand side of Eq. (8) models epithelial cell proliferation, where N_{epi}^* is the surface density of EPCs at confluence i.e. the carrying capacity. The effect of TGF-β1 is to suppress the proliferation of EPCs, thereby reducing their maximum surface density.

MSCs. Consistent with their regenerative potential, MSCs can occupy both the cartilage ring and the submucosal layer, and diffuse passively through both regions. Their proliferation rate is increased by both TGF-β1 and TNF-α, and when they are inside the cartilage ring they differentiate into chondrocytes at a rate that is stimulated by TGF-$\beta 1$. The equation governing the density of MSCs is thus

$$\begin{aligned}\frac{\partial N_{\mathrm{msc}}}{\partial t} &= \frac{1}{r}\frac{\partial}{\partial r}\left(-rvN_{\mathrm{msc}} + rD_{\mathrm{msc}}\frac{\partial N_{\mathrm{msc}}}{\partial r}\right) \\ &\quad + b_{\mathrm{msc}}N_{\mathrm{msc}}\left(N_{\mathrm{msc}}^{*}\left(\frac{C_{\mathrm{tgf}}}{K_{\mathrm{mt}}} + \frac{C_{\mathrm{tnf}}}{K_{\mathrm{mn}}}\right) - N_{\mathrm{msc}}\right) - k_{\mathrm{cm}}f(r)C_{\mathrm{tgf}}N_{\mathrm{msc}},\end{aligned} \tag{9}$$

for $R_l \leq r \leq R_o$. The function $f(r)$ in the third term on the right-hand side of Eq. (9) is equal to 1 inside the cartilage and equal to zero inside the submucosa,

$$f(r) = \begin{cases} 0 & \text{if } R_l \leq r \leq R_c(t), \\ 1 & \text{if } R_c(t) < r \leq R_o. \end{cases} \tag{10}$$

The MSCs enter the trachea through the outer surface, at $r = R_o$, by being applied during implantation surgery. The boundary conditions on the MSCs are therefore, from Eq. (3),

$$\frac{\partial N_{\mathrm{msc}}}{\partial r}(R_l, t) = 0, \quad D_{\mathrm{msc}}\frac{\partial N_{\mathrm{msc}}}{\partial r}(R_o, t) = I_{\mathrm{msc}}g(t), \tag{11}$$

where $g(t)$ specifies the time course of the infiltration, which is taken to be constant for a duration T_{inf}:

$$g(t) = \begin{cases} 1 & \text{if } 0 \leq t \leq T_{\mathrm{inf}}, \\ 0 & \text{if } t > T_{\mathrm{inf}}. \end{cases} \tag{12}$$

Fibroblasts enter the submucosa axially, migrate randomly within it and proliferate according to a purely logistic growth law, hence Eq. (4) is for the fibroblasts

$$\frac{\partial N_{\mathrm{fib}}}{\partial t} = \frac{1}{r}\frac{\partial}{\partial r}\left(-rvN_{\mathrm{fib}} + rD_{\mathrm{fib}}\frac{\partial N_{\mathrm{fib}}}{\partial r}\right) + \frac{2}{h}I_{\mathrm{fib}} + b_{\mathrm{fib}}N_{\mathrm{fib}}(N_{\mathrm{fib}}^{*} - N_{\mathrm{fib}}) \tag{13}$$

for $R_l \leq r \leq R_c$. The numbers of fibroblasts is assumed not to be influenced by macrophages or growth factors; fibrotic activity is attributed only to the quantity of collagen they secrete (see 'Fibroblasts' below). Fibroblasts do not enter or leave the submucosa radially hence the boundary conditions are

$$\frac{\partial N_{\mathrm{fib}}}{\partial r}(R_l, t) = \frac{\partial N_{\mathrm{fib}}}{\partial r}(R_c, t) = 0. \tag{14}$$

Chondrocytes are produced by the differentiation of MSCs. They are assumed to proliferate up to a certain density and undergo apoptosis at a rate that is proportional to the concentration of TNF-α. The equation governing the density of chondrocytes, N_{chn}, is thus

$$\frac{\partial N_{\text{chn}}}{\partial t} = -\frac{1}{r}\frac{\partial}{\partial r}(rvN_{\text{chn}}) + b_{\text{chn}}N_{\text{chn}}(N^*_{\text{chn}} - N_{\text{chn}}) + k_{\text{cm}}f(r)C_{\text{tgf}}N_{\text{msc}} - d_{\text{chn}}C_{\text{tnf}}N_{\text{chn}}, \tag{15}$$

for $R_c \leq r \leq R_o$. Chondrocytes are assumed not to be able to migrate, hence there is no diffusion term in Eq. (15) and no boundary conditions need be specified.

IL-10 is secreted inside the trachea by MSCs, where it diffuses and undergoes degradation:

$$\frac{\partial C_{\text{ilk}}}{\partial t} = \frac{1}{r}\frac{\partial}{\partial r}\left(-rvC_{\text{ilk}} + rD_{\text{ilk}}\frac{\partial C_{\text{ilk}}}{\partial r}\right) + b_{\text{ilm}}N_{\text{msc}} - d_{\text{ilk}}C_{\text{ilk}}, \tag{16}$$

for $R_l \leq r \leq R_o$. The boundary conditions stipulate that IL-10 is secreted into the trachea by EPCs on the lumenal surface at a rate proportional to EPC density, but does not diffuse out of the trachea at $r = R_o$:

$$D_{\text{ilk}}\frac{\partial C_{\text{ilk}}}{\partial r}(R_l, t) = -b_{\text{ilk}}N_{\text{epi}}, \quad \frac{\partial C_{\text{ilk}}}{\partial t}(R_o, t) = 0. \tag{17}$$

TGF-$\beta 1$ is produced by macrophages and MSC-derived chondrocytes and diffuses through the domain while being degraded

$$\frac{\partial C_{\text{tgf}}}{\partial t} = \frac{1}{r}\frac{\partial}{\partial r}\left(-rvC_{\text{tgf}} + rD_{\text{tgf}}\frac{\partial C_{\text{tgf}}}{\partial r}\right) + b_{\text{tch}}N_{\text{chn}}(N^*_{\text{chn}} - N_{\text{chn}}) + b_{\text{tm}}N_{\text{mac}} - d_{\text{tgf}}C_{\text{tgf}}, \tag{18}$$

for $R_l \leq r \leq R_o$. It is assumed that TGF-$\beta 1$ is produced by the chondrocytes only while they are proliferating i.e. until their numbers reach the carrying capacity, hence the rate of production of TGF-$\beta 1$ by these cells is assumed to be proportional to the difference between the current density and the carrying capacity, N^*_{chn}. Zero-flux boundary conditions are applied to the concentration of TGF-$\beta 1$ at the lumenal and outer surfaces:

$$\frac{\partial C_{\text{tgf}}}{\partial r}(R_l, t) = \frac{\partial C_{\text{tgf}}(R_o, t)}{\partial r} = 0. \tag{19}$$

TNF-α is released by macrophages at a rate that is decreased by IL-10, diffuses through the trachea and is degraded. Hence the equation governing the concentration of TNF-α is

$$\frac{\partial C_{\text{tnf}}}{\partial t} = \frac{1}{r}\frac{\partial}{\partial r}\left(-rvC_{\text{tnf}} + rD_{\text{tnf}}\frac{\partial C_{\text{tnf}}}{\partial r}\right) + b_{\text{tnf}}\frac{K_{\text{ti}}^2}{K_{\text{ti}}^2 + C_{\text{ilk}}^2}N_{\text{mac}} - d_{\text{tnf}}C_{\text{tnf}}, \tag{20}$$

for $R_l \leq r \leq R_o$ and with zero-flux boundary conditions at the edges of the domain:

$$\frac{\partial C_{\text{tnf}}}{\partial r}(R_l, t) = \frac{\partial C_{\text{tnf}}}{\partial r}(R_o, t) = 0. \tag{21}$$

Collagen (type I) is secreted and degraded by fibroblasts within the submucosa and, since it comprises large polymeric molecules, is assumed to have negligible diffusivity. The rate of collagen secretion is assumed to be increased by TGF-β1. The equation governing the density of collagen is therefore

$$\frac{\partial C_{\mathrm{col}}}{\partial t} = -\frac{1}{r}\frac{\partial}{\partial r}(rvC_{\mathrm{col}}) + \left(b_{\mathrm{col}}\left(1+\frac{C_{\mathrm{tgf}}}{K_{\mathrm{ct}}}\right) - d_{\mathrm{col}}C_{\mathrm{col}}\right)N_{\mathrm{fib}}, \tag{22}$$

for $R_l \le r \le R_c$. Although collagen (type II) is also secreted within the cartilage ring by chondrocytes, this is not included in the model because thickening of the cartilage region is attributed to the loss of GAGs [26] (see 'ECM remodelling' below).

GAGs are also assumed to have negligible diffusivity, and are produced and degraded by chondrocytes. Also the GAGs are assumed to undergo proteolysis at a constant rate independent of chondrocytes. The equation governing the concentration of the GAGs is thus

$$\frac{\partial C_{\mathrm{gag}}}{\partial t} = -\frac{1}{r}\frac{\partial}{\partial r}(rvC_{\mathrm{gag}}) + (b_{\mathrm{gch}} - d_{\mathrm{gch}}C_{\mathrm{gag}})N_{\mathrm{chn}} - d_{\mathrm{gag}}C_{\mathrm{gag}}, \tag{23}$$

for $R_l \le r \le R_c$.

ECM remodelling. Motivated by the discussion in Sect. 3.2 the dilation, $\nabla \cdot \mathbf{v}$, was formulated so as to be positive when there is a net increase in the collagen concentration in the submucosa and negative when the there is a net decrease in GAG concentration in the cartilage ring. This dependence is formulated in cylindrical polar coordinates as

$$\begin{aligned}\frac{1}{r}\frac{\partial}{\partial r}(rv) = \mathscr{H}\left(A_{\mathrm{vc}}\left(b_{\mathrm{col}}\left(1+\frac{C_{\mathrm{tgf}}}{K_{\mathrm{ct}}}\right) - d_{\mathrm{col}}C_{\mathrm{col}}\right)N_{\mathrm{fib}}\right)\\ + \mathscr{H}\left(-A_{\mathrm{vg}}\left((b_{\mathrm{gch}} - d_{\mathrm{gch}}C_{\mathrm{gag}})N_{\mathrm{chn}} - d_{\mathrm{gag}}C_{\mathrm{gag}}\right)\right),\end{aligned} \tag{24}$$

for $R_l \le r \le R_o$. The function $\mathscr{H}(\theta)$, which relates the net rate of ECM production to the dilation, was chosen to be

$$\mathscr{H}(\theta) = \begin{cases} \theta & \text{if } \theta \ge 0, \\ 0 & \text{if } \theta < 0. \end{cases} \tag{25}$$

This choice was made to ensure that the dilation does not become negative, thereby making stenosis irreversible. Because collagen and GAGs do not occupy the same points in space in the model, the first term in the right-hand side of Eq. (24) will be zero in the cartilage and the second term will be zero in the submucosa. The boundary condition on Eq. (24) is that the outer surface of the trachea remains fixed:

$$v(R_o, t) = 0. \tag{26}$$

Once Eq. (24) is solved to determine v at a given value of t the evolution of the radii of the tracheal lumen and the cartilage ring is calculated using

$$\frac{dR_l}{dt} = v(R_l, t), \quad \frac{dR_c}{dt} = v(R_c, t), \tag{27}$$

with initial conditions, $R_l(0) = R_{l,0}$ and $R_c(0) = R_{c,0}$, which specify the initial radii of tissue boundaries within the decellularised trachea.

3.4 Nondimensionalization

Nondimensionalization was performed to simplify the model by reducing the number of parameters. The general approach used was to scale the variables so as to make the maximum cell densities and chemical concentrations in the non-dimensional model of the order unity, and where possible to scale the production and proliferation rates to unity rather than rates of removal and degradation. To this end the following set of substitutions were made:

$$\left.\begin{array}{lll} \hat{t} = t/\tau, & \hat{r} = r/L, & \hat{v} = \tau v/L, \\ \widehat{R_\times} = R_\times/L, & \widehat{N}_\times = N_\times/N_\times^*, & \widehat{C}_\times = C_\times/C_\times^* \end{array}\right\} \tag{28}$$

where τ is a convenient time scale over which tissue growth can be measured, and L is a characteristic length scale. In Eq. (28) and in what follows the subscript $\times$ denotes any of the subscripts for the species in the model. The following new (dimensionless) parameters are introduced:

$$\left.\begin{array}{lll} \widehat{R}_\times = R_\times/L, & \widehat{D}_\times = D_\times\tau/L^2, & p = N_{\rm msc}^*/N_{\rm chn}^*, \\ N_{\rm mac}^* = 2I_{\rm mac}/(hd_{\rm mac}), & C_{\rm ilk}^* = b_{\rm ilm}N_{\rm msc}^*/d_{\rm ilk}, & C_{\rm tgf}^* = b_{\rm tm}N_{\rm mac}^*/d_{\rm tgf}, \\ C_{\rm tnf}^* = b_{\rm tnf}N_{\rm msc}^*/d_{\rm tnf}, & C_{\rm col}^* = b_{\rm col}/d_{\rm col}, & C_{\rm gag}^* = b_{\rm gch}/d_{\rm gch}, \\ \widehat{b}_{\rm msc} = b_{\rm msc}N_{\rm msc}^*\tau, & \widehat{b}_{\rm epi} = b_{\rm epi}N_{\rm epi}^*\tau, & \widehat{b}_{\rm fib} = b_{\rm fib}N_{\rm fib}^*\tau, \\ \widehat{b}_{\rm chn} = b_{\rm chn}N_{\rm chn}^*\tau, & \widehat{b}_{\rm tch} = b_{\rm tch}(N_{\rm chn}^*)^2\tau/C_{\rm tgf}^*, & \widehat{b}_{\rm ile} = b_{\rm ile}N_{\rm epi}^*\tau/(LC_{\rm ilk}^*), \\ \widehat{k}_{\rm cm} = k_{\rm cm}C_{\rm tgf}^*\tau, & \widehat{I}_{\rm epi} = I_{\rm epi}\tau/(LN_{\rm epi}^*), & \widehat{I}_{\rm msc} = I_{\rm msc}\tau/(LN_{\rm msc}^*), \\ \widehat{I}_{\rm fib} = I_{\rm fib}\tau/(LN_{\rm fib}^*), & \widehat{d}_{\rm mac} = d_{\rm mac}\tau, & \widehat{d}_{\rm chn} = d_{\rm chn}C_{\rm tgf}^*\tau, \\ \widehat{d}_{\rm ilk} = d_{\rm ilk}\tau, & \widehat{d}_{\rm tnf} = d_{\rm tnf}\tau, & \widehat{d}_{\rm tgf} = d_{\rm tgf}\tau, \\ \widehat{d}_{\rm col} = d_{\rm col}N_{\rm fib}^*\tau, & \widehat{d}_{\rm gch} = d_{\rm gch}N_{\rm chn}^*\tau, & \widehat{d}_{\rm gag} = d_{\rm gag}\tau, \\ \widehat{K}_{\rm mi} = K_{\rm mi}/C_{\rm ilk}^*, & \widehat{K}_{\rm ct} = K_{\rm ct}/C_{\rm tgf}^*, & \widehat{K}_{\rm et} = K_{\rm et}/C_{\rm tgf}^*, \\ \widehat{K}_{\rm ti} = K_{\rm ti}/C_{\rm tnf}^*, & \widehat{K}_{\rm mt} = K_{\rm mt}/C_{\rm tgf}^*, & \widehat{K}_{\rm mn} = K_{\rm mt}/C_{\rm tnf}^*, \\ \widehat{A}_{\rm vc} = A_{\rm vc}C_{\rm col}^*, & \widehat{A}_{\rm vg} = A_{\rm vg}C_{\rm gag}^*, & \widehat{T}_{\rm inf} = T_{\rm inf}\tau, \\ \widehat{h} = h/L. & & \end{array}\right\} \tag{29}$$

whereby the equations become in dimensionless form, dropping the carets,

$$\frac{\partial N_{\rm mac}}{\partial t} = \frac{1}{r}\frac{\partial}{\partial r}\left(-rvN_{\rm mac} + rD_{\rm mac}\frac{\partial N_{\rm mac}}{\partial r}\right) + d_{\rm mac}(1 - \left(1 + \frac{C_{\rm ilk}^2}{K_{\rm mi}^2}\right)N_{\rm mac}), \tag{30}$$

$$\frac{dN_{\rm epi}}{dt} = -\frac{v(R_l,t)}{R_l}N_{\rm epi} + \frac{2}{h}I_{\rm epi} + b_{\rm epi}N_{\rm epi}(1 - \frac{C_{\rm tgf}(R_l,t)}{K_{\rm et}} - N_{\rm epi}), \tag{31}$$

$$\begin{aligned}\frac{\partial N_{\rm msc}}{\partial t} &= \frac{1}{r}\frac{\partial}{\partial r}\left(-rvN_{\rm msc} + rD_{\rm msc}\frac{\partial N_{\rm msc}}{\partial r}\right) \\ &\quad + b_{\rm msc}N_{\rm msc}\left(\frac{C_{\rm tgf}}{K_{\rm mt}} + \frac{C_{\rm tnf}}{K_{\rm mn}} - N_{\rm msc}\right) - k_{\rm cm}f(r)C_{\rm tgf}N_{\rm msc},\end{aligned} \tag{32}$$

$$\frac{\partial N_{\rm fib}}{\partial t} = \frac{1}{r}\frac{\partial}{\partial r}\left(-rvN_{\rm fib} + rD_{\rm fib}\frac{\partial N_{\rm fib}}{\partial r}\right) + \frac{2}{h}I_{\rm fib} + b_{\rm fib}N_{\rm fib}(1 - N_{\rm fib}), \tag{33}$$

$$\frac{\partial N_{\rm chn}}{\partial t} = -\frac{1}{r}\frac{\partial}{\partial r}(rvN_{\rm chn}) + b_{\rm chn}N_{\rm chn}(1 - N_{\rm chn}) + pk_{\rm cm}C_{\rm tgf}N_{\rm msc} - d_{\rm chn}C_{\rm tnf}N_{\rm chn}, \tag{34}$$

$$\frac{\partial C_{\rm ilk}}{\partial t} = \frac{1}{r}\frac{\partial}{\partial r}\left(-rvC_{\rm ilk} + rD_{\rm ilk}\frac{\partial C_{\rm ilk}}{\partial r}\right) + d_{\rm ilk}(N_{\rm mac} - C_{\rm ilk}), \tag{35}$$

$$\frac{\partial C_{\rm tgf}}{\partial t} = \frac{1}{r}\frac{\partial}{\partial r}\left(-rvC_{\rm tgf} + rD_{\rm tgf}\frac{\partial C_{\rm tgf}}{\partial r}\right) + b_{\rm tch}N_{\rm chn}(1 - N_{\rm chn}) + d_{\rm tgf}(N_{\rm mac} - C_{\rm tgf}), \tag{36}$$

$$\frac{\partial C_{\rm tnf}}{\partial t} = \frac{1}{r}\frac{\partial}{\partial r}\left(-rvC_{\rm tnf} + rD_{\rm tnf}\frac{\partial C_{\rm tnf}}{\partial r}\right) + d_{\rm tnf}\left(\frac{K_{\rm ti}^2}{{K_{\rm ti}}^2 + C_{\rm ilk}^2}N_{\rm mac} - C_{\rm tnf}\right), \tag{37}$$

$$\frac{\partial C_{\rm col}}{\partial t} = -\frac{1}{r}\frac{\partial}{\partial r}(rvC_{\rm col}) + d_{\rm col}(1 + \frac{C_{\rm tgf}}{K_{\rm ct}} - C_{\rm col})N_{\rm fib}, \tag{38}$$

$$\frac{\partial C_{\rm gag}}{\partial t} = -\frac{1}{r}\frac{\partial}{\partial r}(rvC_{\rm gag}) + d_{\rm gch}(1 - C_{\rm gag})N_{\rm chn} - d_{\rm gag}C_{\rm gag}, \tag{39}$$

$$\begin{aligned}\frac{1}{r}\frac{\partial}{\partial r}(rv) &= \mathscr{H}\left(A_{\rm vc}d_{\rm col}(1 + \frac{C_{\rm tgf}}{K_{\rm ct}} - C_{\rm col})N_{\rm fib}\right) \\ &\quad + \mathscr{H}\left(-A_{\rm vg}\left(d_{\rm gch}(1 - C_{\rm gag})N_{\rm chn} - d_{\rm gag}C_{\rm gag}\right)\right),\end{aligned} \tag{40}$$

$$\frac{dR_l}{dt} = v(R_l,t), \quad \frac{dR_c}{dt} = v(R_c,t). \tag{41}$$

The boundary conditions are

$$\left.\begin{array}{lll}\dfrac{\partial N_{\mathrm{mac}}}{\partial r}(R_l,t)=0, & \dfrac{\partial N_{\mathrm{mac}}}{\partial r}(R_c,t)=0, & \dfrac{\partial N_{\mathrm{msc}}}{\partial r}(R_l,t)=0,\\ \dfrac{\partial N_{\mathrm{msc}}}{\partial r}(R_o,t)=I_{\mathrm{msc}}g(t)/D_{\mathrm{msc}}, & \dfrac{\partial N_{\mathrm{fib}}}{\partial r}(R_l,t)=0 & \dfrac{\partial N_{\mathrm{fib}}}{\partial r}(R_c,t)=0,\\ \dfrac{\partial C_{\mathrm{ilk}}}{\partial r}(R_l,t)\;=-b_{\mathrm{ile}}N_{\mathrm{epi}}/D_{\mathrm{ilk}}, & \dfrac{\partial C_{\mathrm{ilk}}}{\partial t}(R_o,t)=0, & \dfrac{\partial C_{\mathrm{tgf}}}{\partial r}(R_l,t)=0,\\ \dfrac{\partial C_{\mathrm{tgf}}}{\partial r}(R_o,t)\;=0, & \dfrac{\partial C_{\mathrm{tnf}}}{\partial r}(R_l,t)=0, & \dfrac{\partial C_{\mathrm{tnf}}}{\partial r}(R_o,t)=0,\\ v(R_o,t)\;=0. & & \end{array}\right\} \quad (42)$$

3.5 Initial Conditions

At $t=0$ the only cells occupying the trachea will be EPCs on the lumenal surface (if the trachea has been seeded with EPCs during the operation or *ex vivo*). Note that seeding with MSCs is modelled as an influx of cells for $t \geq 0$. The trachea is assumed to initially have uniform distributions of collagen and GAGs representative of a healthy trachea. These are calculated by solving for the steady-state concentrations of Eqs. (38) and (39) which assuming $C_{\mathrm{tgf}}=0$ i.e. no inflammation, and neglecting the small rate of GAG degradation, are both equal to unity. The initial conditions on the variables are therefore

$$\left.\begin{array}{llll} N_{\mathrm{mac}}(r,0)=0, & N_{\mathrm{epi}}(0)=N_{\mathrm{epi},0}, & N_{\mathrm{msc}}(r,0)=0, & N_{\mathrm{fib}}(r,0)=0,\\ N_{\mathrm{chn}}(r,0)=0, & C_{\mathrm{ilk}}(r,0)=0, & C_{\mathrm{tgf}}(r,0)=0, & C_{\mathrm{tnf}}(r,0)=0,\\ C_{\mathrm{col}}(r,0)=1, & C_{\mathrm{gag}}(r,0)=1, & R_l(0)=R_{l,0}, & R_c(0)=R_{c,0}, \end{array}\right\} \quad (43)$$

where $N_{\mathrm{epi},0}=0$ was used to model experiments without seeding with EPCs, and $N_{\mathrm{epi},0}=1$ used where EPCs are applied to the lumenal surface resulting in an initially-confluent density of cells.

3.6 Parameter Values

A typical adult human trachea is approximately 20 mm in diameter, with the wall being 3 mm thick consisting of 2.5 mm of cartilage and 0.5 mm of mucosa, and each ring segment being 4 mm high [27]. Using $L=1$ mm as the characteristic length scale these dimensions give the non-dimensional parameter values $R_{l,0}=7$, $R_{c,0}=7.5$, $R_o=10$ and $h=4$.

An appropriate value for τ, commensurate with the time scale over which recovery of the patient proceeds, is 1 day. Hence to simulate topical application of MSCs for 1 h during surgery a value of $T_{\mathrm{inf}}=0.05$ was used.

Following [9] the diffusion coefficients of IL-10, TGF-β1 and TNF-α were calculated using the formula $D_\times = 6.7\times10^{-10}(M_\times/180)^{-1/3}\,\mathrm{m^2s^{-1}}$ where $M_\times$ is the molecular weight. The values of the diffusion coefficients of the motile cells,

$D_{\mathrm{mac}}, D_{\mathrm{msc}}$ and D_{fib}, were all taken to be equal to $1.6 \times 10^{-13}\,\mathrm{m^2s^{-1}}$ which is at the lower end of the range of values quoted for the neutrophil motility within collagen gels [58]. The degradation rates of the cytokines were estimated from published data of their half-lives measured in vivo: the half-life of IL-10 is quoted as 2 hrs in [63] giving $d_{\mathrm{ilk}} = 8.3\,\mathrm{day}^{-1}$, the half-life of TGF-$\beta$1 as 100 min in [80] giving $d_{\mathrm{tgf}} = 10\,\mathrm{day}^{-1}$, and the half-life of TNF-α as 70 min in [79] giving $d_{\mathrm{tnf}} = 14\,\mathrm{day}^{-1}$. Applying the formulae listed in (44) the non-dimensional parameter values are:

$$\left.\begin{array}{l}
D_{\mathrm{mac}} = D_{\mathrm{msc}} = D_{\mathrm{fib}} = 0.0138,\ D_{\mathrm{ilk}} = 12.5,\ D_{\mathrm{tgf}} = D_{\mathrm{tnf}} = 11,\ d_{\mathrm{ilk}} = 8.3,\\
d_{\mathrm{tgf}} = 10,\ d_{\mathrm{tnf}} = 14,\ d_{\mathrm{mac}} = 1,\ d_{\mathrm{chn}} = 10,\ d_{\mathrm{col}} = d_{\mathrm{gch}} = 0.1,\ d_{\mathrm{gag}} = 0.01,\\
k_{\mathrm{cm}} = 1,\ b_{\mathrm{epi}} = b_{\mathrm{msc}} = b_{\mathrm{fib}} = b_{\mathrm{chn}} = b_{\mathrm{ile}} = b_{\mathrm{tch}} = 1,\ I_{\mathrm{msc}} = 0/0.01,\\
I_{\mathrm{epi}} = I_{\mathrm{fib}} = 0.01,\ K_{\mathrm{mi}} = 0.01,\ K_{\mathrm{et}} = 0.1,\ K_{\mathrm{mn}} = K_{\mathrm{mt}} = K_{\mathrm{ti}} = K_{\mathrm{ct}} = 1,\\
A_{\mathrm{vc}} = 0.9,\ A_{\mathrm{vg}} = 1.5,\ T_{\mathrm{inf}} = 0.05,\ R_o = 10,\ h = 4,\ p = 1,\ N_{\mathrm{epi},0} = 0/1,\\
R_{l0} = 7.02,\ R_{c0} = 7.5\,.
\end{array}\right\} \tag{44}$$

The parameter values that could not be obtained directly from published data were estimated using a heuristic approach. For example based on typical cell division rates of the order $1\,\mathrm{day}^{-1}$ the proliferation rate parameters for all cells i.e. $d_{\mathrm{mac}}, b_{\mathrm{msc}}$ etc. were all taken to be unity. Because ECM has a slower turnover rate than cells the ECM production/degradation rate parameters, $d_{\mathrm{col}}, d_{\mathrm{gch}}$ and d_{gag}, were taken to be 1-2 orders of magnitude smaller. Further, the rate of turnover of cells inside the trachea was assumed to be dominated by cell proliferation, initiated by a relatively small rate of cell infiltration, hence the values of the flux parameter, $I_{\times}$, for all cells was set to 0.01. The parameter values $K_{\mathrm{mi}} = 0.01$ and $K_{\mathrm{et}} = 0.1$ in the equations for the macrophages and EPCs were chosen to give strong bistable behaviour in the submodel for inflammation arising from the interaction between these two cell types. Similarly the value $d_{\mathrm{chn}} = 10$ was chosen to strengthen the deleterious effect of TNF-α on chondrogenesis. The parameter values $A_{\mathrm{vc}} = 0.9$ and $A_{\mathrm{vg}} = 1.5$ were selected to give realistic amounts of stenosis [26] at the end of the simulations.

In the absence of further available information the values of the other parameters were set to unity. It must be emphasised that the parameter values listed in Eq. (44) are tentative and further work is required to identify the values with more precision.

4 Numerical Methods

The approach for solving Eqs. (30–43) numerically was to first split each variable defined over the full radial extent of the trachea i.e. the MSCs, IL-10, TGF-β1 and TNF-α, into two new variables defined exclusively within the submucosa and cartilage. Then the transformations

$$x = \frac{r - R_l(t)}{R_c(t) - R_l(t)}, \quad \text{for } R_l \leq r \leq R_c, \tag{45}$$

$$x = \frac{r - R_o}{R_c(t) - R_o}, \quad \text{for } R_c \leq r \leq R_o, \tag{46}$$

were used to map the moving intervals $R_l(t) \leq r \leq R_c(t)$ and $R_c(t) \leq r \leq R_o$ to the single fixed interval $0 \leq x \leq 1$. This made the equations amenable to the NAG routine d03ph which solves two-point boundary value problems comprising systems of parabolic-elliptic partial differential equations coupled to systems of ODEs.

The substitution (45) transforms the non-dimensional version of Eqs. (4) and (40), appropriate for the variables defined within the submucosa, into

$$\begin{aligned}\frac{\partial u}{\partial t} = &-\frac{1}{r(x,t)}\frac{\partial}{\partial x}\left(\frac{r(x,t)}{R_c - R_l}\left(uv + J_r - \left(\dot{R}_l + x(\dot{R}_c - \dot{R}_l)\right)u\right)\right)\\ &+\frac{s(x,t)u}{R_c - R_l} + \frac{2}{h}I_u(r(x,t), \pm h/2, t) + f(u, r(x,t), t),\end{aligned} \tag{47}$$

$$\frac{1}{r(x,t)}\frac{\partial}{\partial x}\left(\frac{r(x,t)v}{R_c - R_l}\right) = \mathscr{H}\left(A_{vc}d_{\mathrm{col}}(1 + \frac{C_{\mathrm{tgf}}}{K_{\mathrm{ct}}} - C_{\mathrm{col}})N_{\mathrm{fib}}\right), \tag{48}$$

where,

$$J_r = -\frac{D_u}{R_c - R_l}\frac{\partial u}{\partial x}, \tag{49}$$

$$r(x,t) = R_l + (R_c - R_l)x, \tag{50}$$

$$s(x,t) = -\dot{R}_l(R_c - R_l) - (\dot{R}_c - \dot{R}_l)(R_l + 2(R_c - R_l)x), \tag{51}$$

and the dots over the variables R_c and R_l represents the time derivative. Similar equations hold for the variables defined within the cartilage (which are omitted here for brevity). Continuity of the values of u and the flux of u were applied at $x = 1$ (the boundary corresponding to $r = R_c$) to the variables defined in the submucosa and cartilage for the same species.

The transformed equations corresponding to Eqs. (30) and (32–39) and the ODEs (31) and (41) were solved numerically using the NAG routine d03ph. A spatial domain with 40 grid points was used because it was found there was no significant change in the solution values when a finer grid was employed. At each time step v was approximated by solving Eq. (48), with the condition $v(R_o) = 0$, as a matrix equation obtained using a first-order approximation for the derivative.

For the results of simulations shown in Fig. 4b the average chondrocyte density and average GAG concentration were calculated using the formula

$$u_{\text{av}}(t) = \frac{2}{R_o + R_c(t)} \int_0^1 r(x,t)u(r(x,t),t)\,dx, \tag{52}$$

which was evaluated at each time point in the simulations using Simpson's rule.

5 Results

The model equations, defined by Eqs. (30–41) with boundary conditions given by (42) and initial conditions by (43), were solved numerically using the methods described in Sect. 4 with the parameter values listed in Sect. 3.6. To compare different types of therapies, simulations were performed with or without the intra-operative application of MSCs, and with or without preseeding of the tracheal lining with EPCs. Simulations were carried out up to $t = 60$ days, consistent with typical post-operative recovery and monitoring time frames [26]. The main objective is to compare how these different seeding strategies perform with regard to minimising the amount of stenosis and regenerating the cartilage of the trachea, since these are the major determinants of the long term viability of the implant.

Figure 3 shows how the densities of the cells and the concentrations of ECM and cytokines at various times vary radially through the trachea, for the case of application of MSCs only and without seeding of the lumenal surface with EPCs. Spatial profiles are given at $t = 0$ (curves with triangles), at $t = 5$ days (curves with crosses) and at $t = 60$ days (curves with circles) after implantation of the trachea. At $t = 0$ there are no cells present inside the trachea, only collagen and GAGs being present. By 5 days after implantation macrophages have fully infiltrated the submucosa invoking inflammation by secreting large amounts of TGF-β1 and TNF-α through the full radial cross section of the trachea. At this time the concentration of IL-10 and the densities of MSCs and chondrocytes remain low. However by $t = 60$ days, MSCs have infiltrated right the way through the cartilage ring, regenerating the chondrocytes, and have migrated into the submucosa where they inhibit macrophage activity by secreting IL-10. This reduces the levels of TGF-β1 and TNF-α thereby resolving the inflammation. The continued secretion of small amounts of TGF-β1 and TNF-α by the small population of residual macrophages supports the presence of a small number of MSCs at large times.

Figure 4a shows the time courses of the density of EPCs on the lumenal surface for different seeding strategies. In the case where only MSCs are applied to the trachea (see also Fig. 3) there are initially no EPCs present on the lumenal surface but EPCs slowly infiltrate axially from the edges of the graft. However, EPC proliferation is inhibited during the inflammation because of the high concentration of TGF-β1. Through the down regulation of TGF-β1 secretion once the MSCs invade the submucosa, reepithelization of the trachea can occur, as indicated by the solid curve in Fig. 4a. Without application of MSCs to the trachea the epithelial lining fails to regenerate (dashed curve). In the case where EPCs are preseeded onto the lumenal surface and MSCs are applied to the outer surface

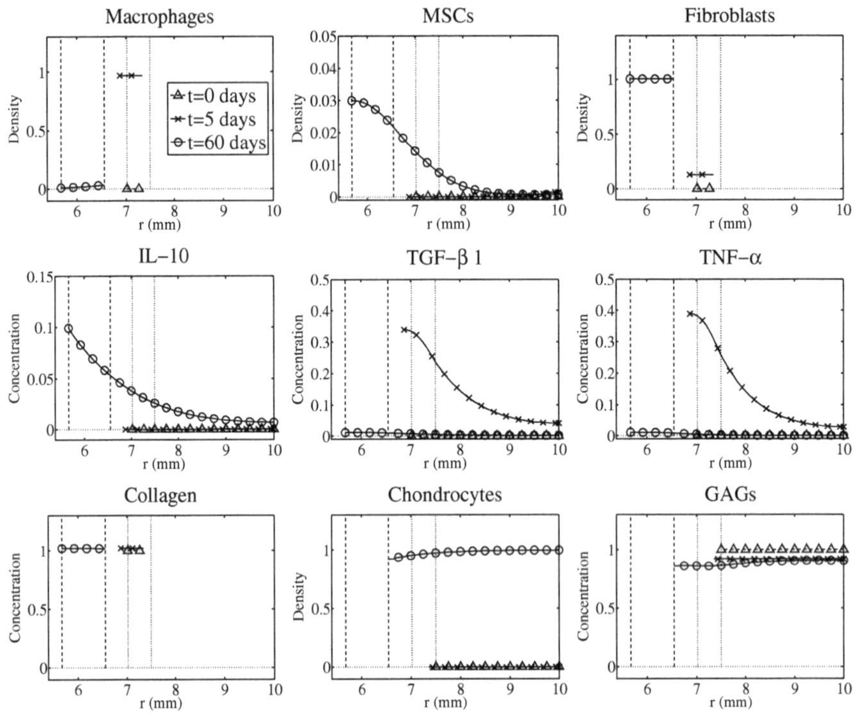

Fig. 3 Spatial profiles of the densities and concentrations of the different species participating in the regeneration of the trachea (note the use of the different vertical axes scalings). Results are shown for seeding of the trachea with MSCs but without seeding with EPCs. Initially, at $t = 0$, there are no cells present inside the trachea (*curves* with *triangles*). At the intermediate time, $t = 5$ days (*curves* with *crosses*), cells have infiltrated and the trachea is in a state of inflammation. By the final time, $t = 60$ days (*curves* with *circles*), the inflammation has been resolved. The vertical *dotted lines* and *dashed lines* indicate respectively the initial and final radial positions of the lumenal surface and the inner radius of the cartilage ring

(dotted-dashed curve), the IL-10 secreted by the EPCs contributes to the suppression of inflammation and the integrity of the epithelial layer remains intact at large time.

In fact the amount of IL-10 secreted by the confluent epithelium is sufficient to damp down inflammation in the model even without exogenous application of MSCs. A small insult to the epithelium in this case will result in a brief transient influx of macrophages analogous to a wound healing response. However results of simulations shows that there is a threshold seeding density of EPCs ($\approx 20\,\%$ confluent) below which the amount of secreted IL-10 is insufficient to suppress inflammation. This accords with experiments that show stenosis can be induced by severely damaging the tracheal lining [53]. However the EPC seeding density required to prevent inflammation occurring need not be very high as suggested by observations of tissue-engineered trachea regeneration that show small islands of mucosa forming 72 h after implantation [P. Macchiarini, private communication].

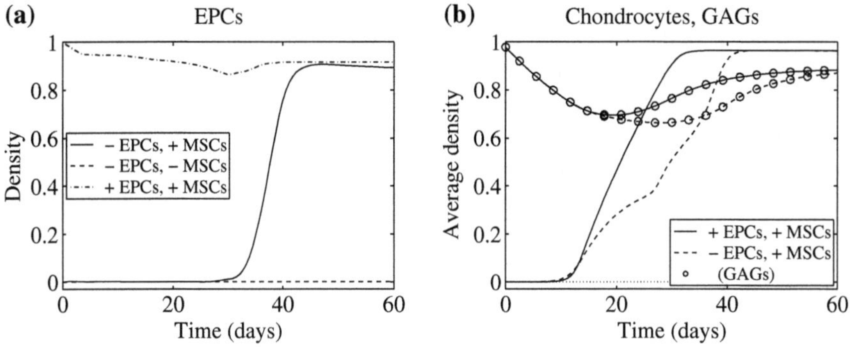

Fig. 4 Graphs showing the time courses of **a** the density of epithelial cells lining the tracheal lumen, and **b** the average density of chondrocytes and GAGs within the cartilage ring, calculated using Eq. (52). Results are shown for different combinations of cell seeding: in the legend of each graph a '+' means the indicated cell type is used in seeding, a '−' means that it is not. In **b** the *curves* with *circles* indicates GAGs, those without *circles* are chondrocytes

This threshold effect of the EPC seeding density is caused by a bistable behaviour in the model resulting from the mutual inhibition between EPCs and macrophages through the action of TGF-β1 and IL-10. Interestingly, Fig. 4a shows a short transient downturn in EPC density after about $t = 30$ days due to the secretion of TGF-β1 by the proliferating chondrocytes. This demonstrates a dual role for TGF-β1 in mediating a pro-inflammatory response in the submucosa but enhancing regeneration of the cartilage through stimulating MSC proliferation and differentiation into chondrocytes. This accords with results of in vitro studies that show that the proximity of chondrocytes can suppress proliferation of EPCs [30].

The model demonstrates that inflammation within the submucosa will also be detrimental to cartilage regeneration. Figure 4b shows the time course of the average density of chondrocytes and GAGs in the cartilage ring for the cases where MSCs are applied and the trachea is seeded, or not seeded, with EPCs. The quantities shown were calculated using (52) with u substituted with N_{chn}or C_{gag}. In the case where there is no preseeding with EPCs the plateauing of the chondrocyte density is delayed and more degradation of GAGs occurs because the TNF-α secreted by the macrophages induces apoptosis of the chondrocytes. If the parameter values in the model are such that inflammation is too severe, e.g. an excessive number of macrophages leading to high C_{tnf} levels, or the chondrocytes are very sensitive to C_{tnf} i.e. $d_{\mathrm{chn}} \gg 10$ in (15), chondrogenesis will be blocked and the GAGs completely degraded.

Figure 5 depicts axial cross sections of the trachea showing how the radii of the tracheal lumen and the inner surface of the cartilage ring vary with time for the different seeding methods. Given in Table 1 are the corresponding values of the degree of stenosis of the trachea, defined as the percentage decrease in lumenal radius, 60 days after implantation i.e. $(R_{l,0} - R_l(60))/R_{l,0} \times 100\,\%$. In the cases where MSCs are not applied to the trachea, shown in Fig. 5a and c, stenosis is more severe ($> 30\,\%$) because no chondrogenesis occurs hence GAG degradation proceeds unabated. In these cases the implant is unstable i.e. the lumenal radius becomes progressively smaller without levelling out. However when MSCs are applied, as depicted in Fig. 5b and d, the chondrocytes, which differentiate from the MSCs, are able to maintain the cartilage and halt the degradation of the GAGs, resulting in stable implants with smaller amounts of stenosis ($<20\,\%$). Where there is no seeding of the trachea with EPCs, shown in Fig. 5 a and b, inflammation occurs, resulting in thickening of the submucosa. In the case of (a) this process is unstable i.e. the thickening is continual whereas in (b) once the inflammation has been resolved with the aid of the MSCs the thickness of the submucosa approaches a constant value. Similar observations were reported for experiments of tissue-engineered tracheas implanted into pigs [26], where the outcome was superior when the tracheas were preseeded with MSC-derived chondrocytes (assuming these cells to contain residual MSCs) and the best results were obtained when the tracheas were also preseeded with EPCs. The model results also agree qualitatively agree with the results of experiments of decellularised tracheas implanted into dogs [54] where it was shown that the intraoperative application of MSCs or bone marrow aspirate reduced the amount of stenosis and, in accordance with Fig. 4a, increased the rate of reepithelisation.

Stenosis in the model is a result of dilation of tissue elements arising from excess secretion of collagen and degradation of GAGs. The increasing width of the tissue domain causes a dilution of the cell and ECM densities (this dilution effect appears explicitly as the second term on the right-hand side of (47), where $s(x,t)$ is negative when high rates of stenosis occur). In the submucosa this dilution locally clears space in the tissue for new cells and ECM to be deposited. Depending on the degree of inflammation this can bring about a positive feedback effect leading to continual expansion of the domain. Dynamical stability analysis techniques can reveal conditions on the parameter values for which this positive feedback arises. In the simulations a key parameter controlling stability in the model was found to be A_{vc}, this being the constant of proportionality relating the rate of dilation of the submucosa to the net rate of production of collagen. It was found that with all other parameters held constant there was a threshold value of A_{vc} below which the dilution was too small to stimulate a sufficient rate of tissue growth to sustain the positive feedback effect. An increase in the value of A_{vc} above the threshold would relate to aberrant morphological changes in the ECM which are associated with fibrosis [18].

In living tissue in the absence of an epithelial lining, inflammation and stenosis may also be brought about by residual immune cell activity resulting from contamination and infection via the tracheal lumen [26]. Fibrotic growth may also be contributed to by EPCs transforming into myofibroblasts as a result of EMT or by differentiation of MSCs, effects that are not considered in the present model. Indeed EMT may explain why the poorest outcomes reported in [26] were for seeding with EPCs and without MSC-derived chondrocytes; chronic inflammation may be causing the transformation of epithelial cells to those with a fibroblastic phenotype that contributes to fibrosis.

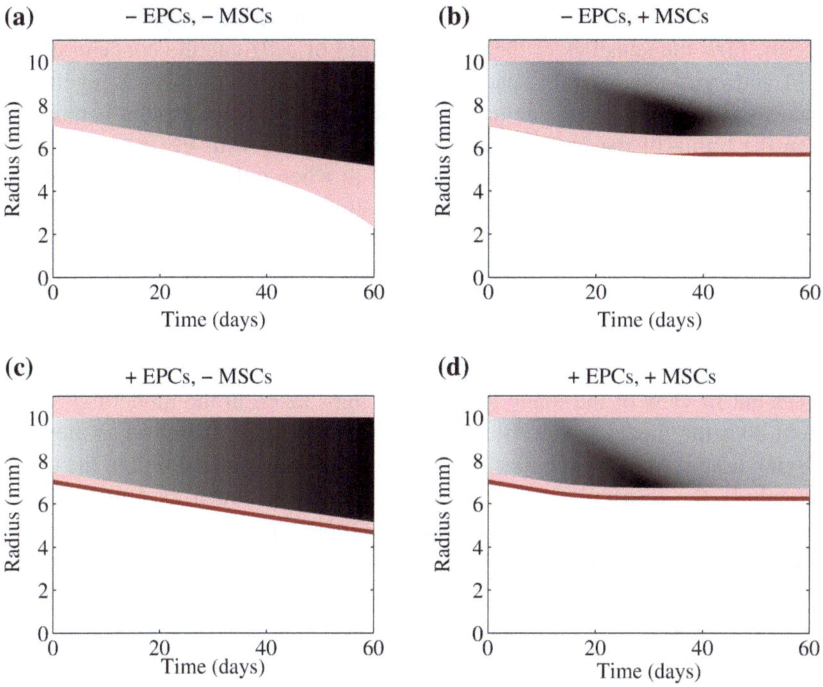

Fig. 5 Graphs showing how the radial cross section of the trachea changes with time during regeneration, for different cell seeding strategies. The *grey* region is the cartilage ring where the *lighter shading* indicates the presence of GAGs and the *darker shading* reveals its absence. The *pink* region below the cartilage is the submucosa. The *dark red* colouring indicates the presence of epithelial cells. In the title of each graph a '+' means the indicated cell type is used in seeding, a '–' means that it is not

Table 1 The final amounts of stenosis of the trachea for the results shown in Fig. 5, these being the percentage decrease in the radius of the tracheal lumen over the 60 days simulation time i.e. $(R_{l,0} - R_l(60))/R_{l,0} \times 100\,\%$. Results are shown for various seeding combinations: a '✔' means the indicated cell type is used in the seeding, a '×' means that it is not

Cell seeding		Stenosis
EPCs	MSCs	(%)
×	×	67
✔	×	33
×	✔	19
✔	✔	11

6 Discussion

This chapter describes the application of mathematical modelling techniques to the study of regeneration of a tissue-engineered trachea in situ. A mathematical model is presented, in the form of a set of coupled partial differential equations (PDEs), to describe how the densities of the different types of cells and the concentrations of ECM and cytokines vary in space and time through the trachea as regeneration proceeds. The model results demonstrate that stenosis of the implant is minimised when the lining of the lumen of the trachea is seeded with EPCs and MSCs are applied to the external surface, as compared to when either or both of these cells are excluded (see Fig. 5 and Table 1). This strategy works because the EPCs and MSCs synergistically damp down the inflammation that causes excess secretion of collagen in the submucosa resulting in stenosis, and the MSCs regenerate the chondrocytes so as to maintain the GAGs in the cartilage thereby preventing airway stenosis due to malacia.

The modelling results accord qualitatively with observed experimental outcomes but in view of the many simplifying steps taken in the modelling, to what extent does the model reflect the processes of tracheal regeneration as they occur in living tissue? Some of the mechanisms crucial to the regeneration of the trachea, particularly the revascularisation, have been excluded. Such an omission means that the model cannot be used to provide a complete prognosis. Instead the model is concerned with the processes of inflammation, particularly how inflammation in the submucosa affects the regeneration of the cartilage and how chondrogenesis and MSC infiltration affect inflammation in the submucosa. However the inflammatory cytokines included in the model (IL-10, TGF-β1 and TNF-α) are only a few of those that form part of the highly complicated signalling network governing inflammation of epithelial tissues.

As is typically the case with complex biological systems, there is an incomplete knowledge of the biology underlying the mechanisms contributing to the regeneration process. This is particularly the case with understanding what factors determine the differentiation fate of MSCs. Although an extensive knowledge has been built up in recent years of the different chemical and mechanical signals that guide MSC migration and their differentiation into chondrocytes, the complete picture is still unclear. Yet such knowledge is key to harnessing effectively the power of MSCs for regenerative medicine, and its absence is a barrier to developing effective mathematical models. The advantage of using donor trachea rather than an artificial matrix is that the former is a more natural environment for guiding the migration of the MSCs into the cartilage ring and for encouraging them to differentiate into chondrocytes. This motivated the simplified approach taken in the modelling whereby MSCs were assumed to differentiate directly into chondrocytes within the cartilage but not within the submucosa. An extension of the model could be to allow MSCs in the submucosa to differentiate into myofibroblasts where they could potentially have a pro-fibrotic effect [16, 72].

The complex and incomplete knowledge of the biology of tracheal regeneration warranted extensive use of such simplifications to keep the number of equations and parameters in the model to a tractable level. The guiding principle was to use the smallest possible set of tissue components that could be assembled to form a realistic model. This represents a top-down approach to the biomedical modelling rather than a more complete systems biology approach that would seek to include as much information about cells and signalling pathways as possible.

A simple radial geometry was chosen to represent the trachea which simplified the description of the transport of species axially. This allowed a 1D formulation to be used thereby reducing the amount of computational effort required to solve the model equations. If a less simplified geometry were used, for example if the posterior wall were to be taken into account by solving the equations on a domain comprising a 2D ring segment (see Fig. 1b), this would give a more realistic description of how a real trachea changes shape under stenosis. However because the model is purely local the mechanisms that give rise to stenosis do not depend critically on the overall extent of the deformation of the trachea. Hence more emphasis has been given in the modelling to the interactions between the different types of cells and cytokines rather than to the details of the tracheal geometry.

For tractability many aspects of the biological mechanisms were given highly simplified descriptions in the modelling. For example the complex processes of homing and migration of macrophages, EPCs and fibroblasts into the trachea from the host tissue was not modelled in detail. Instead their influx was assumed to occur at a constant rate characterised by the parameters I_{mac}, I_{epi} and I_{fib}. Also consideration was not given to the complex mechanisms of interconversion of TGF-β1 between an inactive form bound to the ECM and an active soluble form diffusing throughout the domain [35] and only consideration was given to a small number of specific effects of TGF-β1 on cells pertinent to its inflammatory role. The ability to devise a model that is tractable numerically and analytically, but is also realistic from a biologist's perspective is a key skill of a mathematical modeller.

Continuum modelling approaches to describe cell migration and proliferation such as the reaction-diffusion formulation used in this chapter have been much used in biomedical modelling applications; for other examples see [9, 69]. Such formulations have the advantage over individual-based models of being relatively economical to solve computationally. In some cases these models are also analytically tractable and so mathematical techniques can be used for analysing their solutions. For example because of the mutually antagonistic effects of TGF-β1 (pro inflammatory) and IL-10 (anti inflammatory) in the model, a hysteresis effect arises whereby steady states corresponding to inflamed tissue and healthy tissue can occur for the same values of the model parameters but arise from the use of different seeding conditions. Bifurcation and stability analysis of the equilibrium solutions of the model could yield precise conditions on parameters and initial seeding densities that gives rise to inflammation and stenosis, and in principle be used to guide therapeutic strategies. Relating the bifurcation structure of the solutions of a model to aberrant and normal states in real tissue is a useful concept in biomedical modelling.

It is important to note that the continuum modelling approach adopted in this chapter is essentially a phenomenological description of what is on the spatial scale of individual cells a very complex process. For example, implicit in the diffusion term that models migration of MSCs within the cartilage [in the first term of the right-hand side of Eq. (9)] is a process which involves proteolysis of GAGs by the MSCs so as to clear a path for migration through the ECM [91]. The fact that such mechanisms are not modelled in detail does not necessarily invalidate the use of a continuum modelling approach, though the determination of parameter values should come from in vivo measurements taken at the tissue scale e.g. the thickness of the trachea wall, rather than at the scale of individual cells e.g. motility coefficients derived from in vitro cell migration assays. Much current research in mathematical biology concerns trying to develop continuum models that sensibly link microscopic cell behaviour to macroscopic tissue growth characteristics; however, at present this requires the use of unrealistically simple models of individual cells.

The alternative is to model cells as discrete entities and perform simulations involving large numbers of such cells, such as in the model used by our group to study the growth of blood vessels into porous biomaterials [42]. The complexity of cell migration within the fully 3D tracheal tissue geometry, especially when consideration is also given to angiogenesis, suggests that the modelling of tracheal regeneration would benefit from applying individual-based modelling techniques to complement the PDE methods adopted here. Indeed the current trend in biomedical modelling, boosted by the increasing speed and power of computers, is towards making progressively more detailed and sophisticated multiscale models that can link sub-cellular processes to the outcomes at the level of a whole organism [33]. The trend is in part motivated by the desire to create mathematical and computational models that can represent real tissue faithfully enough to be effective for the development of new drugs and therapies. This *in silico* revolution [21] is spurred on by the establishment of modelling projects involving large multi-disciplinary and cross-institutional collaborations between specialist researchers [77]. Such projects make extensive use of systems biology approaches, which allows models to be obtained formally from existing information contained in databases. Mathematical and computational modelling for tracheal tissue engineering could benefit from the use of such techniques and collaborations in the future.

Although individual-based modelling approaches are appealing, continuum models also have an important role to play in biomedical modelling because they can allow a deeper understanding of the underlying dynamics of a system than is possible by performing numerical simulations alone. The study of continuum models also forms an important interdisciplinary link between traditional areas of mathematics and biology, thereby stimulating research in both areas, and helps scientists to relate similar phenomena in disparate fields. These motivations form a core philosophy of mathematical biology. Alan Turing's model for diffusion-driven instability [74] is compelling because it can explain biological pattern formation in diverse areas including embryology and ecology. His genius established general principles of what gives rise to patterning in tissues; identifying the precise mechanisms in specific cases came later [48].

To the extent that the mathematical model presented in this chapter does encapsulate the main mechanisms guiding tracheal regeneration, a major difficulty is how to determine the values of the model parameters as they pertain to real tissue. Justification for the choice of some of the parameter values listed in (44) was given but others were chosen on an ad hoc basis to demonstrate typical characteristics of the model solutions. Further work needs to be done, using expert knowledge and experimental measurements, to obtain more reliable values for the parameters. If the amount of experimental data available is limited then this will place restrictions on parameter identifiability for such complicated models [64]. In any case, it may be impractical and unethical to obtain parameter values from direct measurements of real tracheal tissue. An alternative is to use in vitro models [52, 59]. Thus in the example of the need to determine the mobility of MSCs within the cartilage, quantified by the parameter D_{msc} in the model, new in vitro experiments could be made of MSCs invading cartilage, the results of which could have much more general significance to the understanding of cartilage regeneration. Just as biology stimulates research in mathematics, the need to quantify key elements in the mathematical model properly can motivate new and insightful biological experiments.

A challenging aspect of developing mathematical models for use in tissue engineering is the need to account for the mechanical stresses that occur within growing tissues. This problem has stimulated previous work that our group has carried out on modelling of tissue growth into porous biomaterials, that takes into account the forces generated between cells and their surroundings [46], and the forces acting on membranes of tissue that grows from host tissue into scaffolds [44]. An interesting aspect of mathematical modelling of tracheal regeneration is the need to link inflammation and changes in ECM to stenosis, this corresponding to an increase in the size of the domain of the model. This was done in the present case by formulating the model equations as a moving boundary problem, with the dilation of tissue elements being related to the rate of ECM secretion and degradation [see Sect. 3.2 and Eq. (24)]. This approach may have applicability to other biological modelling problems, since the mechanisms that regulate the extent of tissue boundaries in an organism are fundamental to the understanding of tissue development and disease.

The work carried out by our group represents only a small step forward in developing of mathematical modelling of tissue-engineered tracheal regeneration. The model currently has limitations to its suitability for guiding the development of therapies, and further work must be done to overcome these limitations. The present model is better suited for generating hypotheses for how regeneration works, identifying key processes that require further elucidation, and finding ways forward towards developing more complex and realistic models. This chapter has highlighted an important use for mathematical models: they can allow knowledge of different biological mechanisms to be integrated within a single theoretical framework, thereby allowing an understanding of how these mechanisms interact within real tissue. Modelling results can yield interesting and unexpected phenomena which can suggest possible explanations for experimental outcomes and can flag possible problems with proposed therapies. An example from the present study is the deleterious effects that the TGF-β1 produced within the cartilage may have on the epithelium (Fig. 4b). This result suggests not to use excessive amounts of exogenous TGF-β1 in an attempt to boost the regeneration of the cartilage. Also, to the extent that TNF-α is responsible for inhibiting chondrogenesis in the trachea, this would suggest using drugs to specifically target the pathways involved in the production of TNF-α.

To summarise, this chapter has demonstrated the potentially important role that mathematical modelling has for tissue engineering of whole organs using stem cells. As the knowledge of the biology underlying regenerative processes improves, particularly with regards to stem cell biology, the effectiveness of mathematical models should develop in parallel. The ultimate goal should be to develop realistic mathematical and computational models that can accurately predict the time course of regeneration and be useful for developing new therapies, thereby contributing directly to improving the prognoses of patients. But even in their nascent form, mathematical models have an important role to play in the formulating of medical hypotheses and in contributing to the development of new modelling approaches to biomedicine.

Acknowledgments The authors wish to acknowledge the financial support of the Biotechnology and Biological Sciences Research Council (BBSRC) and the Engineering and Physical Sciences Research Council (EPSRC).

References

1. Aggarwal, S., Pittenger, M.F.: Human mesenchymal stem cells modulate allogeneic immune cell responses. Blood **105**(4), 1815–1822 (2005)
2. Ahrendt, G., Chickering, D.E., Ranieri, J.P.: Angiogenic growth factors: a review for tissue engineering. Tissue Eng. **4**, 117–130 (1998)
3. Aizawa, T., Kon, T., Einhorn, T.A., Gerstenfeld, L.C.: Induction of apoptosis in chondrocytes by tumor necrosis factor-alpha. J. Orthopaed. Res. **19**(5), 785–796 (2001)
4. Asnaghi, M.A., Jungebluth, P., Raimondi, M.T., Dickinson, S.C., Rees, L.E.N., Go, T., Cogan, T.A., Dodson, A., Parnigotto, P.P., Hollander, A.P., Birchall, M.A., Conconi, M.T., Macchiarini, P., Mantero, S.: A double-chamber rotating bioreactor for the development of tissue-engineered hollow organs: from concept to clinical trial. Biomaterials **30**(29), 5260–5269 (2009)
5. Assoian, R.K., Fleurdelys, B.E., Stevenson, H.C., Miller, P.J., Madtes, D.K., Raines, E.W., Ross, R., Sporn, M.B.: Expression and secretion of type-beta transforming growth-factor by activated human macrophages. Proc. Natl. Acad. Sci. U.S.A. **84**(17), 6020–6024 (1987)
6. Bader, A., Macchiarini, P.: Moving towards in situ tracheal regeneration: the bionic tissue engineered transplantation approach. J. Cell. Mol. Med. **14**(7), 1877–1889 (2010)
7. Baiguera, S., Birchall, M.A., Macchiarini, P.: Tissue-engineered tracheal transplantation. Transplantation **89**(5), 485–491 (2010)
8. Bailey, D.P., Kashyap, M., Bouton, L.A., Murray, P.J., Ryan, J.J.: Interleukin-10 induces apoptosis in developing mast cells and macrophages. J. Leukocyte Biol. **80**(3), 581–589 (2006)

9. Bailon-Plaza, A., van der Meulen, M.C.H.: A mathematical framework to study the effects of growth factor influences on fracture healing. J. Theor. Biol. **212**(2), 191–209 (2001)
10. Balderman, S.C., Weinblatt, G.: Tracheal autograft revascularization. J. Thorac. Cardiovasc. Surg. **94**(3), 434–441 (1987)
11. Barry, F.P., Murphy, J.M.: Mesenchymal stem cells: clinical applications and biological characterization. Int. J. Biochem. Cell B. **36**(4), 568–584 (2004)
12. Boeuf, S., Richter, W.: Chondrogenesis of mesenchymal stem cells: role of tissue source and inducing factors. Stem. Cell. Res. Ther. **1**(4), 31 (2010)
13. Bonfield, T.L., Konstan, M.W., Berger, M.: Altered respiratory epithelial cell cytokine production in cystic fibrosis. J. Allergy Clin. Immun. **104**(1), 72–78 (1999)
14. Bonfield, T.L., Konstan, M.W., Burfeind, P., Panuska, J.R., Hilliard, J.B., Berger, M.: Normal bronchial epithelial-cells constitutively produce the antiinflammatory cytokine interleukin-10, which is down-regulated in cystic-fibrosis. Am. J. Resp. Cell. Mol. **13**(3), 257–261 (1995)
15. Borthwick, L.A., Parker, S.M., Brougham, K.A., Johnson, G.E., Gorowiec, M.R., Ward, C., Lordan, J.L., Corris, P.A., Kirby, J.A., Fisher, A.J.: Epithelial to mesenchymal transition (EMT) and airway remodelling after human lung transplantation. Thorax **64**(9), 770–777 (2009)
16. Brody, A.R., Salazar, K.D., Lankford, S.M.: Mesenchymal stem cells modulate lung injury. Proc. Am. Thorac. Soc. **7**(2), 130–3 (2010)
17. Burchfield, J.S., Iwasaki, M., Koyanagi, M., Urbich, C., Rosenthal, N., Zeiher, A.M., Dimmeler, S.: Interleukin-10 from transplanted bone marrow mononuclear cells contributes to cardiac protection after myocardial infarction. Circ. Res. **103**(2), 203–211 (2008)
18. Cox, T.R., Erler, J.T.: Remodeling and homeostasis of the extracellular matrix: implications for fibrotic diseases and cancer. Dis. Model Mech. **4**, 165–178 (2011)
19. Crampin, E.J., Gaffney, E.A., Maini, P.K.: Reaction and diffusion on growing domains: scenarios for robust pattern formation. B. Math. Biol. **61**(6), 1093–1120 (1999)
20. Curcio, E., Macchiarini, P., De Bartolo, L.: Oxygen mass transfer in a human tissue-engineered trachea. Biomaterials **31**(19), 5131–5136 (2010)
21. Di Ventura, B., Lemerle, C., Michalodimitrakis, K., Serrano, L.: From in vivo to in silico biology and back. Nature **443**(7111), 527–533 (2006)
22. Djouad, F., Delorme, B., Maurice, M., Bony, C., Apparailly, F., Louis-Plence, P., Canovas, F., Charbord, P., Noel, D., Jorgensen, C.: Microenvironmental changes during differentiation of mesenchymal stem cells towards chondrocytes. Arthritis Res. Ther. **9**(2), R33 (2007)
23. Fan, H.B., Liu, H.F., Toh, S.L., Goh, J.C.H.: Enhanced differentiation of mesenchymal stem cells co-cultured with ligament fibroblasts on gelatin/silk fibroin hybrid scaffold. Biomaterials **29**(8), 1017–1027 (2008)
24. Fiorentino, D.F., Zlotnik, A., Mosmann, T.R., Howard, M., Ogarra, A.: IL-10 inhibits cytokine production by activated macrophages. J. Immunol. **147**(11), 3815–3822 (1991)
25. Genden, A. Iskander, J.S. Bromberg, Mayer, L.: The kinetics and pattern of tracheal allograft re-epithelialization. Am. J. Resp. Cell. Mol. **28**(6), 673–681 (2003)
26. Go, T., Jungebluth, P., Baiguero, S., Asnaghi, A., Martorell, J., Ostertag, H., Mantero, S., Birchall, M., Bader, A., and Macchiarini, P.: Both epithelial cells and mesenchymal stem cell-derived chondrocytes contribute to the survival of tissue-engineered airway transplants in pigs. J. Thorac. Cardiovasc. Surg. **139** (2), 437–443 (2010)
27. Grillo, H.C.: Anatomy of the trachea. Surgery of the Trachea and Bronchi, pp. 39–60. B. C. Decker Inc., Hamilton (2004)
28. Halliday, D.A., Clemente, G., Rathjen, D.A., Ferrante, A.: Rapid degradation of articular cartilage proteoglycan by neutrophils: comparison with macrophages and synovial fibroblasts. Inflamm. Res. **49**(9), 441–444 (2000)
29. Hegewald, A.A., Ringe, J., Bartel, J., Kruger, I., Notter, M., Barnewitz, D., Kaps, C., Sittinger, M.: Hyaluronic acid and autologous synovial fluid induce chondrogenic differentiation of equine mesenchymal stem cells: a preliminary study. Tissue Cell **36**(6), 431–8 (2004)

30. Hicks, W., Sigurdson, L., Gabalski, E., Hard, R., Hall, L., Gardella, J., Powers, C., Kumar, N., Lwebuga-Mukasa, J.: Does cartilage down-regulate growth factor expression in tracheal epithelium?. Arch. Otolaryngol. **125**((11), 1239–1243 (1999)
31. Hiraki, Y.u.j.i., Shukunami, C.h.i.s.a.: Angiogenesis inhibitors localized in hypovascular mesenchymal tissues: Chondromodulin-I and tenomodulin. Connect. Tissue Res. **46**(1), 3–11 (2005)
32. Hou, X.Q., Wu, X.J., Ma, J.X., Lv, X.H., Jin, X.: Erythropoietin augments the efficacy of therapeutic angiogenesis induced by allogenic bone marrow stromal cells in a rat model of limb ischemia. Mol. Biol. Rep. **37**(3), 1467–1475 (2010)
33. Hunter, P.J., Crampin, E.J., Nielsen, P.M.F.: Bioinformatics, multiscale modeling and the IUPS physiome project. Brief. Bioinform. **9**(4), 333–343 (2008)
34. Hwang, N.S., Varghese, S., Puleo, C., Zhang, Z.J., Elisseeff, J.: Morphogenetic signals from chondrocytes promote chondrogenic and osteogenic differentiation of mesenchymal stem cells. J. Cell. Physiol. **212**(2), 281–284 (2007)
35. Hyytiäinen, M., Penttinen, C., Keski-Oja, J.: Latent TGF-beta binding proteins: extracellular matrix association and roles in TGF-beta activation. Crit. Rev. Clin. Lab. Sci. **41**, 233–264 (2004)
36. Jungebluth, P., Go, T., Asnaghi, A., Bellini, S., Martorell, J., Calore, C., Urbani, L., Ostertag, H., Mantero, S., Conconi, M.T., Macchiarini, P.: Structural and morphologic evaluation of a novel detergent-enzymatic tissue-engineered tracheal tubular matrix. J. Thorac. Cardiovasc. Sur. **138**(3), 586–593 (2009)
37. Kalathur, M., Baiguera, S., Macchiarini, P.: Translating tissue-engineered tracheal replacement from bench to bedside. Cell. Mol. Life Sci. **67**(24), 4185–4196 (2010)
38. Karagiannidis, C., Velehorschi, V., Obertrifter, B., Macha, H.N., Linder, A., Freitag, L.: High-level expression of matrix-associated transforming growth factor-beta(1) in benign airway stenosis. Chest **129**(5), 1298–1304 (2006)
39. Kertesz, N., Wu, J., Chen, T.H.P., Sucov, H.M., Wu, H.: The role of erythropoietin in regulating angiogenesis. Dev. Biol. **276**(1), 101–110 (2004)
40. Krampera, M., Pizzolo, G., Aprili, G., Franchini, M.: Mesenchymal stem cells for bone, cartilage, tendon and skeletal muscle repair. Bone **39**(4), 678–683 (2006)
41. Kurowska-Stolarska, M., Stolarski, B., Kewin, P., Murphy, G., Corrigan, C.J., Ying, S., Pitman, N., Mirchandani, A., Rana, B., van Rooijen, N., Shepherd, M., McSharry, C., McInnes, I.B., Xu, D., Liew, F.Y.: IL-33 amplifies the polarization of alternatively activated macrophages that contribute to airway inflammation. J. Immunol. **183**(10), 6469–6477 (2009)
42. Lemon, G., Howard, D., Rose, F.R.A.J., King, J.R.: Individual-based modelling of angiogenesis inside three-dimensional porous biomaterials. Biosystems **103**, 372–383 (2011a)
43. Lemon, G., Howard, D., Tomlinson, M.J., Buttery, L.D., Rose, F.R.A.J., Waters, S.L., King, J.R.: Mathematical modelling of tissue-engineered angiogenesis. Math. Biosci. **221**, 101–120 (2009)
44. Lemon, G., Howard, D., Yang, H., Ratchev, S.M., Segal, J.I., Rose, F.R.A.J, Jensen, O.E., Waters, S.L., King J.R.: Tissue growth into a rapid-prototyped model pore system: experiments and mathematical model. Biomech. Model. Mechanobiol. **103**, 372–383 (2011b)
45. Lemon, G., King, J.R.: Multiphase modelling of cell behaviour on artificial scaffolds: effects of nutrient depletion and spatially non-uniform porosity. Math. Med. Biol. **24**, 57–83 (2007)
46. Lemon, G., King, J.R., Byrne, H.M., Jensen, O.E., Shakesheff, K.M.: Mathematical modelling of engineered tissue growth using a multiphase porous flow mixture theory. J. Math. Biol. **52**, 571–594 (2006)
47. Macchiarini, P., Jungebluth, P., Go, T., Asnaghi, M.A., Rees, L.E., Cogan, T.A., Dodson, A., Martorell, J., Bellini, S., Parnigotto, P.P., Dickinson, S.C., Hollander, A.P., Mantero, S., Conconi, M.T., Birchall, M.A.: Clinical transplantation of a tissue-engineered airway. Lancet **372**(9655), 2023–2030 (2008)
48. Maini, P.K., Baker, R.E., Chuong, C.-.M.: The Turing model comes of molecular age. Science **314**, 1397–1398 (2006)

49. Morishima, Y., Nomura, A., Uchida, Y., Noguchi, Y., Sakamoto, T., Ishii, Y., Goto, Y., Masuyama, K., Zhang, M.J., Hirano, K., Mochizuki, M., Ohtsuka, M., Sekizawa, K.: Triggering the induction of myofibroblast and fibrogenesis by airway epithelial shedding. Am. J. Resp. Cell. Mol. **24**(1), 1–11 (2001)
50. Moses, H.L.: TGF-beta regulation of epithelial-cell proliferation. Mol. Reprod. Dev. **32**(2), 179–184 (1992)
51. Murray, J.D.: Mathematical Biology. 2nd edn. Springer, Berlin (1993)
52. Naito, H., Tojo, T., Kimura, M., Dohi, Y., Zimmermann, W-H., Eschenhagen, T., Taniguchi, S.: Engineering bioartificial tracheal tissue using hybrid fibroblast-mesenchymal stem cell cultures in collagen hydrogels. Interact. Cardiovasc. Thoracic Surg. **12**(2), 156–161 (2011)
53. Nakagishi, Y., Morimoto, Y., Fujita, M., Ozeki, Y., Maehara, T., Kikuchi, M.: Rabbit model of airway stenosis induced by scraping of the tracheal mucosa. Laryngoscope **115**(6), 1087–1092 (2005)
54. Nakamura, T., Sato, T., Araki, M., Ichihara, S., Nakada, A., Yoshitani, M., Itoi, S., Yamashita, M., Kanemaru, S., Omori, K., Hori, Y., Endo, K., Inada, Y., Hayakawa, K.: In situ tissue engineering for tracheal reconstruction using a luminar remodeling type of artificial trachea. J. Thorac. Cardiovasc. Sur. **138**((4), 811–819 (2009)
55. Nathan, C., Ding, A.H.: Nonresolving inflammation. Cell **140**(6), 871–882 (2010)
56. Nikolaev, N.I., Obradovic, B., Versteeg, H.K., Lemon, G., Williams, D.J.: A validated model of GAG deposition, cell distribution, and growth of tissue engineered cartilage cultured in a rotating bioreactor. Biotechnol. Bioeng. **105**(4), 842–853 (2010)
57. Oswald, J., Boxberger, S., Jorgensen, B., Feldmann, S., Ehninger, G., Bornhauser, M., Werner, C.: Mesenchymal stem cells can be differentiated into endothelial cells in vitro. Stem Cells **22**(3), 377–384 (2004)
58. Parkhurst, M.R., Saltzman, W.M.: Quantification of human neutrophil motility in 3-dimensional collagen gels—effect of collagen concentration. Biophys. J. **61**(2), 306–315 (1992)
59. Pfenninger, C., Leinhase, I., Endres, M., Rotter, N., Loch, A., Ringe, J., Sittinger, M.: Tracheal remodeling: comparison of different composite cultures consisting of human respiratory epithelial cells and human chondrocytes. In Vitro Cell Dev-An. **43**(1), 28–36 (2007)
60. Ponte, A.L., Marais, E., Gallay, N., Langonne, A., Delorme, B., Herault, O., Charbord, P., Domenech, J.: The in vitro migration capacity of human bone marrow mesenchymal stem cells: Comparison of chemokine and growth factor chemotactic activities. Stem Cells **25**(7), 1737–1745 (2007)
61. Prasse, A., Pechkovsky, D.V., Toews, G.B., Jungraithmayr, W., Kollert, F., Goldmann, T., Vollmer, E., Muller-Quernheim, J., Zissel, G.: A vicious circle of alveolar macrophages and fibroblasts perpetuates pulmonary fibrosis via CCL18. Am. J. Resp. Crit. Care. **173**(7), 781–792 (2006)
62. Quintavalla, J., Uziel-Fusi, S., Yin, J.Y., Boehnlein, E., Pastor, G., Blancuzzi, V., Singh, H.N., Kraus, K.H., O'Byrne, E., Pellas, T.C.: Fluorescently labeled mesenchymal stem cells (MSCs) maintain multilineage potential and can be detected following implantation into articular cartilage defects. Biomaterials. **23**(1), 109–119 (2002)
63. Radwanski, E., Chakraborty, A., Van Wart, S., Huhn, R.D., Cutler, D.L., Affrime, M.B., Jusko, W.J.: Pharmacokinetics and leukocyte responses of recombinant human interleukin-10. Pharm. Res. **15**(12), 1895–1901 (1998)
64. Raue, A., Kreutz, C., Maiwald, T., Klingmuller, U., Timmer, J.: Addressing parameter identifiability by model-based experimentation. IET Syst. Biol. **5**(2), 120–U78 (2011)
65. Reing, J.E., Zhang, L., Myers-Irvin, J., Cordero, K.E., Freytes, D.O., Heber-Katz, E., Bedelbaeva, K., McIntosh, D., Dewilde, A., Braunhut, S.J., Badylak, S.F.: Degradation products of extracellular matrix affect cell migration and proliferation. Tissue Eng. Part A **15**(3), 605–614 (2009)

66. Remlinger, N.T., Czajka, C.A., Juhas, M.E., Vorp, D.A., Stolz, D.B., Badylak, S.F., Gilbert, S., Gilbert T.W.: Hydrated xenogeneic decellularized tracheal matrix as a scaffold for tracheal reconstruction. Biomaterials **31**(13), 3520–3526 (2010)
67. Ribatti, D., Nico, B., Vacca, A., Roncali, L., Presta, M.: Endogenous and exogenous fibroblast growth factor-2 modulate wound healing in the chick embryo chorioallantoic membrane. Angiogenesis **3**, 89–95 (1999)
68. Rojas, M., Xu, J.G., Woods, C.R., Mora, A.L., Spears, W., Roman, J., Brigham, K.L.: Bone marrow-derived mesenchymal stem cells in repair of the injured lung. Am. J. Resp. Cell. Mol. **33**(2), 145–152 (2005)
69. Schugart, R.C., Friedman, A., Zhao, R., Sen, C.K: Wound angiogenesis as a function of tissue oxygen tension: a mathematical model. Proc. Natl. Acad. Sci. U.S.A. **105**(7), 2628–2633 (2008)
70. Strieter, R.M: What differentiates normal lung repair and fibrosis? Inflammation, resolution of repair, and fibrosis. Proc. Am. Thorac. Soc. **5**(3), 305–310 (2008)
71. Sunderkotter, C., Steinbrink, K., Goebeler, M., Bhardwaj, R., Sorg, C.: Macrophages and angiogenesis. J. Leukocyte Biol. **55**, 410–422 (1994)
72. Tomasek, J.J., Gabbiani, G., Hinz, B., Chaponnier, C., Brown, R.A.: Myofibroblasts and mechano-regulation of connective tissue remodelling. Nat. Rev. Mol. Cell Bio. **3**(5), 349–363 (2002)
73. Tsukamoto, Y., Helsel, W.E., Wahl, S.M.: Macrophage production of fibronectin, a chemoattractant for fibroblasts. J. Immunol. **127**(2), 673–678 (1981)
74. Turing, A.M.: The chemical basis of morphogenesis. Phil. Trans. Roy. Soc. Lond. B **237**, 37–72 (1952)
75. Uccelli, A., Moretta, L., Pistoia, V.: Mesenchymal stem cells in health and disease. Nat. Rev. Immunol. **8**(9), 726–736 (2008)
76. van Haaften, T., Byrne, R., Bonnet, S., Rochefort, G.Y., Akabutu, J., Bouchentouf, M., Rey-Parra, G.J., Galipeau, J., Haromy, A., Eaton, F., Chen, M., Hashimoto, K., Abley, D., Korbutt, G., Archer, S.L., Thebaud, B.: Airway delivery of mesenchymal stem cells prevents arrested alveolar growth in neonatal lung injury in rats. Am. J. Resp. Crit. Care. **180**(11), 1131–1142 (2009)
77. Viceconti, M., Clapworthy, G., Jan, S.V.S: The virtual physiological human—a european initiative for in silico human modelling. J. Physiol. Sci. **58**(7), 441–446 (2008)
78. Villa, P., Bigini, P., Mennini, T., Agnello, D., Laragione, T., Cagnotto, A., Viviani, B., Marinovich, M., Cerami, A., Coleman, T.R., Brines, M., Ghezzi, P.: Erythropoietin selectively attenuates cytokine production and inflammation in cerebral ischemia by targeting neuronal apoptosis. J. Exp. Med. **198**((6), 971–975 (2003)
79. Waage, A., Brandtzaeg, P., Halstensen, A., Kierulf, P., Espevik, T.: The complex pattern of cytokines in serum from patients with meningococcal septic shock—association between interleukin-6, interleukin-1, and fatal outcome. J. Exp. Med. **169**(1), 333–338 (1989)
80. Wakefield, L.M., Winokur, T.S., Hollands, R.S., Christopherson, K., Levinson, A.D., Sporn, M.B.: Recombinant latent transforming growth factor-beta-1 has a longer plasma half-life in rats than active transforming growth factor-beta-1, and a different tissue distribution. J. Clin. Invest. **86**(6), 1976–1984 (1990)
81. Wakitani, S., Goto, T., Pineda, S.J., Young, R.G., Mansour, J.M., Caplan, A.I., Goldberg, V.M.: Mesenchymal cell-based repair of large, full-thickness defects of articular-cartilage. J. Bone Joint Surg. Am. **76A**(4), 579–592 (1994)
82. Wang, G.S., Bunnell, B.A., Painter, R.G., Quiniones, B.C., Tom, S., Lanson, N.A., Spees, J.L., Bertucci, D., Peister, A., Weiss, D.J., Valentine, V.G., Prockop, D.J., Kolls, J.K: Adult stem cells from bone marrow stroma differentiate into airway epithelial cells: potential therapy for cystic fibrosis. Proc. Natl. Acad. Sci. U.S.A. **102**(1), 186–191 (2005)
83. Wells, R.G, Fibrogenesis, V.: TGF-beta signaling pathways. Am. J. Physiol-Gastr. L **279**(5), G845–G850 (2000)

84. Westacott, C.I., Barakat, A.F., Wood, L., Perry, M.J., Neison, P., Bisbinas, I., Armstrong, L., Millar, A.B., Elson, C.J.: Tumor necrosis factor alpha can contribute to focal loss of cartilage in osteoarthritis. Osteoarthr. Cartil. **8**(3), 213–221 (2000)
85. Willis, B.C., Borok, Z.: Epithelial-mesenchymal transition: potential role in obliterative bronchiolitis?. Thorax **64**(9), 742–743 (2009)
86. Willis, B.C., Borok Z.: TGF-beta-induced EMT: mechanisms and implications for fibrotic lung disease. Am. J. Physiol. Lung Cell Mol. Physiol. **293**(3), L525–534 (2007)
87. Wynn, T.A.: Common and unique mechanisms regulate fibrosis in various fibroproliferative diseases. J. Clin. Invest. **117**((3), 524–529 (2007)
88. Wynn, T.A., Barron, L.: Macrophages: master regulators of inflammation and fibrosis. Semin. Liver Dis. **30**(3), 245–257 (2010)
89. Yagi, H., Soto-Gutierrez, A., Parekkadan, B., Kitagawa, Y., Tompkins, R.G., Kobayashi, N., Yarmush, M.L.: Mesenchymal stem cells: mechanisms of immunomodulation and homing. Cell Transplant. **19**(6), 667–79 (2010)
90. Yeh, T.H., Tsai, C.H., Chen, Y.S., Hsu, W.C., Cheng, C.H., Hsu, C.J., Lee, S.Y.: Increased communication among nasal epithelial cells in air-liquid interface culture. Laryngoscope **117**(8), 1439–1444 (2007)
91. Zaman, M.H., Trapani, L.M., Sieminski, A.L., MacKellar, D., Gong, H., Kamm, R.D., Wells, A., Lauffenburger, D.A., Matsudaira, P.: Migration of tumor cells in 3D matrices is governed by matrix stiffness along with cell-matrix adhesion and proteolysis. Proc. Natl. Acad. Sci. U.S.A. **103**, 10889–10894 (2006)
92. Zani, B.G., Kojima, K., Vacanti, C.A., Edelman, E.R.: Tissue-engineered endothelial and epithelial implants differentially and synergistically regulate airway repair. Proc. Natl. Acad. Sci. U.S.A. **105**(19), 7046–7051 (2008)
93. Zhang, Q.Z., Su, W.R., Shi, S.H., Wilder-Smith, P., Xiang, A.P., Wong, A., Nguyen, A.L., Kwon, C.W., Le, A.D.: Human gingiva-derived mesenchymal stem cells elicit polarization of M2 macrophages and enhance cutaneous wound healing. Stem Cells **28**(10), 1856–1868 (2010)
94. Zheng, Z.H., Li, X.Y., Ding, J., Jia, J.F., Zhu, P.: Allogeneic mesenchymal stem cell and mesenchymal stem cell-differentiated chondrocyte suppress the responses of type II collagen-reactive T cells in rheumatoid arthritis. Rheumatology **47**(1), 22–30 (2008)

Author Index

Stud Mechanobiol Tissue Eng Biomater (2013) 10: 441–442
DOI: 10.1007/978-3-642-32563-2

MIX
Papier aus verantwortungsvollen Quellen
Paper from responsible sources
FSC® C105338

If you have any concerns about our products,
you can contact us on
ProductSafety@springernature.com

In case Publisher is established outside the EU,
the EU authorized representative is:
Springer Nature Customer Service Center GmbH
Europaplatz 3, 69115 Heidelberg, Germany

Printed by Libri Plureos GmbH
in Hamburg, Germany